Towards Cognitive Autonomous Networks

Towards Cognitive Autonomous Networks

Network Management Automation for 5G and Beyond

Edited by

Stephen S. Mwanje & Christian Mannweiler
Nokia Bell Labs
Munich
Germany

Registered Offices

John Wiley & Sons, Inc., 111 River Street, Hoboken, NJ 07030, USA

John Wiley & Sons Ltd, The Atrium, Southern Gate, Chichester, West Sussex, PO19 8SQ, UK

Editorial Office

The Atrium, Southern Gate, Chichester, West Sussex, PO19 8SQ, UK

For details of our global editorial offices, customer services, and more information about Wiley products visit us at www.wiley.com.

Wiley also publishes its books in a variety of electronic formats and by print-on-demand. Some content that appears in standard print versions of this book may not be available in other formats.

Library of Congress Cataloging-in-Publication Data applied for

HB ISBN: 9781119586388

Cover Design: Wiley
Cover Images: Background stock market © MarsYu/Getty Images,
Artificial intelligence © Yuichiro Chino/Getty Images,
Cityscape with abstract light © Hiroshi Watanabe/Getty Images

Set in 9.5/12.5pt STIXTwoText by SPi Global, Chennai, India

10 9 8 7 6 5 4 3 2 1

Contents

List of Contributors *xix*
Foreword I *xxi*
Foreword II *xxv*
Preface *xxvii*

1   **The Need for Cognitive Autonomy in Communication Networks** *1*
         Stephen S. Mwanje, Christian Mannweiler and Henning Sanneck
1.1   Complexity in Communication Networks   *2*
1.1.1   The Network as a Graph   *2*
1.1.2   Planes, Layers, and Cross-Functional Design   *4*
1.1.3   New Network Technology – 5G   *6*
1.1.4   Processes, Algorithms, and Automation   *9*
1.1.5   Network State Changes and Transitions   *9*
1.1.6   Multi-RAT Deployments   *10*
1.2   Cognition in Network Management Automation   *11*
1.2.1   Business, Service and Network Management Systems   *11*
1.2.2   The FCAPS Framework   *13*
1.2.3   Classes/Areas of NMA Use Cases   *15*
1.2.4   SON – The First Generation of NMA in Mobile Networks   *17*
1.2.5   Cognitive Network Management – Second Generation NMA   *18*
1.2.6   The Promise of Cognitive Autonomy   *18*
1.3   Taxonomy for Cognitive Autonomous Networks   *19*
1.3.1   Automation, Autonomy, Self-Organization, and Cognition   *19*
1.3.2   Data Analytics, Machine Learning, and AI   *21*
1.3.3   Network Autonomous Capabilities   *22*
1.3.4   Levels of Network Automation   *23*
1.3.5   Content Outline   *25*
1.3.5.1 Requirements Analysis   *26*

1.3.5.2   Foundations   *26*
1.3.5.3   Recent Cognitive Solutions   *26*
1.3.5.4   Motivating the Future   *26*
           References   *27*

2   **Evolution of Mobile Communication Networks**   *29*
           Christian Mannweiler, Cinzia Sartori, Bernhard Wegmann, Hannu Flinck,
           Andreas Maeder, Jürgen Goerge and Rudolf Winkelmann
2.1   Voice and Low-Volume Data Communications   *30*
2.1.1   Service Evolution – From Voice to Mobile Internet   *31*
2.1.2   2G and 3G System Architecture   *33*
2.1.3   GERAN – 2G RAN   *35*
2.1.4   UTRAN – 3G RAN   *36*
2.2   Mobile Broadband Communications   *38*
2.2.1   Mobile Broadband Services and System Requirements   *38*
2.2.2   4G System Architecture   *39*
2.2.3   E-UTRAN – 4G RAN   *40*
2.3   Network Evolution – Towards Cloud-Native Networks   *42*
2.3.1   System-Level Technology Enablers   *42*
2.3.2   Challenges and Constraints Towards Cloud-Native Networks   *46*
2.3.3   Implementation Aspects of Cloud-Native Networks   *47*
2.4   Multi-Service Mobile Communications   *49*
2.4.1   Multi-Tenant Networks for Vertical Industries   *50*
2.4.2   5G System Architecture   *51*
2.4.3   Service-Based Architecture in the 5G Core   *54*
2.4.4   5G RAN   *56*
2.4.5   5G New Radio   *59*
2.4.6   5G Mobile Network Deployment Options   *63*
2.5   Evolution of Transport Networks   *69*
2.5.1   Architecture of Transport Networks   *69*
2.5.2   Transport Network Technologies   *70*
2.6   Management of Communication Networks   *72*
2.6.1   Basic Principles of Network Management   *72*
2.6.2   Network Management Architectures   *76*
2.6.2.1   Legacy 3GPP Management Integration Architecture   *77*
2.6.2.2   Service-Based Architecture in Network Management   *78*
2.6.3   The Role of Information Models in Network Management   *79*
2.6.4   Dimensions of Describing Interfaces   *80*
2.6.4.1   Dimension 1: Hierarchy of the Management Function   *80*
2.6.4.2   Dimension 2: Levels of Abstraction   *81*
2.6.4.3   Dimension 3: Layers in Communication   *81*

2.6.4.4   Dimension 4: Meta Data   *81*
2.6.5   Network Information Models   *82*
2.6.5.1   Model of the Dynamic Behaviour   *82*
2.6.5.2   Format of the Data   *84*
2.6.5.3   Semantical Part of the Model   *85*
2.6.6   Limitations of Common Information Models   *85*
2.7   Conclusion – Cognitive Autonomy in 5G and Beyond   *87*
2.7.1   Management of Individual 5G Network Features   *87*
2.7.2   End-to-End Operation of 5G Networks   *88*
2.7.3   Novel Operational Stakeholders in 5G System Operations   *88*
            References   *89*

3   Self-Organization in Pre-5G Communication Networks   *93*
            Muhammad Naseer-ul-Islam, Janne Ali-Tolppa, Stephen S. Mwanje and
            Guillaume Decarreau
3.1   Automating Network Operations   *94*
3.1.1   Traditional Network Operations   *94*
3.1.2   SON-Based Network Operations   *95*
3.1.2.1   Centralized SON   *95*
3.1.2.2   Distributed SON   *96*
3.1.2.3   Hybrid SON   *97*
3.1.3   SON Automation Areas and Use Cases   *97*
3.2   Network Deployment and Self-Configuration   *98*
3.2.1   Plug and Play   *98*
3.2.1.1   Auto-Connectivity   *99*
3.2.1.2   Auto-Commissioning   *99*
3.2.1.3   Dynamic Radio Configuration   *100*
3.2.2   Automatic Neighbour Relations (ANR)   *101*
3.2.2.1   The ANR Procedure   *101*
3.2.2.2   NRT and ANR Limitations   *103*
3.2.3   LTE Physical Cell Identity (PCI) Assignment   *103*
3.2.3.1   PCI Assignment Objectives   *104*
3.2.3.2   PCI Assignment Strategies   *106*
3.2.3.3   PCI Assignment Challenges   *107*
3.3   Self-Optimization   *108*
3.3.1   Mobility Load Balancing (MLB)   *108*
3.3.1.1   Scenarios for Load Balancing and Traffic Steering   *109*
3.3.1.2   Standardization Support for Load Balancing and Traffic Steering   *109*
3.3.2   Mobility Robustness Optimization (MRO)   *111*
3.3.2.1   Optimization Objectives for MRO   *111*
3.3.2.2   Standardization Support for MRO   *114*

3.3.3   Energy Saving Management *115*
3.3.3.1   Scenarios for Energy Saving *116*
3.3.3.2   Standardization Support for Energy Saving *117*
3.3.4   Coverage and Capacity Optimization (CCO) *117*
3.3.4.1   Scenarios for CCO *118*
3.3.4.2   Solution Ideas for CCO *119*
3.3.5   Random Access Channel (RACH) Optimization *120*
3.3.6   Inter-Cell Interference Coordination (ICIC) *122*
3.4   Self-Healing *124*
3.4.1   The General Self-Healing Process *125*
3.4.2   Cell Degradation Detection *125*
3.4.3   Cell Degradation Diagnosis *127*
3.4.4   Cell Outage Compensation *128*
3.5   Support Function for SON Operation *129*
3.5.1   SON Coordination *129*
3.5.1.1   SON Function Conflicts *129*
3.5.1.2   SON Function Coordination *131*
3.5.2   Minimization of Drive Test (MDT) *133*
3.5.2.1   Scenarios and Use Cases for Drive Tests *133*
3.5.2.2   Standardization Support for MDT *135*
3.6   5G SON Support and Trends in 3GPP *136*
3.6.1   Critical 5G RAN Features *136*
3.6.2   SON Standardization for 5G *137*
3.7   Concluding Remarks *140*
           References *141*

4   Modelling Cognitive Decision Making *145*
           Stephen S. Mwanje and Henning Sanneck
4.1   Inspirations from Bio-Inspired Autonomy *146*
4.1.1   Distributed, Efficient Equilibria *146*
4.1.2   Distributed, Effective Management *147*
4.1.3   Robustness Amidst Self-Organization *147*
4.1.4   Adaptability *147*
4.1.5   Natural Stochasticity *148*
4.1.6   From Simplicity Emerges Complexity *148*
4.2   Self-Organization as Visible Cognitive Automation *148*
4.2.1   Attempts at Definition *149*
4.2.2   Bio-Chemical Examples of Self-Organizing Systems *149*
4.2.3   Human Social-Economic Examples of Self-Organizing Systems *151*
4.2.4   Features of Self-Organization – As Evidenced by Ant Foraging *152*
4.2.5   Self-Organization or Cognitive Autonomy? – The Case of Ants *154*

4.3	Human Cognition	*154*
4.3.1	Basic Cognitive Processes	*155*
4.3.2	Higher, Complex Cognitive Processes	*156*
4.3.2.1	Thought	*156*
4.3.2.2	Language	*157*
4.3.2.3	Problem-Solving	*157*
4.3.2.4	Intelligence	*157*
4.3.3	Cognitive Processes in Learning	*158*
4.4	Modelling Cognition: A Perception-Reasoning Pipeline	*159*
4.4.1	Conceptualization	*160*
4.4.2	Contextualization	*160*
4.4.3	Organization	*161*
4.4.4	Inference	*161*
4.4.5	Memory Operations	*162*
4.4.6	Concurrent Processing and Actioning	*162*
4.4.7	Attention and the Higher Processes	*163*
4.4.8	Comparing Models of Cognition	*164*
4.5	Implications for Network Management Automation	*167*
4.5.1	Complexity of the PRP Processes	*167*
4.5.2	How Cognitive Is SON?	*168*
4.5.3	Expectations from Cognitive Autonomous Networks	*168*
4.6	Conclusions	*169*
	References	*170*
5	**Classic Artificial Intelligence: Tools for Autonomous Reasoning**	*173*
	Stephen Mwanje, Marton Kajo, Benedek Schultz, Kimmo Hatonen and Ilaria Malanchini	
5.1	Classical AI: Expectations and Limitations	*174*
5.1.1	Caveat: The Common-Sense Knowledge Problem	*174*
5.1.2	Search and Planning for Intelligent Decision Making	*175*
5.1.3	The Symbolic AI Framework	*176*
5.2	Expert Systems	*177*
5.2.1	System Components	*177*
5.2.2	Cognitive Capabilities and Application of Expert Systems	*177*
5.2.3	Rule-Based Handover-Events Root Cause Analysis	*178*
5.2.4	Limitations of Expert Systems	*179*
5.3	Closed-Loop Control Systems	*180*
5.3.1	The Controller	*180*
5.3.2	Cognitive Capabilities and Application of Closed-Loop Control	*181*
5.3.3	Example: Handover Optimization Loop	*181*

5.4	Case-Based Reasoning	*182*
5.4.1	The CBR Execution Cycle	*183*
5.4.2	Cognitive Capabilities and Applications of CBR Systems	*184*
5.4.3	CBR Example for RAN Energy Savings Management	*185*
5.4.4	Limitations of CBR Systems	*185*
5.5	Fuzzy Inference Systems	*186*
5.5.1	Fuzzy Sets and Membership Functions	*186*
5.5.2	Fuzzy Logic and Fuzzy Rules	*187*
5.5.3	Fuzzy Interference System Components	*188*
5.5.4	Cognitive Capabilities and Applications of FIS	*189*
5.5.5	Example Application: Selecting Handover Margins	*190*
5.5.5.1	Step 1: Fuzzification	*190*
5.5.5.2	Step 2: Apply Fuzzy Operator(s)	*191*
5.5.5.3	Step 3: Apply Weighted Implication	*192*
5.5.5.4	Step 4: Aggregate All Outputs	*192*
5.5.5.5	Step 5: Defuzzify	*192*
5.6	Bayesian Networks	*192*
5.6.1	Definitions	*193*
5.6.2	Example Application: Diagnosis in Mobile Networks	*193*
5.6.3	Selecting and Training Bayesian Networks	*194*
5.6.4	Cognitive Capabilities and Applications of Bayesian Networks	*195*
5.7	Time Series Forecasting	*196*
5.7.1	Time Series Modelling	*196*
5.7.2	Auto Regressive and Moving Average Models	*198*
5.7.3	Cognitive Capabilities and Applications of Time Series Models	*198*
5.8	Conclusion	*199*
	References	*199*
6	**Machine Learning: Tools for End-to-End Cognition**	*203*
	Stephen Mwanje, Marton Kajoa and Benedek Schultz	
6.1	Learning from Data	*204*
6.1.1	Definitions	*205*
6.1.2	Training Using Numerical Optimization	*207*
6.1.3	Over- and Underfitting, Regularization	*209*
6.1.4	Supervised Learning in Practice – Regression	*211*
6.1.5	Supervised Learning in Practice – Classification	*212*
6.1.6	Unsupervised Learning in Practice – Dimensionality Reduction	*213*
6.1.6.1	Factor Analysis	*213*
6.1.6.2	Principal Components Analysis	*214*
6.1.6.3	Independent Components Analysis	*214*
6.1.6.4	Implementations	*215*
6.1.7	Unsupervised Learning in Practice – Clustering Using K-Means	*215*

6.1.8   Cognitive Capabilities and Limitations of Machine Learning   *216*

6.1.9   Example Application: Temporal-Spatial Load Profiling   *218*

6.2   Neural Networks   *219*

6.2.1   Neurons and Activation Functions   *220*

6.2.2   Neural Network Computational Model   *221*

6.2.3   Training Through Gradient Descent and Backpropagation   *222*

6.2.4   Overfitting and Regularization   *224*

6.2.5   Cognitive Capabilities of Neural Networks   *226*

6.2.6   Application Areas in Communication Networks   *226*

6.3   A Dip into Deep Neural Networks   *227*

6.3.1   Deep Learning   *227*

6.3.2   The Vanishing Gradients Problem   *228*

6.3.3   Drivers, Enablers, and Computational Constraints   *229*

6.3.3.1   Computational Power   *230*

6.3.3.2   Timing Constraints   *230*

6.3.3.3   Quantity of Data   *231*

6.3.4   Convolutional Networks for Image Recognition   *231*

6.3.4.1   Convolution Layers   *233*

6.3.4.2   Max Pooling   *234*

6.3.5   Recurrent Neural Networks for Sequence Processing   *235*

6.3.5.1   Long Short-Term Memory   *236*

6.3.6   Combining LSTMs with Convolutional Networks   *237*

6.3.7   Autoencoders for Data Compression and Cleaning   *238*

6.3.8   Cognitive Capabilities and Application of Deep Neural Networks   *240*

6.4   Reinforcement Learning   *241*

6.4.1   Learning Through Exploration   *241*

6.4.2   RL Challenges and Framework   *242*

6.4.3   Value Functions   *243*

6.4.4   Model-Based Learning Through Value and Policy Iteration   *244*

6.4.5   Q-Learning Through Dynamic Programming   *245*

6.4.6   Linear Function Approximation   *246*

6.4.7   Generalized Approximators and Deep Q-Learning   *247*

6.4.8   Policy Gradient and Actor-Critic Methods   *248*

6.4.8.1   Reinforce Algorithm   *249*

6.4.8.2   Reducing Variance   *250*

6.4.8.3   Policy Gradient Algorithm   *251*

6.4.8.4   Actor-Critic   *251*

6.4.9   Cognitive Capabilities and Application of Reinforcement Learning   *252*

6.5   Conclusions   *253*

     References   *253*

7   **Cognitive Autonomy for Network Configuration**   *255*
          Stephen S. Mwanje, Rashid Mijumbi and Lars Christoph Schmelz
7.1   Context Awareness for Auto-Configuration   *256*
7.1.1   Environment, Network, and Function Contexts   *257*
7.1.2   NAF Context-Aware Configuration   *259*
7.1.3   Objective Model   *260*
7.1.4   Context Model – Context Regions and Classes   *263*
7.1.5   Deriving the Context Model   *265*
7.1.6   Deriving Network and Function Configuration Policies   *266*
7.2   Multi-Layer Co-Channel PCI Auto-Configuration   *267*
7.2.1   Automating PCI Assignment in LTE and 5G Radio   *268*
7.2.2   PCI Assignment Objectives   *269*
7.2.3   Blind PCI Auto Configuration   *270*
7.2.4   Initial Blind Assignment   *271*
7.2.5   Learning Pico-Macro NRs   *272*
7.2.6   Predicting Macro-Macro NRs   *272*
7.2.7   PCI Update/Optimization and New Cells Configuration   *273*
7.2.8   Performance Expectations   *273*
7.3   Energy Saving Management in Multi-Layer RANs   *274*
7.3.1   The HetNet Energy Saving Management Challenge   *275*
7.3.2   Power Saving Groups   *276*
7.3.3   Cell Switch-On Switch-Off Order   *277*
7.3.4   PSG Load and ESM Triggering   *278*
7.3.5   Static Cell Activation and Deactivation Sequence   *279*
7.3.6   Reference-Cell-Based ESM   *280*
7.3.7   ESM with Multiple Reference Cells   *281*
7.3.8   Distributed Cell Activation and Deactivation   *283*
7.3.9   Improving ESM Solutions Through Cognition   *285*
7.4   Dynamic Baselines for Real-Time Network Control   *285*
7.4.1   DARN System Design   *286*
7.4.2   Data Pre-Processing   *288*
7.4.3   Prediction   *288*
7.4.4   Decomposition   *289*
7.4.4.1   Adaptation   *290*
7.4.4.2   Baseline Generation   *290*
7.4.5   Learning Augmentation   *290*
7.4.5.1   Knowledge Base   *291*
7.4.5.2   Alarm Generation   *292*
7.4.5.3   Metric Clustering   *293*
7.4.6   Evaluation   *294*
7.4.6.1   Accuracy of Generated Baselines and Clusters   *294*

7.4.6.2   Effect of Baseline Adaptation   *294*
7.4.6.3   Effect of Learning Augmentation   *295*
7.5   Conclusions   *297*
            References   *298*

8   Cognitive Autonomy for Network-Optimization   *301*
            Stephen S. Mwanje, Mohammad Naseer Ul-Islam and Qi Liao
8.1   Self-Optimization in Communication Networks   *302*
8.1.1   Characterization of Self-Optimization   *302*
8.1.2   Open- and Closed-Loop Self-Optimization   *304*
8.1.3   Reactive and Proactive Self-Optimization   *305*
8.1.4   Model-Based and Statistical Learning Self-Optimization   *306*
8.2   Q-Learning Framework for Self-Optimization   *306*
8.2.1   Self-Optimization as a Learning Loop   *307*
8.2.2   Homogeneous Multi-Agent Q-Learning   *308*
8.2.3   The Heterogeneous Multi-Agent Q-Learning SO Framework   *309*
8.2.4   Fuzzy Q-Learning   *310*
8.3   QL for Mobility Robustness Optimization   *314*
8.3.1   HO Performance and Parameters Sensitivity   *314*
8.3.2   Q-Learning Based MRO (QMRO)   *315*
8.3.3   Parameter Search Strategy   *317*
8.3.4   Optimization Algorithm   *318*
8.3.5   Evaluation   *318*
8.4   Fuzzy Q-Learning for Tilt Optimization   *322*
8.4.1   Fuzzy Q-Learning Controller (FQLC) Components   *322*
8.4.1.1   State and Action Fuzzy Variables   *322*
8.4.1.2   Rule-Based Fuzzy Inference System   *323*
8.4.1.3   Instantaneous Reward   *324*
8.4.2   The FQLC Algorithm   *324*
8.4.3   Homogeneous Multi-Agent Learning Strategies   *325*
8.4.4   Coverage and Capacity Optimization   *327*
8.4.5   Self-Healing and eNB Deployment   *327*
8.5   Interference-Aware Flexible Resource Assignment in 5G   *329*
8.5.1   Muting in Wireless Networks   *330*
8.5.2   Notations, Definitions, and Preliminaries   *331*
8.5.3   System Model and Problem Formulation   *332*
8.5.4   Optimal Resource Allocation and Performance Limits   *334*
8.5.5   Successive Approximation of Fixed Point (SAFP)   *335*
8.5.6   Partial Resource Muting   *335*
8.5.6.1   Triggering the Resource Muting Scheme   *336*
8.5.6.2   Detection of Bottleneck Users   *336*

8.5.7   Evaluation   *337*

8.6   Summary and Open Challenges   *340*

       References   *341*

9   **Cognitive Autonomy for Network Self-Healing**   *345*

       Janne Ali-Tolppa, Marton Kajo, Borislava Gajic, Ilaria Malanchini,

       Benedek Schultz and Qi Liao

9.1   Resilience and Self-Healing   *346*

9.1.1   Resilience by Design   *347*

9.1.2   Holistic Self-Healing   *348*

9.2   Overview on Cognitive Self-Healing   *349*

9.2.1   The Basic Building Blocks of Self-Healing   *350*

9.2.2   Profiling and Anomaly Detection   *351*

9.2.3   Diagnosis   *353*

9.2.4   Remediation Action   *354*

9.2.5   Advanced Self-Healing Concepts   *354*

9.2.6   Feature Reduction and Context Selection for Anomaly Detection   *356*

9.2.6.1   Feature Reduction   *356*

9.2.6.2   Context Selection   *357*

9.2.6.3   Feature Reduction and Context Selection in the Future   *358*

9.3   Anomaly Detection in Radio Access Networks   *358*

9.3.1   Use Cases   *359*

9.3.2   An Overview of the RAN Anomaly Detection Process   *360*

9.3.3   Profiling the Normal Behaviour   *361*

9.3.4   The New Normal – Adapting to Changes   *362*

9.3.5   Anomaly-Level Calculation   *364*

9.3.6   Anomaly Event Detection   *365*

9.4   Diagnosis and Remediation in Radio Access Networks   *366*

9.4.1   Symptom Collection   *367*

9.4.2   Diagnosis   *367*

9.4.3   Augmented Diagnosis   *369*

9.4.4   Deploying Corrective Actions   *371*

9.5   Knowledge Sharing in Cognitive Self-Healing   *371*

9.5.1   Information Sharing in Mobile Networks   *371*

9.5.2   Transfer Learning and Self-Healing for Mobile Networks   *373*

9.5.3   Applying Transfer Learning to Self-Healing   *374*

9.5.4   Prognostic Cross-Domain Anomaly Detection and Diagnosis   *374*

9.5.5   Cognitive Slice Lifecycle Management   *375*

9.5.6   Diagnosis Knowledge Cloud   *376*

9.5.7   Diagnosis Cloud Components   *377*

9.5.8   Diagnosis Cloud Evaluation   *378*

9.6   The Future of Self-Healing in Cognitive Mobile Networks   *379*
9.6.1   Predictive and Preventive Self-Healing   *379*
9.6.2   Predicting the Black Swan – Ludic Fallacy and Self-Healing   *380*
            References   *382*

10   Cognitive Autonomy in Cross-Domain Network Analytics   *385*
            Szabolcs Nováczki, Péter Szilágyi and Csaba Vulkán
10.1   System State Modelling for Cognitive Automation   *386*
10.1.1   Cognitive Context-Aware Assessment and Actioning   *386*
10.1.2   State Modelling and Abstraction   *387*
10.1.3   Deriving the System-State Model   *389*
10.1.3.1   The Static-State Model   *390*
10.1.3.2   State Trajectory Modelling and State Clustering   *391*
10.1.4   Symptom Attribution and Interpretation   *392*
10.1.5   Remediation and Self-Monitoring of Actions   *394*
10.2   Real-Time User-Plane Analytics   *396*
10.2.1   Levels of User Behaviour and Traffic Patterns   *396*
10.2.2   Monitoring and Insight Collection   *398*
10.2.3   Sources of U-Plane Insight   *400*
10.2.4   Insight Analytics from Correlated Measurements   *401*
10.2.5   Insight Analytics from Packet Patterns   *402*
10.3   Real-Time Customer Experience Management   *405*
10.3.1   Intent Contextualization and QoE Policy Automation   *406*
10.3.2   QoE Descriptors and QoE Target Definition   *408*
10.3.3   QoE Enforcement   *410*
10.4   Mobile Backhaul Automation   *411*
10.4.1   The Opportunities of MBH Automation   *412*
10.4.2   Architecture of the Automated MBH Management   *413*
10.4.3   MBH Automation Use Cases   *416*
10.5   Summary   *417*
            References   *418*

11   System Aspects for Cognitive Autonomous Networks   *419*
            Stephen S. Mwanje, Janne Ali-Tolppa and Ilaria Malanchini
11.1   The SON Network Management Automation System   *420*
11.1.1   SON Framework for Network Management Automation   *420*
11.1.2   SON as Closed-Loop Control   *421*
11.1.3   SON Operation – The Rule-Based Multi-Agent Control   *422*
11.2   NMA Systems as Multi-Agent Systems   *423*
11.2.1   Single-Agent System (SAS) Decomposition   *423*

11.2.2   Single Coordinator or Multi-Agent Team Learning   *424*

11.2.3   Team Modelling   *425*

11.2.4   Concurrent Games/Concurrent Learning   *425*

11.3   Post-Action Verification of Automation Functions Effects   *426*

11.3.1   Scope Generation   *427*

11.3.2   Performance Assessment   *428*

11.3.3   Degradation Detection, Scoring and Diagnosis   *429*

11.3.4   Deploying Corrective Actions – The Deployment Plan   *431*

11.3.5   Resolving False Verification Collisions   *433*

11.4   Optimistic Concurrency Control Using Verification   *436*

11.4.1   Optimistic Concurrency Control in Distributed Systems   *436*

11.4.2   Optimistic Concurrency Control in SON Coordination   *437*

11.4.3   Extending the Coordination Transaction with Verification   *437*

11.5   A Framework for Cognitive Automation in Networks   *440*

11.5.1   Leveraging CFs in the Functional Decomposition of CAN Systems   *440*

11.5.2   Network Objectives and Context   *442*

11.5.3   Decision Applications (DApps)   *443*

11.5.4   Coordination and Control   *444*

11.5.4.1   Configuration Management Engine (CME)   *444*

11.5.4.2   Coordination Engine (CE)   *445*

11.5.5   Interfacing Among Functions   *446*

11.6   Synchronized Cooperative Learning in CANs   *446*

11.6.1   The SCL Principle   *448*

11.6.2   Managing Concurrency: Spatial-Temporal Scheduling (STS)   *449*

11.6.3   Aggregating Peer Information   *451*

11.6.4   SCL for MRO-MLB Conflicts   *452*

11.6.4.1   QMRO Rewards   *453*

11.6.4.2   QLB Reward Function   *453*

11.6.4.3   Performance Evaluation   *454*

11.6.4.4   Observed Performance   *454*

11.6.4.5   Challenges and Limitations   *456*

11.7   Inter-Function Coopetition – A Game Theoretic Opportunity   *456*

11.7.1   A Distributed Intelligence Challenge   *457*

11.7.2   Game Theory and Bayesian Games   *458*

11.7.2.1   Formal Definitions   *459*

11.7.2.2   Bayesian Games   *460*

11.7.3   Learning in Bayesian Games   *461*

11.7.4   CF Coordination as Learning Over Bayesian Games   *463*

11.8   Summary and Open Challenges   *464*

11.8.1	System Supervision	*464*
11.8.2	The New Paradigm	*465*
11.8.3	Old Problems with New Faces?	*466*
	References	*466*

12   **Towards Actualizing Network Autonomy**   *469*

Stephen S. Mwanje, Jürgen Goerge, Janne Ali-Tolppa, Kimmo Hatonen, Harald Bender, Csaba Rotter, Ilaria Malanchini and Henning Sanneck

12.1	Cognitive Autonomous Networks – The Vision	*470*
12.1.1	Cognitive Techniques in Network Automation	*471*
12.1.1.1	Matching Problem Requirements	*471*
12.1.1.2	Development Effort vs Required Data	*472*
12.1.1.3	Training Time vs Development Effort	*473*
12.1.2	Success Factors in Implementing CAN Projects	*475*
12.1.3	Implications on KPI Design and Event Logging	*476*
12.1.4	Network Function Centralization and Federation	*477*
12.1.5	CAN Outlook on Architecture and Technology Evolution	*478*
12.1.6	CAN Outlook on NM System Evolution	*483*
12.2	Modelling Networks: The System View	*486*
12.2.1	System Description of a Mobile Network	*486*
12.2.2	Describing Performance	*488*
12.2.3	Implications on Automation	*489*
12.2.4	Control Strategies	*490*
12.2.4.1	Configuration vs Goal-Based Control	*490*
12.2.4.2	Command-Based vs State-Based Configuration Management	*491*
12.2.4.3	Benefits and Limitations of Configuration- and Goal-Based Control	*493*
12.2.4.4	Implicit Mix of Strategies	*494*
12.2.5	Two-Dimensional Continuum of Control	*495*
12.2.6	Levels of Policy Abstraction	*497*
12.2.7	Implications on Optimization	*500*
12.2.7.1	Modelling Optimization	*500*
12.2.7.2	Dealing with Uncertainty	*501*
12.2.8	The Promise of Intent-Based Network Control	*502*
12.2.8.1	Definition	*504*
12.2.8.2	Intent-Based Cognitive Autonomy	*505*
12.3	The Development – Operations Interface in CANs	*506*
12.3.1	The DevOps Paradigm	*506*
12.3.2	Requirements for Successful Adoption of DevOps	*508*
12.3.3	Benefits of DevOps for CAN	*509*

12.4   CAN as Data Intensive Network Operations   *510*
12.4.1   Network Data: A New Network Asset   *510*
12.4.2   From Network Management to Data Management   *511*
12.4.3   Managing Failure in CANs   *512*
          References   *514*

          Index   *517*

List of Contributors

Editors

Stephen S. Mwanje,
Nokia Bell Labs, Munich,
Germany

Christian Mannweiler,
Nokia Bell Labs, Munich,
Germany

Other Authors

Janne Ali-Tollpa,
Nokia Bell Labs, Munich,
Germany

Harald Bender
Nokia Bell Labs, Munich,
Germany

Guillaume Decarreau,
Nokia Bell Labs, Munich,
Germany

Hannu Flinck,
Nokia Bell Labs, Espoo,
Finland

Borislava Gajic,
Nokia Bell Labs, Munich,
Germany

Jürgen Goerge,
Nokia Bell Labs, Munich,
Germany

Kimmo Hatonen,
Nokia Bell Labs, Espoo,
Finland

Marton Kajo,
Technical University of Munich

Qi Liao,
Nokia Bell Labs, Stuttgart,
Germany

Andreas Maeder,
Nokia Bell Labs, Munich,
Germany

Ilaria Malanchini,
Nokia Bell Labs, Stuttgart,
Germany

Rashid Mijumbi,
MSD Ireland

Mohammed Naseer-Ul-Islam,
Nokia Bell Labs, Munich,
Germany

Szabolcs Novaczki,
Commsignia, Budapest,
Hungary

Csaba Rotter
Nokia Bell Labs, Munich,
Germany

Henning Sanneck,
Nokia Bell Labs, Munich,
Germany

Cinzia Sartori,
Nokia Bell Labs, Munich,
Germany

Benedek Schultz,
Formerly at Bell Labs, Budapest,
Hungary

Christoph Schmelz
Nokia Bell Labs, Munich,
Germany

Péter Szilágyi
Nokia Bell Labs, Budapest,
Hungary

Csaba Vulkán
Nokia Bell Labs, Budapest,
Hungary

Bernhard Wegmann
Nokia Bell Labs, Munich,
Germany

Rudolf Winkelmann
Nokia, Munich,
Germany

Other Contributors

Levente Bodrog,
WorldQuant Predictive,
Hungary

Szilard Kocsis,
Formerly at Bell Labs, Budapest,
Hungary

Foreword I

It has been over a decade since the concept of Self-Organizing Networks (SONs) emerged from NGMN and was standardized in 3GPP. The subsequent industry success of LTE has greatly benefited from the gradual introduction of SON to enhance mobile network optimization, thereby contributing to cost reductions, improved service reliability, and superior customer experience.

At Deutsche Telekom, we have successfully implemented SON functions and tools in several of our markets for RAN coverage, capacity and handover optimization use cases such as neighbour management and tilt optimization. Despite the adoptions, SON has fallen short of original industry expectations at its inception; implementations have been limited in scope to use cases affecting the RAN and not as part of end-to-end solutions to automate the management and operation of networks. The lack of highly granular network information has been another bottleneck for more efficient SON solutions.

The dawning of the 5G era is a new opportunity to evolve a new generation of SON, enabled by AI/ML methods, that will have the desired impact beyond mobile access network optimizations. Certainly, the challenges introduced by the disruptive deployment of 5G trigger the need for network transformation and a radical change in the way networks and services are managed and orchestrated. Think about the extreme range of requirements that come with new 5G network architectures, including massive capacity, ultra-high reliability, personalized services with dramatic improvements in customer-experience, or support for massive machine-to-machine communication.

Networks are also transforming into disaggregated, programmable, software-driven, service-based and holistically-managed infrastructures. For example, 5G network slicing, which is a critical enabler for operators to establish new types of business models, will impose unprecedented operational agility and higher cooperation across all network and service domains.

Long-term, our vision at Deutsche Telekom is to design, build and run fully autonomous networks. We even envisage that vertical customers will be able to self-provision tailored services based on network slices. Conventional automation techniques will only bring us so far on this path. To address these challenges, the next generation of SON should use AI/ML to achieve a higher level of automation across all processes and network layers.

As this will be a long and costly journey, we need to act smart, reuse as much as possible and start now to build first medium-term solutions that will pay into our long-term targets. At Deutsche Telekom, working with industry partners, we have already started to implement AI/ML for automation tasks in network operations. At the same time, we have identified and started to address the skill transformation required to ensure the right combination of network domain expertise with data science capabilities in our organization.

Today at Deutsche Telekom, we are using an AI/ML solution in the RAN to identify and resolve problems relating to radio interference, congestion or misconfigured handovers. Another project focuses on RAN energy management by targeting energy consumption reductions through an intelligent use of base-station power-saving features.

At our national company in Greece, we are trialling a holistic smart monitoring solution that targets cross domain network performance monitoring (RAN, Core, Transport, and customer service) to speed up issue detection and problem resolution. The solution from Nokia leverages an advanced ML model algorithm to identify and update patterns mapped into a comprehensive knowledge base repository. I look forward to the progression of this test, as well as further direct collaboration between our companies on solutions to push AI-based automation in our networks. On top of these activities, we have several network related AI/ML use cases in our pipeline that will address both domain-specific and end-to-end service management, such as anomaly detection in evolved packet core, automatic capacity planning of IP network interfaces and IMS test automation and validation.

Our industry is at an inflection point. The journey to become true digital service providers requires a complex network transformation based on an aggressive automation roadmap. This book addresses the technical issues that lie at the heart of this transformation. The authors take a long-term view, but also map out a path that will bring us towards cognitive autonomous networks. It is, therefore, a comprehensive resource for anyone interested in network automation strategies, opportunities and challenges.

As such, I recommend it as a guide on your long-term journey towards our common goal of zero touch operations for future networks and services.

Alex Jinsung Choi
SVP, Deutsche Telekom

Foreword II

Radio network management and optimization practices have remained very human centric since the dawn of cellular networks. Network operations including cell site layout selection, RF planning, reuse code planning and access, handover and throughput optimization were carefully planned using static tools by radio experts or recently, through rule-based automation with SONs (Self-Organizing Networks).

5G introduces a paradigm shift that will make such traditional approaches obsolete and, in many cases, unusable. The era of 5G will see an unprecedented mix of users, things and connected devices that will exhibit consumption characteristics that will overburden the available capacity for wireless access. 5G introduces technologies for low latency, mission critical services that will require dynamic management of the network. This enables concepts such as Network Slicing, which will drive unprecedented digitalization. With the sheer volume of industrial and commercial machines accessing the network, static planning, management, and optimization of the wireless network becomes a thing of the past.

At Nokia Software, the need for simplification and automation of network operations across all domains (Radio, Transport, Core, and Applications) drives the foundation of all our products. When working across domains, the orchestration of functions and features across the domains requires complex decoding of working policies and control mechanisms that have been very domain specific until now. Nokia Software has made rapid strides in harmonizing a software blueprint called the CSF (Common Software Foundation) as a common platform for all products. That problem being solved, the next step is to harmonize, interactions, interpretations and insights across Applications. This requires Cognitive models that are not limited by hard coded rules. The desired outcome is a network that self-manages through actions based on derived contexts and insights. In this book, the authors have explored tools for end-to-end cognition which will apply to the entire software tool chain required for 5G network operations.

The authors possess extensive and deep knowledge of the value, utilization and institutionalization of SON practices since its inception to its worldwide acceptance today. The content of this book reflects that vast knowledge and seeks to overcome known SON limitations related to Conflict Management and Objective Mapping by adopting Cognitive Decision making. This book also addresses the need to carefully manage the introduction of Cognitive Autonomy and manage existing relationships and structures in human-centric organizations.

SON will play an important role in managing and optimizing the lifecycle of each Network Slice. SON will collect machine learning based analytics from all domains and transform them into actionable intelligence. SON will understand the context in which the Network Slice is operating and tune the network predictively or in near real-time. Context determination cannot be human driven, it is based on Machine Learning that detects and defines models that can predict and apply proactive modifications to the network.

We are delighted to be working with the team that introduced the world to Self-Organizing Networks through their LTE SON book in 2012. Since then, many of the concepts and potential use of SON that were described in this book have been realized through Nokia Software led projects and products.

Now, Nokia Bell Labs is repeating history by introducing us to cutting edge concepts in this new book which will serve as a blueprint for all Nokia Software products that aim to bring in complete autonomy to Network Operations. The entire Operator community that has benefited from SON is aligned with Nokia in our journey towards Cognitive SON. A journey where Network Operations that will be performed in digital time, with time created for humans to elevate their productivity.

Ron Haberman
Chief Technology Officer, Nokia Software

Preface

This book attempts to motivate the evolution of network management towards a fully cognitive and fully autonomous network. It is premised on the understanding that:

A Cognitive network is one able to reason and formulate recommendations for subsequent behaviour but may, however, require the human operator to approve those recommendations before they are effected.

Autonomous network on the other hand is one able to act on its own i.e. it does not require dictation or rules from anyone else but may, however, be unable to reason based on its environment or even be smart enough to make the best static decisions (e.g. unable to interpret its internal rules).

We hypothesize that the desired eventual scenario is a Cognitive and Autonomous Network that can reason its environment, act independent of (free from) any external direction and learn from its interaction to ever improve its responses. Such a network is far from being achieved but significant steps have been taken thereto. The book attempts to highlight these steps and the eventual path thereof.

Chapter 1 introduces the discussion by presenting the sources of complexity in communication networks and the related network management. It highlights the specific features of networks critical for NM as well as the transitions that communication-networks management systems have undergone. This is the basis for justifying the need for evolving NM techniques with evermore cognitive capabilities. It also lays out the taxonomy used throughout the book to ensure a clear and concise discussion in subsequent chapters.

To ground the discussion in concrete requirements, Chapter 2 presents the hitherto expected evolution of communication networks taking today's 4G networks as the basis. The chapter attempts to evaluate the evolution with consideration of all the different network components – access (for both consumer and enterprise systems), backhaul, transport, core and the internet. It discusses the different ideas that have been presented by the different fora in justifying and proposing how the networks should evolve. Part of it forms a basis for justifying cognitive approaches to network management while the other part presents ideas about how this should be implemented or achieved.

As a baseline for further discussions on what could be possible in 5G and future networks, Chapter 3 reviews Self-Organizing Networks (SONs) in 2/3/4G which is the first generation of network management automation (NMA). Specifically, it discusses the philosophical techniques used therein as well as their application to different network management problems in configuration, optimization and healing of the (mainly) radio networks. This affords it a platform to justify the need to advancing NMA with or through cognitive techniques.

Chapter 4 presents the theoretical foundations of NMA to map out the scope and 'line of thinking' underlying previous, current and future implementation of NMA techniques. It contrasts the concepts of self-organization and cognitive automation and attempts to break down a fully cognitive process into its orthogonal sub-processes as a means of identifying what the available or known cognitive techniques can provide.

Chapters 5 and 6 summarize the available techniques for achieving cognitive automation. Essentially, they present a toolbox of ML and AI techniques that could be useful for automating networks providing, for each tool, an evaluation of the accorded degree of cognition as well as the basis on which the specific tool may be selected for application towards particular network challenges.

The application of the techniques in the cognitive tool box of Chapter 5 towards the network management problems creates the Cognitive Network Management (CNM) paradigm, which is the second generation of NMA after SON. With a core proposal of adding learning and analytics capabilities on existing SON functionality, CNM already advances NMA a significant step although the achieved outcomes cannot claim to have delivered a cognitive autonomous network.

Here, using the philosophical foundation in Chapter 4 and the cognitive tool box in Chapters 5 and 6, Chapters 7–10 present the so-far available achievements of the CNM paradigm. Specifically, Chapters 7, 8, and 9 respectively present CNM solutions that have been developed for self-configuration, self-optimization and self-healing problems for application in both 4G and 5G networks. All three chapters evaluate the developments across all network domains – the access separating the radio and RAN problems from the fixed access problems as well as the core and transport networks which are treated together given their interrelations in the virtualized environments.

Chapter 10 puts the CNM ideas together to highlight the value multi-domain cognitive automation, wherein the solutions introduce concepts that evaluate effects across multiple domains to decide the appropriate actions. This multi-domain challenge is complemented in Chapter 11 with the system-view of the automation challenges – essentially how to ensure the automation system achieves its expected end-to-end objectives. Thus, this addresses the supervision and verification of actions taken by automation functions as well as the challenge of minimizing conflicts amongst the multiple cognitive actors.

Since the book does not claim that network automation challenge has been fully solved, Chapter 12 attempts to look to the future path towards complete Autonomy in Networks. It lays out a clear vision of what would be expected from a network management perspective and relates it other high-level wishes including intent-based operations. It also attempts to give a high-level outline of how this could possibly be achieved and what implications this may have on network operations.

We challenge you, the reader, to realize this vision with us.

Stephen S. Mwanje
Christian Mannweiler
Nokia Bell Labs

1

The Need for Cognitive Autonomy in Communication Networks

Stephen S. Mwanje, Christian Mannweiler and Henning Sanneck

Nokia Bell Labs, Munich, Germany

Communication networks have significantly evolved to the point that they have become very complex to operate. Concurrently, the demands placed thereon by the different stakeholders continue to increase. Users require more diversified, robust yet cheaper services; operators require network operation to be cheap and simple with short lead times to introduce new services while governments demand ubiquitous networks offering reliable services.

At the core of meeting all these demands is network automation – increasing the capabilities of networks to undertake more and more operational tasks which have historically been done manually. This already is an ongoing process, but as we motivate the next level of automation, we must look deep into the structure of networks to identify the areas of greatest promise for automation. This chapter takes this deep evaluation to set the baseline for the subsequent discussion. It seeks to answer these questions: (i) why do we need to pursue the path towards cognitive autonomy in networks? What is the gap between where we are now and the final destination of a Cognitive Autonomous Network (CAN)?

The chapter presents a high-level overview of communication networks and the related Network Management (NM). It highlights the complexity of networks as justification for Network Management Automation (NMA) and discusses specific features of networks that are either the source or the embodiment of complexity in networks. Thereafter, it summarizes the NMA evolution and the existing framework within which NMA solutions have been developed. The discussion includes evaluation of the pressing NM challenges and the justification for evolving NM towards evermore cognitive capabilities. Finally, the chapter gives a taxonomy of the terminology used throughout the book and a preview of the contents presented in the remainder of the book.

Note that although the chapter and the rest of the book are heavily biased towards mobile networks, much of the discussion is generic and equally applies to other network domains. In fact, some chapters have sections that discuss concepts, implications and applications in both core and transport networks.

1.1 Complexity in Communication Networks

A communication network is responsible for getting data from one device (A) to another device (B), each of which may be a mobile device, for example, like a handheld device, a car, a robot or a fixed device like a fixed phone, a personal computer, or a server. In modern networks, the information which may include voice, simple text messages, files such as email/webpages or streaming media is all transmitted as packet data. The different network subparts contain devices and network elements (NEs) of varying size, whether considered in terms of physical size, form factor or available capacity. In the access' part, (where user terminals connect to the network), the typical network will have very many small-sized network elements, i.e. devices with small form-factor and where the maximum number of user terminals that can be concurrently served by the network element is also relatively small. The 'core' part of the network will instead have a small number of large-size network elements which are capable of thousands of concurrently supported users or sessions.

The thousands of network elements that move data around the network create an extremely complex graph of devices, functions, and processes. The sources of this complexity are many and varied – ranging from the system and functional design of the individual nodes of the graph to the inter-dependencies among these nodes, especially in the network's operational processes. Managing this complexity, which is a task delegated to network management, requires innovative approaches to be implemented in the network management systems for the networks to remain operable. The following sections discuss these sources of complexity and their implications for network management.

1.1.1 The Network as a Graph

A communication network may be represented as a graph [1]. It may, however, also be represented by a set of graphs each of which represents a distinct perspective of the network. In simple terms, a graph is an ordered pair $G = (V, E)$ of a set V of vertices and a set E of edges between the vertices. The nature of relationships amongst the vertices and edges characterizes a graph as a simple graph, a multi-graph, or a directed graph, etc. A communication network can be viewed as a graph on the specific view of the network. For example, on a global scale, a whole network can be a single graph if each transmitter and receiver point is considered a node/vertex with the communication links between them as the edges. Another graph, however, can be constructed for the set of neighbouring cells in a mobile network while the set of interfering cells would also be a different graph. These two graphs may also be considered as dynamic graphs if the user terminals are also considered as nodes on the graph, i.e. the terminals are dynamic leaves of the graph. It is, therefore, sensible to study networks in a way that each network is a set of graphs representing different perspectives.

Corresponding to the graph view of networks, network complexity directly translates into complexity of a graph, which may be measured in different perspectives that inherently signal the sources of the complexity. In general, there are three sources of complexity:

- **Network protocol stack in each node**: The abstraction of the physical system into a graph attempts to neglect the internal details of each node to focus on the interactions among nodes. However, each node can be complex with interactions amongst multiple graphs only realized within the node. For instance, individual nodes provide linkage between graphs of network routing and higher layer connected applications, e.g. for network transport dimensioning, network partitioning or for network caching.
- **Scale**: This, which is often the simplest dimension of complexity, is an indication of the number of nodes in the graph and the degree of interconnections among them. Correspondingly, it can be expressed in terms of the simplest measures for describing the structure of the graph, typically expressed in terms of the size or density.
- **Connectivity**: This is the most considered aspect of graphs – a measure of the degree of distribution/centralization and the homogeneity of its connectivity. By default, graphs assume a distributed system where value comes not from the individual nodes but from the complete unit that maximizes the connectivity amongst the nodes. As such, corresponding measures of complexity describe this connectivity in terms of the node or edge centrality measures, which describe the network positions of vertices and edges [2]. The most widely used measure is the edge betweenness, which identifies edges that are most crucial to maintaining a network's connectivity. However, there are many other measures including radius, closeness, etc. [3]

It is evident that these aspects of complexity are all present in networks and an understanding of this graph view is valuable to the process of designing NMA systems capable of addressing that complexity. NMA needs to leverage the multi-graph of a network to address the NMA challenges. Of particular interest is the modelling of the graph view within NMA – either as a single graph representing a single model with a broad view or as a multi-graph representing multiple different models with partial views that are related to each other. The subsequent sections describe, for the different sources of network complexity, the extent to which that complexity may be mapped to the complexity of a network graph.

1.1.2 Planes, Layers, and Cross-Functional Design

The first dimension of complexity comes from the design of network systems. Multiple functions need to be performed to move data from one end of the network to the other including, among others, multiplexing of physical resources among multiple users, finding the most optimal paths to the destination and error free delivery of the data. Given the many functions, it would be complicated to design the communication network devices as single monolithic units. Correspondingly, the underlying design philosophy for networks has been the concept of separation of functionality. This has created the view of networks in terms of planes that separate functions (e.g. packet forwarding from signalling) and layers that address the same function within the plane (e.g. packet forwarding) but separate scope from the functionality of the plane. The purpose of 'planeing' and 'layering' is that functionality in one layer and plane can be accessed and used independently of the functions of other planes and layers.

As illustrated by Figure 1.2, a network has three planes – the data, control and management planes [4] – so designed to separate responsibility and ease the development of network functionality:

- The data plane is responsible for the transfer of user information and includes all functions and processes needed to ensure that are needed to appropriately forwarding of user data across the network. The functionality ranges from formatting signals over the physical medium to managing the end-to-end sessions over which applications data are carried [5].
- The control plane includes all the functions and processes responsible for identifying, allocating, testing, and releasing network resources to the different users and sessions supported by the network, as well as the functions and processes for managing the users and sessions.
- The management plane is responsible for the domain and end-to-end administration of the network system. This includes the administration and management of network elements, functions, and resources; the supervision of resource management procedures and, to a limited extent, the management of user terminals.

Recent discussions e.g. [6], have also proposed the addition of an automation plane as a distinct dimension of network design separate from the other three planes. The major reason for this is simply because automation is no longer an addition to networks, but a core part of today's networks implemented in all the different layers of the network. So, a single automation plane will allow for

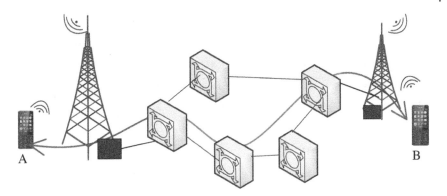

Figure 1.1 Data transfer across a communication network.

harmonization of the automation functionality throughout the network and take on the prior-to human operator functions. This, in a way, is similar to the earlier calls for a Knowledge Plane (see [7]), as the part of the network that relies on the tools of Artificial Intelligence (AI) and cognitive systems to build and maintain high-level models of what the network is supposed to do, to provide services and advice to other elements of the network.

Within the data and control planes, a layered architecture is employed (see Figure 1.2). Note, however, that although it is not explicitly stated, the abstraction of management functionality e.g. to distinguish element management from network management as highlighted in Section 1.2.1, may also be considered a form of layering with a different abstraction that, in the end, achieves the same outcomes. The layering concept in the data and control planes again emphasizes the need to separate responsibility and allow the different functions to be designed separately. As illustrated by the 5G user and control plane protocol stacks in

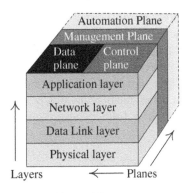

Figure 1.2 The network planes and layers [4].

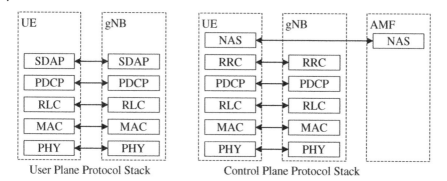

Figure 1.3 5G user plane and control plane protocol stacks [8].

Figure 1.3, for example, by separating the physical (PHY) and the media access control (MAC) layers, different expert teams can be deployed to concurrently and separately specify the two functionalities. The integration activity then simplifies to specifying the interfaces between these functional layers.

However, the separation of responsibility is a source of complexity. First, functionality becomes duplicated in the different layers as is noted to be the case for error control and link recovery mechanisms. The two functions are needed because of different scope (i.e. local link and end-to-end transport respectively) but, in the end, they implement similar capabilities which may lead to redundancy. Moreover, existing models may have to be redesigned owing to new requirements that demand cross-layer or cross-plane designs. For example, for sophisticated fine-tuning/optimization of network resources utilization, the strict separation between the control and management planes begin to get blurred, which further complicates the network. A case in point here is Software Defined Networking (SDN) where an SDN controller takes the responsibility of selecting data paths, a task which has traditionally resided in the physical switches. In this respect, the SDN controller merges the control and management planes or at least blurs the boundaries thereof. In same sense, the challenge of managing neighbour relations amongst cells in a mobile network is a typical management plane challenge yet an implementation that relies on radio measurements from the User Equipment (UE) uses the control plane signalling to relay configuration information. This results in a management plane functionality that is partially implemented in the control plane, again blurring the distinction between the two planes.

1.1.3 New Network Technology – 5G

A new generation of mobile network technology is rolled out every decade or close to a decade. Each generation is typically more complex than the previous one – be it in terms of supported services, or the applicable platforms, protocols, and procedures motivated by the supported services. While the second generation (2G) brought mobile telephony, the third generation (3G) introduced mobile Internet access and the fourth generation (4G) democratized mobile broadband Internet, the fifth generation (5G) will not only support a service type but be an enabler for a fully connected world [8]. 5G will support diverse use cases and applications including, among others, critical and massive machine-type communications (cMTC, mMTC) to enable Internet-of-Things (IoT), industrial Internet, Smart Factories, Vehicle-to-Anything (V2X), Smart Cities, etc.

The new use cases introduce new and often conflicting requirements for throughput, latency, reliability, and cost; which demand a flexible architecture that efficiently combines several radio and core network technologies. MTCs, for example, impose challenging requirements on major network Key Performance Indicators (KPIs), like round-trip time (latency), number of supported subscribers, peak data rates, and reliability. These requirements drive the evolution of the underlying technology which, in turn, drives new approaches to network management. For instance, high reliability requirements and the need to support massive device numbers result in new network optimization use cases that were never considered in earlier systems.

The development of 5G systems not only focuses on the definition of a new radio interface but on a complete End-to-End (E2E) system with a flexible architecture exploiting various underlying technologies to address the different use case and applications. Amongst the new technologies are the introduction of virtualized network nodes and functions, including cloudified Radio Access Networks (RANs) and the support for various vertical scenarios, for heterogeneous user devices and for an even denser deployment of cells. Figure 1.4 depicts the heterogeneity of 5G networks: a multitude of radio access and backhaul/fronthaul technologies is combined with different levels of centralization and virtualization, jointly allowing for a flexible deployment of network functions in various locations of the infrastructure.

Amongst the most critical new technologies introduced with 5G are Network Function Virtualization (NFV) [9] and network slicing [10].

NFV deployed over general-purpose computing and networking hardware (Network Function Virtualization infrastructure, NFVI) decouples 'softwarized' network functions from the underlying hardware resources with prior-to impossible benefits. The joint utilization of NFVI resources leverages multiplexing gains while the decoupling of hardware and software allows for separating resource management from network function management. Then, since network functions can be instantiated in centralized, aggregation, or edge clouds as shown in Figure 1.4, NFV facilitates flexible architecture that enables the protocol stack to be split into what is termed 'front-haul split'. Correspondingly, monolithic base

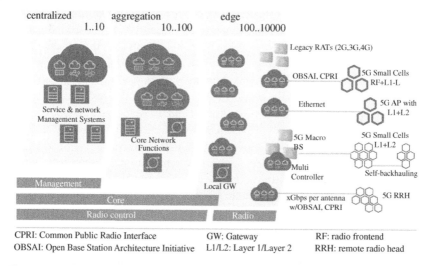

CPRI: Common Public Radio Interface   GW: Gateway   RF: radio frontend
OBSAI: Open Base Station Architecture Initiative   L1/L2: Layer 1/Layer 2   RRH: remote radio head

Figure 1.4 The high-level 5G deployment architecture.

stations with the full 3GPP radio protocol stack are complemented by lightweight access points (APs) that only implement lower protocol layers with their upper layers hosted at the associated macro base stations or edge cloud. While different front-haul splits are feasible, a typical approach is to split the layer 2 between the Radio Link Control (RLC) and Packet Data Convergence Protocol (PDCP) sub-layers [11]. Such splits automatically increase the complexity of the network, not least due to the increased number and variety of managed objects.

Network slicing, on the other hand, is the E2E virtualization of the mobile network. It enables the commissioning of multiple, logically self-contained mobile network instances that are tailored according to the requirements of specific use cases (e.g. a vertical industry); and that operate over shared infrastructure. This will have major effects on network complexity: Firstly, network functionality (particularly control plane functions) could be shared amongst multiple slices, which thus do not operate fully independently. This creates extra edges in the network graph(s) that were not there in earlier technologies. Moreover, UEs (the dynamic leaves on the graph) must be directed to the suitable network slice ('network slice selection') and could attach to multiple slices depending on the concurrent services they should be connected to.

Each of these services, use cases and technologies introduces an aspect of complexity on the network and its associated graph(s). Yet NMA must concurrently address this complexity and improve network operability for 5G to be economically viable for mobile network operators (MNOs). Consequently, novel NM approaches are required to accommodate a considerable share of the additional complexity.

On the other hand, vertical sectors will ask for (limited) network control for customized NM and configuration. Future NM systems must, as such, provide according interfaces for both human operator and software agents as well as appropriate accounting and charging systems thereof.

1.1.4 Processes, Algorithms, and Automation

A network involves very many processes and procedures that control and optimize access, allocation, and utilization of the network's resources. In the case of 5G, even more procedures are expected owing to the desired flexibility of the system. These processes interact with each other, both locally within a network element and, across multiple network elements, e.g. scheduling and resource allocation procedures influence and are influenced by mobility procedures just as UE energy management and network coverage management procedures also influence each other. Network management and NMA need to deal with these underlying procedures and with their interactions. Any increase in the number and complexity of the network procedures directly correlates with increased NM procedures.

Moreover, many of the procedures are evermore moving from being control algorithms and towards being automation algorithms. Recently, there is even a push to provide machine learning based algorithms for fundamental procedures like scheduling [12] and other radio resource management procedures [13]. With such deep automation within the network's signalling system, NM becomes a control system for an underlying automation system and NMA an outer loop automation above the lower level automated signalling system.

This outer loop automation above an automated system gives a different perspective of the network graph. The graph is no longer a single layer multi-graph but a graph of two multi-graphs; one in the lower signalling network and the other in the network management and automation. Processes for NMA must, as such, account for this mixed view.

1.1.5 Network State Changes and Transitions

The network and elements or devices thereof may be considered as state machines which continuously transition from one state to another. The different aspects of the network change both their internal as well as their observed states. For example, UE states change due to the mobility of the UEs while the network's physical environment undergoes short-term changes (e.g. due to moving objects) and long-term changes (e.g. due to changes in weather or the yearly seasons). The network itself also undergoes significant state changes as the aggregate of the individual changes in the network elements. For example, failures in the network elements or in the processes running on these elements may trigger localized or wide area changes in the network.

All these changes in states raise a different dimension of the network management and NMA change. In principle, the network graph or multi-graph cannot be assumed to be static but is a dynamic graph whose structure is dictated by the changes in the individual network elements.

1.1.6 Multi-RAT Deployments

Today's networks are a composite of multiple technologies servicing different applications. Wide area mobility is concurrently served by 2G, 3G, and 4G networks depending on user profile, service type, and coverage. It is expected that many of these earlier technologies will still coexist in the future concurrently with the new technologies, for now, concurrently with 5G. The reasons for this are varied, the simplest being that, typically, the new technology initially provides only patchy coverage, so the existing technologies must continue to serve users in the areas where the new technology has not yet been deployed. However, there are cases where the new technology does not fully address the services expected from users – e.g. it took long after the initial deployment to be able to serve voice over Long-Term-Evolution (LTE), so the existing 2/3G networks had to coexist with LTE to ensure that voice could also be served as well as the mobile data served by LTE.

The strong demand for higher user throughput has motivated deployments of more cells and greater spectrum. Increased cell count has resulted in heterogeneous networks (HetNets), where a number of small cells are deployed alongside macro cells. On the other hand, additional radio spectrum is currently only available at higher frequencies where cell sizes are far smaller than for LTE. For example, within the 5G radio access research, significant effort has been spent on the inclusion of millimetre wave (mmW) spectrum, whose propagation conditions only allow for very small cell sizes. Consequently, 5G deployments will feature many more small cells in the higher bands leading to what have been called Ultra-Dense Networks (UDNs).

In addition, scarcity of licensed spectrum for 3GPP networks has motivated the development of interworking features between 3GPP and un-licensed band technologies like IEEE 802.11 WLAN. Amongst these, is Licensed Assisted Access (LAA) that allows 3GPP 4G networks to also operate in unlicensed spectrum, and wireless local area network (WLAN) interworking that allows offloading of cellular network traffic to WLAN. This support for interworking with other spectrum and technologies will remain a critical requirement for 5G systems in order to support the ubiquitous mobile-broadband requirement as well as other use cases like industrial and vehicular communication.

This deployment of multi-RAT, multi-band, multi-layer network raises another dimension of complexity for the network graph. Although the individual technologies may be different from each other, the services and thus the network and service management functionalities need to be seamless across the layers, bands, or radio access technologies (RATs).

Moreover, as explained in the prior sections, each layer, band, or RAT is itself a complex multi-graph, yet NMA must consistently deal with all this complexity. It is evident that rule-based automation algorithms cannot effectively deal with all these complexities mainly because the human designers who write the rules cannot anticipate and explicitly encode all the situations which may occur in a network. Correspondingly, it becomes very hard to fully describe a network state let alone list all possible states and enumerate the strictly deterministic optimal responses to all such states. This justifies the need to move from NMA solutions and, instead, consider statistical and machine-learning based solutions that may work together to transform mobile communication networks into Cognitive Autonomous Networks.

1.2 Cognition in Network Management Automation

The International Telecommunication Union (ITU) Telecommunication Standardization Sector (ITU-T) recommendation M.3010 [14] describes the concept of a Telecommunication Management Network (TMN) as a framework for service providers to manage their service delivery networks. This is defined as achievable through four logical layers of abstraction: Business Management Layer (BML), Service Management Layer (SML), Network Management Layer (NML), and Element Management Layer (EML). The ITU-T published recommendation M.3400 [15] extending the TMN framework, which introduced the concept of fault, configuration, accounting, performance, and security (FCAPS) as the individual management function that must be undertaken in each of the four logical management layers.

1.2.1 Business, Service and Network Management Systems

Although the network is full of complexity, the individual devices/terminals in the network need to remain simple to configure and operate in bounded time. This directly translates into complexity in the network management which must at all times retain the broader view across the network. Specifically, the complexity, which is evident all through the network's business and operations support systems, translates into complex information models, dependencies, FCAPS processes, etc.

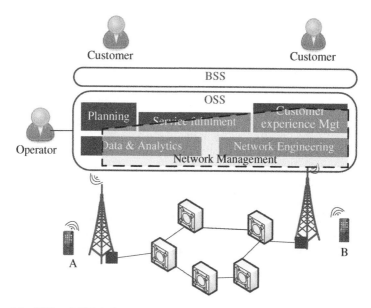

Figure 1.5 BSS and OSS deliver value from/to customers and the network.

For a given network, Business Support Systems (BSS) and Operations Support Systems (OSSs) ensure delivery of value from the network [16]. On the one side, customers expect to get their ordered service as and when ordered, and on the other side, the network operator wants to deliver that service at the highest-possible cost efficiency. The BSS and OSS are responsible for delivering that value to both the customer and the Communication Services Provider (CSP) that operates the network (see Figure 1.5).

The BSS applications are responsible for managing all customer-facing jobs and processes [16]. These will typically include, amongst others, capturing the customer orders in the appropriate detail, triggering the respective service delivery processes; maintaining the customer relationship which includes notifying the customer of important information such as timelines and engineer visits; as well as managing the post-delivery processes like beginning the billing process as soon as the service is activated. Correspondingly, the BSS provides the OSS with a detailed customer order comprising data such as service type, optional service parameters, the customer ID and perhaps the location details or termination equipment ID if it's a fixed-line service. It then tracks the order and fulfilment process making sure to trigger the requisite processes along the way.

The OSS takes responsibility for the operation of the network and the technical aspects of services [16], i.e. OSS aggregates the service and network management functions. The defining line between OSS and the other CSP IT systems is set by the network in that OSS takes responsibility for functions on the network with all the others left to the BSS and IT layers. Correspondingly, OSS is influenced by demands and trends in the customer-facing BSS and the technology of the network layer. The different functions accomplished by an OSS's may be grouped into the following broad areas:

- Planning is responsible for ensuring there are adequate resources and capabilities for the expected demand to guarantee an efficient network and the meeting operational and strategic objectives.
- Service fulfilment is responsible for the timely delivery of an active billable service adhering to technical and business rules whilst minimizing cost and effort.
- Customer experience management (CEM) that is responsible for executing company strategy and defining operational policies to manage the end-to-end service performance as perceived by the individual customer, well before a network alarm is raised.
- Data management and analytics that are responsible for gathering, storing, and exploiting the data about the network, service, and customers for the different OSS processes. It mines the data for insights into customer behavior as well as network performance characteristics and trends.
- Network Engineering facilitates the tasks of deployment, configuration, and monitoring and maintenance of network elements. This is irrespective of complexity of the devices (e.g. with proprietary interfaces and configuration syntax) or their location (be it in an office-block basement, a roadside cabinet, or at the customer's premises).

Network management is the subset of OSS with functions that define the behavior of the network and performance of the respective services [17]. It is responsible for the knowledge, control, and re-configuration of the network's devices to match operational objectives. It defines the set of devices and the location where the devices may be used, the format and customers for which they should be used. Successful implementations of network management will, as such, directly translate into ease of network operability.

1.2.2 The FCAPS Framework

The core scope and responsibilities for network management are defined by the TMN framework [15] in the five functions of Fault, Configuration, Accounting, Performance, and Security Management (FCAPS).

- **F- fault management**: Detection, isolation, and correction of abnormal operation of the network and its environment.
- **C- configuration management**: Control, identify, collect data from and provide data to network elements (NEs). Configure aspects of NEs, such

as configuration file management, inventory management, and software management.

- **A- accounting management**: Measure the use of network services and resources and determine costs to the service provider and prepare charging to the customer for such use.
- **P- performance management**: Gather and analyse statistical data, evaluate, and report upon the behaviour of equipment and the effectiveness of the network or network elements. To maintain overall performance at a defined level.
- **S- security management:** Authentication, access control, data confidentiality, data integrity, and non-repudiation. May be exercised for communications between systems, between customers and systems, and between internal users and systems.

The FCAPS model provides reference functionality for operating a network. However, other models have been developed to support the processes within the network operator that go beyond network management. For example, while the TMN takes a bottom-up approach, the TeleManagement Forum's Business Process Framework enhanced Telecom Operations Map (eTOM) takes a top-down approach that describes and analyses different levels of enterprise processes according to their significance and priority for the service provider's business. The service provider in this case is any entity who takes the network as infrastructure to offer some service either as the business (e.g. the CSPs) or as the platform for their business, e.g. the webscale companies. The eTOM approach provides a blueprint of process directions to service providers and an indication of software components boundaries and interaction points to both the service providers and their suppliers. Another framework that has also influenced the industry, albeit to a much lesser extent, is the IT Information Library (ITIL®).

Derivative operations models have been developed by different entities – both individual organizations and Standard-Defining Organizations (SDOs). The 3GPP model which also underpins the internal models of most CSPs, takes the TMN framework and specifies a model for NM [4]. The specifications include abstractions of the data structures specified as network resource models, the inter-process interfaces called Information Services as well as implementable solutions specified as solutions sets in Common Object Request Broker Architecture (CORBA), Common Management Information Protocol (CMIP), Simple Object Access Protocol (SOAP), etc.

The CSP will deploy a layered set of tools depicted in Figure 1.6 for the different sub-processes of network management. The reader should note that these models are independent of the applied protocols, be it SNMP, CMIS/CMIP or NETCONF but detailed discussion of these and other protocols is outside the scope of this chapter.

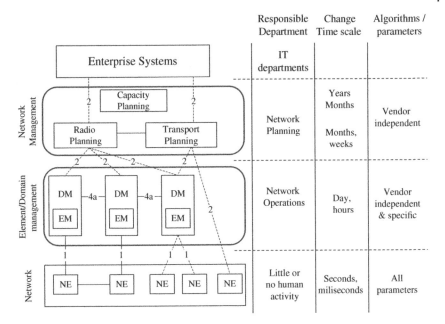

Figure 1.6 The matching of a typical network management tool chain and the 3GPP network management model. Source: Adapted from [19, 20].

1.2.3 Classes/Areas of NMA Use Cases

The push for network automation and specifically for network management automation is not new. There have been previous activities in both industry and academia to implement automation ideas. The most successful ones have been in the network control plane where multiple functions, including user management, resource assignment, etc., have been developed to minimize manual efforts in network control. However, the concepts behind self-organizing networks have also been quite successful in introducing automation in the management plane. Moreover, many publications and studies argue that adding cognition will improve the outcomes of network management automation even further.

In this context, automation focusses on the network configuration, optimization, and healing processes, which directly translate into automation of the configuration, performance, and fault management processes of the FCAPS framework. The security(S) and accounting (A) The FCAPS areas of security and accounting are not developed or discussed within network management and are instead typically affiliated with systems security and BSS businesses respectively. The configuration, optimization and healing processes also directly relate to the network's operations states, respectively reflecting the pre-operational phase, the normal operations state and the degraded operations state.

Configuration is a general term for making changes on one or multiple network devices to achieve some objective. The self-configuration term is, however, used in a much more restrictive way to refer to the required configuration to bring the network device(s) into operation. In the context of this book however, self-configuration is expanded in scope to refer to all configurations which may be triggered by a non-optimization/-healing algorithm to bring the network/device into a specific form of operation, even when it has perhaps been operating in a different state. An example of this could be the case of when a cell is triggered to update its neighbour relations without a specific optimization or degradation event.

Optimization, according to the Merriam-Webster dictionary [18], is the act, process, or methodology of making something (such as a design, system, or decision) as fully perfect, functional, or effective as possible. This applies to networks as well, i.e. after configuring and commissioning the network (or some of its devices or functionality) into commercial operation, some actions may need to be taken to improve the utility of that network's or device's resources or functionality. Thus, optimization in the network's context implies deploying functionality that continuously evaluates the network's state to identify avenues for improved performance. Self-optimization, then, implies that the monitored entity holds such functionality/capability within its fold and triggers such capability without any external indication. In networks, this implies that the network triggers, runs, and controls the optimization algorithms without any intervention from the operator(s).

Self-optimization functions optimize configuration parameters to improve the network performance from a given normal or acceptable network state to a better state. In contrast, self-healing focusses on dealing with failure events in a way that guarantees continuity of service. Self-healing functions ensure the network is still able to serve its customers even in the case of any performance degradation events, for example, network failures or unexpected changes. The typical workflow involves monitoring the symptoms of a fault to evaluate the cause and propose the most appropriate response.

Note, however, that although NMA use cases have been and continue to be grouped among these general areas, there is no distinct delineation of the FCAPS into the NMA use cases. Instead specific use cases addressing one or more functionalities of the FCAPS framework are assigned to one of the three use case areas. For example, the optimization of network tilts cannot be strictly mapped to a specific FCAPS area since tilts may be changed both to optimize performance as well as to recover from network failures like a cell outage. However, tilt optimization is generally classified as an optimization use case.

Besides the three typical network operations states, there has been a push to add an umbrella network automation use case area that is simply referred to as self-operation. This is intended to be similarly applicable in all three stages, i.e. to include use cases with functionality that takes a longer-term view of the network and can recommend changes at each of the three stages of the network. The expectation is that the wider view provides for real-time awareness of end-to-end performance. This is derived using correlated insight on user experience and application state as well as insight on the network state across multiple network domains (including RAN, transport, core, cloud, and other resources).

The self-operation areas may also be considered to include functionality for system-wide supervision of the automation function. Amongst these functions are the verification of action proposed by the automation functions, the coordination of the different functions for system-wide efficacy and the management of the automation function e.g. to reconfigure their control spaces. The contribution of this book to this and other areas is summarized in Section 1.3.5.

1.2.4   SON – The First Generation of NMA in Mobile Networks

The Self-Organizing Networks (SON) [19] concept was introduced to cellular networks to minimize operational complexity and costs, and to improve overall network performance. SON may be considered the first generation NMA as it was the first implementation of large-scale automation in network management. Therefore, individual functions, the so-called Self-Organizing Network Functions (SFs), were proposed to address specific operations, administration, and maintenance (OAM) problems like balancing load among cells [20] or optimizing handover trigger execution [21].

The SFs are state-machines that map observed network behaviour, e.g. changes in KPIs, to (re-)configurations of Network Configuration Parameters (NCPs). Each function is a closed control loop system that acquires data from the network and uses the measurements to determine or compute new configuration values for the network (element) (in the following denominated as 'network configurations') according to a set of algorithm internal rules or policies. In other words, the function is a state machine that exhibits static rule-based behaviour for matching inputs (network KPIs) to outputs (network configurations), in which it is the input – output relationship or the path which is predesigned into the solution through the rules (states and state transitions) of the algorithm.

It was observed that it is necessary to manage and coordinate the multiple SON functions (or instances thereof) [22]. Similar to the design of the individual functions, SON coordination and management have been proposed to be performed by centralized functions according to fixed rules, or through policies that are created based on fixed rules designed by the SON manufacturer with input from MNO (see [22–24]).

However, this static behaviour by the SFs and their management and coordination functions makes the system rigid. The functions are unable to adapt to major environmental or operational changes that result from, e.g. technical upgrades, network architecture modifications, time-variant (de-)activation of small cells, or new operator business and service models. Accordingly, a more flexible and adaptive automated OAM system was required.

1.2.5 Cognitive Network Management – Second Generation NMA

To enable flexibility in SON Functions, there have been many publications that proposed to add cognitive capabilities to the SON functions. This application of cognitive techniques towards network management creates the Cognitive Network Management (CNM) paradigm, which may be considered the second generation of network management automation (after SON). Herein, cognition is defined as the capability to reason over the candidate contexts when formulating recommendations for subsequent behaviour, i.e. consequently enabling context-specific recommendations for action.

With a core proposal of adding learning and analytics capabilities on existing SON functionality, CNM already advances NMA in a significant step. It allows for greater flexibility and adaptability with respect to environmental or operational changes by introducing cognition and turning closed-loop control-type SFs into Cognitive Functions (CFs). In concrete terms, this implies the application of Machine Learning (ML) and data analytics techniques that enable CFs to mimic human cognitive behaviour, i.e. to improve their behaviour based on learned (historical) data and their interaction with the network [8]. The objective of CNM, therefore, is that CFs: (i) explore their environment, (ii) learn optimal behaviour fitting to that environment; (iii) learn from their experience and that of other instances of the same or different CF type; and (iv) learn to achieve higher-level service/application goals and related technical objectives as defined by the network operator or its customers, e.g. through Key Quality Indicator (KQI) targets or Service Level Agreements (SLAs).

1.2.6 The Promise of Cognitive Autonomy

The achieved CNM outcomes cannot be claimed to have delivered a cognitive autonomous network, which may then have been considered the third generation of NMA. Specifically, autonomy requires the following to be achieved: (i) the collection/aggregation of insight about the network states, (ii) proliferation of cognitive decision making to the peripheral network operations functions, (iii) unified control and coordination of the automation system, and (iv) the abstraction of the network operations interface.

Firstly, it is necessary to aggregate knowledge and information from various sources of data to capture the true network state and to ensure that such a state is consistently communicated to all automation functions. Secondly, cognition capabilities need to be extended to the peripheral functionalities of the automation functions. For example, the establishment of baselines for decision making need to be automated and through a cognitive process that ensures the baselines are set per applicable context.

The automation system needs to have a unified control and coordination mechanism which must also leverage the cognitive capabilities of the individual functions when ensuring the system level objectives. In other words, the system-wide functionality must also apply cognitive techniques when deriving end-to-end decisions. This may, for example, include learning the appropriate configurations and rights of the automation functions depending on the available set of active functions. Similarly, the process of setting performance targets to be achieved by the functions need to be automated in a cognitive way. This will allow the human operators to abstract their design or configuration actions on the network and to interact with the network from a higher abstracted level.

Many of these challenges have not been addressed but there are existing works that address some sub aspects of the challenges. This book captures and describes these results and activities, and, most importantly, sets a baseline for identifying the open spaces that need to be filled to achieve Cognitive Autonomous Networks.

1.3 Taxonomy for Cognitive Autonomous Networks

To ensure consistent terminology throughout this book, this section highlights the critical terms that are used.

1.3.1 Automation, Autonomy, Self-Organization, and Cognition

According to multiple English language dictionaries (Merriam-Webster, Oxford, dictionary.com), the terms relating to automation, autonomy, and cognition are as summarized in Table 1.1. Based on these wider definitions, in the context of the book, the following definitions apply:

Automation means an entity, E, is configured and triggered to accomplish a task without external intervention while using some functionality for which the mechanism of accomplishing the task is not of interest. Correspondingly, even when the mechanism is created by the external entity, the entity E is still considered automated even if the external entity intervenes during execution of that mechanism.

Table 1.1 Dictionary definitions for automation, autonomy, and cognition.

Term	Dictionary definition	Synonyms/Examples
Automation	• the use of machines or computers instead of people to do a job, especially in a factory or office • predefining a process to run on its own • the technique, method, or system of operating or controlling a process human intervention reduced to a minimum	—
Autonomy	• the quality or state of being self-governing; especially: the right of self-government e.g. The territory was granted autonomy. • self-directing freedom and especially moral independence, e.g. personal autonomy	• e.g. a self-governing state
Autonomous	• having the right or power of self-government e.g. an autonomous territory • undertaken or carried on without outside control, i.e. self-contained, e.g. an autonomous school system • existing or capable of existing independently e.g. an autonomous zooid • responding, reacting, or developing independently of the whole e.g. an autonomous growth • (of a country or region) having the freedom to govern itself or control its own affairs, e.g. 'the federation included 16 autonomous republics' • having the freedom to act independently, • 'an autonomous republic': (in Kantian moral philosophy) acting in accordance with one's moral duty rather than one's desires	• self-governing, independent, sovereign, free, self-ruling, self-determining, autarchic; self-sufficient • e.g. 'school governors are legally autonomous'
Autonomic	• Equivalent to autonomous • involuntary or unconscious; relating to the autonomic nervous system, e.g. 'the symptoms included gastrointestinal and autonomic disturbance'	—
Cognitive	• of or relating to cognition; concerned with the act or process of knowing, perceiving, etc. cognitive development; cognitive functioning. • of or relating to the mental processes of perception, memory, judgement, and reasoning, as contrasted with emotional and volitional processes.	

Autonomy implies that the entity acts on its own over multiple related or unrelated tasks and where the mechanisms used to accomplish the tasks are of no interest.

A **Self-Organization** implies that an entity is capable of selecting actions without external control as triggered by a signal which may be external or otherwise. The internal mechanisms of such an entity are of no interest to the owner or user of such an entity.

A **Cognitive Entity** is one capable of perceiving a signal and process it into a data element over which the entity reasons to select an action. It conceptualizes and contextualizes the data element and, logically or arithmetically, relates the data element to other data elements to make inferences about the elements, their relations, and subsequently selects the appropriate action.

In effect, a self-organized entity may be cognitive and when it is not cognitive, such self-organization is achieved through a hard-wired decision logic within the entity. Therefore, this implies that all cognitive entities are self-organized but not all self-organized entities are cognitive.

In the context of networks, then, cognition means that the network is capable of reasoning over multiple sources of data to learn context as well as the best actions in these contexts. Correspondingly, cognitive autonomy implies that the network applies its cognition to determine how to behave in different contexts and subsequently independently acts on its own in those contexts.

1.3.2 Data Analytics, Machine Learning, and AI

Although the terms Data Analytics, ML, artificial intelligence and augmented intelligence are widely used, not all of them have been clearly defined. Correspondingly, they are hard to distinguish to the extent that they are, in many cases, used interchangeably. However, in the context of this book, we use them as described below:

Data Analytics refers to the processes and methods used to augment the human experience of the information hidden in large datasets. Such processes and methods include: Data visualization through graphical or other representations and statistical analysis using derived metrics such as averages, percentiles, variance, metrics from linear regression, etc.

ML, which is probably the only term from above that has been formally defined, unfortunately has too many definitions each with a contentious phrase, which then demands for a working definition for this book: ML is the science of getting computers to learn from data on observations and real-world interactions and to autonomously improve their learning over time. ML systems are used to augment human knowledge (or proficiency) through their capacity to improve their perception and understanding of a model hidden in data set. ML uses data analytics but goes a step farther to automate the learning process in that the human does not have to describe the model which the ML algorithm uses to improve its performance.

There is no comprehensible globally acceptable definition of artificial intelligence (AI). A working definition applied in this book is that artificial intelligence is the ability of computer systems to independently perform tasks that would otherwise require human intelligence. Correspondingly, when used as a field of study/work it refers to the study and development of theories, devices and software for systems capable of performing tasks that would normally require human intelligence. Herein, human intelligence refers to higher level human cognitive capabilities such as speech recognition, visual perception, decision making, and language skills.

Augmented intelligence attests to the assistive role of AI and machine learning solutions, i.e. that these techniques are supposed to assist humans in accomplishing tasks as opposed to completely replacing humans.

In this book, we take the view that, in practice, it is hard to distinguish the extents to which augmented intelligence supports human activities and that to which artificial intelligence can be free standing. We therefore use the combined term 'cognition' that includes every capability relating to the process of deriving knowledge, structure and/or understanding from one or more events or data elements. Correspondingly, we refer to cognitive capabilities, techniques, or technologies instead of referring to AI since we are interested in understanding the degree to which a technology can achieve a specific cognitive capability that is of interest for application in networks.

1.3.3 Network Autonomous Capabilities

Networks can be labelled with multiple terms regarding their degree of automation. Although the typically used labels have not been clarified as orthogonal, a network is, in general, expected to accomplish the following seven autonomous capabilities increasing in difficulty:

1) Take a specific routine action e.g. download a file
2) **Detect an event**: Instrument a parameter and run a level crossing detector on the instrumented data
3) **Correlate events/data**: Match multiple instrumented events
4) **Diagnose events**: Distinguish amongst multiple correlated events i.e. create cause–effect relationships amongst events/data
5) **Contextualize events/data**: Add context to the detection and correlation of events, i.e. a certain event is true under certain contexts and untrue otherwise
6) **Anticipate standalone events**: Detect a series of events and correlate them to determine the probability of occurrence of another event, e.g. the gradual but

continuous increase in cell traffic in a certain area may indicate a major event (say a sports event) from which one can anticipate heavy data upload during and subsequent to the event.

7) **Anticipate correlated events:** Detect multiple unrelated multi-contextual events and correlate them to derive the probability of occurrence of another event, e.g. be able to anticipate an increase in traffic in one part of the city because of a planned human event – such as a music show by relating such an event to other previous events.

The reader should note that the autonomous capability consideration is only one dimension (amongst many) of a network's automation space. Other dimensions include the network scope, the use case, and method of control.

Network scope implies that automation is possible at different levels of the network – e.g. element or domain, inter/intra domain. The use case dimension implies that the same or a different autonomous capability can be applied for different use cases – e.g. resource management, mobility management, quality assurance, etc. Similarly, the capabilities are applicable to different forms of control including intent-based, policy-based, or any other form of network control method.

1.3.4 Levels of Network Automation

As stated, owing to the multi-dimensional nature of network effects and the different dimensions in which automation may be described, it is practically impossible to describe an entire network in terms of its degree/level of automation. Correspondingly, even although orthogonal descriptions could be imagined for a given use case, a network's degree of automation can only be delineated in the following intuitive non-orthogonal descriptions with their differences relative to automation capabilities illustrated by Figure 1.7.

First, automation is generally evaluated relative to a manually controlled network, which is one where operations personnel evaluate network state and decide the course of action. They may automate their evaluation and actions in form of scripts which are executed in a controlled manner to ensure that undesired actions are avoided or as much as possible minimized.

The network may be considered automated, typically for a specific functionality, if the operator has pre-set a certain action to be undertaken without the operator explicitly triggering that action. This is typically implemented through scripting, i.e. the network allows scenario-specific execution of automation scripts. By and large, the network is still human-controlled, but allows the scripts to be run autonomously for those specific and very routine tasks. Such tasks may for example include scheduling execution of macros to collect network statistics and mining the statistics for specific events with distinguishable fingerprints.

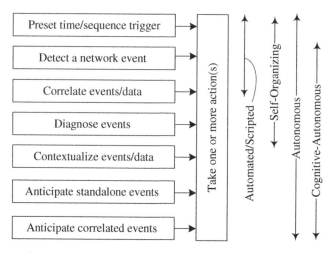

Figure 1.7 A delineation of the levels of autonomy in networks.

Scripts may however also go beyond evaluation to include configuration or control actions like scheduling network element resets.

A SON may be considered as the superset for automated or scripted networks in that it does not only automate the selection and execution of actions but also interprets events under different context to determine the cause-effect relations of the events under the different contexts.

An autonomous network is one able to act on its own. It does not require dictation or rules from anyone else but may however not be able to reason based on its environment or even be smart enough to make the best static decisions (e.g. it may be unable to interpret its internal rules).

A cognitive network is one able to reason and formulate recommendations for subsequent behaviour but may however require the human operator to approve those recommendations before they are executed.

From the above definitions, it is evident that a SON, at least as it has been so far developed, is closer to an autonomous network since it is capable of acting independently only that the derivation of such decisions is locked into the algorithm. On the other hand, CNM may be seen as the addition of cognitive skills to SON i.e. SON is made smarter by including cognitive capabilities in the SON's mechanisms of deriving actions.

Cognition would, however, need to be expanded if the network is to also be autonomous. In other words, a cognitive autonomous network holds both cognitive and autonomous abilities in that it is capable of reasoning when formulating recommendations for behaviour and then subsequently independently executing such derived recommendations. This requires the Cognitive Autonomous

Network (CAN) to have a much wider view of the network for any action that is taken and to have a deeper understanding of the operator's expectations.

Moreover, the cognitive functions need to have a deeper cognitive capability beyond reasoning. In the SON framework, the SON designer fully understands the cause-effect model for a given problem (such as antenna tilt changes) and so he designs the rules for selecting actions based on what is observed in the network. The CNM framework expands the capability of the functions to allow them to learn this cause-effect model for the specific problem. The CAN framework will require the function to be able to perceive insights from a wider set of data so as: (i) to learn a larger context-space within which the function should select actions, and (ii) to build a deeper and wider cause-effect model for the function in all those learned contexts. Essentially, the CAN framework adds a dynamic brain to all Self-x functions as illustrated in Figure 1.8.

1.3.5 Content Outline

This book is intended to be a single resource articulating what is and what could be possible in as far as Network (Management) Automation is concerned. It summarizes the evolution of networks and network management and discusses the expectations for cognitive capabilities in networks; the available tools for cognitive automation and the so far available recent solutions thereof. The book may be considered to have four major themes described here.

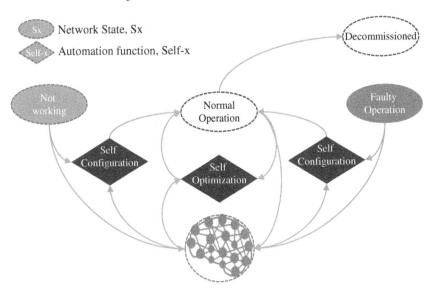

Figure 1.8 AI-/ML-based cognition on configuration, optimization, and healing network processes.

1.3.5.1 Requirements Analysis

The first part is a statement of requirements for NMA in the form of a description of the structure of networks and how they are expected to evolve in 5G and beyond. The corresponding discussion, with its foundations given in the first part of this chapter and the details covered in Chapter 2, takes today's 4G networks as basis to present the expected evolution in all the different parts of the network. With no specific limitation what has or will be standardized, we show what changes and advancements are expected, be it in the access (for both consumer and enterprise systems), backhaul, transport, core and the internet. The discussion summarizes the ideas presented by different consortia (like research projects) and those being standardized in different SDOs. All this is critical in justifying the need to advance network management and its automation, so the chapter specifies the conditions that need to be fulfilled by advanced NMA systems.

1.3.5.2 Foundations

The second part of the book, presented in Chapters 3–6, describes the inputs of the NMA advancement project. First, Chapter 3 summarizes SON and the concepts behind the developed SON algorithms. As opposed to simply listing the state-of-the-art SON solutions, the chapter focusses on delineating the reasoning or 'way of thinking' behind the development of today's SON functions. Secondly, Chapter 4 reviews the broad ideas on automation, with special emphasis on self-organization and bio-influenced decision making and a hypothesis on what cognition is and what it can achieve. Chapters 5 and 6 then summarize the available tools for cognitive automation starting in Chapter 5 with the simple rule-based/expert systems that attempt to mimic the solutions of an expert and then discussing the more involved deep learning concepts in Chapter 6.

1.3.5.3 Recent Cognitive Solutions

The third part of the book discusses the recent solutions that attempt to apply the cognitive concepts summarized in Chapters 5 and 6 towards solving telco problems. Since the majority of NMA challenges are in configuration, performance, and fault management, the core discussion is separated into three Chapters 7–9, respectively on Cognitive Self-configuration, optimization, and healing with emphasis on a domain-centric view. However, solutions that provide a wider view across multiple domains are discussed in Chapter 10. The solutions presented in the four chapters are responsible for mitigating most of the complexity challenges raised in Section 1.1 of this chapter. In particular, they allow NM processes to be expected at scale amidst the complexity.

1.3.5.4 Motivating the Future

The last part of the book motivates the continued evolution of NMA solutions towards advanced cognitive techniques. First, Chapter 11 discusses the system-wide challenges and the respective solutions which mainly address the concern of interdependence among NMA solutions. It highlights available solutions and their limitations, based on which it proposes a candidate approach that may be able to provide an ultimate solution for coordinating cognitive functions. Then the last chapter presents the way forward and the open questions that must be addressed to get there. Besides showing where and how the cognitive concepts discussed in Chapter 5 and 6 are best suited for telecom network related challenges, it highlights the gap between current NMA systems and the desired CAN. It also attempts to show the links to other major ideas shaping the industry, including, among others, intent-based network control and DevOps concepts. All this is wrapped up in the long-term vision of making networks cognitive and autonomous.

We challenge the reader to realize this vision.

References

1 Birand, B., Zafer, M., Zussman, G. et al. (2011). Dynamic graph properties of mobile networks under levy walk mobility. IEEE Eighth International Conference on mobile ad-hoc and sensor systems, Valencia.

2 Marsden, P.V. (2015). Network centrality, measures of. In: *International Encyclopedia of the Social & Behavioral Sciences*, 2e (eds. P. Baltes and N. Smelser). Pergamon.

3 Hernández, J.M. and Mieghem, P.V. (2011). *Classification of Graph Metrics*, 1–20. Mekelweg, The Netherlands: Delft University of Technology.

4 Fall, K.R. and Stevens, W.R. (2011). *TCP/IP Illustrated, Volume 1:The Protocols*. Addison-Wesley.

5 Zimmermann, H. (1980). OSI reference model-the ISO model of architecture for open systems interconnection. *IEEE Transactions on Communications* 28 (4): 425–432.

6 Ziegler, V., Theimer, T., Sartori, C. et al. (2016). Architecture vision for the 5G era. IEEE International Conference on Communications Workshops (ICC), Kuala Lumpur.

7 Clark, D.D., Partridge, C., Ramming, J.C. et al. (2003). A knowledge plane for the internet. Proceedings of the conference on Applications, technologies, architectures, and protocols for computer communications, Karlsruhe.

8 Mwanje, S., Decarreau, G., Mannweiler, C. et al. (2016). Network Management Automation in 5G: Challenges and Opportunities.Proceedings of the 27th IEEE International Symposium on Personal, Indoor and Mobile Radio Communications (PIMRC), Valencia.

9 ETSI GS NFV-MAN (2014). *Network Functions Virtualisation (NFV)*. Valbonne: Management and Orchestration.

10 3GPP TR 22.891 (2016). Feasibility Study on New Services and Markets Technology Enablers, Stage 1 (Release 14).

11 NGMN Alliance (2015). *Further Study on Critical C-RAN Technologies*. Frankfurt: NGMN Alliance.

12 Comsa, I.-S., De-Domenico, A., and Ktenas, D. (2017). Qos-driven scheduling in 5g radio access networks-a reinforcement learning approach. In: *GLOBECOM 2017–2017 IEEE Global Communications Conference*, 1–7. IEEE.

13 Calabrese, F.D., Wang, L., Ghadimi, E. et al. (2016). Learning radio resource management in 5G networks: Framework, opportunities and challenges. *IEEE Communications Magazine (September)*: 138–145.

14 ITU-T (2000). *Principles for a telecommunications management network*. ITU-T, M.3010.

15 ITU-T (2000). *TMN management functions*. ITU-T, M.3400.

16 Open Networking Foundation (2016). *Impact of SDN and NFV on OSS/BSS-ONF Solution Brief*. Palo Alto, CA: Open Networking Foundation.

17 3GPP TS 32.101 (2017). *Telecommunication management, Principles and high level requirements*. Sophia Antipolis: ETSI.

18 The Merriam-Webster.com Dictionary. "optimization (*n.*)". https://www.merriam-webster.com/dictionary/optimization (accessed 7 February 2019).

19 Hamalainen, S., Sanneck, H., and Sartori, C. (2011). *LTE Self-Organising Networks (SON): Network Management Automation for Operational Efficiency*. Wiley.

20 Mwanje, S.S. and Mitschele-Thiel, A. (2013). A q-learning strategy for lte mobility load balancing. IEEE 24th International Symposium on Personal Indoor and Mobile Radio Communications (PIMRC), London.

21 Mwanje, S.S. and Mitschele-Thiel, A. (2014). Distributed cooperative Q-learning for mobility-sensitive handover optimization in LTE SON. IEEE Symposium on Computers and Communication (ISCC), Madeira.

22 Schmelz, L.C., Amirijoo, M., Eisenblaetter, A. et al. (2011) A coordination framework for self-organisation in LTE networks. IFIP/IEEE International Symposium on Integrated Network Management (IM), Dublin.

23 Bandh, T., Romeikat, R., Sanneck, H. et al. (2011). Policy-based coordination and management of SON functions. *IFIP/IEEE International Symposium on Integrated Network Management (IM)*, Dublin.

24 Frenzel, C., Lohmuller, S., and Schmelz, L.C. (2014). *Dynamic, context-specific SON management driven by operator objectives*. Krakow: IEEE Network Operations and Management Symposium (NOMS).

2

Evolution of Mobile Communication Networks

Christian Mannweiler[1], Cinzia Sartori[1], Bernhard Wegmann[1],
Hannu Flinck[2], Andreas Maeder[1], Jürgen Goerge[1] and Rudolf Winkelmann[3]

[1]*Nokia Bell Labs, Munich, Germany*
[2]*Nokia Bell Labs, Espoo, Finland*
[3]*Nokia, Mobile Networks, Munich, Germany*

In 2019, the Third Generation Partnership Project (3GPP) officially completed the specification of Release 15, also referred to as the Phase 1 of the fifth generation (5G) of 3GPP mobile networks. This release has defined a completely new radio system complemented by a next-generation core network and includes the specifications for both standalone (SA) and non-standalone (NSA) operation of 5G New Radio (NR).

This chapter presents the evolution of mobile communication networks from GSM to the 5G System (5GS). The ever more demanding performance requirements are motivated by describing the evolution of communication services to be supported by mobile networks. Important technical milestones of each system generation are described, focusing on fundamental system design choices, overall system architecture, individual network features in different domains including radio, core, and transport networks, as well as selected technological and protocol enhancements. The chapter also describes the supported communications services and the mobile network architectures and technologies that have been developed from 2G to 5G in order to address the requirements of the listed services. The evolution of important network technologies, concepts, and functional architectures in the domains of radio access network (RAN), CN, and transport networks across network generations are analysed with respect to their relevance for network management.

The chapter then presents 5G system design as well as individual capabilities. 5G mobile networks are introduced by highlighting the key technological enablers. Starting from an overall system perspective, the chapter also presents the most

Towards Cognitive Autonomous Networks: Network Management Automation for 5G and Beyond,
First Edition. Edited by Stephen S. Mwanje and Christian Mannweiler.
© 2021 John Wiley & Sons Ltd. Published 2021 by John Wiley & Sons Ltd.

important novelties in the 5G RAN and features of NR. The discussion is then extended to evaluate the impact on transport networks.

The chapter concludes with an evaluation of the network management approaches for pre-5G networks, detailing the management system architecture, interactions between managing functions and managed objects, and the approach for modelling network elements and network data. The shortcomings of traditional network management approaches are described: while specific components, such as, Self-Organizing Network (SON) functions, will remain a crucial part of 5G management systems, novel capabilities are required to address the need for automation and cognitive behaviour of the network management procedures.

2.1 Voice and Low-Volume Data Communications

Mobile communication systems are typically classified into network generations. The **first generation** of mobile communication networks was based on circuit switched analogue technologies. It became available in the 1980s with the aim of providing ubiquitous voice services. This was not widely adopted owing to low inter-operability amongst the technologies, an issue which was addressed starting from the second generation. Table 2.1 gives an overview of the sequence of generations over the last decades, starting with the second generation, the first standard of the series to be developed by the 3GPP, which today is the leading organization in the specification of mobile communication systems. The Global System for Mobile Communications (GSMs) is the most prominent representative of 2G mobile communication networks. In Europe, GSM networks have been deployed starting from the early 1990s, offering voice and low-volume

Table 2.1 Overview of 3GPP releases and mobile network generations.

Network generation	Year	First 3GPP release
2G GSM	1992	GSM Phase 1
'2.5G' GPRS	1997	GSM Phase 2+/Rel. 97
3G UMTS	2000	Rel. 99
'3.5G' HSDPA	2002	Rel. 5
LTE	2008	Rel. 8
4G LTE-A	2011	Rel. 10
'4.5G' LTE-A Pro	2016	Rel. 13
5G new radio	2018	Rel. 15

data service to the subscribers. The design of the third generation (3G) mobile communications system, Universal Mobile Communications System (UMTS) started in the late 1990s and 2000s to enhance the data services provided by mobile networks.

2.1.1 Service Evolution – From Voice to Mobile Internet

Pre-5G mobile network generations already had to serve a continuously increasing variety of applications characterized by different traffic profiles and the associated Quality of Service (QoS) requirements. While second-generation networks were mainly designed to carry voice traffic and low-volume data service, the major novelty in 3G networks constituted mobile Internet access, particularly web browsing, access to mail service and streaming services with medium data rates.

In the 1990s, the **second generation** (2G) GSM [1] replaced the first generation. GSM is based on Time-Division Multiple Access (TDMA) and were the first digital wireless technologies which extended voice services to a vast population. In addition to voice, GSM introduced services such as text messages Short Message Service (SMS), fax services, voice mail, and High-Speed Circuit-Switched Data (HSCSD), with data rates of up to 57.6 kbps by aggregating four radio channels of 14.4 kbps. SMS has always been very popular, reaching its peak volume in 2011 and only slowly decreasing thereafter. SMS was the first coexisting data service in a circuit-switched system and used a control channel of GSM. A single text message comprises a maximum of 160 characters with best effort QoS profile. Multimedia Messaging Service (MMS), allowing the transmission of multimedia content up to 300 kB, was introduced in 2002, but never reached a comparable level of popularity. In contrast, SMS still plays a role today, e.g. IoT services with low data rates and relaxed latency constraints or for online payment systems (short message service-transaction authentication number (SMS-TAN)) or in developing countries for interaction with authorities.

As a circuit-switched system, GSM is particularly tailored for voice communication. Initially, the utilized voice codecs (full rate, FR and half rate, HR) required a transmission data rate of up to 13 kbps and allowed a trade-off capacity (HR) for quality (FR). FR is very robust even in the case of fluctuating radio channel quality since it allows for the identification of the more important parts within the encoded audio signal, so that those parts can be prioritized accordingly for radio transmission, e.g. using stronger channel coding. In 1997, the enhanced full rate (EFR) codec was introduced, which required a marginally lower data rate of 12.2 kbps while providing improved speech quality. EFR uses the same algorithm as the 12.2 kbps mode of the Adaptive Multi-Rate Narrowband (AMR-NB) speech codec. AMR-NB, with a total of eight-bit rate modes, was adopted as the standard codec by 3GPP in 1999.

GSM only provides very limited capacity for data transmission. UEs can only be allocated a single channel and the effective data rate is limited to 9.6 kbps. To cope with the growth of the Internet market, GPRS (General Packet Radio Service) was specified in the late 1990s, which achieved data rates of up to 172 kbps by allocating multiple time slots of a GSM frame to a single UE. Later, Enhanced Data Rates for GSM Evolution (EDGE) increased the data rate to around 230 kbps by introducing a more efficient modulation scheme, which allows for a maximum data rate of 59.2 kbps per time slot, and by bundling of four time slots per UE. In 2016, 3GPP standardized 'Extended Coverage GSM (EC-GSM)', a low-power Internet of Things (IoTs) protocol based on EDGE/GPRS for support of cellular IoT.

In the late 1990s, 3GPP developed the UMTS, the **third generation** (3G) global standards [2]. The radio technology changed to Wideband Code Division Multiple Access (WCDMA) to fulfil the performance requirements of International Mobile Telecommunication-2000 (IMT-2000). IMT-2000 was required to support multi-media (voice, data, video) traffic with data rates of up to 2 Mbps in stationary conditions and 384 kbps while moving at moderate speeds. The main goal of 3G was to increase of user bit rates on the radio interface to satisfy the expected demand for more data-intensive services, such as mobile video telephony or faster downloads. To support different services, four traffic classes with different latency sensitivity and throughput were identified: (i) conversational, (ii) streaming, (iii) interactive, and (iv) background class. The distinguishing characteristics of the classes are data rate and delay sensitivity. Table 2.2 depicts the UMTS QoS classes, their fundamental characteristics, and example applications.

Table 2.2 Quality of Service classes for the 3G system [3].

Traffic class	Fundamental characteristics	Example application
Conversational class	• Preserve time relation (variation) between information entities of the stream • Conversational pattern (stringent, low delay)	Voice, video telephony, video games
Streaming class	• Preserve time relation (variation) between information entities of the stream	Streaming multimedia
Interactive class	• Request response pattern • Preserve data integrity	Web browsing, network games
Background class	• Destination dos not expect the data within a certain time • Preserve data integrity	Background download of emails

In the conversational QoS class, the AMR-Wideband (AMR-WB) speech codec offers nine source data rates between 6.6 and 19.85 kbps. The bit rate is controlled by the RAN depending on load conditions and channel quality. In terms of end-to-end delay, a maximum value of 400 ms shall not be exceeded.

The maximum (downlink) data rate increased from 384 kbps in the original release (Release 99) to typically 7.2 Mbps (for the most common terminals of category 8) and 42.2 Mbps for MIMO-capable terminals, respectively. In uplink direction, maximum data rates increased from 384 kbps to 11.5 Mbps. To cope with the continuous surge in data traffic volume, 3GPP standardized High-Speed Downlink Packet Access (HSDPA) in Rel. 5 and later enhanced this to Evolved High-Speed Packet Access (HSPA+) in Rel. 7 [4]. Multi-carrier HSPA+ specified in further 3GPP releases, allows a theoretical data throughput of 337 Mbps in the downlink and 34 Mbps in the uplink.

2.1.2 2G and 3G System Architecture

The gradual transition from 2G to 3G networks also marked the shift from predominant voice services to increasingly important data services. Therefore, support for packet-based data transfer in the core network has already been introduced in GPRS and EDGE. Since the air interface still behaved like a circuit-switched call, part of the efficiency associated with a packet-based connectivity was lost. The main network elements (NEs) of the 2.5G GSM/(E)GPRS system are the Base Station Subsystem (BSS), which form the GSM EDGE Radio Access Network (GERAN), and the Core Network (CN). The Packet Control Unit (PCU) introduced for GPRS-EDGE is normally placed in the BSS, although the standard also foresees the possibility of having it in the Serving GPRS Support Node (SGSN) of the CN.

While the 3G/3.5G RAN is very different from the 2G/2.5G RAN and required operators to deploy new radio networks, the CN architecture substantially remained unchanged. The CN of the 2.5G and 3G/3.5G systems is composed of two domains: the circuit-switched (CS) domain providing services like voice or SMS, and the packet-switched (PS) domain providing IP-based services, cf. Figure 2.1. The 3G CN is an evolution of the 2G CN architecture, complemented with elements for new functionalities. The main NEs of the system architecture comprise:

- The **Mobile Station** (**MS**) is a combination of a terminal equipment (also referred to as Mobile Equipment, ME) and a separated module, the 2G Subscriber Identity Module (SIM)/3G Universal Subscriber Module Identity (USIM) where the subscriber data are stored.
- The RAN comprises of the 2G **GSM EDGE Radio Access Network** (**GERAN**) and the 3G **UMTS Terrestrial Radio Access Network** (**UTRAN**), which are described in Sections 2.1.3 and 2.1.4, respectively.

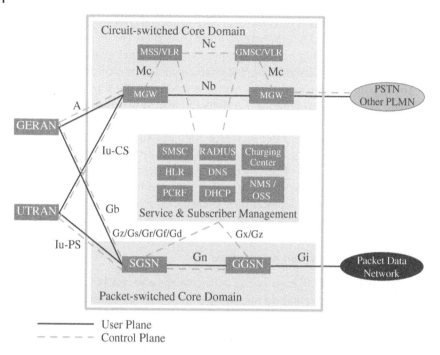

Figure 2.1 Architecture of 2G and 3G mobile networks.

- The **Home Location Register** (**HLR**) contains all subscriber data and their last known location and the **Visitor Location Register** (**VLR**) contains temporary subscriber information. The **Authentication Centre** (**AuC**) is a protected database holding the same secret keys as in the user's (U)SIM card.
- **Circuit-switched core**: the **Mobile Switching Centre Server** (MSS) carries circuit-switched data while the **Gateway MSC** (**GMSC**) terminates call control interfaces to external networks. From Rel. 4, the MSS was split into control and user plane functions, the MSS and **Media Gateway** (MGW) respectively.
- **Packet-switched core**: the **Serving GPRS Support Node** (SGSN) provides, amongst others, user authentication, mobility management, and session management by setting up PDP connections through which data are sent to external Packet Data Networks (PDNs). Since the SGSN performs the same functions as the MSS for voice traffic, they are typically collocated. The **Gateway GPRS Support Node** (GGSN) serves as the gateway towards an operator's IMS or to external PDNs.

While Rel. 4 still supports both the circuit-switched and packet-switched domains in the CN, the CN of Rel. 5 is fully IP-based, including voice services. Both SGSN and GGSN are upgraded to support circuit-switched services like voice. The Call State Control Function (CSCF) provides call control functionality for multimedia sessions and the Media Gateway Control Function (MGCF) controls media gateways. The Media Resource Function (MRF) supports features such as multi-party conferencing.

2.1.3 GERAN – 2G RAN

The second generation (2G) of mobile communication was almost completely voice-dominated, the architecture and system design followed the circuit switched approach with setting up and release of dedicated channels for voice calls [5]. Robustness against call drops was the main driver for an interference avoiding cellular network deployment. Cell and frequency planning were the most important operations before the installation of the system, i.e. substantial offline optimization effort took place during the planning phase. Frequency reuse is the core concept of cellular GSM network, where the same frequency is simultaneously used in geographical areas with sufficient distance to eliminate co-channel interference between adjacent cells.

The GERAN architecture consists of a base transceiver station (BTS) and base station controller (BSC). As depicted in Figure 2.2, several BTSs and one BSC build the so-called base station subsystem (BSS) using the *Abis* interface between BTS and BSC. The BSC manages the radio resources for one or more BTSs, it handles radio channel setup, frequency hopping, and handovers. The BSS also includes a transmission and rate adaptation unit (TRAU) which transforms GSM speech coded signals into Integrated Services Digital Network (ISDN) format used in network and switching subsystem (NSS). A high-speed line (T1 or E1) is used to connect the BSS with the Mobile Switching Centre (MSC) linked via the A interface, while mobile devices are linked via the so-called Um interface. The MSC

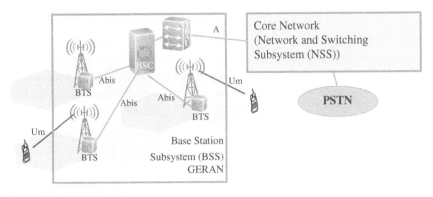

Figure 2.2 GERAN architecture.

belongs to the NSS which includes other core network elements like HLR, VLR, and AuC. To connect to another network, a GMSC is required.

2.1.4 UTRAN – 3G RAN

The UTRAN resembles the GERAN architecture consisting of multiple radio network subsystems (RNSs), where an RNS consists of one radio network controller (RNC) which controls and serves, depending on the manufacturer, up to several thousands of base stations called Node B (NB). Each Node B can provide service to multiple cells. A cell is defined as a certain sector with a specific carrier frequency.

As depicted in Figure 2.3, the newly standardized interfaces of the UTRAN architecture are the *Iub* interface between Node B and the RNC and the *Iur* between two RNCs. In the first UTRAN Release (Rel. 99), these logical interface connections are established by means of Asynchronous Transfer Mode (ATM)-transport links. The *Iu* interface connects the UTRAN with the Core Network (CN) which can be instantiated as *Iu-CS* for circuit-switched connection with MSC (like the *A* interface in GSM) and as *Iu-PS* to connect with the packet-switched part of the CN. From 3GPP Rel. 4, ATM transport links were replaced by Internet Protocol (IP) connectivity with mechanisms like Multi-Protocol Label Switching (MPLS) as a suitable alternative to manage different traffic classes with guaranteed QoS (cf. Section 2.5).

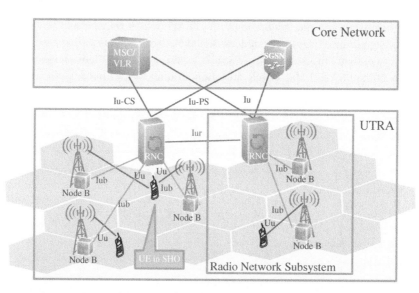

Figure 2.3 UTRAN architecture with interfaces to the core network.

The UMTS air interface connecting the UTRAN with the user equipment (UE) is called *Uu* interface and is the most considerable technology change between GSM/GPRS networks and UMTS networks. While GSM/GPRS used TDMA technology with multiple narrow-band 200 kHz carriers as multi-user access technology, for UMTS networks WCDMA was chosen with 5 MHz carrier bandwidth. Furthermore, from spectrum usage perspective, UMTS is offering both duplex modes, namely frequency division duplex (FDD) and time division duplex (TDD), while GSM was only operating in FDD mode combined with a quasi TDD component to support the duplex separation by means of a fixed time shift of the uplink and downlink slots.

Another remarkable difference is the frequency reuse factor of 1 in UTRAN. This is a result of the soft capacity philosophy of CDMA. However, it also requires new RRM methods like soft handover (SHO) or sometimes also called macro diversity, where the UE is simultaneously connected to two or more different cells. This guarantees smooth and failure-free cell changes on the expense of cell capacity, since resources in several cells are assigned to transmit the same information. The UE stays in the same carrier, while in GSM, the terminal must re-tune the frequency along with the cell change resulting in a hard handover with short disconnection. UMTS no longer needs the labour-intensive frequency reuse planning on expense of the (less crucial) scrambling code planning. Staying with mobility robustness example, the 3G approach with soft and softer (intra-frequency) handover is rather robust per se. However, the cell breathing effect is increasing the network planning complexity to balance between too many SHO links and failures due to intercell interference.

Introducing high-speed packet access (HSPA) to UMTS supported improvement in end-user experience with cost-efficient mobile broadband (e.g. mobile Internet), seen as an interim generation step called 3.5G [6]. It started with HSDPA (high speed downlink packet access) in 3GPP Rel. 5 and Rel. 6 introduced a new transport channel in the uplink, known as high speed uplink packet access (HSUPA) which allows for sending of large e-mail attachments, pictures, video clips, etc.

There has only been a limited number of evolutionary architectural changes for 3G, mainly related to radio resource management (RRM) and scheduling. While in Rel. 99, scheduling and retransmissions were handled by the RNC, this functionality was moved to the Node B in HSPA in Rel. 5. This placement closer to the air interface ensured faster reaction with Hybrid Automatic Repeat reQuest (HARQ) and lower latency. Another remarkable function of HSPA is the fast channel-dependent scheduling in a TDM manner providing a gain referred to as multi-user diversity.

2.2 Mobile Broadband Communications

Similar to 3G, the development of the **fourth generation** (4G) of mobile networks was driven by the subscribers' and the industry's demand for increased throughput due to new services requiring even higher data rates. Accordingly, the requirements of the International Mobile Telecommunications-Advanced (IMT-Advanced) report, issued by ITU-R in 2008, mandate, among others, a nominal downlink throughput of 100 Mbps for mobile users, increased spectral efficiency, and a scalable channel bandwidth. In general, 4G systems shall provide an adequate QoS for the most important services, amongst them mobile broadband access, video chat, mobile TV, and (at that time new) services like high-definition television (HDTV).

2.2.1 Mobile Broadband Services and System Requirements

To cover these QoS requirements, the most widely used system for 4G mobile networks, Evolved Packet System (EPS) of the 3GPP implements several mechanisms to provide the appropriate QoS in both the radio access (E-UTRA, Evolved Universal Terrestrial Radio Access) and the core network (EPC, Evolved Packet Core). Most importantly, for 4G, 3GPP has introduced the notion of an EPS bearer that is associated with a set of QoS parameters, namely QoS Class Identifier (QCI), Allocation and Retention Priority (ARP) as well as Maximum Bit Rate (MBR) and Guaranteed Bit Rate (GBR) values for uplink and downlink. The EPS bearer defines the end-to-end QoS granularity of the system, allowing for a fine-grained definition of QoS parameters for the service to be delivered by the bearer. The QCI, in an integer from 1 to 9, indicates nine different QoS performance characteristics, such as resource type (GBR or Non-GBR), priority (1–9), packet delay budget (50–300 ms), and packet error loss rate (10^{-2} down to 10^{-6}). ARP defines a bearer's priority level (1–15) and its pre-emption capability and vulnerability. For example, for conversational voice service, the following standardized QCI characteristics are defined: QCI = 1, priority level = 2, packet delay budget = 100 ms, and allowed packet error/loss rate = 10^{-2}. For buffered streaming and TCP-based services, two QCI values with different priority levels are defined [7]. The lowest standardized packet delay budget for the EPS, defined for, e.g. V2X, augmented reality, or discrete automation traffic, is 10 ms.

Regarding maximum data rates, the level of spatial multiplexing used on the radio link is a major factor. In a fundamental single input and single output antenna setup (SISO), the maximum downlink bit rate is 75 Mbps with 256 QAM modulation scheme. In 4 × 4 multiple input, multiple output antennas (MIMO) setups, approximately 300 Mbps can be achieved. Additionally, with different levels of carrier aggregation up to 100 MHz across different frequency bands, latest

UE categories even support up 1.2 Gbps downlink speed (sometimes referred to as 'Gigabit LTE'). Typical peak data rates in uplink direction are approximately 75 Mbps or, by using two component carrier aggregation, 150 Mbps.

2.2.2 4G System Architecture

Despite the high data rate provided by 3G systems, the continuous increase of data traffic led to the standardization of the fourth-generation mobile networks. The Long-Term-Evolution-Advanced (LTE-A) is based on orthogonal frequency-division multiplexing (OFDM) radio access and delivers more capacity for faster and better mobile broadband experiences, enabled by the 'always on' principle. In contrast to the previous network generations, where circuit-switched technology concepts built the baseline for the network architecture, 4G has been designed to support only packet-switched services [8]. It aims to provide seamless end-to-end IP connectivity between UE and the PDN. Accordingly, the RAN is called evolved UTRAN (E-UTRAN), which is accompanied by the System Architecture Evolution (SAE) including the EPC network. Together, LTE and SAE comprise the EPS.

The EPS adopted a rather flat architecture for improved scalability and better data throughput performance. The overall architecture implements an all-IP-based CN to which different RANs, 3GPP as well as non-3GPP, can connect through gateway interfaces. In addition, the EPC supports increased flexibility, allowing several mobility protocols, QoS, and Authentication, Authorization, and Accounting (AAA) services with support for multiple access technologies. Since the circuit-switching domain from earlier generations has been discontinued, voice is handled by the so-called 'Voice over LTE' (VoLTE) service. Figure 2.4 [9] depicts the overall architecture of the EPS. Key components include:

- **Home Subscriber Server** (HSS): replaces the previous HLR and contains subscription data, user information, registration, authentication (including keys for authentication and encryption), and maintains current gateway addresses.
- **Mobility Management Entity** (MME): manages most aspects of mobility including serving GW selection, roaming (with information from home HSS), idle mode and traffic data management, paging.
- **Policy and Charging Rules Function Server** (PCRF): provides service definition and gating for multiple data flows, QoS, billing, and application related functions.
- **Serving Gateway** (S-GW): is the local mobility anchor as the mobile moves between base stations. It is the visited gateway during roaming. It also provides lawful interception features, e.g. to support the requirements of US Communications Assistance for Law Enforcement Act (CALEA).

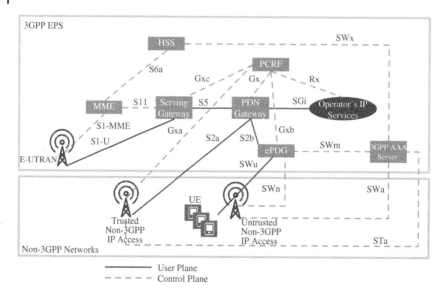

Figure 2.4 Network architecture of the Evolved Packet System (EPS).

- **Packet Data Network Gateway** (**PDN-GW** or **P-GW**): provides external IP interconnectivity, IP addresses, routing and anchors sessions towards fixed networks. It enforces data flow policies and provides lawful interception features CALEA.
- **Evolved Packet Data Gateway** (**ePDG**): provides inter-working functionality with untrusted non-3GPP access networks, particularly security mechanisms such as IPsec tunnelling of connections with the UE over an untrusted non-3GPP access. Trusted non-3GPP accesses can directly interact with the EPC.

2.2.3 E-UTRAN – 4G RAN

The E-UTRAN architecture abandons the hierarchical architecture of previous generations consisting of a controlling node (BSC/RNC) towards a flat architecture with a single powerful access node called enhanced Node B (eNB) which combines the base station (Node B) functions to terminate the radio interface and the functions of the RNC to manage radio resources.

The E-UTRAN illustrated by Figure 2.5 consists only of eNBs interconnected with each other by a new interface called *X2*. The inter-connection with the EPC is realized by means of the *S1* interface, which can be instantiated as user plane (UP) interface *S1-U* to the S-GW and as control plane (CP) interface *S1-MME* to the

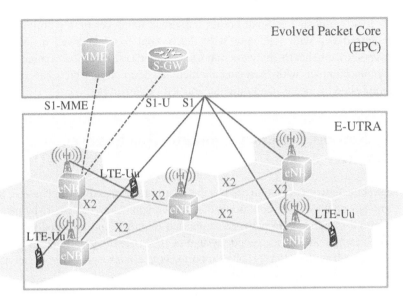

Figure 2.5 E-UTRAN architecture.

MME of the EPC. The *LTE-Uu* is the radio interface that connects the UE with the eNB.

The radio interface shows significant differences to the preceding UMTS. Multi-user access of UMTS is based on WCDMA, using fixed 5 MHz carriers, while LTE is using Orthogonal Frequency-Division Multiple Access (OFDMA) in downlink (DL) and Single-Carrier Frequency-Division Multiple Access (SC-FDMA) in uplink (UL) as multi-user access with scalable carrier bandwidth between 1.4 and 20 MHz. For spectral efficiency reasons, the tight frequency reuse of one has been kept in LTE irrespective of missing SHO capability resulting in considerable cell edge interference and strong impact on cell edge throughput and on timing of mobility management. Since the target cell identified for handover becomes the dominant interferer, very accurate timing of the cell change is crucial. And since radio conditions are different for each cell edge, mobility robustness optimization (MRO) was one of the first SON use cases standardized in LTE.

SON was introduced with the very first LTE standard release (3GPP Rel. 8), starting with features like Automatic Neighbour Relation (ANR), Automatic Physical Cell Identifier (PCI) assignment and Inter-Cell Interference Coordination (ICIC). ICIC is a good example of more complex network configuration: the need for sophisticated ICIC resulted from choosing OFDMA to enhance spectral efficiency while keeping frequency reuse of one.

More network optimization related SON features like Coverage and Capacity Optimization (CCO), Mobility Load Balancing (MLB) and as discussed above

MRO to accomplish a cell-pair border individual handover timing optimization were introduced with following 3GPP Rel. 9, and more followed in the later releases showing that the network management and configuration becomes more and more complex with local and momentary adaptation which can only be accomplished by means of automation and self-organization.

2.3 Network Evolution – Towards Cloud-Native Networks

The mobile network evolution can briefly be summarized as follows: The 2G GSM technology enabled voice communication to everyone and everywhere by means of a circuit-switched core network, supported roaming, and provided increased capacity compared to the analogue system of 1G. Since the year 2000, GPRS and EDGE, as well as 3GPP Rel. 99 (3G), supported both circuit-switched voice services as well as packet-switched data services.

3GPP Rel. 8 and 3GPP Rel. 10 (4G) have introduced a further design break by being a fully IP-based system. They feature a flat, scalable architecture designed to manage QoS for high-throughput services (mobile broadband). All services are packet-switched and no longer circuit-switched. The IP Multimedia Subsystem (IMS) provides a fully Session Initiation Protocol (SIP)-based control architecture.

For 5G, the technology evolution to cloud services has motivated the next major review of the network architecture. In the design of 5G, the 4G CN functionality, comprising rather monolithic network elements (such as, MME), has been further decomposed into smaller functions. Moreover, a service-based interface (SBI) design has complemented the traditional reference points between network functions. Before the details of the 5G system architecture are described in Section 2.4, this section will elaborate on the underlying technologies that will facilitate a transformation to cloud-native mobile networks.

2.3.1 System-Level Technology Enablers

Traditionally, telecommunication applications were designed and operated as proprietary solutions of tightly coupled hardware and software. Yet, the global development in Communication Service Provider (CSP) networks shows a shift from traditional voice and message services to web-centric data services. This fundamentally changes the architecture of mobile networks. Most of today's data services rely on cloud technologies which have been established in the IT industry over the last decades and helped to reduce costs, increase efficiency, improve business agility, and enable new business models. CSPs are looking for similar benefits by evolving their infrastructure towards cloud-native mobile networks:

1) **Increased agility in introducing new services**: the traditional delivery model required months from ordering to delivery, but the telco cloud allows resources for new functions to be available within minutes. Self-service interfaces and deployment automation allow rapid introduction of new services. So CSPs can quickly test new services and remove them with low cost if they do not fulfil expectations. Finally, 'pay per use' models can be supported more easily.

2) **Reduced hardware complexity**: Hardware consolidation and commoditization allows the same hardware to be shared between functions of different generations or in different phases of their lifecycle.

3) **Operational efficiency**: Through unified infrastructure, a single set of operations, administration, and maintenance (OAM) capabilities enables management of all VNFs. Consequently, higher levels of automation in service and network operations can be achieved.

Cloud service models outline how MNOs or, more generally CSPs, can evolve their networks towards providing cloud-based and web-centric service offering. The three basic cloud service models, Software as a Service (SaaS), Platform as a Service (PaaS) and Infrastructure as a Service (IaaS), are shown in Figure 2.6 and compared with a traditional 'on-premises' deployment.

SaaS is the type of cloud service with the highest level of abstraction. It enables users to connect to and use cloud-based applications, e.g. over the worldwide web. Exemplary SaaS offerings include Salesforce, Google's Apps and Gmail, Microsoft's Office365, Cisco's WebEx, or Dropbox.

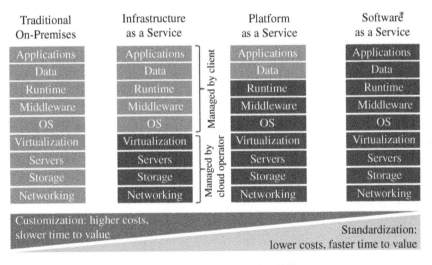

Figure 2.6 Cloud service models – technology mapping [10].

PaaS operates at a lower level of abstraction and usually comes with a platform on top of which software can be created and run. One of the benefits of this model is that all applications built using the platform tools PaaS can exploit the platform runtime services, such as scalability, high-availability, and multi-tenancy services. Examples include Google App Engine, Microsoft (Azure) Service Fabric, or Red-Hat OpenShift.

IaaS provides users access to computing, networking, and storage resources such as servers, I/O ports, and databases. Clients use their own platforms and application software within an IaaS infrastructure. Examples include Amazon Web Services, Microsoft Azure, Google Compute Engine, OpenStack, or VMware vCloud.

From the MNO perspective, the two initially relevant cloud service models comprise IaaS and PaaS, that correspond to different phases for moving mobile networks evolving networks towards a telecommunication cloud [11], cf. Figure 2.7.

Many telecommunication operators currently find themselves amid the first phase, where virtualized networks achieve a strict separation of software-based functionality from the underlying general-purpose hardware by introducing a virtualization layer (e.g. hypervisors). For this transitive phase, which roughly corresponds to adopting the IaaS model, many vendors have converted their existing products into software versions able to run in a cloud environment at relatively low cost without great impact on the product architecture and functionality. The goal has been to remove vendor-specific hardware and replace it with commodity IT hardware. However, for many NFs, this has proven to be quite inefficient and therefore required some performance optimizations (e.g. Intel Data Plane Development Kit 'DPDK') which again introduced hardware dependencies.

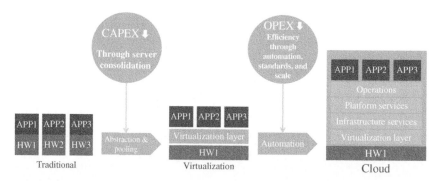

Figure 2.7 From traditional to cloud-native telecommunications networks [11].

Table 2.3 Comparison of virtualized and cloud-native networks.

Characteristic	Virtualized networks	Cloud-native networks
Level of automation	Manual operations; limited work-flow/script-driven automation	Extreme automation, model and policy- driven, 'post DevOps'
Investment strategy	Investment driven by standard refresh cycles	Investment driven by new business models, e.g. digitization of service delivery, new novel 5G/IoT service
Deployment characteristics	COTS + hypervisor + VNF (often single vendor) with limited orchestration capabilities	Multi-vendor, horizontal, interoperable cloud components
Key technology components	COTS servers, Linux, hypervisors, OVS/VPP, OpenStack	PaaS, machine learning, hyper-converged servers, compact DCs (e.g. edge clouds), hardware acceleration
Modularity	Handful of VNFs (limited scaling)	Dozen to hundreds VNFs or containers (dynamic, ephemeral networks)
Design and operation methodology	Software design mainly based on physical devices characteristics; reliability/redundancy/recovery schemes (frequently) part of application software	Software design built based on cloud principles; cloud reliability/redundancy/recovery mechanisms are provided by platform services

Source: Adapted from [12].

To fully harvest benefits and opportunities of the telco cloud, a system and product architecture transformation is required. Therefore, the second phase increasingly applies the design principles for cloud-native applications to mobile networks functions. The rather monolithic NFs and telco applications are decomposed into smaller building blocks which will not only use IaaS but are developed and operated using higher-level services and abstractions offered by a PaaS model. The network infrastructure as a whole evolves towards a set of platforms that offer automated and unified approaches to manage and operate applications. Exemplary platform services include procedures such as state management, resource scaling, load balancing, and similar application agnostic operations, thus relieving the telco application from these tasks. The most important differentiators between virtualized and cloud-native networks are summarized in Table 2.3.

2.3.2 Challenges and Constraints Towards Cloud-Native Networks

While the long-term shift towards cloud-native networks must be acknowledged, it is important to recognize the constraints of currently available commercial off-the-shelf (COTS) infrastructure components. Many telco applications, such as, delay-sensitive RAN functions, often have more challenging and diverse performance requirements than web-based IT applications. These must be considered when designing cloud-native telco networks, the key considerations for cloud-native mobile networks being:

1) **Mix of services and requirements**: Traditional telco services like voice require low and almost jitter-free latency, but limited throughput and to be considered telco-grade, they require 99.999% availability [13]. In addition, 5G networks aim at providing telco services to vertical markets for applications such as Industrial Internet of Things (IIoTs), Industry 4.0, smart grid, or car-to-car communications. These require e2e connectivity concurrently fulfilling high throughput, low latency, low jitter, and high reliability [14, 15]. Cloudified systems have yet to prove the capability to fulfil such demanding requirements at large scales. As such, networks may continue to mix of legacy and telco-cloud implementations.

2) **Network Function Virtualization** (**NFV**): Purely software-based telecom functionality shall run on non-proprietary, commoditized hardware, but specific functionality continues to rely on specific hardware, e.g. for performance reasons. Evolution to first virtualization and then cloud-native implementation remains challenging and ,in some cases not even possible. IaaS model as adopted in IT must be complemented with a dedicated cloud platform and classical network management for legacy systems.

3) **Programmability and dynamic reconfigurability of mobile networks**: Software-defined networking (SDN) splits NFs into the decision logic hosted in a control application and the controlled NF in the network that executes the decision. This decouples the control application from the controlled NF, allowing, for example, full separation of the control and the forwarding function within the gateway and hence more flexible service chaining. Another example comprises SON functions that can dynamically optimize and reconfigure RAN and transport functions for better network utilization.

4) **Cloud-native radio access**: The RAN is a particularly challenging environment for NFV and cloud implementations, due to strict real-time requirements and latency constraints of wireless communication. The RAN will continue to require a distributed topology, with dedicated hardware to support signal processing in the base stations. This topology inherently limits the expected benefits in terms of both statistical multiplexing and improved redundancy levels without significant additional deployments. On the other hand, cloud RAN

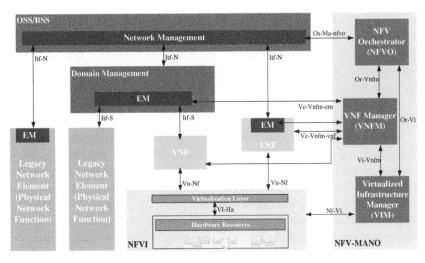

Figure 2.8 ETSI NFV MANO reference architecture. Source: Adapted from [16].

deployments are frequently accompanied by the introduction of Multi-access Edge Computing (MEC). Many applications, services, and contents may benefit from being executed at the access network.

5) **End-to-end management and orchestration**: NFV and, later, cloud-native networks were expected to exchange application-specific and frequently proprietary hardware with less expensive commodity components that would harmonize and thus reduce the operational effort for infrastructure management. Initially however, NFV has instead increased the operational effort by adding an extra stack of functions for orchestration and lifecycle management (LCM) of virtualized resources, namely the ETSI NFV MANO functions NFV Orchestrator (NFVO), VNF Manager (VNFM), Virtualized Infrastructure Manager (VIM). A further evolution of NFV MANO has defined the interfacing with the traditional 3GPP OAM functions, i.e. Element and Domain Managers as well as Operation Support Systems (OSSs), cf. Figure 2.8. It is the next evolutionary step that will achieve the seamless integration of LCM and FCAPS procedures through related industry efforts such as the Linux Foundation's Open Network Automation Platform (ONAP) project or ETSI's Zero-touch Network and Service Management (ZSM) specifications.

2.3.3 Implementation Aspects of Cloud-Native Networks

In cloud environments, microservices provide a scalable and efficient means to enable delivery of confined functionality as independent software building blocks which can be upgraded separately and in more frequent and smaller steps, e.g. continuous integration, delivery, and deployment. As a specific realization of service-oriented architectures (SOA), microservices expose well-defined interfaces to allow for lightweight service discovery, composition, and consumption, for example, the 5G Access and Mobility Management Function (AMF) could be further separated in smaller sub-functions like UE registration, mobility management, etc.

In addition to microservices, a cloud native system supports the implementation of stateless applications that can scale without the need to transfer, e.g. session context information across processing units. In such implementation architectures, session context states are kept in an external database that is accessed by all micro-services. This is realized by a so-called Shared Data Layer (SDL) architecture that splits the data storage, both subscriber and session data, from the service logic enabling stateless virtualized network functions (VNFs). The basic principle is depicted in Figure 2.9. This means that VNFs no longer need to manage their own data and will only run the required business service logic, making them easier and faster to develop [17]. As outlined above, historically, the 3GPP network architecture has been composed of self-contained network elements, each element storing and processing subscriber and service data. With the increase of subscribers and the demand for supporting new services and functions, this design faces scalability issues. Initial solutions, including standalone NFV, comprised scaling up individual network elements, which added complexity and limited optimization possibilities. Therefore, the shift to truly cloud-native system design has to follow.

The heterogenous compute, I/O networking, and storage capabilities of a telco cloud infrastructure are distributed over numerous nodes and need to be interconnected on demand. The **Smart Network Fabric** is a multi-layer, scalable, and dynamically reconfigurable IP/optical cloud interconnection to seamlessly process and transport data according to the service requirements. It connects central, metro, and edge nodes of public mobile network operators, datacentres of Internet

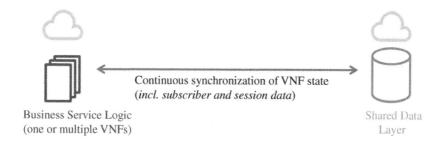

Continuous synchronization of VNF state
(*incl. subscriber and session data*)

Business Service Logic
(one or multiple VNFs)

Shared Data
Layer

Figure 2.9 Concept of the shared data layer.

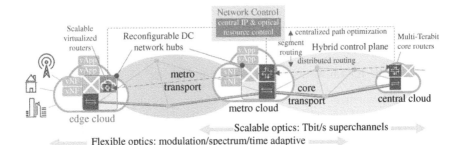

Figure 2.10   Key principles of the smart network fabric. Source: Adapted from [18].

service and cloud computing providers, private nodes in enterprises or residential homes, and hub nodes of the Internet backbone.

To ensure both efficiency and reliability, the scaling and allocation of resources amongst clouds needs to be dynamic. The result is a unified cloud infrastructure that is dynamically distributed over a mesh of locations interconnected by means of the Smart Network Fabric. Figure 2.10 illustrates the Smart Network Fabric connecting edge, metro, and central clouds. While IP continues to be the enabling convergence protocol, Segment Routing could be a potential solution for providing a 'hybrid control plane', i.e. exploiting the efficiency of centralized path optimization and retaining the scalability of distributed routing.

The outlined technology enablers and their role in building a cloud-native mobile network have had a major impact on the design of the 3GPP 5G system Rel. 15 and continue to shape the design of upcoming releases. This is described on the following sections for both the overall 5G system and the associated RAN.

2.4   Multi-Service Mobile Communications

Mobile networks of the fifth generation are becoming an important enabler for the ongoing digitization of business and daily life and the transition towards a connected society. Services in high priority fields for society (e.g. education, health, government services) or business areas (e.g. smart grids, intelligent transport systems, industry automation) will rely on a versatile mobile network infrastructure. In these areas, machine-type devices will increasingly contribute to mobile traffic, introducing highly diverse characteristics and requirements significantly different from today's dominant human-centric traffic. Hence, 5G networks must combine high performance with the support for service-specific functionality. For economic reasons, a common infrastructure platform that is shared amongst multiple communication services is needed. 5G networks will exploit the so-called 'network slicing feature' to enable multi-service and multi-tenant capable systems.

2.4.1 Multi-Tenant Networks for Vertical Industries

For 5G development, three fundamental communication service classes have been introduced, (i) massive broadband, characterized by very high data rates, (ii) critical machine-type communications (cMTCs), also referred to as ultra-reliable, low-latency communications (URLLCs) and characterized by low latency and high reliability, and (iii) massive machine-type communications (mMTCs), characterized by highly scalable and cost-efficient network operations. A detailed analysis on the requirements per service class is given in [14], For functional requirements, however, multi-service mobile networks shall incorporate numerous capabilities, some of which are already supported by the EPS including:

- Service-specific traffic detection and network performance monitoring rules;
- Service-specific network control: settings for QoS, mobility, security, etc. can be dynamically derived based on detected services and network context, for example, mobility management configurations may be specified for UE type and class, UE mobility pattern, the detected service characteristics in terms of reliability and continuity, and the capabilities of the radio access technology (RAT);
- Adaptation of NF configuration to enable service differentiation, i.e. NF selection, placement, and configuration is adapted to the supported communication service. For instance, virtualized NFs may be located at the central cloud for services with relaxed latency requirements or placed at the network edge for low latency services. Furthermore, multi-connectivity may be configured for increased throughput for broadband services, or it could be configured to increase reliability for URLLC services.

For multi-tenancy support, each tenant's business requirements, service level agreement (SLA) policies, and resource isolation requirements, need to be considered:

- Network resources such as networking capacity and storage processing are divided into different pools for different resource commitment (sharing) models.
- Different resource management policies are defined per tenant. In particular, each tenant can require different QoS levels which may be fulfilled by means of, for example, particular NF chains, function configurations, or radio scheduling policies.

- Tenants, depending on their level of expertise, may request different service monitoring and configuration capabilities via a GUI or APIs. Therefore, selected system functionality may be exposed to allow tenants supervision and control of their communication service in a limited fashion.
- Network operators and infrastructure providers require tools and processes to prioritize tenants and their respective services. As mediator and admission controller, they need to prioritize and admit tenants' service requests according to an objective function that reflects overall operator targets and utility.

2.4.2 5G System Architecture

3GPP foresees several deployment options for the fifth generation (5G) of mobile networks, the so-called 'standalone' (SA) option and several variations of 'non-standalone' option. In SA ('Option 2'), 5G UEs use 5G New Radio (NR) for both user plane and control plane traffic. gNBs connect to the 5G core network (5GC), which could also interact with legacy (e.g. LTE) base stations. Options 3 and 7 combine a mixed 4G–5G RAN with either 4G or 5G CN. 5G NR is used as an additional carrier for user plane traffic whereas control plane traffic is handled by LTE. In NSA option 3 'LTE assisted, EPC connected', a 4G core network is used, while option 7 'LTE assisted, 5GC connected' assumes a 5GC. Both options have two variants: either gNBs directly connect to the core network or via the LTE base station (note: the latter case is not depicted in Figure 2.11). Option 4 is similar to option 7: control and user plane traffic is now handled by 5G NR while LTE only serves as an additional carrier for user plane traffic. Table 2.4 summarizes the major characteristics of SA and NSA.

The substantial increase of mobile broadband traffic volume as well as the objective to support a wide range of diverse services for vertical industries requires

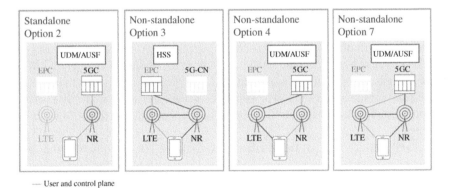

— User and control plane
— User plane only

Figure 2.11 Selection of 5G deployment options.

Table 2.4 Comparison of 5G NR standalone and non-standalone options.

	Standalone (SA)	Non-standalone (NSA)
NR radio cells	Directly used by 5G device for control and user planes	5G used as secondary carrier, under the control of LTE eNB ('Option 3' or 'Option 7')
Choice of core network	5G core network (5GC, 'Option 2'), which may also anchor inter-RAT mobility with LTE	4G EPC ('Option 3') or 5GCN ('Option 7')
CSP perspective	Simple, high performance overlay	Leverages existing 4G deployments
Network vendor perspective	Independent RAN product	Requires tight interworking with LTE
End user experience	Peak throughput set by NR; Dedicated low latency transport	Peak throughput is up to the sum of LTE and NR; Latency impacted when routed via LTE

a flexible, elastic, and programmable network. To achieve this objective, 3GPP specified a new system architecture for 5G. The 5GC is a cloud native CN, that leverages on NFV and SDN as the enabling technologies for abstracting network functions from the underlying hardware.

In terms of architecture, the 5GS is an evolution of the 4G system. It introduces distributed User Plane Functions (UPFs), formerly S-GW and P-GW, and maintains Control and User Plane functions Separation (CUPS). Several use cases will benefit from having flexible UPF deployment options, depending on the communication service requirements. For example, while low latency services require UPFs close to the radio, the best option for basic Internet access continuous to be rather centralized UPFs. In the control plane, the 5GC further decomposes 4G functions, e.g. by introducing separate functions for Session Management (SMF) and Access and Mobility Management (AMF). In practice, this means that smaller software components can be deployed independently (e.g. as microservices), configured, and scaled to support the cloud-centric operations and the LCM model of the 5GS.

3GPP Rel. 15 system architecture specifications [19–21] describe the 5GC architecture in two ways with the same set of network functions: (i) a reference point-based architecture, reflecting a traditional 3GPP architecture (not shown here), and (ii) a service-based architecture (cf. Figure 2.12) which is fully aligned with cloud principles. Option 2 reflects the 'Network Cloud OS' concept where

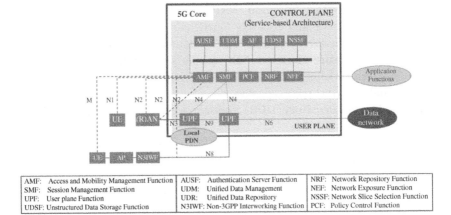

AMF: Access and Mobility Management Function	AUSF: Authentication Server Function	NRF: Network Repository Function
SMF: Session Management Function	UDM: Unified Data Management	NEF: Network Exposure Function
UPF: User plane Function	UDR: Unified Data Repository	NSSF: Network Slice Selection Function
UDSF: Unstructured Data Storage Function	N3IWF: Non-3GPP Interworking Function	PCF: Policy Control Function

Figure 2.12 Architecture (service-based view) of the 5G system.

network services are 'composed' using a library of functions hosted in a cloud and 'chained' together to create the end-to-end service. The key components of the 5GC architecture include:

- **Access and Mobility Management Function** (**AMF**): The single control plane function that terminates the interface from the access networks and from the UE, manages access control and mobility, and plays a key role in network slice functionality by serving all slices a UE is accessing
- **Session Management Function** (**SMF**): This establishes and manages sessions for all access types according to the network policy. In EPC, this functionality is spread across the MME and S/PGWs
- **User Plane Function** (**UPF**) is equivalent to the user plane of the EPC serving/packet data network gateway (S/P-GW), but is enhanced to support flow-based QoS and a new session and service continuity mode that allows a make-before-break function for URLLC. Operators can deploy multiple UPF instances in distributed and centralized locations, according to service type for example
- **Policy Control Function** (**PCF**) delivers a common policy framework by exposing policies as a service that are consumed by any authorized client. Besides QoS and charging, it supports network slicing, roaming, and mobility management policies
- **Authentication Server Function** (**AUSF**) provides a common authentication framework for all access types
- **Unified Data Management** (**UDM**) stores subscriber data and profiles, as is done by the HSS or HLR. It is envisioned that it integrates subscriber information for both fixed and mobile access in 5GC

- **Network Repository Function** (NRF) provides new 5G functionality. NRF provides registration and discovery functionality enabling network functions and services to discover each other and communicate directly via open APIs
- **Network Exposure Function** (NEF) supports the external exposure of capabilities of network functions, such as, monitoring capability, provisioning capability, and policy/charging capability
- **Unified Data Repository** (UDR) represents a common backend for the UDM, NEF, and PCF
- **Network Slice Selection Function** (NSSF) provides the new 5G functionality of assigning UEs to network slice instances
- **Unstructured Data Storage Function** (UDSF) provides the new functionality allowing control plane functions to store their session data and to become session stateless
- **Non-3GPP Interworking Function** (N3IWF) is a 5GC function for the integration of the stand-alone untrusted non-3GPP access to the access agnostic, universal core. N3IWF terminates IPSec and IKE v2 and exposes towards the common core N2, N3 similar to the RAN.

Procedures as well as the Policy and Charging Control Framework for the 5G system are described in [20, 21], respectively.

Figure 2.12 depicts the service-based architecture (SBA) design. It comprises a major change to the reference-point-based design of previous generations and is also expected to be the more relevant deployment option.

2.4.3 Service-Based Architecture in the 5G Core

The 3GPP SBA largely follows SOA software design principles. SOA-based systems comprise functions that offer services that are exposed to service consumers through a well-defined interface. The high level of modularity allows flexible composition of complex services by mixing and matching fine-grained services. The Open Group determines four defining properties of a service [22]: (i) It logically represents a business activity with a specified outcome; (ii) it is self-contained; (iii) it is a 'black box' for service consumers; (iv) it may be composed of other services.

The SBA of the 5GC defines services that any service consumer with proper authorization can use. A service consumer can consume multiple services and it can create its service offerings that are exposed to other service consumers. The relationship between service consumer and service producer is not fixed by the type of logical entity, as was the case of the earlier used reference point-based architecture versions, where reference points between logical entities were defined together with the protocols. Instead, in SBA service, consumers do not have static relationships with the service instance they are using, but a service consumer discovers those services it needs by a service discovery procedure offered by NRF. This service-based approach results in more flexible and dynamic architecture where new services can be introduced without impacting other service producers or service consumers. Only a new service interface needs to be defined and made available to the needing service consumers. The approach matches well with the cloud computing concept where the location and capacity of a service can change dynamically.

In the 3GPP Rel. 15 SBA model, services are implemented by Network Functions (NFs) and exposed through SBI. In contrast to a reference-point-based architecture, an NF exposes interfaces that are neither bound to a specific reference point nor to a specific consumer. Rather, the binding takes place through service discovery, selection and well-defined authentication and authorization means. Each of the services provided by an NF exposes a separate SBI for any service consumer to use. For example, the AMF is defined to contain and offer four different services to any authorized service consumer, typically to NFs such as SMF and Policy and Charging Function [21]. Any of the following AMF services, can be upgraded without impacting other interfaces and their consumers [23]:

1) Namf_Communication service used to communicate with the UE and the RAN
2) Namf_EventExposure service providing notifications based on mobility events
3) Namf_MT Service which provides paging
4) Namf_Location service which is a servicing positioning request.

Each of these services are self-contained and their life cycle can be managed separately.

The roles of NF service producer and NF service consumer closely follow the traditional client-server model of the Web-services in their interaction with each other. The service APIs are defined by using REpresentational State Transfer (REST)-ful [24] service operations where standard HTTP-methods are applied whenever possible. According to the REST model, the application-level state is contained in the responses to the requests. Based on the state in the response message from the service producer, the service consumer learns how to interact with the service producer in subsequent transactions. The service producer doesn't need to maintain any state for a given service consumer. The service consumers and producers use the HTTP-protocol to exchange the representations of resource states and operations on them. A URI consist of Path-part and Query-part [25]. All SBI APIs and services are defined by the supported resource URI with possible parameters and by the allowed HTTP-operations (e.g. GET, PUT, POST, PATCH, etc.) against the resources.

All service producers register and deregister their services with the NRF. Service consumers use the services of NRF to discover and select service instances

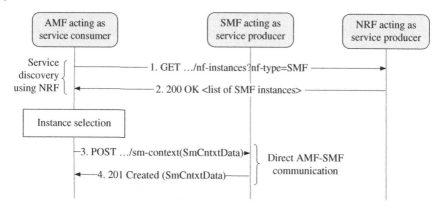

Figure 2.13 Example of SBA interactions between AMF and SMF.

they would like to consume. NRF authorizes a service consumer to access and use service instances as part of the service discovery process. NRF also provides system level services such as monitoring the load situation of the services and their liveliness. Therefore, NRF is the most critical and central component of the whole SBA.

Figure 2.13 shows an example how the NFs of the 5GC SBA interact with each other. In the depicted scenario, an AMF contacts an SMF to create a context for a UE for subsequent requests to establish PDU sessions. This may have been triggered by NAS-signalling at the time when UE registered to the network. At step 1, the AMF needs to discover an SMF instance that has capacity to serve the UE. It sends a service discovery message to NRF in a form of 'HTTP-GET' message with proper parameters. As a response, the NRF returns a list of matching SMF instance identifiers if there is more than one SMF instance available in the '200 OK' message at step 2. The AMF completes the instance selection by using the list of possible SMF instance identifiers and sends in a 'POST' message containing information about the UE to the selected SMF at step 3. If the context creation was successful, the SMF responds with '201 Created' message to the AMF at step 4.

2.4.4 5G RAN

The design of the fifth-generation radio network architecture had to consider a diverse set of requirements, both functional as well as performance related, making it a challenge for system design and standardization [26]. These – partially interdependent – main requirements are:

- As far as possible, integration of new radio (NR) and E-UTRA into a common architecture allowing seamless handover and interworking between both RATs
- Allow for flexible centralization of radio functions depending on the infrastructure capabilities and deployment scenario

- Scalability for different use cases and scenarios ranging from ultra-broadband to massive IoT
- Support for virtualization and softwarization of RAN functions.

The first requirement in this list – integration of NR and E-UTRA – has an impact to virtually all aspects of the architecture as it is eminent in the different options for stand-alone (SA) and NSA system architecture, cf. Table 2.4. For the RAN architecture, the main impact is the integration of RAN nodes with NR and E-UTRA air interface via common interfaces towards 5GC, and between RAN nodes, as shown in Figure 2.14. Similar to the LTE-A, NG-RAN defines a 'flat' architecture (see [27, 28]) where RAN nodes are interconnected via *Xn* interface and connected to the 5GC via the *NG* interface, similar to the 4G EPS. In NG-RAN, a gNB is a RAN node hosting NR radio, while an ng-eNB hosts E-UTRA radio. The RAN-level interface between different RATs enables a variety of interworking options, including:

- Hand-over between NR and E-UTRA with user-plane data forwarding in RAN
- Dual Connectivity (DC) between NR and E-UTRA, with gNB or ng-eNB as Master Node (MN) or Secondary Node (SN)
- E-UTRA-NR co-existence including supplementary uplink (SUL), where a single TDD band is shared by LTE and NR for uplink
- ANR between NR and LTE.

The common RAN-CN interface enables access-agnostic core network as far as possible, reducing the need for RAT-specific functions. Flexible centralization,

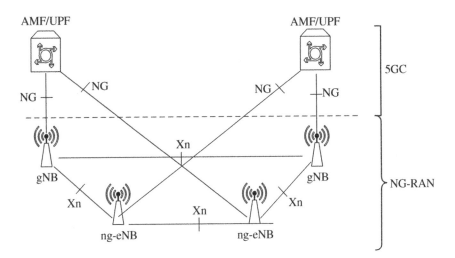

Figure 2.14   NG-RAN architecture [27].

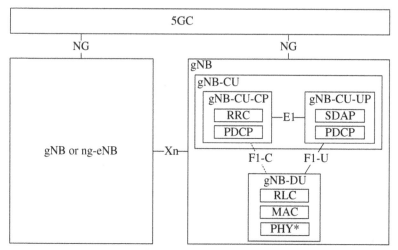

* -lower-layer(intra-PHY) split defined e.g.in CPRI Forum

Figure 2.15 Logical split RAN architecture [28].

scalability, and virtualization is supported by means of a logical split architecture of RAN nodes, which allows more independent scaling and instantiation of UPFs as well as dislocated placement of RAN functions. NG-RAN Rel. 15 [29, 30] defines a logical fronthaul interface (*F1-C* and *F1-U*, cf. Figure 2.15) for 'higher-layer' split and a split between control-plane and user-plane. Higher-layer split in this context, denotes a split between Packet Data Convergence Protocol (PDCP) and Radio Link Control (RLC) protocol, but still in OSI layer 2. Lower-layer split options denote options where the split is in the physical layer. Those options are not defined by 3GPP, but. by CPRI forum, for example, as *eCPRI* which specified a transport for an intra-PHY split point [31]. This option has a significantly lower bandwidth requirement than the original CPRI split point designed for quantized symbol transport.

It is important to note that RAN logical split is transparent to the network and to the UE. This allows realization of RAN nodes both as monolithic (and distributed) as well as centralized (but with fronthaul split) deployments [26], as well as instantiation of RAN functions according to service or network slice requirements.

The split logical RAN architecture of NG-RAN Rel-15 is shown in Figure 2.15. The *F1* logical interface interconnects the Centralized Unit of a gNB (gNB-CU) and the distributed unit (gNB-DU). It is further distinguished in *F1-C* for control plane and application protocol, as well as *F1-U* for user-plane data. Furthermore, the *E1* control interface provides CP/UP separation in the case of an existing *F1* interface.

Table 2.5 Scaling limits of gNB logical entities.

Logical entity	Hosting entity	Maximum allowed logical entities within hosting entity	Limiting element	Source
gNB-CU	gNB	1	By definition	[28]
gNB-DU	gNB	2^36	gNB-DU ID	[32]
gNB-CU-CP	gNB-CU	1	By definition	[28]
gNB-CU-UP	gNB-CU	2^36	gNB-CU-UP ID	[33]
NR cell	gNB	1 024–16 384	NR cell identity and gNB ID	[34]
NR cell	gNB-DU	512	maxCellIngNBDU	[32]
TNL association	gNB-CU	32	maxNoOfTNLAssociations	[32]

One gNB-CU-CP can be associated to several gNB-DU and gNB-CU-UP entities, allowing independent scaling and redundancy for user plane. There are some limitations on the number of logical entities possible within a gNB, as shown in Table 2.5.

2.4.5 5G New Radio

The new 5G RAN (denoted as 'NR') features several key technology components which allow fulfillment of the quite challenging and diverse requirements on latency, reliability, and throughput. One of the main features of NR is the extremely flexible support for carrier frequencies from 450 MHz up to 40 GHz, with a supported channel bandwidth in the range of 5 up to 400 MHz. The higher frequency ranges, in combination with carrier aggregation enable configurations with up to 1200 MHz aggregated channel bandwidth, allow 20 Gbps throughput for eMBB to reach use cases. The lower frequencies are ideal for wide coverage scenarios, potentially with building penetration such as required for massive IoT. Table 2.6 lists a few of the most important NR features and their benefits. This section elaborates on selected features in more detail.

This flexibility is enabled by a scalable carrier numerology with sub-carrier spacing ranging from 15 kHz for lower channel bandwidth up to 120 kHz for 400 MHz channels. In total, a sub-carrier spacing in the set of {15, 30, 60, 120} kHz is supported for data channels, where the scaling with $2^n \cdot 15$ kHz allows for efficient FFT processing for the OFDM symbol generation as well as compatibility with LTE subcarrier spacing (15 kHz).

Table 2.6 Selection of NR features and their benefits.

Feature	Benefit
Usage of spectrum above 6 GHz	10× to 100× more capacity
Massive MIMO and beamforming	Higher capacity and coverage
Lean carrier design	Low power consumption, less interference
Flexible frame structure	Low latency, high efficiency, future-proofness
Scalable OFDM air interface	Address diverse spectrum and use cases
Scalable numerology	Support of multiple bandwidth and spectrum
Advanced channel coding	High robustness for short blocks, large data blocks with low complexity

NR supports time-slot durations between 0.125 and 1 ms depending on the sub-carrier spacing and the corresponding OFDM symbol length in time. One slot always comprises 14 symbols, meaning that one radio frame of 10 ms length can comprise up to 160 slots for higher sub-carrier spacings. To meet ultra-low latency requirements, NR further enables flexible configuration of downlink and uplink slots within one frame.

5G NR needs to support different type of services in a radio cell, meaning that the frame design needs to be sufficiently flexible to support configurations for 0.5 ms latency of layer 2 packets on the one side, but on the other side longer and more power-efficient configurations for low-complex, massive IoT devices. This design objective is achieved by introducing bandwidth parts (BWPs), which serve as a self-contained structure in the resource grid which enables different configurations of sub-carrier spacing, slot formats, and correspondingly, mapping to quality of service requirements within a single carrier.

Figure 2.16 illustrates the basic principle of the flexible frame structure in 5G NR. Each of the different parts of the frame annotated with different use cases could be configured such that efficiency, e.g. in terms of reaching QoS KPIs, resource utilization, device power consumption, etc. is optimized. Devices which are configured for BWPs only need to decode common control information, and control information for the corresponding BWP. This feature can be also be applied to the end-to-end concept of network slicing, such that BWPs and corresponding configurations are associated with one network slice or potentially several slices. It should be noted that the flexibility comes with a price related to schedule and implementation complexity in the network. On the device side, low

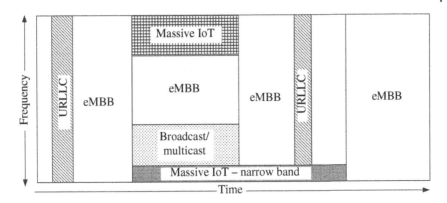

Figure 2.16 Flexible frame structure in 5G NR.

complex design would require conscious choices of configuration options which need to be agreed by industry stakeholders.

A further key pillar in the 5G NR framework is the extensive support of massive MIMO and beamforming. Massive MIMO in the context of NR denotes support of antenna arrays with a large number of controllable antennas (e.g. 32 controllable transmit antennas [32 Tx], or more). Beamforming denotes a specific transmit/reception strategy where massive antenna panels are used to form directional beams with high antenna gains. This is specifically required to overcome the challenging propagation characteristics of higher frequencies above 6 GHz. Nevertheless, beamforming is also useful for frequency deployments below 6 GHz, e.g. in the case if LTE carriers in 1.8 GHz bands should be complemented, at the same site, with NR resources in 3.5 GHz bands. In addition to the coverage advantages of beamforming, higher-order spatial multiplexing can be applied by re-using the same time-frequency resources in different directional beams.

Three types of antenna array architectures are supported. The one most capable in terms of flexibility and granularity of beams is digital baseband beamforming. In this case, each antenna element or antenna port has a dedicated transceiver unit which can be controlled in terms of gain and phase individually. This allows dynamic frequency-selective beamforming where different beam patterns can be generated for different frequency ranges, thus allowing the combination of spatial, time, and frequency multiplexing. However, digital beamforming is subject to high power consumption and cost characteristics when bandwidth increases.

As a further concept, analogue beamforming denotes an architecture where beam patterns are generated by adapting Tx/Rx weights in RF (i.e. in the analogue domain at the antenna). This architecture allows the implementation of (semi-) static beam patterns with pre-defined Tx/Rx weights which can be switched at different time instances. For example, beam switching is used to generate a 'rotating' beam pattern in time with one directional high-gain beam for coverage optimization. For control channels and channel acquisition, a wide coverage beam can be generated. In this architecture, only one transceiver unit is needed, reducing cost and complexity especially for high-frequency scenarios.

Finally, hybrid antenna arrays combine digital with analogue beamforming. Several RF units are combined and controlled by a single transceiver unit, such that a limited number of beams/beam patterns can be generated at the same time, but with less granularity and flexibility than with fully digital beamforming. This enables CCO also in higher frequencies but with less complexity if, for example, up to eight transceiver units are supported.

From a network management and operational perspective, it is important to understand that each of the antenna array architectures require different sets of parameters and also exhibit different operational behaviours in terms of resource allocation, parameter optimization e.g. for mobility, and other management aspects.

5G NR is the first 3GPP Release where beamforming is an integral design objective of the air interface. For this purpose, NR provides a beam management framework to optimize performance using beamforming. Beam management consists of the following phases:

- beam indication, which helps the UE to select beams in uplink and downlink directions both for initial access and in connected state
- beam measurement and reporting, including reporting of suitable downlink and uplink beams for a UE to the network
- beam recovery for rapid link reconfiguration against sudden blockages
- beam tracking and refinement to optimize beam parameters at UE and gNB.

Beam management procedures are anchored mainly at MAC and PHY layers. However, cell level mobility is related to beam management in the sense that the same procedures are applied, but additionally the serving cell is changed and, possibly, a path switch for user plane data from the core network is performed.

Multi-connectivity is one of the key features of NG-RAN (including E-UTRA and NR RATs), which is not only used for bandwidth aggregation and throughput enhancements, but also for reliability enhancements both on control- and user-plane by duplicating packets on radio bearers anchored in physically disjunct locations. Multi-connectivity in NG-RAN is denoted as Dual Connectivity (DC – similar to LTE), but with more flexibility and capabilities. In general, DC user plane operation (traffic splitting, aggregating, duplicating, and discard) is anchored in the PDCP layer. The following types are differentiated according to the air interfaces involved and connectivity to the core network, following the standalone and non-standalone architecture options described in Table 2.4:

- **MR-DC**: A generic term for Multi-RAT Dual Connectivity where air interfaces of the master and the secondary node are not the same
- **NR-NR DC** (short: NR-DC): Dual Connectivity where both radio nodes are gNBs hosting NR radio interfaces
- **E-UTRA-NR-DC** (short: EN-DC): Dual Connectivity where the master node is an LTE eNB connected to an EPC, and the secondary node is a gNB with NR air interface
- **NR-E-UTRA-DC** (short: NE-DC): Dual Connectivity where the master node is a gNB with NR air interface, and the secondary node is an ng-eNB, i.e. an eNB with E-UTRA (LTE) air interface connected to a 5G core
- **NG-E-UTRA-NR DC** (short: NGEN-DC): Dual Connectivity where the master node is an ng-eNB, and the secondary node is a gNB.

In practice, the variants which will presumably gain the highest market relevance will be NR-DC and EN-DC since many NR deployments will start as a 'capacity booster' for an existing LTE network.

To enable URLLC services, 3GPP has also defined a set of features including:

- Robust channel coding and repetition for more reliable transmission
- shorter transmission time intervals down to 0.125 ms to achieve 1 ms overall latency over the air
- packet duplication on PDCP layer
- multi-connectivity for increased reliability, including RRC diversity via signalling radio bearer (SRB) split
- flexible function placement with local breakout and edge computing
- HARQ enhancements, i.e. flexible and faster HARQ round-trip time (RTT) as well as asynchronous and automatic HARQ
- MAC scheduling enhancements for reduced transmission latency, including pre-reserved, restricted, and pre-emptive schemes.

2.4.6 5G Mobile Network Deployment Options

Today, cloud-based IT and, although to a lesser extent, telecommunications infrastructures are largely based on centralized topologies where users' traffic is sent to a small number of datacentres in a central place. This model is well suited for mobile Internet services and has the great advantage of keeping an operator's total cost of ownership (TCO) low.

However, there are several use cases that greatly benefit from having their application components closer to the user, i.e. placed in the access network. For example, many 5G use cases require low latency with round trip times of a few milliseconds, amongst them industry automation and tactile Internet services. Other use cases such as autonomous vehicles and drones benefit from minimizing

V2X communication delay by placing applications closer to the user. In addition, transporting large traffic volumes to a central place, as some of content delivery network (CDN) use cases require, may cause transport bottlenecks. Those factors, low latency applications and proper scaling of CDNs lead to a paradigm shift in cloud architecture design.

Multi-tier edge cloud deployments form the basis of the emerging infrastructure paradigm that features distributed datacentres at the edge of the network. Those edge clouds are located at different aggregation levels between the radio periphery and the core networks. The exemplary three-tier network infrastructure (see Figures 2.1–2.4) consists of a very large number (100s up to a few thousands) of edge sites and a considerable number (up to 100) of aggregation sites, which are connected to a few (up to 10) large centralized datacentres. Such distributed edge sites create the basis for introducing flexible deployment of softwarized network functions, cloudified RANs, MEC platforms, e.g. [35, 36], and related applications.

As outlined in Section 2.3, decomposition and hardware-independent design of 5G applications allow for a flexible and adaptive commissioning of network functions in different locations of the network. Virtualized NFs can even be re-located during operation with limited impact on performance. Figure 2.17 depicts an example of network evolution exploiting decomposition and flexible allocation of NFs, including both fixed and mobile access NFs. Those access NFs are decomposed into smaller software functions hosted on computation platforms in the edge clouds together with converged core network functions and relevant applications and services. Mostly cloud-based centralized RAN and optical line terminal (OLT) functions of passive optical networks (PONs) are deployed in the edge cloud in high-density areas, while more distant and moderately populated areas are served by small access hub locations containing real-time (RT) baseband and/or OLT functions. Such flexible deployment options are essential to adapt to the foreseen traffic variations of 5G networks.

In a more detailed breakdown, Figure 2.18 differentiates four basic regions that span over six aggregation levels to form a topological view of 5G cloudified networks:

- The **antenna site region** usually hosts functions that are characterized by a tight coupling of software and hardware (physical network functions (PNFs)) due to performance reasons. PNFs are most commonly located at the antenna or RF premises site. However, they can also reside in other locations of the network infrastructure.
- The **edge cloud region** covers the peripheral sites with transport hubs and packet-optical aggregation points, comprising the first two aggregation levels, 'photonic backhaul' and 'metropolitan area networks (MANs)'. RAN nodes located rather close to the antenna site comprise a very limited amount of general-purpose compute, storage, and networking hardware (e.g.

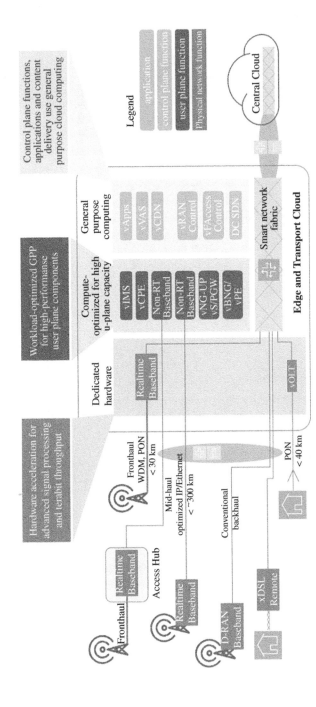

Figure 2.17 Network evolution exploiting flexible deployment options.

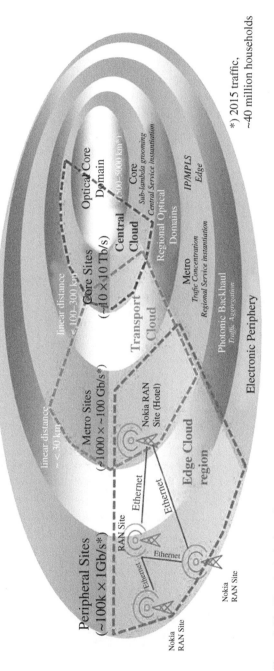

Figure 2.18 Topological view of a large size mobile network infrastructure.

NFVI resources) or application-specific hardware (e.g. baseband units). At metropolitan-level, depending on the deployed RAN split architecture, a single site has to process between 100 Gbps and several Tbps.

- The **transport cloud region** comprises medium-sized transport hubs and datacentres at major aggregation sites of the network. It spans metropolitan area sites, the so-called 'regional optical domain', and the 'core' aggregation level. For the latter, single sites must process approximately 10 Tbps of traffic.
- The **central cloud region** refers to a limited (typically a low double-digit) number of large-scale data-centres. They comprise the sites of the optical core domain where total traffic volume can grow up to a few hundreds of Tbps.

The adaptive decomposition, disaggregation, and (re-)allocation of network functions to these regions, depending on the service requirements and deployment constraints, is a significant novelty over legacy networks. It allows realization of the same functional architecture by instantiating the network functions in different physical deployment scenarios, or even multiple deployment scenarios in parallel. For example, if multiplexing gains are exploited, a higher level of centralization is preferred, which is suitable for services such as eMBB. On the other hand, if traffic needs to be processed closer to the network edge, e.g. due to latency requirements or limited transport capacity, network functions can be executed in the edge cloud region. This more distributed deployment scenario can be useful for URLLC-type traffic.

Beyond the flexible deployment of VNFs to multiple and varying cloud locations, a further factor impacting the mobile transport networks comprises the increasing number of small cell deployments in 5G when compared to earlier cellular network generations. Therefore, the overall backhaul and – depending on the used technology – the midhaul, i.e. the links between distributed units (DU) and centralized units (CU), will be more that branching and network topology become more interconnected, e.g. mesh network between the small cells, such that the number and overall length of links will dramatically increase. Second, 5G NR air interface will allow for far higher data rates. Together with the expected use of massive MIMO, this will dramatically increase the data rates on the fronthaul links between radio units (RU) and DUs.

In the era of 5G, mobile networks will undergo a fundamental change. Future RAN solutions will be deployed in a most cost efficient and agile way, serving any kind of clients' or end-customers' service pattern needs (eMBB, URLLC, IoT) or mobile network densification. Traditional monolithic BTS/RAN deployment principles will gradually dissolve into more distributed 'split RAN' architectures (cf. Section 2.4.4), which will require a more sophisticated transport underlay, clearly beyond traditional backhauling.

It is expected that such split RAN architectures will first be applied more locally by decomposing and connecting BTS baseband unit (BBU) functions into separate entities of RUs, DUs, and CUs. However, in next RAN deployment stages, the decomposed BBU functions are supposed to be spread wider over a much larger deployment footprint, in any kind of (integrated) RU, DU, and CU assemblies (and combinations) between different RAN or central office (CO) locations and connected by means of an external transport network. Some mobile operators with fibre-rich assets, particularly in the last miles, have already started these stages with first rollouts.

3GPP has defined certain RAN split categories [28], mainly

- between RUs and DUs, based on real-time-sensitive low layer split (LLS) options along with corresponding fronthaul connection profiles for the transport network
- between DUs and CUs, based on less time-stringent (and less bandwidth-hungry) high layer split (HLS) options, which use a midhaul-type transport connection profile.

Each split category may have further detailed split options, e.g. 3GPP LLS options 8, 7.1, 7.2X, 7.3, or 6, each having its own benefits and drawbacks [37]. Lastly, split RAN options may have specific ways of implementation:

- DUs and particularly CUs can be further disaggregated into a control plane (CU-CP) and several distributed user plane (CU-UP) instances with new mid-haul type interfaces, cf. Figure 2.15.
- Most recently, additional new RAN entities are envisioned to be introduced, like the RAN intelligent control (RIC) in ORAN specification work [38].
- Moreover, DUs and CUs can be implemented using either compact/bare-metal or COTS/cloud driven approaches. The latter virtualizes native BBU functionalities into virtualised functions (vBBU).

In general, the entirety of transport connection profiles, including those for the new split RAN interfaces, are referred as x-haul profiles, i.e. they combine traditional backhauling with the new mid- or fronthauling needs. For the different RAN deployment options, transport-relevant KPI indicators include bandwidth, latency, or site distance.

Due to these developments, mobile traffic transport profiles on the backhaul, midhaul, and fronthaul will have to rely on a mix of transport technologies and deployment options. In some areas, due to cost reasons, backhaul of small cells will be based on wireless multi-hop technology (microwave and millimetre-wave in non-access frequency bands) or even on self-backhauling, thus reducing capacity in the access. However, in areas with higher traffic demand, a 'fibre-to-the-X' (FTTX) optical wireline solution will be preferred, e.g. eCPRI [31].

Overall, these developments will lead to an increase in the number of transport connection points as well as the dynamicity of the topological adjustments. Hence, mobile traffic control and management planes need to be upgraded to technologies such as SDN to manage the rapidly changing traffic mixes and connectivity configurations.

2.5 Evolution of Transport Networks

Transport networks transfer bit streams between different types of accesses over varying distances. They form the underlying infrastructure of a communication network. Transport networks are structured into tiers as shown in Figures 2.18 and 2.19. The evolution of each of the hierarchy levels has been driven by very different capacity, reliability, and cost requirements of the network domains. The growing amount of bursty traffic patterns, transport network unification, and convergence has made packet-based technologies preferable across all transport network hierarchies. Since transport networks are strategic and imply significant TCO expenditures, their renewal is mid- to long-term oriented, with more gradual deployment steps than most other parts of the network infrastructure.

2.5.1 Architecture of Transport Networks

Access networks at the periphery (e.g. radio network, FTTX, PON, hybrid fibre-coaxial [HFC], DSL, Ethernet) collect individual traffic flows from the end users and aggregate the traffic towards the backhaul networks. The topology of access networks is typically a star, a tree, or a chain that connects individual customer premises equipment (CPE) or base stations to a last mile(s) hub node that feeds the next level of the aggregation network. The number of equipment and links in the access network is very high in comparison to the other transport tiers.

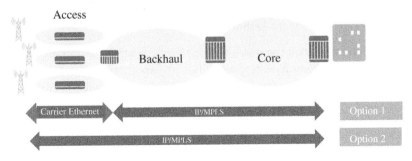

Figure 2.19 Carrier Ethernet and IP/MPLS options.

Link capacities typically vary from a few hundred kbps (e.g. IoT-type devices) up to a few Gbps. Last miles access links may either be shared among access points (e.g. PON, cable, fibre ring) or have dedicated point-to-point connections to the user or customer premises. Consequently, the cost of physical deployments in the access network (such as deploying fibre ever closer to customers – 'fibre deep') is a dominant factor when considering upgrading the access.

The topology of higher transport tiers is based on rings that offer redundant paths in case of link or node failures. A traditional mobile backhaul network typically covers several aggregation switches or routers that connect to the access links. Here, packet-based technology is applied, e.g. to gain benefits from statistical multiplexing, or to unify future transport infrastructure. Backhaul networks further aggregate the access tier traffic towards the (packet-optical) core network that connect various backhaul networks. Backhaul networks link capacities up to 100 Gbps. Between backhaul networks and the core network, there could be additional levels of hierarchies if networks are interconnected to form so-called MANs or optical domain networks that provide regional network capacity without the core network, Figure 2.18. Both, core backhaul and (ultra) long-haul networks are native wide area networks (WANs) as opposed to local area networks (LANs) because of their wide geographical span.

The role of future transport networks in interconnecting the RAN entities, as well as mobile core functions and premises will dramatically change to allow better techno-economic scalability (in terms of RF-site density, higher throughput and tighter latency figures) of any future RAN solution. While this will help to cope with newly introduced service patterns, it will also result in a sharp increase of transport connection points to be managed.

As outlined in Section 2.4.6, traditional RAN deployment practices will undergo a major architectural change with regards to functional decomposition, spatial distribution, disaggregation, virtualization and openness. Likewise, the transport domain will follow exactly the same trends, with its subdomains of traffic aggregation, switching, routing, and transmission.

2.5.2 Transport Network Technologies

The evolution of mobile backhaul networks that connect base stations with the core network has followed the evolution of the traffic volumes and the types of the cellular network service. For 2G and 3G networks, voice was the main traffic type and data rates were limited. Circuit-switched backhauls matched the traffic and capacity needs well. In addition, circuit switching provided accurate timing needed for the base stations. Data service volumes exceeded the voice service volumes only after HSPA and LTE/4G networks were taken in use. This pushed the backhaul networks to adopt packet-based technologies. The

LTE network has a flatter architecture than its predecessors and it places more functionality into the base stations without introducing an intermediate RNC that has acted as logical point of traffic aggregation in 3G. In 4G, base stations can be connected to each other (*X2* interface) and to different core networks (network sharing, eDECOR [39]), which lead to more complex network configurations. The dominant backhaul technologies for the LTE/4G networks have been carrier Ethernet and IP/MPLS over optical networks that consolidates the traffic of voice, video, and other data services.

Transport networks introduce a demarcation line between enterprise or customer networks and network provider networks. This demarcation line is between customer premises equipment (CPE) and provider edge equipment (PE). PE equipment tunnel the customer premises traffic across the operator network, so that the routing protocols and addressing schemes used in the customer networks do not interfere with corresponding schemes of provider network infrastructure. In mobile networks, this demarcation line can be located at the boarder gateways connecting the cell sites with the backhaul transport as well as in the gateways between the core network datacentre and the transport network.

Carrier Ethernet extends native Ethernet technology, which originally stems from enterprise networks and LANs, to MANs and WANs by introducing a set of new capabilities. It defines both the physical and the data link layer frame structure, which contains global destination and source MAC addresses. Carrier Ethernet extends the data link layer frame structure and adds bandwidth profiles and three connectivity service types: 'E-Line', 'E-LAN', and 'E-Tree'. E-Line service is meant for leased line replacement that connects two customer Ethernet ports over a WAN. E-LAN emulates a multipoint-to-multipoint LAN service with MAC learning and bridging across WAN. E-Tree service supports multiple point-to-multipoint trees based on Ethernet Virtual Connections. Carrier Ethernet services can be implemented over many different types of layer 1 transport technologies. The most relevant ones for cellular backhaul include Ethernet over Synchronous Optical Networking (SONET)/Synchronous Digital Hierarchy (SDH), Ethernet over MPLS, Ethernet over WDM, and Ethernet over microwaves. Carrier Ethernet specification are developed and maintained by Metro Ethernet Forum (MEF) and IEEE 802.

MPLS [40, 41] provides means for ensuring isolation and QoS to different label paths. Older SONET/SDH transport systems can be emulated as an overlay to IP/MPLS using circuit emulation service of Pseudo Wire [42]. MPLS introduces virtual circuits between layer 3 and layer 2. Packet forwarding is based on labels rather than IP addresses. MPLS labels can be distributed among the participating MPLS nodes by interior routing protocols (Intermediate System-to-Intermediate System [IS-IS] and Open Shortest Path First [OSPF]), Resource Reservation Protocol – Traffic Engineering (RSVP-TE), or a specific Label Distribution Protocol [43].

Moreover, Border Gateway Protocol (BGP) has extensions for label distribution [44]. The ingress gateway that receives a packet from the customer encapsulates it with MPLS labels (called *PUSH* operation) that determine the complete path across the transport network. The other nodes on the path solely operate on the MPLS labels by swapping or removing (*POP* operation) them. The egress gateway *POPs* the labels and forwards the original packet. The MPLS labels can convey other information than QoS and path switching which leads to different variants of MPLS, such as MPLS-Transport Profile [45] and MPLS Source Routing that fits well with SDN-type of control and traffic engineering.

Use of Ethernet and IP/MPLS in the backhaul is not mutually exclusive but can co-exist so that the interfaces to the base stations are Ethernet based, and the traffic flows are mapped to IP/MPLS in the aggregation switches at the boarder of the access and backhaul (Figure 2.19).

2.6 Management of Communication Networks

The required flexibility of 5G networks demands network functions and services to be dynamically deployed and configured. So, orchestrating entities must communicate with many different managed entities across multiple interfaces. Network management algorithms, especially for self-organization and self-optimization, depend on the communication capabilities between the different management functions as well as between the management system and the network elements.

Frequently, the integration of the different parts is cumbersome owing to the incompatible interfaces among the different entities. As such, the idea of 'common models' has been raised from time to time, but the industry was not able to agree to implement such a single common model.

This section describes the development of network management approaches considering the technological, architectural, and functional evolution of communication networks as well as the related modelling that underlies the management. It illustrates basic principles and paradigms for network management in general, presents most common architectures and technologies. Then, to illustrate the scope of the modelling task, the section discusses in a general, systematic way the various dimensions of 'modelling' to motivate the different possible information models. It also highlights the benefits of such common models as well as inherent limitations that then justify a different approach.

2.6.1 Basic Principles of Network Management

Network management is a fundamental component within the network's OAM system. Such systems comprise '*the whole of operations required for setting up and maintaining, within prescribed limits, any element entering into the setting up of a connection'* [46]. International Telecommunications Union – Telecommunication Standardization Sector (ITU-T) refers to the set of functions and procedures to manage a telecommunication network as a Telecommunication Management Network (TMN). Recommendation M.3000 [47] defines a TMN as providing '*the means used to transport, store and process information used to support the management of telecommunication networks and services'*. Formally, ITU-T differentiates between *Telecommunications Managed Areas* and *TMN Management Services*. The former denotes a set of resources that are logically and/or physically involved with one or more telecommunications services and that shall be managed as a whole. Examples include switched data networks, mobile communications networks, or switched telephone network. The latter refers to the integrated set of processes (management services) of a company in order to achieve the business and management objectives, such as quality to the telecommunications service customers or operational productivity [48].

To support these objectives, [49] proposes to categorize network management into five coarse-grained functional areas, referred to as FCAPS:

- **Fault Management**: the functions to enable detection, isolation and correction of abnormal operation of the telecommunication network and its environment
- **Configuration Management**: the functions that exercise control over, identify, collect data from and provide data to network elements (NEs). They configure such NE aspects as configuration file management, inventory management, and software management
- **Accounting Management**: the functions that enable the measurement of the use of network services and resources; the determination of costs to the service provider and charges to the customer for such use and support the determination of prices for services
- **Performance Management**: a set of functions that gather and analyse statistical data, evaluate and report upon the behaviour of telecommunication equipment and the effectiveness of the network or network elements so that overall performance can be maintained at a defined level
- **Security Management**: a set of functions that provide authentication, access control, data confidentiality, data integrity, and non-repudiation and that may be exercised in the course of any communications between systems, between customers and systems, and between internal users and systems

Further, [50] defines three different TMN architectures and their fundamental elements:

TMN functional architecture is the structural and generic framework of management functionality that is subject to standardization and is described by using four fundamental elements, namely function blocks, Management

Application Functions (MAFs), TMN Management Functions (and sets thereof), as well as reference points. The list of function blocks comprises Network Element Function (NEF), Operations System Function (OSF), Transformation Function (TF), Workstation Function (WSF). A MAF represents part of functionality of one (or several) TMN management services. TMN Management Functions refer to a set of cooperating MAFs to realize a TMN management service. A TMN reference point defines the service boundary of a functional block, i.e. it describes an external view of its functionality and represents the interactions between a pair of function blocks.

The TMN functional architecture is further divided into four logical layers, each focusing on particular aspects of management, categorized by their level of abstraction. The hierarchical grouping into business, service, network, and element Management is depicted in Figure 2.20. On the lowest layer, the scope of the element management layer comprises control and coordination of a subset of network elements. The network management layer has wider geographical scope; it comprises control and coordination of the network view of all network elements within its scope or technological domain. The (network technology-agnostic) tasks of the service management layer comprise contractual aspects regarding service provisioning to customers, such as, service assurance and fulfilment or order handling and invoicing. Finally, the business management layer includes completely proprietary functionality dedicated to goal setting, investment decisions, and other business strategy aspects.

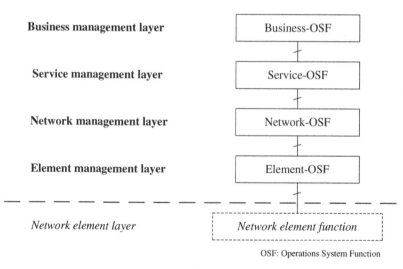

Figure 2.20 ITU-T model for layering of TMN management functions [27].

TMN information architecture: TMN Management Services require the exchange of management information between the elements of managing and managed systems. Standardized information models consist of information elements that are abstractions of resource types, usually modelled as objects. Further, they contain an abstraction of the management and management support aspects of a network resource. Information models are mapped to reference points of the TMN functional architecture. Thus, the resulting *information model of a reference point* unifies the TMN functional and information architectures, defines the minimum scope of management information to be supported by the associated function blocks, and usually corresponds to the physical interface of the TMN physical architecture.

TMN physical architecture depicts the configuration of the physical equipment of a TMN. It comprises two fundamental elements: physical blocks and physical interfaces. Physical blocks are subdivided into Network Element (NE), Operation System (OS), Workstation (WS), and transformation devices that host function blocks providing conversion between different protocols or data formats. The physical architecture model mandates that each physical block at least host the accordingly named function block of the functional architecture. Nevertheless, the recommendation does not exclude additional associations between function blocks and physical blocks. The OS physical block can typically be further broken down into the four logical layers, i.e. Business (B-OS), Service (S-OS), Network (N-OS), and Element E-OS, each containing the associated function block.

The general scoping of telecommunications management by ITU-T has been utilized by industry organizations to define norms, standards, industry specifications, frameworks, and best practices on different management layers. For example, the TM Forum has defined the so-called *Frameworx* concept, consisting of

(1) the Business Process Framework or enhanced Telecom Operations Map (eTOM) [51], which defines service operations procedures, such as, fulfilment, assurance, billing/revenue management, or operations support functions
(2) the Information Framework [52], describing data definitions and according information models required for service operations
(3) the TM Forum Application Framework (TAM) [53], describing a service provider's ecosystem of OSS/BSS applications.

As a whole, *Frameworx* covers the service and (parts of) the business management layers as defined in TMN. The next subsection focuses on the network and element Management layers and how they have been further standardized or specified in selected network domains.

2.6.2 Network Management Architectures

For the network and element management layers, different TMN management areas have defined their dedicated realization of the generalized TMN functional architecture. For example, 3GPP working group SA5 has specified the management reference model and interfaces for mobile networks as depicted in Figure 2.21 [54].

Functions of the network and element management (NM/EM) layers are summarized as the Operations System, which interfaces with the network element layer and other enterprise (service management) systems in southbound and northbound direction, respectively, and with peer-level functions in other organizations in the horizontal direction. Exemplary functions of the NM layer include management and orchestration of network services, network planning (including radio planning), network configuration management, alarm and event correlation, network supervision, and centralized SON automation. On the EM/DM layer (3GPP adds domain management to this layer), typical functions include element- and domain-level configuration management, network performance monitoring, alarm correlation, and distributed SON automation.

Further, seven types of management interfaces have been defined in the 3GPP management reference model. The most important ones include Type 1 management interface (also referred to as *Itf-S*) between an NE and an Element/Domain

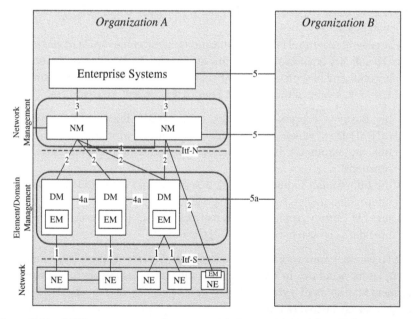

Figure 2.21 3GPP management reference model. Source: Adapted from [54].

Manager of a single PLMN organization and Type 2 management interface (also referred to as *Itf-N*) between the Element/Domain Manager and the Network Manager of a single organization. In rare cases, Network Managers can also have a direct interface with the NE. Horizontal management interfaces (Type 5/5a) between two different organizations usually only exist on a peer level and are typically restricted to higher (network and above) management layers, cf. Figure 2.21. For 5G networks, these horizontal interfaces across administrative domains will become of higher relevance, because of private mobile networks and increasing levels of network sharing, e.g. through network slicing, a 5G concept for deploying multiple logical networks with own management functions on a shared infrastructure. Both will require a tighter interaction between the network management functions of the involved networks and organizations. More details on these topics are presented in Section 2.7

2.6.2.1 Legacy 3GPP Management Integration Architecture

In 3G and 4G mobile networks, 3GPP adopted a top-down, process-driven interface modelling approach referred to as the *Integration Reference Point* (IRP) concept [55]. IRPs comprise the so-called IRP categories and IRP levels. IRP categories include

- **Interface IRP:** this category defines operations and notifications for a specific telecom management domain, such as, alarm management or configuration management. It describes how information is exchanged on an interface
- **Network Resource Management (NRM) IRP:** this category defines the managed objects, i.e. the so-called Information Object Class which represents the manageable aspect of the resource of this IRP.
- **Data Definition IRP:** this category includes the data definitions commonly used by Interface IRPs and/or NRM IRPs.

A three-level approach is used to define each IRP category, i.e. each category is partitioned into the following specification levels:

- **Requirements:** this level includes the conceptual definitions of management interfaces based on selected use cases and subsequently defines concrete requirements for the IRP. The specifications of this level should be rather stable over a longer period of time
- **Information Service (IS):** based on the requirements level, the IS-level provides technology-agnostic definitions for each IRP category. The specifications of this level should only change by means of additions and extensions
- **Solution Set (SS):** the SS-level provides a mapping of each IS definition to technology-specific solutions, e.g. to a management protocol, such as, CMIP, NETCONF, CORBA, or SOAP. The specifications of this level may change if new or better technologies become available.

Using this fundamental approach, 3GPP specifies the procedures for each of the management domains of the FCAPS model and, beyond those, for the following domains: roaming management, fraud management, software management, subscription management, subscriber and equipment trace management, service level trace management, and management of the TMN itself. For further details, the reader is referred to the 3GPP specification series 28 and 32 on telecommunications management [56, 57].

2.6.2.2 Service-Based Architecture in Network Management

Similar to the 5GC control plane, 3GPP has defined the Service-Based Management Architecture (SBMA) for the management plane [58]. A management service offers management capabilities and can combine three so-called management service component types which correspond to the IRP categories defined in Section 2.6.2:

- **Type A** is a group of management operations and notifications independent of managed entities. This type corresponds to the 'Interface IRP' category. Examples include single or bulk configuration management interfaces or interfaces for synchronous and asynchronous performance management data exposure.
- **Type B** comprises the management information represented by information models of the managed entities (managed object). This type corresponds to the 'NRM IRP' category. Examples include NRMs defined for the different NFs of the 5GS, such as, AMF, SMF, PCF, or the gNB.
- **Type C** comprises performance and fault information of the managed entities ('managed data'). This type corresponds to the 'Data Definition IRP' category. Examples include counters for ping-pong or too-late/too-early handovers, radio link failures, and dropped calls.

A management service combines components of either Types A and B (i.e. what are the operations/notifications associated with a specific managed object) or Types A, B, and C (i.e. what are the operations/notifications associated with a specific managed object and which managed data is processed). Service capabilities include, but are not limited to, Configuration (also referred to as Provisioning), Performance, and Fault Management (CM, PM, FM, respectively) capabilities. In the case of a CM service, the service producer is the entity being configured, i.e. it receives CM parameters from the service consumer. Management services can be offered on any management layer, cf. the ITU-T model for layering of TMN management functions in Figure 2.20. Typically, service consumption happens in a 'bottom-up' manner, i.e. services produced on lower layers, e.g. element management, are consumed by consumers on a higher layer, e.g. domain management. Another example comprises the consumption of a Network Function Management Service (NFMS), e.g. a PM service, by an entity on the network slice subnet management level which, in turn, can offer a PM-related Network Slice Subnet Management Service (NSSMS) to an entity on the network slice management level. For further details on 3GPP SBMA, see [58].

2.6.3 The Role of Information Models in Network Management

It requires a huge number of management applications on the different layers of the network management architecture (cf. Figure 2.21) to manage all aspects of the overall network, which comprises the RAN with several thousand network elements deployed across the country (including the necessary site equipment such as tower mount amplifiers, active antennas, and routers), as well as the many hundreds of network elements of the core and transport networks. The communication between all these entities is based on a huge variety of interfaces, by which the systems expose their internal behaviour in an abstract way to their communication partners. Each of these interfaces uses protocol stacks with multiple layers, which require models on all levels of abstraction that can be described by different meta data possibly in many different languages. As a consequence, over the years, the industry implemented a Babylonian language problem.

Information models in the management of telecommunication networks have become extraordinary heterogeneous over the last decades for a variety of reasons. On the one hand, the different network elements differ in scaling and deployment which results in different management applications and requirements for the corresponding interfaces. On the other hand, changing existing interfaces imposes costs and risks for both vendors and MNOs, such that the inertia of changing existing, working interfaces for the sake of harmonization must not be underestimated.

Generally, the description of different aspects of each interface causes many misunderstandings in discussions owing to differences in backgrounds and assumed context for the different stakeholders. Moreover, the multiple interfaces with different levels of abstraction and layers imply a huge complexity and cost in real-life systems, since all these interfaces along with all their aspects must be documented, implemented, and maintained in the long term. Each newly arriving generation of RAT increases the number interdependencies and deployment options such that the number of interfaces and the related effort to maintain them also increases. Given these challenges, it appears obvious to require common management interfaces and common information models to reduce the complexity and allow for generic or vendor-independent solutions.

So far, although the industry has developed and standardized common interfaces for specific layers and types of network elements, (see for example [56, 57]), it was not able to agree on an overall, common 'lingua franca' that harmonizes all the interfaces. Specifically, the idea of starting from the very different entities to derive a common model has not been successful. Instead, an alternative candidate approach is to derive common models – even of different managed elements – based on the requirement that it should be possible to handle common use cases in a common way. For example, the different managed entities all require means to manage the inventory, software versioning and faults. Common defined models based on this requirement would become enablers to handle common aspects of the systems by common applications. However, this approach immediately reveals the inherent limitations of common models: If there is no common use case for a certain aspect of two managed entities, then there is no chance to cover these aspects by a common information model. So, multiple information models have been developed for use in different contexts as discussed in the subsequent sections.

2.6.4 Dimensions of Describing Interfaces

Network management involves several management functions with different focus, scope, and levels of abstraction. Each corresponding interface can be described on different levels of abstraction, and each interface is based on a stack of protocols. Each perspective of describing interfaces may be considered as a distinct dimension of network modelling, which also has distinct requirements. The following paragraphs briefly describe the dimensions, based on which models may be developed to manage communication networks.

2.6.4.1 Dimension 1: Hierarchy of the Management Function

As shown in Section 2.6.1 the different tasks related to operating and managing a telecommunication network can be grouped into a hierarchy of management functions comprising of Business Management, Service Management, Network Management, Element Management, and the Network Elements/Network Functions. Each layer has a specific scope in terms of managed network elements, a specific level of abstraction of those elements and a specific time scale of its operations. Consequently, the overall management system forms a 'Telecommunication Management Network' [50] with multiple different interfaces to support a broad variety of use cases, which leads to multiple models.

In many operators' networks, these graphs of management functions and their interfaces have been specifically designed to support the organizational structure and workflows of individual operators. These environments, which have gradually grown over decades, comprise several dozens of network management applications that are partly customized products from different vendors while the rest have been individually, or even internally, developed applications. Together, they represent a considerable investment and are business critical for the operator, as any changes to the functions or their interfaces imposes costs and risks, which the operator would rather avoid. This is, in many cases, the reason not to adopt standardized interfaces when they would be available. ITU-T recognizes that 'functional and information architecture provide a framework that allows requirements to be documented about *what* a TMN implementation should do' and that 'TMN implementations are not currently a subject for standardization', because 'TMN implementations have to blend and balance a number of divergent constraints such as cost, performance, and legacy deployments, as well as new functionality being delivered' [50].

However, with each new generation of technology, the number and complexity of tools for management, optimization, and automation has considerably increased, which has led to a significant burden to develop and maintain the corresponding interfaces. This is the one motivation for. 3GPP [56, 57] and ZSM, for example, to drive harmonized reference architectures with standardized interfaces between the management functions.

2.6.4.2 Dimension 2: Levels of Abstraction

Any function and each interface between them can be described by several levels of abstraction. A product manager's view on the business requirements is more abstract than a systems engineer's view, which is also different from that of a developer who must deal with the very details of the code. Accordingly, many standardization bodies adopted these levels of abstractions, such as the concept of 3GPP to separate 'Requirements', implementation agnostic 'Information Service', and multiple implementation specific 'Solution Sets' [55].

2.6.4.3 Dimension 3: Layers in Communication

Interfaces between software systems are based on protocol stacks (e.g. ISO/OSI) and are often based on sets of protocols (e.g. CORBA might be used to control the exchange of files, while the actual transfer of files is performed by ftp). Successful communication requires the sender and receiver to use the same sets of protocols and protocol stacks. Peers need to use an agreed dynamic behaviour to exchange data, a well-defined format to structure information elements, and a common understanding on the semantics of the different information elements. All these aspects are equally as important for each layer.

2.6.4.4 Dimension 4: Meta Data

Meta data is used to describe data. Each of the above-mentioned aspects of the interfaces (hierarchy, abstraction or layer) can be described by meta data, while different target groups use different languages to describe the interfaces. While product managers might use plain text, system engineers may prefer the Universal Modelling Language [59], an implementation-independent language used to model various static and dynamic aspects of systems, such as software systems or telecommunication systems. Similarly, software designers may use implementation-specific interface definition languages (IDL) to generate artefacts to ease the implementation of the software in a specific technology.

2.6.5 Network Information Models

Considering the many dimensions, it is obvious that many different kinds of models may be developed to integrate the management functions and managed entities in a communication network. This section analyses how far common protocol stacks and common meta data might enable further automation and will emphasize some inherent limitations of common management interfaces.

2.6.5.1 Model of the Dynamic Behaviour

To integrate a managed entity into a management system, communication between them depends on a common dynamic behaviour, i.e. a common notion on how to organize the exchange of data. The model of the dynamic behaviour of an interface describes how the entities exchange the data, independently of the format or the semantics of the data.

As an example, the dynamic behaviour of internet protocol v4 (IPv4) is passive, i.e. the IPv4 sender puts up to 65 535 logical 1 or 0 on the wire but does not care about any receiver. Multiple receivers might read the data from the wire, and if a receiving interface detects that it is the destination of the data, it will handle the data, while other interfaces shall ignore the data. In contrast, the dynamic behaviour of TCP/IP comprises additional handshaking mechanisms to ensure the correct transmission of the data and to address processes within a server. Protocols on top of TCP/IP utilize the mechanisms provided by TCP/IP, which they augment by additional message sequences to perform specific tasks resulting in different dynamic models. For example, both, ftp and http are meant for file transfer, but they are based on fundamentally different dynamic behaviour: ftp establishes a logical connection (session), transfers one or multiple files in either direction, and then closes the connection. In contrast, http is connection-less, in the sense that the client sends a request for a file to the server and, in turn, receives the requested file as the answer –without the need for the client to open and close an explicit session.

In general, an interface has to mediate between three dynamic behaviours: the internal behaviour of the managed entity, the dynamic behaviour of the interface itself, and the internal behaviour of the managing entity. These dynamics might be very different. For example, the configuration of the managed entity may require the complete configuration database to be exchanged, while the management entity and the interface are designed to deliver individual commands to modify single parameters. There are mismatched expectations if the management system's Performance Management (PM) database is optimized to write many records of the same measurement type while the interface is optimized to transfer big files. The two would match if a file sent by a managed entity contains blocks of measurements of same type. This is, however, rarely the case. There is considerable overhead in communication on the interface if the managed entity sends small files containing only one record. Most importantly, the database has to handle each record by an individual transaction and so the managed entity may have to sequentially cycle through hundreds of different measurement types.

Besides mismatches, the internal behaviour of network elements might be restricted due to the requirements of their functionality in the network, which then demands for different approaches in dealing with these network functions. For example, a BTS cannot change certain parameter values without dropping ongoing calls. As a consequence, a managing entity cannot assume an instant change for such parameters if continuity of the service for the end customer is required.

For different network elements to be connected by a common interface with common dynamic behaviour, the dynamic behaviour of the network element must be compatible to the interface. Otherwise, the network elements have to implement a mediation component to map the internal dynamics to the dynamics of the interface. While this is doable, it is very expensive in terms of R&D, risk, and runtime resources.

The interfaces enable the management systems to perform a transformation in scale and capacity. One management system might use its interfaces to integrate hundreds of thousands of managed entities, each providing small amounts of data. However, from a management system point of view these many small amounts add up to a huge amount. Besides the amount of data, the management system must also handle the load caused by the dynamic behaviour of hundreds of thousands of interfaces towards the managed entities, e.g. to open, maintain, and close long-lived sessions and, especially, to handle exceptional conditions if connections die unexpectedly. Depending on the managed entities, requirements for scaling might be very different. Harmonization of interfaces should consider such differences in scaling.

As shown, there is definitely no 'one interface fits all' scenarios. Managed entities with similar dynamic behaviour and similar requirements from a management system point of view regarding scale and capacity should use the same model of the dynamic behaviour (protocol) to transfer the data. However, a management system that needs to integrate different categories of managed entities must be prepared to support multiple protocols. In other words, it may sometimes be necessary to introduce new protocols/interfaces to the management entity. Since the dynamic behaviour of managed entities is usually very stable across releases of the managed entity, the integration of a new protocol into a management system is typically a one-time exercise. This addition of interfaces is, however, an expensive and high-risk exercise that is also prone to new errors and problems. As such, new interfaces should be introduced with caution as a fundamental mismatch between the dynamic behaviours of the interface and the managed entity increases the chance of failure. Simply stated, a sub-optimal interface that is working is better than an interface that would be perfect in theory but failing in reality due to implementation issues.

2.6.5.2 Format of the Data

Besides the dynamic behaviour, integration of a managed entity into a management system also depends on the commonality of the formatting of the data, i.e. a common notion of the structure of the data and its partitioning into information elements. The model of the format describes how the data, irrespective of the mechanisms to transfer the data or its semantics, is partitioned into information elements. For example, IPv4 defines packets with up to 65 535 binary bits, has a mandatory header section of 5×32 bit containing 13 header fields, and an optional header of up to 4×32 bits, but does not impose any format or structure to the data in the payload.

An interface must transform between three different formats: The internal format of the managed entity, the inherent format of the interface, and the internal format of the managing entity. Usually, the data transfer is organized in stacks of protocols. For a given managed entity, the lower layers of the protocol stack usually do not change with new releases of the managed entity. This is also true for the format of the data of such lower layers, e.g. the coding of the management data into file formats like XML, json, csv, asn.1, xdr, or protobuf is usually very stable. Also, the notion of alarms that are modelled by well-defined fields or the concept to organize configuration data as objects which contain parameters with values is usually stable across releases of network elements. Mapping these formats into a common model is a one-time task, which is not usually complex and thus technically feasible.

In contrast, the specific set of information elements exposed by a managed entity on its management interfaces does depend on the release of the managed entity, therefore, the corresponding management systems must be enabled to handle the additional information elements. This preparation might be complex, since the management system might need to adapt its internal storage system and data access functions, its internal communication systems, and its user interface. In addition, the management system must cope with multiple versions of managed entities in parallel due to the fact that a network-wide upgrade might take several weeks where parts of the network have already been upgraded while other parts are still running the old release.

If an upgrade of a managed entity requires manual work to adapt (i.e. code) and to release a new version of the management systems, which might also potentially require involvement of third-party-suppliers, then the upgrade will be both expensive and slow. Typically, operators require lead times of three to six months to adapt network management tools to new network elements releases, which is incompatible with the flexibility required by 5G. In contrast, a management system that uses generic, metadata-driven tools for data parsing and processing is able to read the description of the updated format and to automatically adapt its internal structures (including database tables, internal communication, and user interface). Although the development of a metadata-driven tool is much more complicated than a statically coded tool, the automated adaptation enables much faster introduction of new features, which might be more important than the costs of the initial implementation.

The upgrade process can further be automated using an interface that enables the managed entity and the management tool to exchange the metadata. Such an interface may use any of the known metadata transfer languages e.g. XMI for UML, CorbaIDL for Corba, XML Schema, WSDL, or YANG. However, since such additional metadata interfaces all have concerns and dimensions as for any other interface, dependencies and long-term maintenance effort might increase considerably.

2.6.5.3 Semantical Part of the Model

Besides the dynamic behaviour and data format models, an interface requires the sender and receiver to agree on common semantics of the information elements. In IPv4, for example, the octets 12–15 indicate the source IP address and octets 16–19 the destination IP address while IP does not prescribe any semantics for the payload, i.e. the serviced data unit. The common understanding of the data semantics allows the communicating entities to accurately interpret the data. Similarly, for management systems, proper handling of the operability data requires the management system and the managed entity to have a common understanding of the meaning of the individual information elements. In this respect, the operability data represents the behaviour of the managed entity, so proper interpretation on both sides is needed to accordingly and appropriately define the behaviour.

2.6.6 Limitations of Common Information Models

As discussed, the degree to which the network can be uniformly controlled, depends on the common understanding of the information elements. Today's networks are characterized by both standard and proprietary features and information elements.

The well-known standardization bodies (3GPP, IETF, ITU-T, etc.) define the call processing procedures between the network functions, e.g. the messages that must be exchanged between the network elements to provide the telecommunication service, e.g. to establish a connection between a user equipment (UE) and the RAN or to perform a proper handover. This guarantees interoperability of network functions and devices from different vendors. Each network function that fulfils a given standard must represent the corresponding properties and parameters with the standardized semantics (e.g. frequency of a cell, cell-Id, system information on broadcast channels, etc.). These have been the causes for the huge success of standardization in telecommunication and the big differentiator compared, say, to the highly non-interoperable solutions of the internet applications.

From the management point of view, the operator can implement common use cases for all these standardized parameters of the corresponding network elements regardless of the vendors, e.g. the planning and assignment of frequencies, cell-Id, and broadcasted system information can be performed for any vendor's cell in the network. In these cases, it is possible to map the standardized parameters into a common information model with agreed semantics of the individual information elements that can be handled by common applications.

By intention, 3GPP does not define the internal algorithms of the network function. For example, each vendor has the freedom to define how and when a BTS decides to perform a handover, to which target cell, and how to consider traffic steering and load balancing. This freedom leaves room for differentiation and technical evolution and will persist as long as vendors are required to differentiate by features.

The air interface of the RAN especially requires very complex algorithms that apply many rules and policies to optimize utilization of the spectrum and to ensure the best possible quality to users. A real-world cell in UTRAN or E-UTRAN requires approx. 1000 parameters, while approx. 10% of these parameters are standardized and 90% relate to proprietary algorithms.

However, from the management point of view, this freedom has dramatic consequences, because these proprietary, internal algorithms do not have any analogy on implementations of other vendors. There is no common use case to set or to optimize the corresponding vendor/feature-specific parameters. The semantics of these vendor/feature-specific parameters cannot be mapped to a common information model. These vendor/feature-specific parameters cannot be handled by common applications that depend on the semantics.

The situation gets even more complex due to the many dependencies within a network function between standardized parameters and vendor-specific parameters. For example, in most cases, it is not possible to configure a real-life cell without touching the vendor-specific parameters, which then implies that generic configuration tools are not possible for even the most basic configuration use cases.

As long as the external behaviour of network elements will be standardized while, at the same time, technical evolution and differentiation of network elements is required, the information models will comprise information elements with both standardized and vendor-specific semantics. As a consequence, management systems must be able to cope with such heterogeneity.

It follows that common information models are not possible, at least in foreseeable future. However, other common models could be developed. This, however, requires a detailed re-evaluation of the approach to network modelling to identify the areas where such models could be possible. The next section presents such an approach and studies the challenges from a systems theory perspective.

2.7 Conclusion – Cognitive Autonomy in 5G and Beyond

MNOs and CSPs have always looked to network automation to improve operational and cost efficiency. Past efforts focused on small steps delivering incremental benefits. SON, automated LCM and APIs allow for programmatic change of network parameters have been proof points for such incremental automation attempts that have, so far, not materialized the vision of cognitive autonomous networks. Dependencies in the management of physical radio networks and of virtualized infrastructure components are major obstacles. Further drivers of MNOs' need for cognitive autonomy in 5G mobile networks include explosion of parameter space, user-and service-specific customization of networks and network elements, variations of network deployment flavours, functional decomposition, and management of multitudes of logical network instances. Overall, four major challenges for 5G mobile network management automation can be identified:

1. Management of novel 5G network features which entails management of feature flexibility requiring dynamic configurability and management of feature complexity with the related vast configuration space and dependencies
2. End-to-end operation of 5G networks
3. Vertical sectors as novel operational stakeholders in 5G system operation.

2.7.1 Management of Individual 5G Network Features

5G mobile network design introduces an extensive infrastructure heterogeneity and flexibility for implementing the network domains (RAN, CN, etc.). In the RAN, this manifests through novel architecture, allowing for a large variety of network function deployments, and a number of flexible and configurable radio features, e.g. the application of a broad range of frequencies, multiple RAT numerologies [60] or flexible beamforming. In the core, formerly monolithic network elements are decomposed into more fine-grained and frequently virtualized

NFs that can be executed in different locations, cf. Section 2.4.6. Moreover, the architecture must integrate various cell types and sizes from classical macro-cells down to pico cells for ultra-dense network deployments, the use of different radio technologies, diverse transport topologies (fronthaul/backhaul), and the (joint) use of licensed and unlicensed spectrum. The introduced flexibility and means for dynamic re-configurations of network features requires new capabilities of the management and orchestration systems.

2.7.2 End-to-End Operation of 5G Networks

On the end-to-end system level, the major challenge arises from the identification and enforcement of an appropriate network operating point, i.e. a comprehensive and reconciled configuration of all network features. Generally, network operations entail the adjustment of the operating point in case of, for example, changes in service requirements, traffic and user distribution or the network context and, finally, its optimization according to the required performance and operational targets. In 2G/3G/4G network deployments, it is possible to determine an appropriate operating point with manageable effort, as the influencing variables are well-known, and their number is limited. In order to keep or adapt the – rather stationary – operating point in an automated way, the industry has developed quasi-static closed-loop control systems called SON functions (cf. Chapter 3). These SON functions have been applied for the configuration, optimization, and failure recovery [61]. In contrast, 5G networks will have considerably more influencing variables resulting from the highly configurable network implementation, i.e. the technological advances in radio technology, overall system heterogeneity and deployment modularity, as well as the mixed physical and virtualized environment. Different communication services will require different operating points owing to the different performance targets, which are likely to change more frequently than today. 5G management systems should address such different and dynamic performance targets by identifying and virtually configuring different operating points simultaneously. The existing SON approaches will probably not provide a sufficient degree of flexibility and dynamicity for implementing an appropriate cognitive 5G network management.

2.7.3 Novel Operational Stakeholders in 5G System Operations

The final challenge is brought in by the requirement of serving multiple vertical industries via virtual network instances, so-called 'network slices', each of which may serve a dedicated 'vertical' company, a so-called tenant. Each instance can be considered as a self-contained network from a tenant's point of view. For each instance, the CSP not only has to identify an operating point and derive according network configurations, but may also have agreed on a contract-specific level of network management features delegated to the tenant. This exposure of former operator-internal functionality can range from a rather simple monitoring of the slice's Quality of Service (QoS) or Quality of Experience (QoE) performance, according to targets agreed in an SLA, to a rather deep control in terms of re-configuring a (virtualized) network function within certain boundaries. Besides exposing selected management features in such a way as to preserve overall security and operational safety of the network, the challenge for the CSP comprises the resolution of conflicting configurations made by different tenants but affecting the same network segment or network function.

In summary, there is a strong requirement for cognitive methods and tools that allow the network to autonomously exploit the emerging flexibility and opportunities provided with the new 5G features. The objective for the network management system is to largely autonomously configure network resources and functions in such a way that each network slice instance reaches its optimal operating point to meet the performance and operability requirements. Cognitive autonomous management needs to map these multiple operating points to the available infrastructure resources in an SLA-conforming and non-conflicting manner.

References

1 3GPP TS 23.002 (2018).*Network architecture*. Sophia Antipolis: 3GPP.

2 3GPP TS 23.401 (2018). *General Packet Radio Service (GPRS) enhancements for Evolved Universal Terrestrial Radio Access Network (E-UTRAN) access*. Sophia Antipolis: 3GPP.

3 3GPP TS 23.107 (2018) *Quality of Service (QoS) concept and architecture*. Sophia Antipolis: 3GPP.

4 3GPP TS 25.308 (2017). *High Speed Downlink Packet Access (HSDPA)*. Sophia Antipolis: 3GPP.

5 Eberspächer, J., Bettstetter, C., Vögel, H.-J., and Hartmann, C. (2008). *GSM - Architecture, Protocols and Services*, 3e. Wiley.

6 Holma, H. and Toskala, A. (eds.) (2010). *WCDMA for UMTS: HSPA Evolution and LTE*, 5e. Wiley.

7 3GPP TS 23.203 (2019). *Policy and charging control architecture*. Sophia Antipolis: 3GPP.

8 Holma, H. and Toskala, A. (eds.) (2011). *LTE for UMTS: Evolution to LTE-Advanced*, 2e. Wiley.

9 3GPP TS 23.402 (2018). *Architecture enhancements for non-3GPP accesses*. Sophia Antipolis: 3GPP.

10 IBM (2016). *A Practical Approach to Cloud IaaS with IBM SoftLayer*. IBM Redbooks.

11 Nokia Networks (2015). Reinventing telcos for the cloud. https://resources .nokia.com/asset/200238 (accessed 20 January 2020).

12 Brown, G. (2017). Designing Cloud-Native 5G Core Networks. https://www .juniper.net/assets/us/en/local/pdf/whitepapers/2000667-en.pdf (accessed 20 January 2020).

13 Taylor, M. (2016). Telco-Grade Services From An IT-Grade Cloud. https://www .metaswitch.com/knowledge-center/white-papers/telco-grade-services-from-an-it-grade-cloud. (accessed 31 January 2019).

14 3GPP TR 22.804 (2018). *Study on Communication for Automation in Vertical domains (CAV)*. Sophia Antipolis: 3GPP.

15 3GPP TR 22.821 (2018). *Feasibility Study on LAN Support in 5G*. Sophia Antipolis: 3GPP.

16 ETSI GS NFV-MAN 001 (2014). *Network Functions Virtualisation (NFV); Management and Orchestration*. Sophia Antipolis: ETSI.

17 Nokia Networks (2019). Creating new data freedom with the Shared Data Layer. https://resources.nokia.com/asset/200238 (accessed 24 January 2020).

18 Nokia Networks (2017). Mobile anyhaul. https://resources.nokia.com/asset/ 201272 (accessed 24 January 2020).

19 3GPP TS 23.501 (2018). *System Architecture for the 5G System*. Sophia Antipolis: 3GPP.

20 3GPP TS 23.502 (2019). *Procedures for the 5G System*. Sophia Antipolis: 3GPP.

21 3GPP TS 23.503 (2018). *Policy and charging control framework for the 5G System*. Sophia Antipolis: 3GPP.

22 The Open Group (2019). Service Oriented Architecture (SOA). https:// collaboration.opengroup.org/projects/soa/pages.php?action=show&ggid=1575. (accessed 18 June 2019).

23 3GPP TS 29.518 (2018). *Access and Mobility Management Services*. Sophia Antipolis: 3GPP.

24 3GPP TS 29.501 (2018). *Principles and Guidelines for Services Definition*. Sophia Antipolis: 3GPP.

25 Fielding, R.T. (2000). Architectural Styles and the Design of Network-based Software Architectures. Doctoral Dissertation. University of California.

26 Maeder, A., Ali, A., Bedekar, A. et al. (2016). A scalable and flexible radio access network architecture for fifth generation mobile networks. *IEEE Communications Magazine* 54 (11): 16–23.

27 3GPP TS 38.300 (2019). *NR and NG-RAN Overall Description*. Sophia Antipolis: 3GPP.

28 3GPP TS 38.401 (2019). *NG-RAN; Architecture description*. Sophia Antipolis: 3GPP.

29 3GPP TS 38.470 (2019). *NG-RAN; F1 general aspects and principles.* Sophia Antipolis: 3GPP.

30 3GPP TS 38.460 (2019). *NG-RAN; E1 general aspects and principles.* Sophia Antipolis: 3GPP.

31 Ericsson A.B. (2018). *Common Public Radio Interface: eCPRI Interface Specification.* Huawei Technologies Co. Ltd, NEC Corporation, Nokia.

32 3GPP TS 38.473 (2019). *NG-RAN; F1 Application Protocol (F1AP).* Sophia Antipolis: 3GPP.

33 3GPP TS 38.463 (2019). *NG-RAN; E1 Application Protocol (E1AP).* Sophia Antipolis: 3GPP.

34 3GPP TS 38.423 (2019). *NG-RAN; Xn Application Protocol (XnAP).* Sophia Antipolis: 3GPP.

35 ETSI MEC ISG (2019). ETSI MEC: An Introduction. https://www.etsi.org/images/files/technologies/ETSI-MEC-Public-Overview.pdf (accessed 24 January 2020).

36 Linux Foundation (2019). LF Edge, Akraino Edge Stack. https://www.lfedge.org/projects/akraino (accessed 24 January 2020).

37 3GPP TR 38.801 (2017). *Study on new radio access technology; Radio access architecture and interfaces.* Sophia Antipolis: 3GPP.

38 O-RAN-WG1.OAM (2019). *O-RAN Operations and Maintenance Architecture.* O-Ran Alliance.

39 3GPP TR 23.711 (2016). *Enhancements of Dedicated Core Networks selection mechanism.* Sophia Antipolis: 3GPP.

40 RFC 3031 (2001). *Multiprotocol Label Switching Architecture.* IETF.

41 RFC 3032 (2001). *MPLS Label Stack Encoding.* IETF.

42 RFC 3985 (2005). *Pseudo Wire Emulation Edge-to-Edge (PWE3) Architecture.* IETF.

43 IETF, RFC 5038, *"The Label Distribution Protocol (LDP) Implementation Survey Results"*, Oct. 2007.

44 RFC 3107 (2001). *Carrying Label Information in BGP-4.* IETF.

45 RFC 5921 (2010). *A Framework for MPLS in Transport Networks.* IETF.

46 CCITT (1992). *Recommendation M.20. Maintenance philosophy for telecommunication networks.* ITU.

47 CCITT (2000). *Recommendation M.3000. Overview of TMN Recommendations.* ITU.

48 CCITT (1997). *Recommendation M.3200. TMN management services and telecommunications managed areas: overview.* ITU.

49 CCITT (2000). *Recommendation M.3400. TMN management functions.*ITU.

50 CCITT (2000). *Recommendation M.3010. Principles for a telecommunications management network.* ITU.

51 TM Forum (2018). GB921. *Business Process Framework (eTOM) Suite.* Release 18.5.

52 TM Forum (2019). GB922. *Information Framework (SID).* Release 18.5.

53 TM Forum (2018). GB929. *Application Framework (TAM) Suite.* Release 18.5.

54 3GPP TS 32.101 (2017). *Telecommunication management; Principles and high level requirements.* Sophia Antipolis: 3GPP.

55 3GPP TS 32.150 (2018). *Integration Reference Point (IRP) concept and definitions.* Sophia Antipolis: 3GPP.

56 3GPP. Specification Series 28, Telecom Management. http://www.3gpp.org/DynaReport/28-series.htm (accessed 24 January 2020).

57 3GPP. Specification Series 32 Telecom Management. http://www.3gpp.org/DynaReport/32-series.htm (accessed 24 January 2020).

58 3GPP TS 28.533 (2019). *Management and orchestration; Architecture framework.* Sophia Antipolis: 3GPP.

59 OMG (2015). *Unified Modeling Language TM (OMG UML).* Object Management Group.

60 Pedersen, K., Berardinelli, G., Frederiksen, F. et al. (2016). A flexible 5G frame structure design for frequency-division duplex cases. *IEEE Communications Magazine* 54 (3): 53–59.

61 Hämäläinen, S., Sanneck, H., and Sartori, C. (eds.) (2011). *LTE Self-Organising Networks (SON): Network Management Automation for Operational Efficiency.* Wiley.

3

Self-Organization in Pre-5G Communication Networks

Muhammad Naseer-ul-Islam, Janne Ali-Tolppa, Stephen S. Mwanje and Guillaume Decarreau

Nokia Bell Labs, Munich, Germany

The concepts of Self-Organizing Networks (SONs) for mobile networks were introduced to cater for the increasing network management complexities. This complexity arises from the simultaneous operation of multiple cellular mobile technologies as well as multiple layers of cells within each of them. With the evolution of each mobile technology, the network elements (NEs) have also become more complex in terms of number of tuneable parameters they provide to meet the diverse service requirements. SON aims to minimize this complexity by automating most of the laborious repetitive network management tasks like NEs' configuration, optimization, and troubleshooting, both during the network's deployment and throughout its operational life. This helps to reduce the amount of manual effort needed for network management and ultimately also reduces the possibilities of human error.

The shift from manual operation of mobile networks makes SON the first generation of network management automation. And, by eliminating the lengthy manual analysis of network problems, SON makes mobile networks more responsive to the changes in their operating environment including, amongst others, appearance of new buildings, weather changes, and mobility changes. This chapter presents the state-of-the-art concepts on SON.

Firstly, it compares the traditional manual network management workflows with the SON-based network management and discusses different architectural options for the implementation of SON solutions. It then turns to the discussion of SON functions as having been designed for the different phases of a mobile network lifecycle, such as planning, deployment, and operations. Various SON use cases, for example, mobility robustness optimization (MRO), etc. are then

Towards Cognitive Autonomous Networks: Network Management Automation for 5G and Beyond,
First Edition. Edited by Stephen S. Mwanje and Christian Mannweiler.

discussed to illustrate how SON functions help to minimize the management complexity of these phases. To present the overall system view of how SON integrates into the whole network management environments of mobile networks, supporting functions like SON Coordination and Minimization of Drive Tests (MDTs) are also discussed.

To set a baseline for further network management automation and the inclusion of cognitive techniques, this chapter includes a section that summarizes the state of the standardization work on SON in 3GPP. This is critical as it highlights the expectation of the industry, especially the operators, and the requirements that have been set for the advancement of network management and its automation. Finally, the chapter concludes by highlighting the challenges that are still open in the full automation of network management as well as the current design approaches for SON functions.

3.1 Automating Network Operations

The introduction of automation functionality in networks is transforming network operations. Initially, specific tasks could be automated through the use of scripting that were executed either at specific times or when specific events are registered in the network. Then, the introduction of self-organizing functionality availed more advanced modules which also further transformed the network operation processes.

3.1.1 Traditional Network Operations

Traditional network operation is based on centralized Operations, Administration, and Maintenance (OAM) architecture, as also described in Section 2.6. The major network management tasks, usually collectively referred to as FCAPS (Fault/Configuration/Accounting/Performance/Security management), are performed from a central Operational and Maintenance Centre (OMC). Most of these tasks are human operator driven with some tool support for network planning and optimization. These tools are semi-automated and need to be heavily supervised by the human operator. Especially, correlating information from different sources like Performance Management PM/Fault Management (FM)/Configuration Management (CM) to analyse particular network problems and to define corrective configuration changes consume a lot of network operator resources. This makes the network management a resource extensive, time-consuming, expensive, error-prone task, and requires a high degree of human expertise [1]. Moreover, the efficiency of these expensive manual processes still depends on the expertise of the human operator.

The resource extensive nature of network management processes both in terms of specialized human resources and software tools as well as their heavily manual execution mean that network planning and optimization can only be done on longer time scales of weeks and months. This means that the network cannot optimally react to short-term variations in the network environment like sudden traffic load variation or changes in the propagation conditions. To compensate for this, network operators often need to over-provide their network resources so that the user experience degradation is minimized.

3.1.2   SON-Based Network Operations

To manage the increasing complexity of network management and to overcome the inefficiencies of the traditional manual process, the concept of SON was introduced. SON introduced closed loop algorithms that can fully automate different use cases of network management related to different phases of a mobile network lifecycle.

SON-based network operation brings automation to many of the resource-extensive and time-consuming network management tasks. This also allows mobile networks to respond to changes in the NEs and their environment more quickly, even in the order of minutes instead of days and weeks compared to the traditional operational setup. The added agility also allows the introduction of new optimization and healing use cases that would not be possible with traditional manual operations, for example, real-time coverage outage detection and compensation.

Automation functions, so called SON Functions, are introduced at different parts of the network to undertake automation for the specific use cases. 3GPP has defined three different architectural options, as described in the next sections, for the implementation of SON functions related to the general 3GPP network management architecture.

3.1.2.1   Centralized SON

Centralized SON refers to SON function implementations, where SON function logic resides in a central entity. In 3GPP network management architecture, such a central entity could be a Network Manager (NM) or a Domain Manager (DM) as shown in Figure 3.1. Centralized SON is a natural choice for SON use cases that have a wider network scope such as a domain or even a full network as information from different NEs can be accessed at that management level. This allows SON functions to analyse the performance of several NEs and coordinate the requirements of those NEs while calculating the network parameter reconfigurations. Additionally, at NM level, SON functions can also be designed for a multi-vendor environment because of the availability of standardized Key

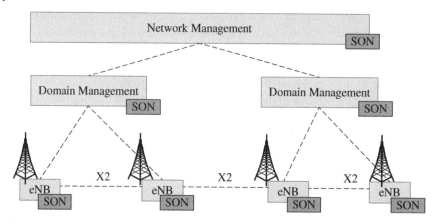

Figure 3.1 Location of SON functions in the 3GPP OAM architecture.

Performance Indicators (KPI)s and measurements via the north-bound interface (Int-N).

Performance KPIs and counters are generated by individual NEs, so, for centralized SON implementations all data needs to be transported to the central entity. To avoid signalling overload, the performance data is usually aggregated and transported to DM and NM in discrete intervals of 15 minutes to an hour. However, this, aggregation loses the short-term dynamics of the performance data and lengthens the SON function response time. For this reason, centralized SON functions are good for long-term optimizations that do not need to respond to instantaneous changes in NE performance and that are not sensitive to the delays involved in data transfer.

From a computational complexity point of view, centralized SON functions can benefit from more compute and storage capacity of central entities. But on the other hand, it also introduces a single point of failure and thus can compromise the performance of large parts of networks if the SON function execution becomes somehow affected.

3.1.2.2 Distributed SON

Distributed SON functions are implemented at NE level, e.g. at the enhanced Node B (eNB). They can still use information from other neighbouring NEs via direct interfaces like X2. This implementation option is good for SON use cases that focus on a limited scope of either a single cell or a cluster of neighbouring cells. As SON function logic is implemented close to the source of NE performance data, so these SON functions can also react to instantaneous changes in performance KPIs and counters. SON use cases that cannot tolerate the data transfer delays to a central entity and the aggregated form of performance data are suitable for distributed implementations. Although most of the parameters at NE level are vendor specific, some multi-vendor SON solutions have also been defined by 3GPP that can be implemented in distributed manner. For this purpose, 3GPP has defined the data elements and interfaces to exchange information amongst neighbouring NEs in a standardized way.

NEs usually have very limited compute and storage resources. Therefore, SON functions can only be implemented in a distributed manner if they are not computationally extensive. For the algorithms, as several neighbouring NEs can potentially be working on same or mutually-dependent optimization issues, special attention should also be given in SON function design to avoid any oscillation behaviour, race conditions and deadlocks amongst the actions of different distributed SON functions. Distributed SON functions also provide an intrinsic redundancy characteristic that avoids any single point of failure for the whole network.

3.1.2.3 Hybrid SON

Hybrid SON combines the benefits of both the centralized and distributed SON implementations. In this approach, some SON functionality is implemented at central entities like NM or DM and some functionality is implemented at the NE level. The central functionality focuses on the more long-term aspects, whereas, the distributed functionality focuses on more instantaneous and short-term optimization goals. One example case could be that the central functions define the policies and boundaries for the configuration parameter modification and the distributed functions perform the actual optimization and reconfigurations within those policies and boundaries.

3.1.3 SON Automation Areas and Use Cases

There are many use cases where automation (and SON) can be applied. Priority use cases were identified by NGNM and subsequent specification taken up by 3GPP. The use case falls into three major categorizations depending on the operational phase of the network and the focus on the automation capability. Specifically, the use cases are listed as part of Self-Configuration, Self-Optimization, or Self-Healing. There were, however, other use cases that were highlighted as important although not necessarily falling in any of these three categories. These were handled separately as support functions for SON operation.

The following sections differentiate these automation areas including the critical features of their respective use cases and their automation algorithms defined by 3GPP as part of the Long-Term Evolution (LTE) specification work. The presented use cases include MDT and SON coordination, which are two of the Support Functions for SON Operation.

3.2 Network Deployment and Self-Configuration

NE deployment and initial configuration is a resource-extensive task because of the involved network (re-)planning efforts and physical site visits for the actual deployment and configuration. Considering the HetNet deployment of present network operators with multiple radio access technology (RATs) and multiple layers of cells even within one RAT, this initial configuration via a physical site visit becomes too expensive. Automation of different tasks in this phase of the network deployment is therefore crucial for network operators as identified by NGNM [2]. Self-configuration use cases deal with the pre-operational phase i.e. the time between the physical deployment of the NE and the time when its fully ready to handle live network traffic.

The tasks relevant for this phase, which are also sometimes collectively referred as '*Plug-and-Play* (*PnP*)', are, for example, establishing the basic connectivity between the new NE and its OAM system, the commissioning of the NE as well as some initial configuration of the radio parameters.

3.2.1 Plug and Play

PnP or Self-configuration deals with automated integration of newly deployed NEs or their parts into the network to make it operational and ready to take live traffic with minimal human involvement. The process works in pre-operational state i.e. the time between the NE power-up and when it is ready to process live network traffic. The tasks involved in this phase can be divided into three main categories:

1) **Auto-connectivity** that deals with the automatic establishment of secure connection between the NE and the networks' OAM system
2) **Auto-commissioning** procedures that deals with the automated provisioning and testing of software and its configuration based on the site-specific data and hardware configurations
3) **Dynamic radio configuration** (*DRC*) that deals with the adjustment of radio parameters of the eNB according to its current network environment.

While 3GPP does not define the exact details and sequence of individual steps there is still a logical sequence for the steps involved i.e. a secure connection needs to be established before any configurations can be done. To supervise the whole procedure and give control to the network operators, 3GPP standardized a Self-Configuration and Software Management Integration Reference Point (IRP) in 3GPP TS 32.501. This allows the introduction of some 'stop points' between different tasks of the self-configuration process. These stop points help to halt the automation workflow and give control to the supervising authority to either resume or stop the process at those stop points.

3.2.1.1 Auto-Connectivity

Auto-connectivity provides a means to automatically connect the newly deployed NE to its OAM system and to acquire the required configuration to initiate auto-commissioning. This helps to minimize the need for 'off-the-shelf' software provisioning and its configuration with the new NEs. Once the NE is deployed and connected to its OAM, the actual NE-to-site mapping can be performed, and it can be configured in a site-specific manner. This also allows for a flexible rollout, as the NE parameter configuration can be changed even up until the time it is physically deployed in the network. The auto-connectivity makes use of Dynamic Host Configuration Protocol (DHCP, IETF RFC 2131) and an Auto-Connection Server (ACS) to establish the connectivity between the NE and its OAM system in the following steps:

- **Setup of basic transport network connectivity:** As a first step, the NE acquires an initial IP address from the operator network using DHCP. Depending on the IP auto-configuration service settings, the NE also gets the IP configurations for communication with the ACS, the operator's Certification Authority (CA) and optionally a Security Gateway (Sec-GW).
- **Secure connection setup:** The NE enrols to the CA using the vendor provided certificate and gets the operator certificate for further authentication and secure connection setup in the operator network environment.
- **Site identification and NE configuration download:** After establishing the security credentials, the NE sends an announcement message to the ACS with its HW ID and optionally the measured GPS coordinates. The ACS can then perform a site identification i.e. the mapping between the HW-ID provided by the NE and the site-ID in the Configuration Management Database (CMDB). The ACS can then update the CMDB with the HW-ID of the new NE as well as the network topology database with the new NE. The ACS can also retrieve site-specific configuration data from CMDB for further (auto-)commissioning and forward it to the NE within the reply acknowledge (ACK) message of the announcement message.
- **Secure connection setup with DM:** if the NE communicated with the ACS using a secure connection which is different from what is provided in the ACK message from ACS, the NE tears down that connection and establishes a new secure connection with its DM using the configurations from the ACK message. The NE is now ready for the actual commissioning.

3.2.1.2 Auto-Commissioning

Once the auto-connectivity is successfully completed, the NE is ready for further commissioning, which consists of the following tasks:

- **Inventory update:** The HW configuration of the new NE such as the processing boards, antenna, etc. is automatically identified and forwarded to the OAM systems to update the CMDB. Additionally, the provided information can also be validated against the network planning for the specific site.

- **Software download:** Because of the automation of self-configuration tasks, it is possible to deploy NEs with very basic SW and the actual SW for NE full operation is only downloaded and validated at this stage.
- **Configuration data download:** After successful download of the latest NE software, the configuration data, prepared during the radio and transport planning for the particular site, is also downloaded to integrate the new NE with other NEs in the network. Additionally, some of the radio parameters may also be configured on the fly using the DRC as explained in the next section.
- **License management:** After the inventory check and SW update the license, management procedures are performed for the planned and installed features.
- **Call processing interfaces:** This process involves the establishment of the call processing interfaces and performs some final self-tests by the NE to validate the successful commissioning.
- Finally, the NE transits into an operational state either automatically or with a human involvement that clears the acceptance of the new NE into the operational network.

3.2.1.3 Dynamic Radio Configuration

In legacy network operations, deployment of new eNodeB follows a comprehensive planning phase covering aspects like the location of the new eNodeB, the required HW and SW configurations as well as detailed radio parameters configuration. This makes it a resource-extensive task in terms of both the specialized human resources and tools needed. This detailed planning requirement becomes too expensive for an incremental deployment especially where eNodeBs are gradually added to the existing network deployment as it needs to be done with every new deployment. Moreover, planning tools assume some characteristics of the environment in which the eNodeBs need to be deployed, which may change between the time of planning and actual deployment. To compensate for any mismatches, additional drive tests are therefore performed after the deployment and parameters are fine tuned to better match the actual network environment.

To reduce planning efforts in the pre-deployment phase, the concept of DRC was introduced as part of the 3GPP self-configuration work [3]. DRC allows dynamic configuration of crucial radio parameters like Physical Cell ID, Neighbour Cell Relation (NCR), Initial Power, and Antenna Tilt setting during the deployment. Although dimensioning of new eNodeB still needs to be done to select the site location as well as the HW and SW features, the detailed radio planning can be partly replaced with the DRC. Apart from making the pre-deployment planning less expensive, this also allows the consideration of the actual network environment when configuring the radio parameters. DRC functionality defined by 3GPP only defines the NE features and interfaces to perform the radio configuration on the fly but the actual algorithms to find the best configuration parameters are still left for vendor implementation. However, two features – Physical Cell ID Allocation and Automatic Neighbour Relation – are defined in greater detail by 3GPP because of the dependencies on multiple neighbouring cells/eNodeBs and are discussed in greater detail in the following sections.

3.2.2 Automatic Neighbour Relations (ANR)

To enable handovers (HOs) and other cell-relationship processes, each cell in the network broadcasts the list of its neighbour cells via various System Information Blocks (SIBs). However, this list of neighbour cells requires to be populated first. In 2G and 3G systems, cell neighbour relations were manually set and only optimized if new NCRs needed to be added or if unnecessary neighbours (those that led to HO failures) needed to be removed. This is a very labour-intensive process which may even be prone to errors since the operator can never be fully aware of the coverage overlap amongst all cells.

In LTE, automation through the Automatic Neighbour Relations (ANR) procedure [4, 5] was introduced with the aim of automatically adding and/or removing neighbours to a cell's Neighbour Cell Relations Table (NRT). For each cell of an eNB, the eNB keeps an NRT, similar to Table 3.1, which tracks each neighbour cell and the procedures that are allowed for such a neighbour cell.

Specified in 3GPP's E-UTRAN Network Resource Model (NRM) [6], the NRT contains an identifier for the target cell, called the Target Cell Identifier (TCI). This identifier corresponds to the target cell's E-UTRAN Cell Global Identifier (ECGI) and Physical Cell Identifier (PCI). The NRT provides a binary field for each process, e.g. HO or load balancing (LB), which identifies whether the specific process is allowed for each of the neighbours.

3.2.2.1 The ANR Procedure

Through the ANR process illustrated by Figure 3.2, unknown neighbour relations for a given cell S can be learned using the User Equipment (UE) measurements within cell S. In the serving cell S, the ANR process adds to S's NRT, any cells which can be measured by the UEs in S. Any added cells are not removed unless specifically (manually) black listed e.g. for causing HO failures.

The ANR procedure (see Figure 3.2) relies on the UEs' ability to read all neighbouring cells. The end-to-end process is as follows:

1) Network send Measurement Command (Measurement Control) to UE to perform the detection/measurement of cells around it.

Table 3.1 3GPP specification of the NRT [5].

Attribute name	Support qualifier	isReadable	isWritable	isInvariant	isNotifyable
Id	M	M	–	M	–
tCI	O	M	M	–	M
isRemoveAllowed	CM	M	M	–	M
isHOAllowed	CM	M	M	–	M
isICICInformationSendAllowed	CM	M	M	–	M
isLBAllowed	CM	M	M	–	M
isESCoveredBy	CM	M	M	–	M
qOffset	CM	M	M	–	M
cellIndividualOffset	CM	M	–	–	M
Attribute related to role					
adjacentCell	M	M	M	–	M

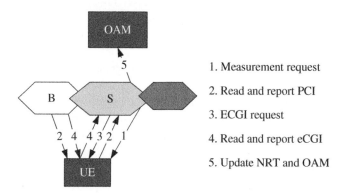

1. Measurement request

2. Read and report PCI

3. ECGI request

4. Read and report eCGI

5. Update NRT and OAM

Figure 3.2 The ANR procedure.

2) UE in cell S reads the PCI of the new cell (B) and UE updates serving cell S with new cell B's PCI via the via Radio Resource Control (RRC)-Reconfiguration message.
3) The eNB requests UE to report E-UTRAN Cell Global ID (ECGI).
4) UE reports the ECGI by reading the BCCH channel.
5) Serving cell S updates NRT and OAM with new cell relationship.

ANR can be performed for Intra-LTE and intra-frequency as well as for Inter-RAT or Inter-frequency cells. In the Inter-RAT/Inter-frequency case, the eNB configures the UE with measurement gaps through which the UE may measure and detect the non-active RAT and frequencies. The ANR function then instructs the UE in connected mode to detect cells on other RATs/frequencies. Then, following the UE's report of PCI of the detected cells in the target RATs/frequencies, the eNB updates its inter-RAT/inter-frequency NRT.

The default ANR procedure does not assume OAM support in identifying new neighbours. It is possible, however, to leverage OAM in which case, every new eNB registers to OAM and downloads PCI/ECGI/IP information with which it updates all NCRs. The default ANR procedure is then used to detect unknown neighbour relationships.

3.2.2.2 NRT and ANR Limitations

The biggest challenge with this procedure is that relationships are documented using a binary field, i.e. a cell is either a neighbour to another cell or not. This translates into very large NRTs as even those cells that barely overlap will easily be considered as important relationships. There are proprietary solutions that have attempted to enrich the NCR description among cells in terms of 'buckets' of overlap, where each bucket identifies a kind of overlap. Three examples of such NCR buckets could for instance be:

a. One cell fully encircles another cell at a different cell site.
b. For two co-located cells (same cell site), one cell fully encircles the other cell.
c. All other cells with partial overlap with one another.

This bucket system improves the description but is not fully optimal for all kinds of cell relations. It is necessary as such that new mechanisms for describing and identifying neighbour relations are identified. Cognitive techniques could be of use here as described in the subsequent chapters.

3.2.3 LTE Physical Cell Identity (PCI) Assignment

The PCI is the primary configuration parameter for the cell and helps in differentiating the signal amongst cells. There are 504 unique PCIs grouped into 168 unique physical-layer cell identity groups (PLIGs, N_{ID}^1), each group having three unique physical-layer cell identities (PLIs, N_{ID}^2) [7]. Thus, a cell's PCI is the combination of the Cell's PLIG, and its PLI, i.e.

$$PCI = 3 \cdot N_{ID}^1 + N_{ID}^2 \tag{3.1}$$

Although seemingly simple, PCI assignment is not a trivial problem, owing to the limited number of PCIs [7] and the need to minimize PCI conflicts amongst cells.

3.2.3.1 PCI Assignment Objectives

The PCI has a one-to-one mapping with the cell's synchronization signals; reference signals (RSs) and their pseudorandom position in frequency, as well as with the scrambling codes for most of the physical channels [8]. The cell's PCI is related to the synchronization signals used by the UEs for cell search. Every 5 ms, the cell transmits two synchronization signals – the primary synchronization signal (PSS) and the secondary synchronization signal (SSS). The PSS is generated from a frequency-domain Zadoff-Chu sequence whose root index has a one-to-one mapping to the physical-layer cell identities N_{ID}^2. The SSS is a concatenation of two sequences, both of which are characterized by two indices m_0 and m_1. The indices are derived from the physical-layer cell identity groups according to (3.2) [7]:

$$m_0 = m' \bmod 31$$

$$m_1 = \left(m_0 + \left\lfloor \frac{m'}{31} \right\rfloor + 1\right) \bmod 31$$

$$m' = N_{ID}^1 + q\frac{(q+1)}{2}$$

$$q = \left\lfloor \frac{N_{ID}^1 + q'\frac{(q'+1)}{2}}{30} \right\rfloor \quad ; \quad q' = \left\lfloor \frac{N_{ID}^1}{30} \right\rfloor \tag{3.2}$$

Besides the synchronization signals, the PCI is also related to the cell's RS and serves as a resource allocator parameter for both the downlink (DL) and uplink (UL) signals. Downlink RSs are allocated in a time-frequency grid, always transmitted in the same OFDM symbol in the time domain. In the frequency domain, however, each cell has a different RS frequency shift whose index u_d given by Eq. (3.3).

$$u_d = \begin{cases} PCI \bmod 3 & ; \ SISO \ scenario \\ PCI \bmod 6 & ; \ MIMO \ scenario \end{cases} \tag{3.3}$$

If neighbouring cells' PCIs have different shifts (u_d), the cells' RSs do not overlap in frequency resulting in less interference on UE channel estimation. This is critical for intra-eNB cells. The UL demodulation RS sequence is defined by a cyclic shift of a base sequence ru;v. Sequences ru;v are divided into 30 groups (u = 0; 1, ..., 29) each having one (v = 0) or two (v = 0; 1) sequences [7]. To minimize RS interference, neighbouring cells should be assigned different base sequences. This requires that PCI mod 30 is different amongst such cells, although more complex schemes have been proposed [9].

Owing to the limited PCIs, some PCIs must be reused in different cells. However, the reuse must seek to achieve the following objectives:

1) **Minimize the number applied PCIs:** which, due to the one-to-one mapping between the reference symbols and the PCIs, ensures that the initial detection of PCIs during the UE's cell search is easier.

2) **Avoid PCI collision, i.e. ensure that** no two neighbouring cells A and B are assigned the same PCI as shown in Figure 3.3a, which would otherwise cause a UE coming from cell A towards the second cell B not being able to detect the new candidate cell.

3) **Avoid PCI confusion, ensure that** two cells C1 and C3 that are both neighbours to a cell C2 do not have the same PCI (Figure 3.3b), which would otherwise make HO measurements in C2 ambiguous and increase HO failures.

4) **Avoid (or minimize) m_0 and m_1 confusion**, as having the same m_0 or m_1 makes a part of the SSS will be similar. This may make a UE in low SINR conditions unable to differentiate the SSS (and the PCIs), resulting in a long synchronization time.

5) **Avoid (or minimize) RS Interference, ensure that** PCI mod 3/6/30 (Figure 3.3c) are dissimilar amongst cells between any two potentially interfering cells, especially for PCI mod 3/6 for cells on the same Base Station which otherwise will have increased co-interference between each other.

The degree of occurrence of the conflicts, especially PCI collisions and confusions, can be reduced by ensuring an adequate separation of cells with the same PCI. A safety margin (SM), shown in Figure 3.3d, defines the number of cells between two cells C1 and C2 that are assigned the same PCI. For example, SM = 0 implies that the same PCI is allocated to two direct neighbour cells. Meanwhile SM = 2, which is the minimum required to guarantee confusion-free assignment (at least within one layer), leaves a space of two cells between the cells C1 and C2. One could imagine that an SM = 2 is adequate in all scenarios, but it is sometimes necessary to have a bigger SM. For example, in scenarios where more cells are expected to be added to the network after the initial deployment, which will

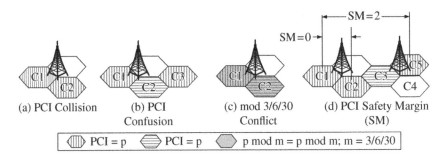

(a) PCI Collision (b) PCI Confusion (c) mod 3/6/30 Conflict (d) PCI Safety Margin (SM)

| PCI = p | PCI = p | p mod m = p mod m; m = 3/6/30 |

Figure 3.3 Critical PCI conflicts and the safety margins (SMs).

be the case for most UDNs, a bigger SM allows a PCI to be assigned to the new cell without changing the assignments of the existing cells.

3.2.3.2 PCI Assignment Strategies

The PCI assignment problem has been widely studied both for single layer scenarios, where it is fairly trivial, and for the more challenging Het-Net scenarios. The majority of the solutions, including [8, 10, 11], apply some degree of graph colouring to solve the problem. However, there exists variations such as [12, 13], which typically focus on the specific problem of introducing a new cell/eNB in an already operational network. In general, the proposed solutions show that PCI assignment is a graph colouring problem. For Het-Net scenarios however, the solutions can be generalized into two strategies shown in Figure 3.4.

1) **Single PCI range:** In this case, the entire PCI range, as shown in Figure 3.4a, is used to assign PCIs in every layer. This strategy includes all approaches that were studied in single layer networks such as those in [9, 10, 13] as well as Het-Net approaches like in [8]. To accurately assign the PCIs requires that each layer has full information about the PCI assignment in the other layer(s). This, however, is in practice not desirable as each layer may be provided by a different vendor and the small cells may typically not have X2 interfaces to directly query their direct neighbours for the outer neighbour relations (NRs). Nevertheless, allocating each layer independently would be inaccurate as the same PCI may be assigned to two cells that are neighbours to one another. Therefore, although it is not desirable, we assume the case of full information across layers, even if only as a reference case against which the other approach is measured.

2) **Range separation:** This approach, proposed in [11], splits the PCI range *a priori* into subranges, allocating a subrange to each layer (Figure 3.4b). All cells in a given layer can only be assigned PCIs from the specific layer's subrange regardless of the applied assignment scheme. The main advantage here is that PCIs can be independently assigned in the different layers without sharing any knowledge across the layers.

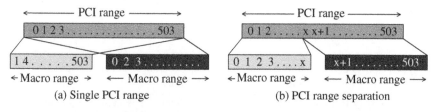

(a) Single PCI range   (b) PCI range separation

Figure 3.4 PCI assignment strategies.

3.2.3.3 PCI Assignment Challenges

As summarized here, different studies on the Het-Net PCI assignment problem have concluded that it is possible to assign PCIs in an automated and conflict free manner. However, in dense urban environments, the small cell density is expected to continue growing at least in the foreseeable future. This results in extremely dense cell deployments, generally called Ultra Dense Networks (UDNs). Then, the original assumptions made about the network deployment (and subsequently used in the PCI assignment studies) cease to be true.

The shortcoming with all the solutions so far is that they have not considered potential UDN scenarios for the PCI assignment. They either considered the traditional single layer scenarios (e.g. in [9, 10, 13]) or the currently deployed Het-Net scenarios, (e.g. [11]) where a macro-cell-network underlays a few small cells (up to 3 small cells/macro) typically in a few hot-spots. Other solutions, for example, [10] consider PCI assignment in the pico layer but with the macro layer completely ignored. Moreover, even where the macro layer is considered, the assumed density of macro cells is also low – typically three sectors per macro e.g. in [8]. In practice, the realistic deployment scenario is such that the macro network is densified with up to 6 cells/macro before the small cell layer is introduced.

The two generic PCI assignment approaches represent the two conflicting desires in the PCI assignment problem for a Het-Net environment. On the one hand, we would like to assign PCIs in each layer independently, i.e. without any concern as to which PCIs have been assigned in the other layer. This, addressed by range separation, would be important in the scenario where each layer is supplied by a different vendor and the vendor, for example, has a different SON solution for PCI assignment. This is also desirable as the macro network is likely to be stable for a long time whilst the pico layer is likely to change over time. On the other hand, independently assigning PCIs leaves the possibility that PCIs assigned in one-layer conflict, to some degree, with the PCIs assigned in the other layer. The resulting effects are expected to be more pronounced in the UDN scenario and would thus require each layer to have full knowledge of the other layer. However, this is not always possible since the layers may be supplied by different vendors.

Also, using the range separation solution for the different (vendor-) layers is not a solution. Firstly, the PCI ranges cannot be adjusted at runtime (i.e. the value of x in Figure 3.4b cannot be adjusted at runtime). This implies that PCIs could easily be exhausted in one layer while they are underutilized in the other layer. Moreover, even when the assignment is confusion free in each layer, there is no guarantee of the same confusion freeness across layers.

The ultimate solution needs to apply some cognitive techniques to learn the cell relations and build a multi-vendor or multi-layer graph of the cells with which to allocate the PCIs. This would also be the case for 5G which is expected to have even more cell density albeit with increased PCI space. Such a solution is discussed Chapter 7.

3.3 Self-Optimization

Self-optimization deals with (re-)configuration of network parameters in the operational stage of mobile network, i.e. while the NEs are handling live network traffic. Network configuration parameters need to be optimized to cater for the changes in the network and its environment. For example, the network environment and therefore the propagation conditions might change due to factors such as construction of new buildings and streets, and seasonal effects like change in vegetation in spring and autumn. The user traffic patterns also change over time, for example, due to the addition of new subscribers or change in the mobility patterns of the users resulting in new hotspots. The network itself can also change because of the addition or removal of cells in the network. Even without these changes, network configuration parameters also need to be adjusted in operational phase to overcome the shortcomings of planning and configuration phases. As the initial configurations are derived from planning tools, which make assumptions about the environment in which the network will be deployed, this operational phase optimization tries to align the network parameters to the actual network environment using the measurements from the deployed network. In this process, data from different network sources like PM/FM/CM are continuously analysed to assess current network performance for which SON algorithms propose configuration changes to improve the performance according to the network operator objectives.

The following sections summarize the critical self-optimization use cases that have been standardized as part of the LTE SON standards.

3.3.1 Mobility Load Balancing (MLB)

User traffic in mobile networks is not uniformly distributed. Ability to balance the load amongst different NEs (cells) is therefore crucial for optimal resource utilization as well as better user experience. This becomes especially important in Het-Net scenarios, where multiple layers of cells on different frequencies and RATs provide coverage for the same geographical areas. Load Balancing (LB) and Traffic Steering (TS) mechanisms, therefore, provide a means to direct traffic to the best available cell layer and RAT. Most of these mechanisms rely on mobility parameters to divert traffic to a specific NE, therefore, care must be taken not to compromise the HO performance when doing such TS and load balancing amongst different NEs.

3.3.1.1   Scenarios for Load Balancing and Traffic Steering

The main target for LB and TS is optimal network resource utilization but depending on the available deployment scenarios there can be different approaches to meet that target.

Intra-frequency neighbour load balancing: User traffic and mobility patterns often lead to hot-spots, where some cells become overloaded whilst their neighbours are still lightly loaded. Mobility Load Balancing (MLB) provides a means to shift UEs from an overloaded cell to a neighbouring cell that still has capacity available. This is done by changing the border between the neighbouring cells and care should be taken that the offloaded UEs can still be served by the neighbouring cells.

Het-Net utilization: In Het-Net deployments with cells on different frequencies and RAT covering the same geographical areas, it is desirable to utilize all the available cell layers as much as possible. Especially, in case of pico cells deployed within the coverage of macro cells, traffic should be offloaded to the pico cells as much as possible, which would otherwise be overshadowed by the macro cell because of its high transmission power.

Het-Net mobility optimization: For fast moving UEs, it is desirable to avoid HO to small cells to minimize HO signalling and throughput impacts for the UE. Therefore, fast moving UEs can be restricted to macro cell layer using the TS mechanisms.

Energy saving: Energy saving (ES) uses cell switch off to save power. ES may trigger load balancing to move load to a neighbour cell if the original serving cell has been identified as a candidate for cell switch. Details on the ES-related decision are discussed in Section 3.3.3.

3.3.1.2   Standardization Support for Load Balancing and Traffic Steering

As for other 3GPP SON use cases, the LB and TS work in 3GPP focused on providing a means to exchange the load information in neighbouring cells to access their offloading capability as well as the means to force certain traffic to a particular RAT or frequency layer by changing the mobility parameters. The actual logic to decide the extent of offloading to be done and which UEs to move to which layer are left for vendor implementation. 3GPP features useful for LB and TS are as follows:

Load information exchange: Neighbouring cells can exchange their load information via 'Resource Status' messages on X2 interface to assess the most suitable neighbours for offload as shown in Figure 3.5. Release-8 introduced reporting of '**Physical Resource Block (PRB)**' usage but it cannot accurately measure the current load of the cell. For example, even a single user with background traffic can use all the PRBs if not needed by other users. But the cell

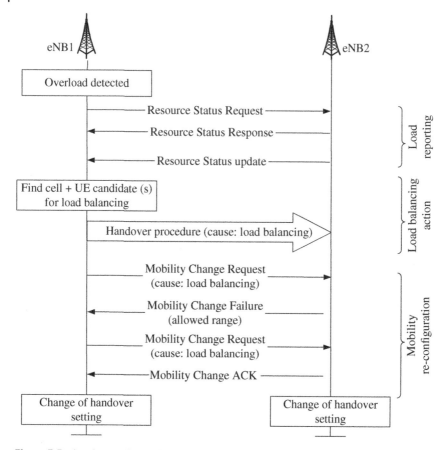

Figure 3.5 Load reporting and mobility change procedures for LB. Source: Adapted from [4].

is free to reduce the PRB allocation to this user if needed by other users without compromising the Quality of Service (QoS) guarantee for the background user. For this reason, Release-9 introduced a new load information element (IE) called '**Composite Available Capacity (CAC)**', which indicates the percentage of capacity available for offloading from a neighbour cell. 3GPP does not define exactly how to calculate this CAC value but the basic assumption is that the reporting cell should be able to accommodate the traffic equivalent to the reported available capacity in CAC. This also allows CAC calculation to include eNB characteristics even beyond the radio usage, for example, QoS requirements for current PRB usage, computation load, transport backhaul load, any reserved resources for other incoming UEs due to HO, etc.

Mobility setting change: For connected mode UEs, offloading is achieved by changing the cell boundary via mobility parameters. Release-9 introduced 'Mobility Change Procedure' on X2 interface to request a shift of cell boundary by a certain delta value to the current HO trigger points. Figure 3.5 presents the message flow of load reporting and mobility change procedures to achieve load balancing via MLB.

Basic biasing: For idle mode UEs, load balancing can be achieved by modifying the cell-pair offset in the cell-reselection trigger conditions. This is applicable for both intra-/inter-frequency cell re-selection.

Absolute priority: Cell priorities can also be defined to modify the chance of a cell to be selected by an idle mode UE for both inter-frequency and inter-RAT scenarios.

3.3.2 Mobility Robustness Optimization (MRO)

MRO tries to ensure a proper mobility behaviour for users moving between cells of the network i.e. HO in connected mode and cell re-selection in idle mode. The main targets for MRO are to minimize the call drops, Radio Link Failures (RLFs) and unnecessary HOs as they affect the service continuity.

3.3.2.1 Optimization Objectives for MRO

Improper setting of HO parameters can lead to many problems which result in either non-optimal network resource utilization because of extra signalling or, in worst case, in service interruption due to RLFs. The specific optimizations targets which are also reflected in Figure 3.6 are:

- **Too late HO:** An HO failure is characterized a Too Late HO failure if the serving Cell A initiates the HO too late or even not at all. The UE experiences an RLF shortly after the HO and then tries to connect to the target Cell B.
- **Too early HO:** A Too Early HO failure happens if a UE experience an RLF shortly after the HO and tries to reconnect to the original serving cell.
- **Wrong cell HO:** A Wrong Cell HO happens if the UE experience an RLF shortly after a successful HO but then tries to connect to another cell.
- **Ping-pongs:** A Ping-Pong HO happens if a UE makes a second HO to the original serving Cell A shortly after a first successful HO from Cell A to Cell B.
- **Short stay:** A Short Stay HO happens if a UE experience a second successful HO to a third Cell C shortly after a successful HO from Cell A to Cell B.
- **Unnecessary HO from high priority frequency layer to lower priority frequency layer:** In multi-RAT and multi-carrier frequency deployments, wrong mobility parameters can also lead to unnecessary HO from one layer to another. Apart from affecting the user experience, this also leads to non-optimal usage of different frequency layers.

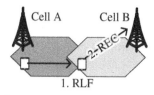

Too Late HO: Connection failure in Cell A followed by connection re-establishment in Cell B

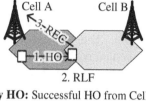

Too Early HO: Successful HO from Cell A to Cell B followed immediately by an RLF in Cell B and finally re-establishment in Cell A

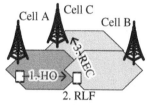

Wrong Cell HO: Successful HO from Cell A to Cell B, followed immediately by an RLF and finally re-establishment in Cell C

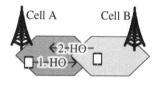

Ping Pong HO: Successful HO from Cell A to Cell B followed by another successful HO back to Cell A within a short time interval

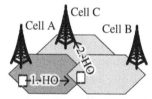

Short Stay HO: Successful HO from Cell A to Cell B, followed by another successful HO from Cell B to Cell C within short time interval

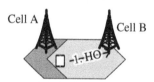

Unnecessary HO: Successful HO high priority frequency layer Cell A to lower priority frequency layer Cell B even when the UE could have been served by Cell A

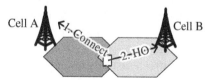

HO Right After Connection: UE establishes a connection to Cell A (transmits from idle mode to connected mode) and immediately performs a HO to Cell B

Figure 3.6 Types of HO failures.

Table 3.2 Measurement events associated with MRO for connected mode mobility.

Mode	Source	Target	Event	Comment
Connected mode	LTE f1	LTE f1	A3: Neighbour becomes Q_{offset} better than serving	Intra-frequency cell-specific
	LTE f1	LTE f2	A5: Serving becomes worse than threshold1 and neighbour becomes better than threshold2	Inter-frequency cell-specific
	LTE	2G/3G	B2: Serving becomes worse than threshold1 and inter-RAT neighbour becomes better than threshold2	Inter-RAT not cell-specific
Idle mode	LTE f1	LTE f1 (or f2 with same priority)	Neighbour becomes Q_{offset} better than serving	Intra-frequency, inter-frequency
	LTE f1	LTE/3G/2G f2 higher priority	A cell on f2 becomes better than $Thresh_{x,high}$	Inter-frequency, inter-RAT
	LTE f1	LTE/3G/2G f2 lower priority	Serving cell becomes worse than $Thresh_{serving,low}$, and a cell on f2 becomes better than $Thresh_{x,low}$	Inter-frequency, inter-RAT

- **HO right after connection:** Misalignment of idle mode and connected mode mobility parameters can lead to situations, where once a UE switches from idle mode to connected mode and immediately afterwards is handed over to a neighbour cell.

In LTE, multiple parameters are defined to influence the mobility behaviour of cells. Idle mode mobility decisions are done by UEs based on cell reselection parameters broadcasted by the eNB. Whereas, connected mode mobility decisions are made by the eNBs with the help of measurements made by the UEs.

HO measurements by the UEs are assumed to be event driven and eNBs can configure parameters to define when those events are triggered. The relevant events are given in Table 3.2 for both connected and idle mode mobility. eNBs can modify the threshold/offset values in those event definitions to adjust the mobility behaviour or UEs in its coverage area. The connected mode mobility measurements are configured individually to each UE via RRC signalling, whereas, the idle mode parameters are broadcast in system information and are, therefore, similar for all the UEs in a cell.

To overcome instantaneous fluctuations in signals, two averaging mechanisms are also introduced as part of mobility parameters. A measurement report is not immediately sent once a trigger condition if fulfilled, as defined in Table 3.2. However, the trigger condition has to remain valid for a period of time in order for the UE to report it to the eNB. This time period is called **Time-to-Trigger (TTT)** and can be configured by the eNB. In idle mode a similar timer, *Treselection*, is also applied to the event trigger conditions.

The measurements done at the physical layer are also subject to averaging to smooth the effects of fast fading. Additionally, for measurement event detection, the UE also has to apply recursive averaging to the physical layer measurements. The eNBs can influence this averaging behaviour by configuring the **Filter Coefficient (FC)**.

Both TTT and FC affect how measurements are done and reported to the eNBs. Greater values help to smooth instantaneous fluctuation in radio signals and avoid making HO decisions for unstable neighbour coverage. However, this also means that the HO trigger is delayed and care must be taken not to configure too large values that negatively affect the HO performance, for example for fast moving UEs.

3.3.2.2 Standardization Support for MRO

3GPP work on MRO mostly focused on defining measurements and messages for the root cause identification for the different mobility problems. The logic to modify the mobility parameters based on the root cause analysis is left up to eNB implementation and is therefore vendor specific. Both centralized and distributed implementations are possible. For centralized implementation, SA5 has defined relevant KPIs to be reported to the NM via Interface-N and can also be used in a multi-vendor manner. However, as many of the mobility characteristics like UE speed is not known at the NM level, therefore, distributed solution is more elaborated in the 3GPP standardization. The benefit of distributed solution is that each eNB can make its own decisions and the complexity does not grow with the network size. Additionally, eNBs can also include information that is either unavailable at the DM/NM level or gets lost in the aggregation needed for the KPI calculation over Itf-N. The main measurements and messages associated with MRO are as follows:

Re-establishment request: Introduced already in Release-8, a re-establishment request allows a UE to reconnect to the strongest cell after an RLF. The messages sent by the UE contains useful information like the PCI of previous cell in which it experienced the RLF as well as the C-RNTI of the UE in that cell. However, it is only possible for intra-LTE RLF scenarios.

RLF report: Introduced in Release-9, RLF report is generated by UE that experienced an RLF and contains information like the UE C-RNTI and radio measurements of serving and neighbouring cell just before the RLF. Once a UE makes a successful re-establishment after the RLF, it can send the RLF report to the current serving cell. The current serving cell can then identify the original serving cell in which the RLF occurred and can send an RLF Indication message to the original cell. This message is defined over X2 interface and assumes that the two cells are neighbour to each other.

HO report: Also introduced in Release-9, HO Report can be used by the target cell to report any problems that the incoming UE experienced during or shortly after the HO. The cell receiving the RLF indication or re-establishment request combines this information with its own information to analyse the root cause of the mobility problem. If it is not responsible for the RLF, it sends the HO report to the cell from which the UE made an HO to it. The assumption is that the actual problem was in the mobility parameters for that HO, for example, in case of Too Early HO or Wrong Cell HO. The HO report contains the root cause as determined by the cell sending the HO report, as well as associated information like the RLF report for further analysis by the original source cell.

T_storecontext: To ensure the original source cell can identify the UE involved in the mobility problem as reported in HO report and RLF report, eNBs must store UE context for a specific time even after the UE is successfully handed over to a neighbour cell or the UE runs out of synchronization. The time period (T_storecontext) is configurable by the network.

3.3.3 Energy Saving Management

ES has become an important network management and optimization target in the recent past along with legacy use cases of QoS assurance and network resource optimization. To meet the increasing traffic demands network operators are forced to densify their network deployments with smaller cell sizes as well as the overlay of different frequency and RAT layers to provide required capacity in the same geographical areas. Combined with the rise in energy prices, this densification leads to an exponential increase in operators OPEX and they are demanding features to save their network energy consumption. Additionally, the environmental impact of high energy consumption is also an important factor for increasing awareness and the demand for ES mechanisms.

Apart from designing energy efficient NEs as the basic of ES, traffic profiles can also be exploited to optimize network configuration to minimize energy consumption. Traditionally, mobile networks are dimensioned for Busy Hour (BH) traffic and remain operating at that configuration even during off-peak hours. For this reason, the main idea of 3GPP work for ES is to look at a means to optimize network configuration based on the current load, for example, by switching off some of the cells during off peak hours. This is especially relevant for Het-Net deployments with multiple layers of cells. But all the ES mechanisms need to make sure that they do not create any coverage holes in the network and affect the UE performance negatively.

3.3.3.1 Scenarios for Energy Saving

3GPP mainly focused on ES SON solutions that utilize switching-off cells in low traffic demand times. The work considered two main deployment scenarios, which differ in the sense of how different neighbouring cells are deployed and how they can compensate for the coverage loss from the switched-off cell. The details were captured in a Technical Report (3GPP TR 36.927). A summary of which is given below:

Overlay cell network coverage: This scenario considers network deployments with overlay cells as shown in Figure 3.7a. This type of deployment is helpful for localized hot-spots, where high capacity small cells enhance the overall capacity and the overlay macro cell provides coverage assurance as well as the basic capacity throughout the cell coverage area. As the overlay cell fully covers the small cell coverage area, a complete switch-off from the small cell layer is also feasible. The scenario can be further categorized into two sub-cases:
- Intra-LTE Inter-Frequency case, where both the macro and small cells use LTE technology but operate on different frequencies to minimize the inter-cell interference. This would be a valid scenario, for example, for mature LTE deployments, where small cells are deployed to enhance the overall network capacity in high traffic demand areas. An example would be the deployment of Home eNBs for residential users or femto cells for commercial venues.
- Inter-RAT case, where the coverage layer is provided by a legacy RAT and LTE small cells enhance capacity in specific locations. This is important for operators who want to enhance their overall capacity whilst also using their legacy networks.

Single cell layer network coverage: This scenario considers network deployment where no overlay coverage is available as shown in Figure 3.7b. In this

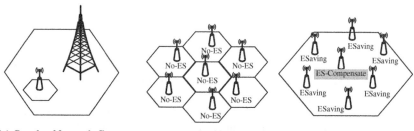

(a) Overlay Network Coverage   (b) Single Layer Network Coverage

Figure 3.7 Energy saving scenarios.

scenario, switching-off a cell for ES would result in coverage holes and therefore one or more neighbouring cells need to compensate for this coverage loss. This can be done, for example, by changing the antenna and transmitting power configurations of the compensating cells. The dynamic change of cell density, however, makes it more complicated than the overlay scenario. Apart from making sure that no coverage holes are created, the compensating cells need to consider other impacts on, for example, PCI allocation, Neighbour Relations, mobility settings, etc. for smoother network operation during ES phase.

3.3.3.2 Standardization Support for Energy Saving

Detailed studies have been done in 3GPP about the support functions needed for ES and recommendations have been provided in different technical reports. Like other SON functions, the optimization logic is left for NE vendor implementation. To simplify the coordination amongst different neighbouring cells, three ES states were identified for the cells involved in ES and corresponding state transitions were introduced in 3GPP TR 32.826 as:

- **No-ES (notParticipatinginEnergySaving) state:** Not performing ES related measures
- **ESaving state:** NE switched off or restricted in resources
- **ES-Compensate (compensatingForEnergySaving) state:** Compensating coverage.

ES-related parameters have also been added to the SON Policy NRM (3GPP TS 32.522) to enable control of ES SON functions and their behaviour by NM. The parameters included the ES state of the cell as well as thresholds and timers to measure the cell load before making any ES actions. Information exchange amongst neighbouring eNBs is also enabled via X2 interface (3GPP TS 36.423). Messages related to ES are, for example:

- **Load information and request:** to switch on/off a neighbouring cell
- **Deactivation indication information element:** Included in ENB CONFIG-URATION UPDATE to notify neighbours about cells currently switched-off for ES
- **Cell activation:** Message from a neighbouring eNB to request activation of a previously switched-off reported cell.

3.3.4 Coverage and Capacity Optimization (CCO)

Coverage and Capacity Optimization (CCO) aims to maximize the network capacity while making sure that no coverage holes exist in the coverage area. Network coverage and capacity is affected by several environmental and network changes, including amongst others: (i) seasonal changes with their effects on vegetation that modify signal propagation from one season to another, (ii) changes in the terrain such as construction or destruction of buildings which can also affect signal propagation, (iii) capacity changes due to variable diurnal patterns in user mobility, e.g. movement from residential areas to commercial areas in the morning and vice versa in the evening, and finally, (iv) network operator initiated changes like deployment of new NE or upgrade of existing ones also impact the overall performance of the network.

A particular snapshot of NE configurations is mostly optimal only for a subset of scenarios experienced by the NE. Considering all the changes, as given above, a static NE configuration would lead to sub-optimal configuration and ultimately wastage of expensive network resources as well as bad QoS for the end users. To avoid this, network parameters need to be dynamically optimized to adjust the network coverage and capacity to the current demands. Additionally, this dynamic optimization can also help to overcome the shortcomings of the network planning phase. As all network planning is done on some assumptions about the network environment, there is always some gap in the actual performance when the network is deployed. Using measurements from UEs in real deployments, CCO can also try to optimize the network configuration to match the actual environment.

3.3.4.1 Scenarios for CCO

The CCO specification in 3GPP 28.627 has identified four scenarios where CCO may apply:

1) **E-UTRAN Coverage holes with 2G/3G coverage: In this Het-Net deployment a** legacy RAT like Universal Mobile Communications System (UMTS)/Global System for Mobile Communication (GSM) provides the overlay coverage and LTE small cells are deployed to enhance network capacity. Imperfections in network planning for the small cell layer might lead to coverage holes for the small cell coverage layer as shown in Figure 3.8. This impacts the UEs ability to utilize the high capacity LTE layer as well as increase the IRAT mobility HOs at the edges of coverage holes.

2) **E-UTRAN Coverage holes without any other radio coverage:** There is no overlay coverage in this scenario and, therefore, it results in a more serious problem of complete network outage in the coverage holes as depicted in Figure 3.8.

3) **E-UTRAN Coverage holes with isolated island cell coverage:** Island cells may be deployed at, say, special venues or other commercial areas. Network planning discrepancies or changes in the environment might affect the radio propagation and ultimately lead to coverage holes and complete network outage in those areas.

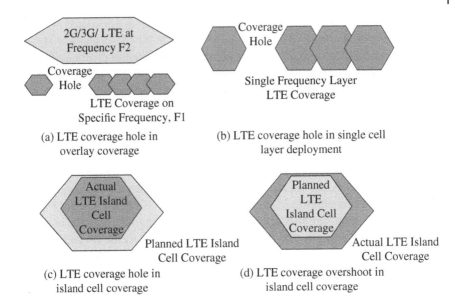

Figure 3.8 Coverage problem scenarios.

4) **E-UTRAN cells with too large coverage:** Changes in the environment might also impact the radio propagation in a way that the cell coverage extends to areas that were not planned to be covered originally. In isolated cells this might not be a big problem but in a neighbourhood of several cells this might lead to severe inter-cell interference problems and QoS issues for UEs.

3.3.4.2 Solution Ideas for CCO

As modifications to cell coverage impact the cell boundaries and overall network capacity, CCO actions need to be performed over relatively larger time intervals. 3GPP recommends performing CCO in a centralized SON architecture (see 3GPP 28.627 section 6.4.3 [28].). The actual optimization can be done at the centralized OAM system such as at NM or DM level. On the other hand, the detection of CCO problems can be done either via KPI/counters reported via Itf-N or even as a result of other SON function analysis. For example, the results of MDT, MRO, MLB, etc. might also indicate some coverage and capacity issues that cannot be solved by adjusting the mobility parameters. In this case, the CCO algorithms need to anal-yse and modify the actual coverage plans of different cells. The centralized CCO can also benefit from network planning tools in an automated manner to analyse some NE configurations.

Cell coverage can be modified adjusting antenna parameters. Antenna tilt and azimuth adjustments have been widely studied as an effective way to adjust cell coverage and capacity. To perform these adjustments in an automated manner, Active Antenna Systems (AASs) and Remote Electrical Tilt (RET) play a vital role, as they eliminate the need for manual adjustments of antenna mount and tower climb for any antenna modification. Using RET, antenna tilts can be changed remotely from OAM systems and AAS allows the dynamic change of the antenna patterns to achieve a particular coverage of a cell.

Additionally, Tx Power can also be used to modify cell coverage and capacity. This can be used to reduce the inter-cell interference when, for example, the cells are transmitting too deeply into the neighbouring cells. However, power modification impacts the whole cell coverage area and not only the cell border areas. This might impact the received signal strengths in the whole coverage area negatively and ultimately the QoS of users might also be affected negatively. Tx power optimization can also be combined with antenna adjustments, but they need to be coordinated so that they try to achieve same targets. Un-coordinated optimization might lead to undesired results, for example, antenna optimization trying to reduce the cell coverage but Tx power trying to increase the power at the same time.

3.3.5 Random Access Channel (RACH) Optimization

Random Access Channel (RACH) is used by UEs to access the radio cells when they are not UL synchronized with the cell or there are no scheduling grants already allocated to the UEs (3GPP TS 36.300). It can be used for initial access in RRC_IDLE state or for RRC Connection Re-establishment after the UE gets out of synchronization. In connected mode, RACH can also be used in case of data arrival with no prior scheduling grant available for the UE.

As RACH resources are shared by all UEs in the cell coverage, proper dimensioning is critical to allow sufficient random-access opportunities for the UEs. Congestion on the random-access resources can severely impact the UE QoS either by blocking the connection attempts or delaying the data transmission.

RA can be performed in two different ways:

1) In the **contention-based** case, the UE performs random-access using a randomly selected preamble from a set of preambles. The eNB may detect the access using correlation of the preambles and furthermore measures the timing of the UE transmission. The eNB then respond to the UE using the same preamble. The eNB assigns a temporary C-RNTI and informs the UE about the scheduled resources and timing advance measurement. Afterwards, the UE responds with its ID based on its current state. The UE in idle state provides the Non-Access Stratum (NAS) information, whereas, the UE in connected state provides an Access Stratum (AS) ID like the C-RNTI. This helps in contention resolution and the eNB can address the UE with its unique ID.

2) The **non-contention-based** access is collision-free as the UE performs the random-access using a dedicated preamble instead of a randomly selected one. The UE is informed about the dedicated preamble, for example, as part of the HO preparation, where the target eNB temporarily reserves a preamble for incoming UE and it is signalled to the UE via the source eNB. In this case, no contention resolution is needed as the reserved preamble is not used by any other UE.

RACH optimization can be implemented as a distributed SON function operating at each eNB. To achieve this, eNBs rely on random access statistics reported from the UEs in their coverage areas as well as the RACH configurations of the neighbouring eNBs as shown in Figure 3.9. UEs can report the random-access attempts as part of the RLF report (3GPP [30]), which indicates how many attempts the UE made before successfully connecting to the eNB. The report includes the number of preambles sent as well as the Tx power used for the unsuccessful attempts.

The UE reports on random-access help the eNB to evaluate the RACH performance. The overall RACH performance of a cell could also be affected by the RACH configuration of neighbouring cells. Therefore, RACH configuration-related information exchange between the neighbouring eNBs is supported via X2 interface. This enables the RACH optimization SON functions to consider the RACH configurations of the neighbouring eNBs into the optimization algorithms. The parameters that are useful for RACH optimization and exchanged over X2 interface are defined in 3GPP, TS 36.211 and are briefly described here:

Root sequence index: Random-access preambles are generated using Zadoff-Chu sequences. Each cell generates the preambles using one of the 838 logical groups of root-sequences and broadcast the index of that in its coverage

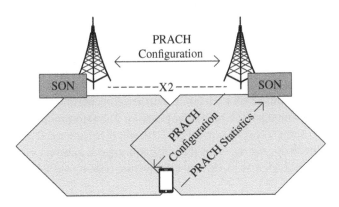

Figure 3.9 RACH optimization cycle.

area. Neighbouring eNBs can exchange this information to avoid scheduling of overlapping sequences.

Zero correlation zone configuration: Each cell also broadcasts a zero-correlation zone configuration to guarantee orthogonality of generated preambles irrespective of UE timing advance and transmission delay. The configuration depends on, for example, cell size and synchronization source.

High speed flag: UE speed can also affect the correlation between the cycles of sequences. Therefore, for high-speed UEs the length of cyclic shift selected with given zero correlation zone is further restricted. Each cell broadcast this 'high speed flag' to indicate to the UEs whether to use the restricted or unrestricted lengths.

PRACH frequency offset: In frequency division duplex (FDD) mode, each cell broadcasts a PRACH frequency offset to indicate which PRB is available for RACH access. This can also be used by the neighbouring eNBs to avoid scheduling of RACH access on same PRBs.

PRACH configuration index (TDD mode only): PRACH configuration index is also needed in addition to PRACH frequency offset to drive the PRB for RACH access. Each cell also broadcasts it and exchanges with the neighbouring eNBs to avoid overlapping PRACH configuration.

3.3.6 Inter-Cell Interference Coordination (ICIC)

Inter-Cell Interference Coordination (ICIC) aims to minimize the interference amongst neighbouring cells. This is required to improve the received SINR situation as it directly impacts the experienced QoS and achievable data rates. ICIC, tries to achieve this by coordinating the usage of radio resources amongst neighbouring cells. For example, in GSM, frequency planning was done to avoid use of the same frequency channels in adjacent cells. This helps to reduce the inter-cell interference but at the cost of poor resource utilization as not all the expensive frequency channels can be used in all cells.

LTE is a frequency reuse one system, meaning the complete frequency band is used in all cells. This makes ICIC extremely important for improved network performance. ICIC can be achieved by intelligent grouping of users and reservation of certain radio resources for each group. Furthermore, power allocations can also be coordinated amongst different neighbours for different groups of these users for ICIC. An example is shown in Figure 3.10, where the inter-cell interference is reduced in the dark grey region as the neighbouring cell does not schedule users in the light grey region with the same frequency resources (PRBs) with which

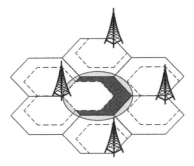

Figure 3.10 LTE ICIC example.

the users in the dark grey region are scheduled. Two implementation options are possible:

1) **Hard frequency split** as in GSM, to be used for scheduling in the central and edge areas of the cells. This lowers the interference but negatively impacts the cell and user performance, due to the reduced number of available PRBs in each region.

2) **Soft frequency reuse pattern**, where the frequency band is still divided into central and edge area resources, but the edge frequencies can be used in the central region if they are not used in the cell edge region.

ICIC can be performed in both UL and DL for data channels and UL control channel. Because the DL control channels are spread across the full frequency band and several PRBs, ICIC cannot be performed on DL control channels.

ICIC can be performed with different levels of dynamicity for network configuration changes. **Fully dynamic** ICIC adjust network parameters frequently and require regular exchange of information amongst the neighbouring cells via X2 interface. Alternatively, **Static ICIC** means fixed configuration of resources to be used for central and edge resources amongst the neighbouring cells and they do not change unless done by the OAM system. A **Semi-Static ICIC** scheme implements slow adaptation of the parameters based on the analysis of measurements of neighbouring cells on a relatively longer time scale.

To support dynamic ICIC schemes, information needs to be exchanged amongst neighbouring cells. Several information elements and messages have been standardized for X2 interface in Release 8 (3GPP TS36.423, 2011):

UL high interference indicator: is a bitmap, where each position represents a PRB. A value of 1 represents high interference sensitivity and 0 represents low interference sensitivity.

UL interference overload indication: this provides a report on interference overload like high, medium, or low interference for each PRB.

Relative Narrowband Transmit Power (RNTP): this provides an indication on DL power restrictions on PRBs in a cell. It contains a bitmap representing each PRB and other related information needed for interference aware scheduling in neighbouring eNBs. On the PRB bitmap, a value of 1 represents 'no promise on the Tx power is given' and a value of 0 represents 'Tx power not exceeding RNTP threshold'.

3.4 Self-Healing

Self-Healing also deals with the operational state but focuses on the failure detection and recovery, for example, of outage detection and NE failure recovery. Due to the distributed nature and large size of mobile networks, FM is a challenging task, which is further accentuated by the fact that in many parts of the network, especially in the radio access network (RAN), there is often little redundancy. Therefore, special SON functions have been created for automating the detection, diagnosis, and recovery of network performance degradation, which are collectively called self-healing SON functions [1].

Self-Healing is an autonomous property that aims to maintain the system in a healthy state, as opposed to states that are considered faulty or degraded [2]. This is realized by continuously monitoring the system, detecting any unexpected degraded states, diagnosing the possible root causes, determining the set of possible corrective actions and a plan for deploying them and finally monitoring the outcome of the recovery. Unlike self-optimization, the self-healing functions can typically only monitor the symptoms of network performance degradation and the root causes are not *a priori* known.

The self-healing concepts and requirements are defined in 3GPP technical specification 3GPP TS 32.541 [14]. In addition, there are four documents specifying the Information Service (IS) [15] and Solution Sets (SSs) for the SON Policy NRM IRP [16]. 3GPP identifies three main use cases: software faults, hardware faults, and cell outage detection and compensation.

For software faults, the corrective actions include software initialization and restarts on different levels and in case this is not able to solve the problem, re-installing, and possibly reverting either to a backup of an earlier software version or an activation of a fall-back software load. Alternatively, software issues can be attempted to be resolved with reconfigurations, but this requires a reliable diagnosis of the problem [3]. In case of faulty hardware, the recovery depends on whether there is a redundant backup unit for the failing resource. If there is no redundancy, the corrective actions can include isolating the faulty resource and trying to circumvent or work around the failure by reconfiguring the other working resources. If there is redundancy, then the recovery would be executed with a switchover using the necessary reconfigurations [3].

The subsequent sections discuss the basic concepts on self-healing for cell degradations. Note, however, that all the concepts originally presented as part of SON can be further improved by using more sophisticated anomaly detection methods as are discussed in Chapter 9.

3.4.1 The General Self-Healing Process

As defined by 3GPP [3], the self-healing process has two parts: the monitoring and the healing, as show in Figure 3.11. The monitoring part continuously monitors the network state and in case a Trigger Condition of Self-Healing (TCoSH) is reached, the healing process is triggered. The healing process gathers any additional information required for diagnosing the root cause of the degraded network state and determining any available recovery actions. In case there are no available corrections, the process stops, and an FM alarm may be raised. Otherwise, the recovery actions are deployed and subsequently the network state is monitored to evaluate the result of the self-healing actions.

3.4.2 Cell Degradation Detection

As outlined in the previous section, the first step in self-healing is the monitoring of the network and detection of the trigger conditions. The triggers are based typically on the indicators available in network management systems:

- **Raw counters or measurements** that provide either the counts of certain events in the managed NE or measurement values of certain monitored parameters, respectively.
- **Key Performance Indicators (KPIs):** KPIs use one or more counters/ measurements as input and calculates a value according to a well-defined (often standardized) formula e.g. in [29]. KQIs usually take KPI aggregates to provide a broader view of the network performance and are rarely useful to assess cell level performance.
- **Alarms:** An alarm is an event triggered by the NE towards the OAM system to notify it on a fault it has detected. HW and SW faults are typically reported as alarms and violation of certain KPI thresholds may also trigger one.

In order to detect degradations, acceptable (or unacceptable) value ranges need to be defined for the indicators. How this is defined, depends on the particular indicator and can be at least of the following type:

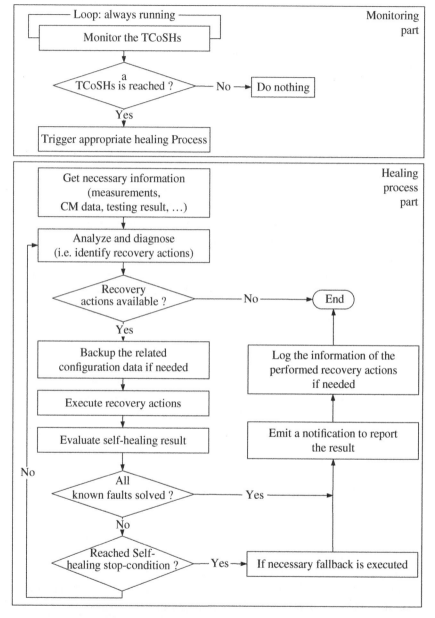

Figure 3.11 Overview of the self-healing process [3].

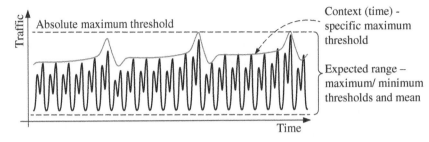

Figure 3.12 Context (time) dependent profile of cell's traffic.

- **An absolute threshold:** For many indicators, it is possible to define thresholds, either maximum, minimum, or a range, within which it should always remain. Typical examples of such indicators are, for example, failure counters, where certain maximum values could be already defined in the network objectives or Service Level Agreements (SLAs). For example, the number of dropped calls should not exceed 1% of all calls. The same threshold would apply to all NEs in the network.
- **Statistical profile:** For certain category of indicators, the acceptable thresholds depend on the managed element and its context and therefore one static absolute threshold cannot be applied to all of them. To avoid having to configure the thresholds individually to all the managed elements, the normal behaviour of each such indicator in any given element can be learned into statistically defined *profiles*. Next, the acceptable ranges can be defined relative to the recorded profiles, for example, as the acceptable number of standard deviations from the mean in a given direction.
- **Context-specific profiles:** As many indicators change with context, it is often necessary to track the indicators in a context specific way. The most used context is time, mainly because human behaviour typically exhibits time varying patterns, e.g. diurnal or daily patterns. These are also reflected especially in the network traffic related indicators, for which the normal and acceptable values may vary heavily with hour of the day or season of the year. For these indicators, profiling is typically done by learning the daily pattern, for example, by creating a statistical profile for each hour of a day for that indicator as illustrated by Figure 3.12.

3.4.3 Cell Degradation Diagnosis

Once a degradation is detected, the next step is to diagnose what caused it to determine if there are recovery actions that can be deployed. This is a much harder problem than degradation detection and typically relies on the analysis done by troubleshooting experts, who use their experience and intuition to drill deeper and find the root causes. Naturally, the aim of self-healing is to automate this step as well and there are some methods already in use to achieve this or at least augment the human expert in making the diagnosis [4].

The most prevalent solutions are rule-based solutions that collect the expert knowledge in form of IF-THEN rules that are typically sequentially executed to infer the possible root cause. Alternative and more advanced solutions have applied **Bayesian Networks (BNs) to statistically make the inference.** A BN consists of a set of variables or nodes and a set of directed edges between these variables. Directed edges normally reflect cause-effect relations within the domain and the strength of an effect is modelled as a conditional probability. When applied to troubleshooting, the nodes represent certain conditions, the faults (i.e. root causes) and their symptoms. Conditional probability tables are used to calculate the probability of the actual state (i.e. the probabilities of the state of the dependent nodes) given the evidence (i.e. proven states of parent nodes).

The major drawback of Bayesian Networks is that it is relatively difficult to build the models, especially specifying the probabilities of the node transitions. Therefore, a balance needs to be found between the complexity and the expressiveness of the model. In self-healing, this is made more difficult by the fact that faults (by definition) are rare and very diverse events. Statistical models based on this information can be very unstable and sensitive to the minor changes in the input variables. Solutions that are proving to be robust to these challenges are the cognitive approaches that apply machine algorithms as discussed in Chapter 9.

3.4.4 Cell Outage Compensation

A cell outage means that a cell has become completely unavailable and inaccessible. No UEs are able to connect to it. This can be due to a hardware or software fault. If there is no redundancy as for example an additional network layer overlay, a cell outage can lead to a complete denial of service, which is why being able to react quickly to cell outages is of highest importance [4].

Cell Outage Compensation (COC) methods aim to quickly restore the service in affected areas in case of a cell outage. They typically do this either by increasing the transmission power or, if adaptive antennas are available, by using the beam-steering and beam-shaping capabilities of the neighbouring network cells. A simple example of the latter is to use a Remote Electric Tilt (RET) unit in a neighbouring cell to tilt it up to increase the coverage area. This can be preferable to adjusting the transmission power, as the cells may often already operate on full power.

The aim of the COC function is a fast restoration of the service and therefore it may do a coarse change of the antenna parameters. Later, they can be optimized further by triggering the Coverage and Capacity Optimization (CCO) function. The CCO function updates the antenna parameters as the COC, but unlike the COC, uses an iterative process to find the best settings. It may, for example, first find the optimal transmission power and afterwards update the antenna tilt. Additionally, after COC, the neighbour relationships may need to be updated using the Automatic Neighbour Relationship (ANR) function.

3.5 Support Function for SON Operation

3.5.1 SON Coordination

As discussed in the above sections, 3GPP SON specification so far focused on specific use cases and provides solutions only for those individual use cases. In network deployments with multiple of those SON functions, they must operate on the same network resources and its environment. Because of this shared network environment, different SON functions may affect the performance other SON functions. For stable network operation, it is, therefore, important to have overview of configuration changes done by different SON functions and to prioritize between different actions in different network scenarios. For example, in case of a network coverage outage, a network operator might want to prioritize the SON functions related to troubleshooting compared to the regular optimization SON functions. These challenges have been discussed by earlier research: [17, 18] and the need for automated methods for the governance of individual SON functions have also been highlighted [19–21]. These methods are broadly termed as SON Coordination and aim to provide a system level overall supervision of simultaneously operational different SON functions and to coordinate their actions for a stable and efficient overall network operation. In the following sections, we first discuss the types of conflict that different SON functions can have and then we elaborate on the coordination mechanisms that can be used either to avoid or minimize the impact of those conflicts.

3.5.1.1 SON Function Conflicts

A SON function conflict refers to a situation where the configuration changes made to one SON function (A) impact the performance of another SON function (B). The performance impact might, for example, be that the SON function A changes the network environment in a way that the impact of changes done by SON function B are minimized and SON function B is not able to achieve its intended targets. Such a SON function conflict can occur if two SON function operations overlap either in spatially or temporal dimensions.

Two SON functions can spatially overlap if the configuration changes performed by them affect the same set of NE(s). This is also referred as the 'Spatial Scope' of a SON function, which includes all the cells where a configuration change is applied as well as the neighbouring cells that could also be impacted due to those changes [4].

SON functions also have a temporal characteristic i.e. they become active for certain network events and take some time to gather all required inputs and process them to obtain the final configuration change recommendation. Once that change is implemented at the concerned NE(s), the respective SON function still needs to monitor the environment of those NEs to access the impact of introduced configuration changes, this is referred as the 'Impact time' of a SON function. Therefore, if another SON function is activated at the same time that also impacts the same network environment, it might disturb the first SON function execution.

Because of these spatial and temporal characteristics of a SON function execution the following types of conflict can arise in different network scenarios:

Network parameter conflicts: SON functions need different network parameters as input to their analysis and to modify the network configuration for better performance. Stability of those parameter values is therefore crucial for correct SON function behaviour. If one SON function (A) changes the input parameters of another SON function (B) while it is still performing the required analysis, the output configuration change generated by B might not be valid anymore and may lead to performance degradation instead of improvement. On the other hand, two SON functions might have an output parameter conflict if they both want to change the same network parameter but using different values. Output parameter conflict can also occur if the second SON function wants to change a network parameter during the impact time of the first SON function that also modified the same parameter.

Measurement conflict: SON functions rely on network measurements to assess network performance and trigger the configuration changes. There is always some delay between the time when configuration change is applied to the network and when its impact is visible in the network measurements. This delay might introduce conflicts between different SON functions if their configuration changes modify the network environment while other SON functions are still working on outdated measurements.

Characteristic conflict: A characteristic conflict appears between SON functions when they want to modify the same characteristic of a NE. For example, antenna tilt and Tx power both change the cell size and MRO and MLB both change the HO trigger point between neighbouring cells. This type of conflict is not directly measurable as conflicting SON functions are operating on different configuration parameters. Therefore, this needs to be carefully analysed at the design phase of a SON function.

3.5.1.2   SON Function Coordination

Coordination between SON functions can basically be achieved in two ways – SON function co-design and SON function coordination.

SON function co-design: The first coordination approach is to carefully analyse the conflicts already at the design phase of SON functions with the aim that SON functions are designed, as far as possible, in a way that they operate on disjoint input and output parameters. To achieve this, SON function design needs to follow some guidelines:

- Reduce the number of shared parameters between different SON functions with clear responsibility of each SON function
- Combine the SON functions into one SON function that operate on the same optimization targets or try to achieve different optimization targets but within the same time period using similar configuration parameters.
- Group SON functions and by-design introduce a run-time behaviour for conflict avoidance.

SON function co-design is a feasible approach to avoid runtime conflicts when there are only a few SON functions. The co-design increasingly becomes impossible as the number of SON functions increase. This approach also needs to analyse all the SON functions together meaning if a new SON function needs to be introduced, the whole process needs to be done across all the SON functions. Even with this high complexity, it is not always possible to design the SON functions with fully dis-joint parameters. Additionally, network operators want to have the flexibility to deploy SON functions from different vendors, so co-design for such an environment is not possible.Although SON function co-design in not applicable to all types of conflicts, it can still be used to minimize the number of potential run-time conflicts and integrate some conflict management features already in the SON function design. This helps reduce the complexity of runtime conflict resolution.

Runtime SON coordination function: To handle runtime conflicts between different SON functions, a SON Coordination Function can also be implemented. This helps detect potential conflicts and ultimately resolve them at runtime. Apart from its main task of conflict resolution, a SON coordination function can also provide the necessary governance capability to the network operator to supervise the autonomously running SON functions (see [31]). For example, a network operator might be able to define a conflict resolution policy or to define a policy to prioritize the execution of different SON functions. To achieve this, the design of SON coordination function needs to follow some principles:

- SON function actions are fully under control of the coordination function
- SON function actions that could potentially lead to any conflicts must be approved by the coordination function before they can be implemented in the network
- The coordination function can prioritize the execution of different SON functions and even pre-empt already running SON functions, if needed
- The coordination function provides an interface to the network operator for the definitions of related policies and the actions of the coordinator for conflict resolution follows those policies
- The coordination function needs to minimize the conflicts and resolve them but should also support the positive interactions between the required SON functions to reach a certain optimization goal
- The coordination logic needs to consider both the spatial and temporal characteristics of the different SON functions as explained above.

Based on the targets of such coordination functions as well as the above-mentioned design principles, a SON coordination function can generate different decisions for the execution of multiple conflicting SON functions. Some examples of such decision options are described below but they are only a representative set and different implementation can also have other criteria to resolve a conflict.

- **Acknowledge:** This is the simplistic case, when there is no conflict with any other operational SON function. As such, the coordinator acknowledges the SON function request and the desired configuration changes are executed on the NE(s).
- **Reject:** The SON Coordination function can also reject the SON function request forcing the SON function not to execute any configuration changes. The main reason for such a decision could be that another SON function with similar spatial or temporal characteristics is currently active in the network. Alternatively, a SON function with higher priority is currently active.
- **Reschedule:** A SON function request can also be rescheduled to another time. The coordinator can decide to reschedule if, for example, the impact time of the other conflicting SON function is about to finish.
- **Pre-emption:** In case of different priorities assigned to different SON functions, the coordinator might also pre-empt the execution of a SON function if there is a configuration change request from a higher priority SON function. For example, an operator might prioritize the failure recovery or outage compensation SON functions over regular optimization functions.
- **Rollback:** When a configuration change by a SON function leads to performance degradation instead of improvement, a logical first step for the SON coordinator is to rollback those changes to the previous stable configuration.

3.5.2 Minimization of Drive Test (MDT)

Mobile network operators rely heavily on drive test to monitor and assess their network performance. Drive tests utilize special equipment to collect network measurements and the sequence of related test call events using specialized equipment. The measurement logs then need to be further analysed for root cause analysis and verification of network performance. All these activities are resource intensive both in terms of human effort and equipment. Moreover, because of the requirement of extensive equipment, drive tests are mainly done in outdoor environments. This limits the effectiveness of drive tests, even with huge resource expenditure, as large portion of indoor traffic remains unaddressed.

Driven by these inefficiencies of drive test procedures, automation of drive test was included as one of the top priorities in the NGMN requirements for cellular network evolution [2]. This started the 3GPP work on a standardized feature, 'Minimization of Drive Tests (MDTs)' in March 2009 [22]. The initial solution was completed as part of 3GPP Release 10 specification [23] and then extended in the later releases.

The focus of 3GPP MDT specification was to use regular subscribers' UE to collect measurement data. Apart from minimizing the network operator efforts for drive test, using subscriber UE has the added advantage that they also provide measurements from areas like indoor coverages, that cannot be covered by normal drive tests. Regular UEs and cells already provide several measurements, for example, for connection establishment and maintenance. Therefore, the MDT work focused on how to efficiently collect those measurements at a common network node for further analysis and optimization.

For the ultimate objective of improved network performance at reduced operational effort and cost, MDT nicely complements the SON paradigm for mobile network operation. SON aims to automate the network management tasks where MDT can serve as a means to collect the relevant network performance measurements and identification of problematic areas. Different SON functions can then perform root cause analysis as well as the necessary parameter configuration for optimized network performance.

3.5.2.1 Scenarios and Use Cases for Drive Tests

Drive testing has been identified as being critical in the following scenarios and use cases:

New base station deployment: Drive tests are used during deployment of new base stations and cells to verify the performance of network planning and fine-tuning of configuration parameters. The measurement collection mainly focuses on DL/UL coverage measurements and mobility between neighbouring cells.

Construction of new infrastructure: Changes in infrastructure (highways/railways/buildings) can alter the radio propagation in cell coverage areas. Drive tests need to be performed in this case to verify that coverage and throughput is still consistent with the actual operator requirements.

Customer's complaint: Drive test might also be needed for root cause analysis in case of customer complaints in specific areas.

Periodic drive tests: Mobile network environments keeps on changing due to several factors, such as changes in subscriber-base as well as their traffic profiles, seasonal changes with their effects on radio propagation, etc. Regular drive tests help to monitor the actual network performance and to identify areas for improvement.

Coverage optimization: The fundamental objective of the drive test is to verify that basic coverage is available in all the planned areas and therefore it was also the focus of early work of MDT in Release 10. The idea was to collect coverage related performance indicators like signal strength values of different cells at each physical location to visualize the overall coverage map of the network. This helps identify the relative coverage quality of different areas and classify areas with weak coverage or coverage holes. Additionally, coverage of different cells can also be analysed to see if there are high interference areas due to coverage overshoot of neighbouring cells. UL coverage can also be analysed to detect areas with poor UL coverage and any mismatch between DL and UL coverages of a cell.

Mobility optimization: Measurement related to serving and neighbouring cell during mobility events can also help to identify mobility problems like high HO or RLF rate. The mobility measurements can also reveal insights to general mobility characteristics of subscribers and can help to further improve the neighbour relations and mobility parameter setting of neighbouring cells. For proper analysis, however, it should first be ensured that the coverage problems are already rectified before looking into mobility problems.

For **Capacity optimization** to maintain a certain QoS, the network operator needs to make sure a minimum network capacity in addition to the network coverage. Measurements, for example, related to load, throughput, etc. help to identify relative load distribution of different areas and the problematic areas where network capacity does not match with the required operator targets. Detailed analysis of these measurements can help to either modify network parameter configuration or to identify the need for new cells deployment. To cater for this, measurements like throughput and data volume for both DL and UL were also included in MDT enhancements for Release-11.

Parameterization for common channels: Mis-match between radio coverage and common channels parameterization can degrade connection setup performance. The target of this use case is to identify the problematic areas where UEs can detect a cell but are unable to successfully setup a connection with in either the UL or DL direction. In such a scenario, the drive test analysis of coverage optimization is not sufficient as coverage is already available in those problematic areas. Therefore, the root cause analysis and optimization procedures focus on adjusting the parameters of common channels instead of cell antenna pattern, Tx power, etc. which are mostly done for coverage optimization.

3.5.2.2 Standardization Support for MDT

3GPP networks support several measurements at the UE and network side for network performance evaluation and management as well as Radio Resource Management (RRM) procedures. MDT features, therefore, focus on reusing those measurements to reduce the UE impact in terms of complexity and power consumption. To provide the geographical distribution of UE measurements, location information was, however, added to the measurement reports as part of MDT feature. Additionally, logging of UE measurements was also introduced for idle mode UEs, as previously only connected mode UEs reported measurements to the network. Therefore, the focus of MDT feature development was not on defining new measurements for UE but to define the procedures to configure and report the measurements to a central entity and to allow the collection of spatial and temporal distribution of those measurements.

MDT is always network triggered, i.e. the OAM decides when and how to configure the MDT measurements. The OAM then sends this configuration to the RAN elements, which then configure the UE accordingly to send the required measurements. This signalling between OAM and RAN elements is supported by extending the Trace procedures from previous 3GPP releases. Additionally, on the radio interface specific procedures and information elements were also defined to support MDT configuration and reporting by UEs.

From a network configuration perspective, two options for MDT were defined:

i. **Signalling-based MDT**, which is used to collect measurements from a specific UE, and
ii. **Management/Area-based MDT**, which is used to collect measurements from a group of UEs in a certain geographical area.

The OAM notifies the RAN nodes in the selected area to start MDT measurements, which then select the specific UEs for MDT reports. The selection is based on parameters received from OAM, UE radio capabilities, and the 'MDT Allowed Flag' given by the CN during call setup. Due to power consumption impacts of MDT, user consent is necessary to include it in the MDT measurements. Therefore, 'MDT Allowed Flag' can be set by Core Network (CN) in a way that the users that have not agreed or are roaming in another network can be excluded from the MDT measurement configuration.

From measurement collection and reporting perspective, two modes were also defined:

1) **Logged MDT**, which is used to store the MDT measurements for a certain period before reporting to the network. This applies to UEs in Idle mode having no active connection to the network. Once a UE transmits from Idle to Connected mode, it indicates the availability of logged measurements. The RAN node can then retrieve the measurements, at any time, as long as the UE remains in the connected mode. Therefore, the UE needs to store the logged measurements for 48 hours after the expiry of logging period or when the logs memory exceeds its allocated MDT storage capacity.

2) **Immediate MDT**, which can be used to immediately report back the measurements. This is done by Connected mode UEs, which have an active connection with the network. This mode allows the collection of UE measurement reports (e.g. serving and neighbouring cells quality); network measurements such as throughput as well as the collection of Failure messages sent by the UE.

The standard does not restrict any combination of above-mentioned MDT types, i.e. both signalling, and management/area-based MDT can configure either logged or immediate MDT based on the connection state of the UEs.

3.6 5G SON Support and Trends in 3GPP

The 5G system is designed to support a very flexible and highly configurable mobile network that can support multiple communication services like eMBB, ultra-reliable, low-latency communication (URLLC), and mIoT. This flexibility, however, also introduces additional network management complexity in terms of explosion of configuration parameter space, user- and service-specific customization of networks and NEs, variations of network deployment flavours, functional decomposition, and management of a multitude of logical network instances, cf. Chapter 2. Automation of various network management tasks is, therefore, essential to overcome the increased complexity.

3.6.1 Critical 5G RAN Features

The legacy framework of SON can be taken as a baseline for the first step towards automated operation and optimization of 5G networks. However, evolving SON for 5G also needs to consider key new 5G features that may impact how SON functions are designed and operated in a 5G network. The key 5G New Radio (NR) features to consider are:

Lean physical layer: The physical layer of 5G NR is very flexible and allows a great variety of configurations, leading to the presence of hundreds of configurable parameters. For example, contrary to LTE, there are no permanent synchronization signals sent over the air, instead they are sent only when needed (when there is data to be transferred).

Beamforming: In 5G NR, beam-based cell sector coverage is used, which increases the link budget and overcomes the disadvantages of the mm-wave channel. In other words, all data transmissions and key signalling transmissions are beam-formed (directional transmission). The Beam Management, i.e. the allocation of beams to the UE when the UE changes position in the cell, also raises new optimization requirements. For example, the UE can very quickly change the beam it is connected to and this needs to be considered for the mobility optimization both within the cell and between neighbouring cells.

Flexible spectrum usage: Another new functionality introduced in 5G NR is the flexible usage of spectrum with 'Bandwidth Parts'. This allows the cell, for example, to allocate a small bandwidth part when the required throughput is small and to allocate a larger bandwidth part when the required throughput is higher. Additionally, this feature also allows to simultaneously support different UE categorizes (narrowband and wideband) within the cell. This flexible management of the spectrum raises questions about the load of the cell, for example, which bandwidth part would count to calculate the available load of a cell that contains multiple bandwidth parts.

Network slices: This is a key 5G feature that enables a flexible way to use the resources of the network for different tenants, and different kind of services. With each network slice supporting different services with their specific performance targets there is also a need to perform network slice specific optimization as well as coordination amongst the requirements of multiple network slices operational on common NEs.

Functional split: In 5G NR, the Node B (gNB) can be split into several parts, and 3GPP has developed the corresponding interfaces. The NR allows the Node B to be split into two parts: CU (Centralized Unit) and DU (Distributed Unit). The CU is in charge of the higher layers of radio protocol stack (Packet Data Convergence Protocol [PDCP], SDAP) and DU is in charge of the lower layers (Radio Link Control [RLC], media access control [MAC], and physical layer).

3.6.2 SON Standardization for 5G

The importance of SON was realized from the beginning of 5G standardization work. For this reason, even the first release of 5G RAN specifications (Rel-15) already included two SON features related to Automatic Neighbour Relation (ANR) and Automatic Setup of NG/Xn/X2 interfaces [24]. Because of the tight schedule of Rel-15 specification and the time required to fully define the baseline physical layer procedures, other SON use cases were deferred to Rel-16 specification work.

In Rel-16, a full Study Item (SI) was defined to analyse the 5G SON requirements in details. In addition, to study the legacy SON use cases that could not be incorporated into Rel-15 specification, one key differentiating aspect of the SI proposal was to consider data analytics driven approaches for 5G management automation and SON, which was also reflected in the name of the SI i.e. 'RAN Centric Data Collection And Utilization for LTE and NR' [25]. The expectations from the analytics capability was high as seen in the motivation for the SI:

> By collecting massive RLF reports via MDT, the time taken to detect the network problems is shortened from 3 months to 4 days and the consistency is more than 88% compared with traditional drive test. Another viable paradigm for RAN Data utilization is positioning, based on MR data in a 7 sq.km field with more than 160 base stations and 600 cells, the CDF of positioning accuracy within 50 m can achieve about 70%. And, with more data collected and used, the higher accuracy can be achieved.

The SI proposal highlighted two main use cases for this data driven optimization as:

- **The problem detection:** the use of MDT allows the identification and solution of problems in the network much faster than using regular Drive Test
- **UE positioning:** The use of radio measurements and their processing can greatly enhance the UE positioning accuracy.

The SI analysed use cases of SON and MDT and identified potential solutions for these use cases. The study uses LTE solutions as a baseline and takes the NR new architecture and features into account, e.g. MR–Dual Connectivity (DC), CU–DU split architecture, beam, etc. The feasibility check of L1/L2 measurements specified in SA5 TS 28.552 was also accomplished. The SI was completed with recommendations to support SON and MDT features for NR, and further established detailed recommendations for the legacy SON use cases as given in the SI Report [26]. The recommendations are summarized below:

Capacity and Coverage Optimization (CCO): The SI recommends enhancing the CCO for NR with measurements related to signal strengths from serving/target cells and beams; measurements related to both successful and unsuccessful RACH access; measurements related to RACH access on both the normal UL and supplementary UL; measurements related to interference and cell load. To comply with the 5G NR flexible functional split, the SI also recommends supporting CCO functionality both at OAM and CU level. In the latter case, the CU is responsible for providing the DU with all required information but leaves it up to the DU to decide how to overcome the problem.

PCI selection: The flexible functional split (CU–DU) architecture of NR gNB also introduces multiple options of where to detect and resolve the PCI conflicts and the SI recommends using the Rel-15 solution as a baseline and further refines it for the split gNB case in the normative work also considering the procedures introduced on F1 interface between the CU and DU.

Mobility optimization: The new proposal of SI for Mobility optimization compared to LTE is the collection of data related to beams and Random Access to allow the UE and network to be able to collect data on the source and target beam as well as also having information about where the UE tried to perform Random Access before accessing the target cell or before the HO failure. Additionally, it is recommended to have an HO Report from the UE, even in the case of a successful HO. This report could include information about the procedure that UE followed prior to the success of the HO, for example, where RACH access failed before being successful for the HO completion.

Load sharing and load balancing optimization: The SI recommended support for both intra-RAT and inter-RAT load sharing and load balancing. For this purpose, it is recommended to support load reporting procedures over X2, Xn, F1, and E1 interfaces that could also potentially consider the load information related to beams, network slices and hardware in addition to the radio load.

RACH optimization: As 5G allows multiple RACH attempts on different RACH resources (beam), the SI concluded that it should be possible to log and to report the RACH processes when successful and when it fails as well as exchange of PRACH parameters related to both the Normal UL and Supplementary UL over the Xn and F1 interfaces.

Energy saving: The SI observed that intra-system ES can already be supported by the cell activation/deactivation features of Rel-15, whereas, inter-system ES needs to be further analysed.

Minimization of Drive Test (MDT) use cases: MDT is one of the major items for 5G SON in 3GPP. The main part is the update of MDT designed for LTE and its implementation to NR. The new NR functionalities like beam forming needs to be added to the measurements that UE and network collects. Also, 5G needs to collect UE and network measurement when the UE is configured with dual connectivity, which could be added to the MDT measurements.

Enhancements of MDT for LTE are also discussed as part of the 5G SON work, which includes the collection of UE measurements and L2 measurements collected at the base station. The measurement enhancements need to consider the new architecture of NG-RAN (Functional Split) and the context of Dual Connectivity.

At the time of writing, the SI work has been completed and a new Work Item has been approved to start the 5G SON specification work [27]. The work item will take the recommendations of the SI work as the baseline and further develop the standardized solutions for the above-mentioned SON use cases. There is a strong interest in developing the 5G SON framework in a data analytics and machine learning driven manner as evident from the SON SI description. However, so far there has been little progress made in this regard in the 3GPP SON specification work due to tight 3GPP schedules and the need to have a first baseline support for the legacy SON use cases in 5G. It is expected that the inclusion of data analytics and Machine Learning (ML) techniques for RAN management automation will be discussed in more detail in the next 5G releases.

3.7 Concluding Remarks

This chapter has summarized the state of the art on Self-organizing networks. Specifically, it gave a brief discussion of the most critical use cases and the features that have already been standardized in support of these use cases. However, there are challenges with the SON framework.

The SON functions have been developed to target individual problems, as is visible amongst the presented use cases, without considering their impacts on other network management tasks and SON functions. Correspondingly, it was identified that SON management is critical to deal with the simultaneous operation of multiple SON functions. This, however, is not an easy task as has been identified in the SON coordination-related studies.

Secondly, SON functions are mostly designed as expert systems that try to mimic the behaviour of network engineers. Thus, it is difficult to adapt those SON functions for changing environments and contexts. This directly motivates the need for more adaptable algorithms (SON functions) that can adapt according to their context as well as their interaction with other SON functions. Additionally, since the other functions may be from a different vendor, the new paradigm must be capable of dealing seamlessly with a Multi-vendor environment, i.e. Multi-vendor function coordination must be a default capability in advanced network automation solutions.

Moreover, the advancement of network automation must consider new use cases that were initially not critical or not required. Specifically, new 5G features like beams, slicing, virtualization, multi-tenancy, unlicensed-band (co-)operation, etc. must be addressed. Additionally, difference in operating environments must be considered in such developments, e.g. to account for new scenarios like industrial environment, URLLC, virtual reality, etc. Finally, the advanced automation functions (or at least the framework thereof) must provide capabilities of end-to-end automation. For example, the automation of network slicing management should be done optimally in a non-piece-wise manner to maximize the value of service-specific logical networks.

References

1 Jaana, L., Wacker, A., and Novosad, T. (eds.) (2006). *Radio Network Planning and Optimisation for UMTS*. Wiley.

2 NGMN (2010). *A Deliverable by the NGMN Alliance: NGMN Top OPE Recommendations*. Frankfurt: NGNM.

3 3GPP TS 32.501 (2008). *Telecommunication management; Self Configuration of Network Elements; Concepts and Integration Reference Point (IRP) Requirements (Rel. 8)*. Sophia Antipolis: 3GPP.

4 Hamalainen, S., Sanneck, H., and Sartori, C. (2011). *LTE Self-Organising Networks (SON): Network Management Automation for Operational Efficiency*. Wiley.

5 3GPP TS 25.484 (2011). *Technical Specification Group Radio Access Network; Automatic Neighbour Relation (ANR) for UTRAN*. Sophia Antipolis: 3GPP.

6 3GPP TS 32.762 (2016). *Technical Specification Group Services and System Aspects; Telecommunication management; Evolved Universal Terrestrial Radio Access Network (E-UTRAN) Network Resource Model (NRM) Integration Reference Point (IRP); Information Service (IS)*. Sophia Antipolis: 3GPP.

7 3GPP TS 38.211 (2018). *Technical Specification Group Radio Access Network; NR; Physical channels and modulation, (Release 15)*. Sophia Antipolis: 3GPP.

8 Teyeb, O., Mildh, G., and Furuskr, A. (2012). Physical Cell Identity Assignment in Heterogeneous Networks. Vehicular Technology Conference (VTC Fall), Québec City.

9 Salo, J.J., Nur-Alam, M., and Chang, K. (2010). *Practical Introduction to LTE Radio Planning – A White Paper on Basics of Radio Planning for 3GPP LTE in Interference Limited and Coverage Limited Scenarios*. Espoo, Finland: European Communications Engineering (ECE) Ltd.

10 Ahmed, F., Tirkkonen, O., Peltomaki, M. et al. (2010). Distributed graph coloring for self-organization in LTE networks. *Journal of Electrical and Computer Engineering* 2010: 1–10.

11 Szilagyi, P., Bandh, T., and Sanneck, H. (2013). Physical cell ID allocation in multi-layer, multi-vendor LTE networks. *Mobile Networks and Management*: 156–168.

12 Kallin, H. and Moe, J. (2013). Method for automatically selecting a physical cell identity (PCI) of a long term evolution (LTE) radio cell. US Patent 8,494,526, filed 19 March 2013 and issued 23 July 2013.

13 Amirijoo, M., Frenger, P., Gunnarsson, F. et al. (2008). Neighbor cell relation list and physical cell identity self-organization in LTE. IEEE International Conference on Communications Workshops, Beijing.

14 3GPP TS 32.541 (2017). *Telecommunication management; Self-Organizing Networks (SON); Self-healing concepts and requirements.*Sophia Antipolis: 3GPP

15 3GPP TS 32.762 (2014). *Telecommunication management; Evolved Universal Terrestrial Radio Access Network (E-UTRAN) Network Resource Model (NRM) Integration Reference Point (IRP); Information Service (IS).* Sophia Antipolis: 3GPP.

16 3GPP TS 32.766 (2014). *Telecommunication management; Evolved Universal Terrestrial Radio Access Network (E-UTRAN) Network Resource Model (NRM) Integration Reference Point (IRP); Solution Set (SS) definitions.* Sophia Antipolis: 3GPP.

17 Dottling, M. and Viering, I. (2009). Challenges in mobile network operation: Towards self-optimizing networks. IEEE International Conference on Acoustics, Speech and Signal Processing, Tapei.

18 Baliosian, J. and Stadler, R. (2007). Decentralized configuration of neighboring cells for radio access networks. IEEE International Symposium on a World of Wireless, Mobile and Multimedia Networks, Helsinki.

19 Schmelz, L.C., Amirijoo, M., Eisenblaetter, A. et al. (2011). A coordination framework for self-organisation in LTE networks. 12th IFIP/IEEE International Symposium on Integrated Network Management (IM 2011) and Workshops, Dublin.

20 Barth, U. and Kuehn, E. (2010). Self-organization in 4G mobile networks: Motivation and vision. 7th International Symposium on Wireless Communication Systems, York.

21 Sanneck, H.B.Y. and Troch, E. (2010). Context based configuration management of plug & play LTE base stations. IEEE Network Operations and Management Symposium-NOMS, Osaka.

22 3GPP TS 36.805 (2009). *Evolved Universal Terrestrial Radio Access (E-UTRA); Study on Minimization of Drive-Tests in Next Generation Networks.* Sophia Antipolis: 3GPP

23 3GPP TS 37.320 (2011). *Universal Terrestrial Radio Access (UTRA) and Evolved Universal Terrestrial Radio Access (E-UTRA); Radio measurement collection for Minimization of Drive Tests (MDT); Overall description; Stage 2.* Sophia Antipolis: 3GPP.

24 3GPP, TS 38.300 (2018). *NR and NG-RAN Overall Description, V15.4.0, Dec. 2018.* Sophia Antipolis: 3GPP.

25 3GPP, RP-182105 (2018). *Study on RAN-centric data collection and utilization for LTE and NR.*, Sophia Antipolis: 3GPP.

26 3GPP, TR 37.816 (2019). *Study on RAN-centric data collection and utilization for LTE and NR*. Sophia Antipolis: 3GPP.

27 3GPP, RP-191594 (2019). *New WID on SON/MDT support for NR*. Sophia Antipolis: 3GPP.

28 3GPP TS 28.627 (2018). *Telecommunication Management; Self-Organizing Networks (SON) Policy Network Resource Model (NRM) Integration Reference Point (IRP); Requirements (Release 15)*. Spphia Antipolis: 3GPP.

29 3GPP TS 32.410 (2018). *Telecommunication Management; Key Performance Indicators (KI) for UMTS and GSM (Release 15)*. Sophia Antipolis: 3GPP.

30 3GPP TS 25.331 (2011). *E-UTRA Radio Resource Control (RRC) Protocol specification (Release 8)*. Sophia Antipolis: 3GPP.

31 Frenzel, C., Lohmuller, S. and Schmelz, L.C. (2014). Dynamic, context-specific SON management driven by operator objectives. IEEE Network Operations and Management Symposium (NOMS), Krakow.

4

Modelling Cognitive Decision Making

Stephen S. Mwanje and Henning Sanneck

Nokia Bell Labs, Munich, Germany

The need for automating network management has been articulated for at least a decade [1, 2]. Successive work has studied the different aspects of the Network Management Automation (NMA) challenge and different solutions for the specific challenges. This has resulted in what is known in the mobile network industry as Self-Organizing Networks (SONs) [1, 2]. The core idea in SON (as discussed in Chapter 3) was the implementation of closed-loop control mechanisms that evaluate radio network state and propose the appropriate configurations or reconfigurations of the network parameters. For some key use-cases (e.g. auto-connectivity, commissioning, and automatic neighbour relationship setup) the automation gains have been proven for many operators in the field. Still, those use cases concern relatively simple workflows and parameter types. Their automation gain comes more from the fact that a very frequent, but simple use case is automated. For use cases which are less frequent but more complex to solve, the degree of automation in deployed systems is still low.

At the same time, the challenges underlying the need for network automation have continued to expand. The complexity of network operations exponentially increases with new technologies and the related density of cells. This has motivated for better mechanisms for decision making when deriving actions taken by the autonomous functions. In other words, NMA needs to evolve to a more cognitive system, i.e. one capable of reasoning over all possible contexts when formulating recommendations for subsequent behaviour. The current (second) generation of NMA that is principally referred to as Cognitive Network Management is thus currently underway to allow for this higher degree of automation. The core idea therein is applying cognitive techniques (mainly machine learning techniques) on to or over the earlier designed SON functions. However, to take full advantage of these cognitive techniques, we need to characterize their possible offers

Towards Cognitive Autonomous Networks: Network Management Automation for 5G and Beyond, First Edition. Edited by Stephen S. Mwanje and Christian Mannweiler.

in comparison to the automation requirements of the networks. This requires an end-to-end understanding of cognition especially as evidenced by the most cognitive entity known – the human.

Therefore, this chapter discusses the foundational principles behind *self-organization and cognition, as demonstrated by biological agents*. It contrasts the concepts of self-organization and cognitive autonomy and attempts to break down a fully cognitive process into its (quasi-)orthogonal sub-processes as a means of identifying the capabilities which the available or known cognitive techniques should provide.

4.1 Inspirations from Bio-Inspired Autonomy

Biology has for long been the source of inspiration for the design of technological artefacts. From early aircrafts mimicking birds to neural networks attempting to model the human brain, biological systems always provide a good starting point for developing new technical systems. The designers seek to create artefacts capable of efficient and autonomous operation in previously unknown and unpredictable environments. Typically, the design attempts to create an entity that displays elements of lifelike skills in design, control, behaviour, and intelligence.

Biological and communication networks have similar characteristics that make it necessary for communication networks to learn from biological networks. Both consist of many relatively autonomous entities interacting with each other in simple hierarchical organizations that result in extremely complex systems. Yet biological systems exhibit self-organization behaviour in that autonomous individuals (e.g. cells, insects, birds, etc.) group or act together in certain structures without any centrally controlled, globally explicit expressed rules leading to the visible patterns. Even with the self-organizing behaviour, they demonstrate characteristics that make them of great interest in communication networks – specifically the efficient and effective distributed control; robustness amidst self-organization; adaptability; natural stochasticity; and the simplicity of units and rules.

4.1.1 Distributed, Efficient Equilibria

Biological systems can self-organize in a fully distributed fashion, for example, creating a complex higher organism from many simpler homogeneous organisms through a self-enhancement process but without exploding the self-enhancement, i.e. with limits to how much the organisms self-enhance. The higher organism develops out of a single fertilized egg in what are known as activator-inhibitor systems. Cells which carry identical genetic code, become different from each other through the combination of a self-enhancement process and longer ranging confinement of the locally self-enhancing process [3].

Moreover, even as distributed systems, biological systems can collaboratively achieve efficient equilibria. For example, the nervous system maintains homeostasis (the stable state of an organism and of its internal environment) by controlling and regulating the other parts of the body through a negative feedback mechanism. A deviation from a normal set point (like a temperature drop) acts as a stimulus to a receptor (e.g. the skin and abdominal organs), which sends nerve impulses to a regulating centre in the brain to take the necessary counter action.

4.1.2 Distributed, Effective Management

Biological systems demonstrate very effective management of constrained resources, yet with a globally amplified intelligence in the emergent outcomes. Insect colonies are, for example, very efficient in both task allocation process (the engagement of workers in specific tasks and numbers appropriate to the prevailing situation) as well as the task partitioning process (the division of one task into sequential actions to be done by more than one individual).

4.1.3 Robustness Amidst Self-Organization

Biological systems are robust to harsh conditions with inherent and sufficient redundancy. This can be observed in how epidemics work – the transmission of infectious diseases between individuals. Given a contact between the individuals, something is communicated and so the degree of connectedness of the individuals underlies the spread of the epidemic. Moreover, the group activity is resilient to failures by internal or external factors. These systems are robust in that a catastrophe befalling one individual does not necessarily translate into failure of the group [4]. Correspondingly, the robustness allows scalability since group activity will be achievable regardless of the size of the group.

4.1.4 Adaptability

Biological systems are adaptive to the varying environmental circumstances, as is demonstrated in firefly flight synchronization. Male fireflies flash to attract females without following the leader yet with a constant period between flashes [5]. With each firefly only exposed to flashes of its neighbours, the fly resets the timing of flashes depending on the differences with their neighbours. With this mechanism, even when their flashing is reset, e.g. by an external light, they will quickly regain synchronization as soon the external light source disappears.

Moreover, a biological system can learn and evolve itself under new conditions. An example here is Somatic Hypermutation in the immune system – the cellular mechanism by which the immune system adapts to the new foreign elements (e.g. microbes) that confront it. Somatic Hypermutation diversifies B-cell (a type of white blood cell) receptors used to recognize foreign elements (antigens) and allows the immune system to adapt its response to new threats during the lifetime of the organism [6].

4.1.5 Natural Stochasticity

Biological systems are naturally stochastic (i.e. non-deterministic) [4] and yet the stochasticity component does not make the system weak or fail. Amongst cells of the same type, for example, the values describing these cells will vary from one cell to another at any given time and will also vary with time for a single cell [7]. Biological systems are thus able to deal with stochasticity and yet generate deterministic or predictable outcomes. Similarly, Brownian motion (the random motion of particles immersed in fluid) is described, as in Eq. (4.1), in terms of a continuous time stochastic process (the Wiener process) in which the position of the particle has a noise term added to the product of its velocity and the viscous coefficient of the fluid.

$$\frac{dv}{dt} = -\gamma v + \eta \tag{4.1}$$

4.1.6 From Simplicity Emerges Complexity

Biological systems can develop complex behaviours based on and using a limited set of simple rules. Ant colonies for example, can find short and efficient routes to food sources using simple mechanism of leaving marks (pheromones) on previously used routes. Thereby, the higher the intensity of pheromones the greater the chance that the respective path is short.

The global outcome in such scenarios is emergent behaviour, i.e. the outcome of the group depends more on the collection of the individual behaviours and not on each individual task they perform.

4.2 Self-Organization as Visible Cognitive Automation

Historically, automation has been likened to self-organization with the discussion typically turning to biologically inspired autonomy. The characteristics of biological systems (as described in Section 3.1) justify the pursuit of imitating said systems in an effort to realize self-organization. Correspondingly, definitions, characterization, and solutions for self-organizing systems tend to look to nature to derive self-organization and this has also been the pattern in communication networks. This presents an external view of self-organization, i.e. a system is considered to be self-organized if is observed to always exhibit consistent probably deterministic or predictable outcomes even when perturbed. Correspondingly, the goals of the individual entities therein or of the whole system are of no significant regard in such a view, especially since they are subjective for each entity. Instead the emergent outcomes, (specifically consistence, predictability, or even determinicity of outcomes) become the measure of the degree of self-organization.

4.2.1 Attempts at Definition

There have been multiple definitions for self-organizing systems amongst them being:

F. Heylighen (1997): 'Self-organization is a process where the organization (constraint, redundancy) of a system spontaneously increases, i.e. without this increase being controlled by the environment or an encompassing or otherwise external system.' [8]

Yates (1987): 'Technological systems are organized by external commands,... human intentions lead to the *building of structures* or machines.... natural systems *become structured* by their own internal processes:... the *emergence of order* within them is a complex phenomenon that intrigues scientists from all disciplines.' [9]

Bonabeau et al. (1999): Self-organization is a set of dynamic mechanisms whereby structure appears at the global level of a system from interactions amongst its lower level components [10]. The rules specifying the interactions amongst the systems' constituent units are executed on the basis of purely logical information, without reference to the global pattern which is an emergent property of the system rather a property imposed upon the system by an external ordering influence.

Camazine (2003): 'Self-organization is a process in which *pattern at the global level* of a system *emerges* solely *from numerous interactions among the lower-level* components of a system.... the rules specifying interactions... are executed using only local information without reference to the global pattern.' [11]

It is evident from all these definitions that a system is defined as self-organizing from an external perspective. The observing entity monitors the behaviour of the system and its interactions with the environment. Then, where it is evident that the environment does not explicitly define how the system responds to stimuli, the system is considered self-organized. However, this does make any conclusion as to how the internal structure of the system is organized.

4.2.2 Bio-Chemical Examples of Self-Organizing Systems

Natural systems provide a long list of self-organized systems examples, so it is no wonder then that nature is the primary source of inspiration when developing SO systems. Many biological entities exhibit Self-Organization (SO) characteristics in their life systems. And the same is observed in chemical and biochemical processes underlying the biological systems.

One of the most common examples of self-organizing systems in biology is the formation and management of colonies in social insects, e.g. haplometrosis in ants. For example, at formation, during the annual mating season, queen and male ants migrate from their colonies to look for mates. When successful, the male dies and the queen finds a good place to start a new colony. The queen maintains male sperm after being inseminated and uses this to fertilize all future eggs. The colony does not however expand indefinitely – mainly because of the migration of virgin queens. Although the behavioural, social, and chemical cues that trigger the migration of virgin queens and males from a colony are not well understood, it has been observed that they are quite literally pushed out of the colony by a plethora of workers [12]. It is through this process that new colonies are formed but there are processes for replacing a dying queen in the different ant species. Therein either the original queen lays a virgin queen shortly before its death or workers battle to become the candidate queen (called a gamergate) which then goes through a series of epigenetic changes in order to become the new queen for that colony.

On the other hand, many of the collective activities performed by social insects result in the formation of complex spatio–temporal patterns. Without centralized control, workers can work together and collectively tackle tasks far beyond the abilities of any one individual. The resulting patterns produced by a colony are not explicitly coded at the individual level, but rather emerge from nonlinear interactions between individuals or between individuals and their environment [13]. A good example here is the foraging for food by ant colonies through which ants can establish efficient routes to and from the nest to the food sources [14].

Self-Organization is also observed in swarms of birds flying in some joint and well-coordinated way over very long distances, e.g. the seasonal migration of swarms for more than 10 000 km [15]. The movement of such a swarm is a SO system since there are no external triggers that control how the swarm maintains direction or how it responds to obstacles in its path.

An example of Self-Organization in physical chemistry is Gerhard Ertl's dynamics in catalyst reactions on surfaces [16]. In one such reaction, it was shown that when exposed to oxygen, a copper surface exhibits long-range Spatial Self-Organization forming distinct stable patterns.

An example of Self-Organization in chemistry is pattern formation in the Belousov-Zhabotinsky (BZ) reaction [17]. The BZ reaction is a family of oscillating chemical reactions, during which, ions of transition-metals catalyse oxidation of various, usually organic, reductants by bromic acid in an acidic water solution. In the simplest form of such an oscillating reaction, the products of one reaction become the reactants of another that regenerates the original reactants.

The BZ reaction makes it possible to observe development of complex patterns in time and space by the naked eye on a convenient human timescale of dozens of seconds and space-scale of several millimetres. The classic BZ reaction involves potassium bromate, cerium(IV) sulphate, and propanedioic acid (aka malonic acid) in dilute sulphuric acid, in which case the colour changes are due to the oscillating oxidation state of cerium–respectively from colourless for Ce(III) to yellow of Ce(IV). The BZ reaction can generate up to several thousand oscillatory cycles in a closed system, which permits studying chemical waves and patterns without constant replenishment of reactants.

4.2.3 Human Social-Economic Examples of Self-Organizing Systems

Besides the observable biological systems, there are also many Self-Organizing human systems that are visible today – mainly in the social-economic human interactions. The formation of human societies can also be considered a SO system. As seen from outside, in such societies no specific force defines how groups of humans should respond to certain events. Instead, groups internally coordinate their activities towards forming the respective communities. This is evident in the formation of countries, organizations, communities, associations, unions, clubs, etc.

Consider, for example, the selection of narcotic drug trade routes as evidenced in Figure 4.1a. Clear patterns of the drugs flow are quite evident, yet no single organization designed these plans for all to follow. However, each concerned group considered the prevailing conditions in each country or region of the world and selected the countries which best fulfilled the organization's business objectives. The outcome of which is a clearly a SO system with no control from an external force.

Similarly, the capitalistic economic system prescribes minimal rules yet supports many individuals and entities with different ideas and goals representing

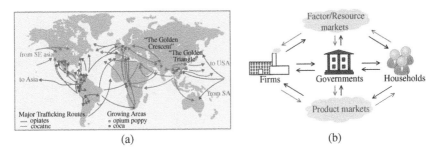

(a) (b)

Figure 4.1 (a) CIA Map of international illegal drug connections (*Source:* http://PBS.org [18]) (b) the circular flow of economic activity in a typical market economy.

local algorithms. Other human systems that have been identified to have stable self-organized and emergent outcomes include, amongst others, the characteristics of traffic jams on the road [19], the flow of persons in evaluation dynamics [20] and the creation of paths over the lawn in parks [21].

4.2.4 Features of Self-Organization – As Evidenced by Ant Foraging

To capture the features of self-organizing systems, let us consider the foraging behaviour of ants, which has long been established as a self-organizing process. Through the foraging mechanism, ants can find the shortest paths between food sources and their nest. Whilst walking from food sources to the nest and vice versa, ants deposit a substance called pheromone on the ground, in this way, forming a pheromone trail. Ants can smell pheromone and, when choosing their way, they tend to choose, in all probability, paths marked by strong pheromone concentrations. The pheromone trail allows the ants to find their way back to the food source (or to the nest). Also, this can be used by other ants to find the location of the food sources found by their nestmates.

Experimentally, it has been shown that this pheromone trail following behaviour can give rise, once employed by a colony of ants, to the emergence of shortest paths [14]. That is, when more paths are available from the nest to a food source, a colony of ants may be able to exploit the pheromone trails left by the individual ants to discover the shortest path from the nest to the food source and back. In an experiment with a double bridge of equal length between food source and nest (Figure 4.2a), it is shown (see [14]) that the ants converge on a single path. Then, in an experiment in which the bridge's branches are of different length as in Figure 4.2b, the shortest branch is most often selected.

This can all be explained by the pheromone laying behaviour in that at every point of choice, an ant will choose the path with the highest pheromone intensity. This has a multiplicative effect where the higher intensity path also has a higher probability of being chosen by subsequent ants further increasing its intensity relative to the lower intensity path.

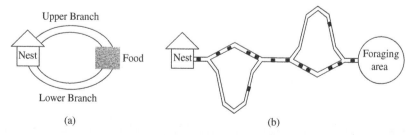

Figure 4.2 Foraging of ants in (a) single branch (b) the original experiment [14].

For an ant that needs to find a path either to the food source of the nest, there is a higher density of pheromone on the short path so the ant chooses to take the shorter route. Eventually, all ants almost exclusively use the shorter path, an emergent behaviour that highlights ants as a good example for a Self-Organized system. Specifically, the SO is observed from the fact that simple rules lead to a good result. Correspondingly, many solutions in distributed control and optimization [7] have been developed on this basis.

Evaluating the SO examples, and specifically ant foraging, several features become evident when contrasted to non-SO systems. These core features, as described in Table 4.1, include the control of the system, the exchanges of information in the system, the resulting structure and complexity of the system as well as its level of scalability.

Table 4.1 Features of Self-organizing systems.

Feature	Property	Description
Control	Distributed	There is no global control for the whole system. Instead, the constituent subsystems perform at least partially autonomous from each other.
Interaction	Local	There is no form of global information exchange but only local interaction amongst the subsystems, e.g. the pheromone trail in the ant case. Another example here is that in a swarm of birds (even where they follow one leader), only local interaction amongst neighbouring birds in a flock is possible when deciding if one is following the right path or not.
Structure	Emergent	The system's global behaviour or functioning emerges as observable patterns or structures but is not externally prescribed, i.e. the final path selected by the ants is not prescribed externally or *a priori*.
Complexity	High	Even when the individual subsystems and their basic rules are simple, the resulting overall system becomes complex and often unpredictable. However, the complexity of the system may also allow for complex decisions to be made.
Scalability	High	Typically, the system will have no significant change in performance even as the number of subsystems increases e.g. Birds can migrate over long distances both in small flocks (less than 50 for the *Merops apiaster*) and as large swarms (thousands for *red-billed quelea* and *starlings*). Similarly, a capital market can function well both in a single-small market like Switzerland or in a multi-country or multi-state market like Europe or the USA.

4.2.5 Self-Organization or Cognitive Autonomy? – The Case of Ants

The natural systems are described above as self-organizing, but this is typically the case when the systems are evaluated based on their emergent behaviour, i.e. as evaluated by an external system. Thus, it is imperative to question the underlying mechanisms for decision making that are applied therein. Specifically, we should consider the possibility that the decision-making process is cognitive, as we do here in the case of ants.

New studies have proven that the behaviour of colony animals (especially ants) is more than random or deterministic self-organization. '... each foraging process of an animal is also a learning process. With foraging repetition, long-term memory continues to accumulate, an animal's knowledge about the environment of its nest gets richer, and the region that the animal is familiar with continues to enlarge' [22].

The observed emergent behaviour is due to an internal cognitive process in which the ants process information to make decisions. Internally, within each insect, there are sub-processes that the insect uses to compute its decision that eventually leads to the observed outcome, i.e. the animals use their intelligence and experience to guide them [22]. Thus, SO is more than a deterministic decision-making state machine.

4.3 Human Cognition

The discussion of cognitive systems usually begins with the most cognitive being so far proven to exist – the human. As defined in Chapter 1, cognition relates to the collection of mental processes and sub-processes that support process of acquiring knowledge and understanding through sensory stimuli, experience, and thought. The brain continuously executes these cognitive processes to process all the information we receive from the environment.

The processes need to harmoniously interact amongst themselves for us to adequately analyse a situation and flexibly adapt to its reality, demands and changes. They, however, also need to be able to effectively act separately on their specialty functions. For example, people with language disorders should still be able to accurately perceive stimuli or solve mathematical problems.

The complete set of mental processes and subsequent skills can be grouped into two major categories [23]: the basic processes that are independently fundamental to the functioning of the brain and the higher or complex processes that are built up from a combination of basic and other complex processes. To distinguish these processes, let us consider a simple mental exercise:

Jon is listening to music whilst driving a car, i.e. some brain power is devoted to listening whilst the rest is devoted to driving. Through his peripheral vision, he notices an object appearing on the street. His brain will focus more attention on the possible danger posed by the appearing object. He then realizes that it is a cyclist riding into the driving lane. Jon will entirely focus his attention on the driving (almost eliminating the sensation of hearing something) and will coordinate a series of motor movements to eliminate the danger (e.g. hoot, brake and/or manoeuvre). Then, if the driving-related stimuli disappear afterwards (e.g. when the car is stopped), the music may even appear louder and more disturbing as the brain quickly switches back to the processing of the audio signal (potentially the only remaining stimulus).

4.3.1 Basic Cognitive Processes

The basic processes that are independently fundamental to the functioning of the brain are sensation, perception, attention, and memory.

The first process is **sensation** which refers to our awareness and capture of the various stimuli in our environment. The stimuli in the form of various forms of physical energy reach our sense organs where the receptors therein are stimulated, and nerve impulses are transmitted in the brain. The sensory organs capture all the available stimuli from the outside world and leave it to the brain to filter that information.

Following sensation, **perception** enables us to process the received impulses and make out the meaning from the stimuli. For example, Jon perceives that what he sees is not a static roadside object but a moving object which may end up in his path (even before working out what kind of object it may be). We continuously use our perception without even noticing it. We are conscious of other people's movements, messages we receive on our phone, food flavours, our posture, etc. What one perceives depends on selection, organization and interpretation of stimuli, i.e. an individual selectively attends to certain stimuli and not to others. However, it also depends on the functioning of sense organs and the brain, previous experiences, psychological state, and interest as well as our motivation and behaviour. Considering previous experiences and interest, for example, if Jon is a 16-year old novice driver listening to his favourite latest hit song, the combination of a short experience and high interest in the song may imply that he may only perceive the possible danger far later than his mother who may even be in the car's back seat.

Attention is the mental process that enables us to filter the sensations. The environment provides a lot of concurrent stimuli, yet we only have limited processing capacity. As such, we must voluntarily and/or involuntarily choose what we focus on. Some actions such as walking, and chewing require little attention and so we automate their execution to proceed without attention. Others, such as speaking and body language, however, require focus, especially when we are giving a lecture.

The degree of automation (reduction in attention) increases with the familiarity of the task. For example, on first learning to drive a car, focusing on everything one must do is difficult but after a whilst it is done naturally and without thinking. And our attentional focus can be influenced by the intensity and character of the stimuli (colour, clarity of visual signal or pitch, timbre, intensity of auditory signal as well as the perceptual dependencies).

It is evident that attention and perception are not completely orthogonal or decoupled but need to proceed in parallel. For example, Jon can perceive something likely to move into his path even whilst attentive to the music but completely switches to fully attend to the moving object as soon as he considers its possible degree of danger.

Memory allows us to encode the data we receive from the environment and the data that we mentally create so that we can consolidate and retrieve it later. Memory is characterized in different types depending on the characteristics and subsequently on usage. The first is sensory memory which allows information that briefly reaches the senses to be registered in the sensory store. Some of this information is successfully passed to the short-term memory store. For example, the visual image obtained when looking at the telephone number will enter the sensory store and be selected from other incoming information for processing. Short-term memory, also described as working memory, is the temporary storage required for such tasks as mental arithmetic, reasoning, or problem solving. After a short period (estimates from a few seconds to a few minutes) the information is either lost from memory or passed on to the more permanent store (long-term memory). Long-term memory, which has been suggested to have virtually unlimited capacity, allows storage of information over very long periods. However, the retrieval of such information is not guaranteed to be fast or perfect. All these types of memories interact together but they don't all depend on the same brain areas. For example, people with amnesia still remember how to walk but can't remember their partner's name.

4.3.2 Higher, Complex Cognitive Processes

Higher processes are those processes that presuppose the availability of knowledge and put it to use [24]. Amongst these are thought, language, and intelligence as well as combinations of these leading to problem solving and learning.

4.3.2.1 Thought

The complexity and heterogeneity of our thoughts are unbounded. This higher mental process is responsible for tasks related to problem-solving, reasoning, decision-making, creative thinking, divergent thinking, etc. To simplify these functions, the brain creates concepts. We need to group and relate ideas, objects, people, or any other kind of elements that come to mind. Usually, this helps us streamline our cognitive processes. However, sometimes we try to be logical and often ignore how irrational we can be. We take shortcuts to process information faster without considering details that might be important. This can lead to cognitive biases, which are deviations from the normal process of reasoning. For example, we sometimes believe that we can figure out what is going to happen in a game of chance. On occasions, cognitive biases can lead to cognitive distortions which are extremely negative with irrational thoughts like 'the world hates me'.

In the case of thinking, we use our stored knowledge to solve various tasks. We logically establish the relationships amongst various objects in our mind and take rational decisions for a given problem. We also evaluate different events of the environment and accordingly form an opinion.

4.3.2.2 Language

Astonishingly, we can produce and comprehend different sounds and words, combine different letters and phrases and, with precision, express what we want to communicate, even in different languages. We even use our body language to communicate.

Language development is produced throughout our lifetime. The communicative skills of each person vary significantly and can be improved through practice. Some language disorders make it especially difficult to communicate for different reasons, although it is also possible to help people with these problems.

Decision making: the thought process of selecting a logical choice from the available options. When trying to make a good decision, a person must weigh the positives and negatives of each option and consider all alternatives. For effective decision making, a person must be able to forecast the outcome of each option as well, and based on all these items, determine which option is the best for that specific situation.

4.3.2.3 Problem-Solving

Problem solving is a mental process that involves discovering, analysing and solving problems. The ultimate goal of problem solving is to overcome obstacles and find a solution that best resolves the issue. The best strategy for solving a problem depends largely on the unique situation. In some cases, people are better learning everything they can about the issue and then using factual knowledge to come up with a solution. In other instances, creativity and insight are the best options.

4.3.2.4 Intelligence

There are many definitions of intelligence, all relating to the amalgamation of multiple cognitive skills leading to learning, understanding, reasoning, and problem solving. In simple terms, however, intelligence has been defined by the *Oxford*

Dictionaries as the ability to acquire and apply knowledge and skills. There are multiple kinds of intelligence (see Gardner's multiple bits of intelligence) but all of them are manifestations of a higher cognitive process. For example, intrapersonal intelligence, linguistic intelligence, logical-mathematical intelligence and musical intelligence all require complex combinations of the basic processes. Similarly, this is true for the recently popular emotional intelligence which involves the ability to monitor one's own and others' emotions, to discriminate amongst them, and to use the information to guide one's thinking and actions [25]. This also requires a complex combination of the basic processes in an unusually unique way that captures unusual characteristics of human emotions in different day-to-day difficulties

4.3.3 Cognitive Processes in Learning

The different theories of learning, except for associative learning, are based on all the cognitive processes together and no cognitive process acts on its own in the learning process. Instead, we integrate all our resources to improve our study habits and achieve meaningful learning.

When faced with a book, we must recognize the letters, avoid distracting ourselves with irrelevant stimuli, remember the words we are reading, associate what we read with other contents that we learned previously, etc. We use our cognitive processes throughout the learning exercise, although the number of cognitive processes used will vary. For example, we will not process the information in the same way if we only want to find a fragment that interests us other than reading the whole book.

Learning also influences and is influenced by writing, which itself is as much a mental process as it is a physiological process. We need to ignore the noises that make it difficult for us to write, to make our writing readable, to remember what we have written in the previous paragraphs, to worry about our spelling, etc. In addition, we also need to properly plan what we want to write, avoid colloquialism, ensure others understand what we mean, and ensure use of proper symbols (e.g. that zero does look like an *o*).

Correspondingly, the learning process is not a single uniform process and requires different cognitive capabilities depending on the desired outcome. Most importantly, different outcomes can be achieved from a learning process, e.g. as described by Bloom's Taxonomy [26] on the Levels of Cognition or Thinking (Table 4.2).

In conclusion, therefore, a cognitive entity must demonstrate a set of capabilities each matched to a different objective. Moreover, it must be able to combine multiple basic capabilities to achieve complex functions which it uses to solve complex problems. These are also the expected outcomes from cognitive autonomous functions in network management systems. The next section will describe how such functionality can be modelled.

Table 4.2 Bloom's Taxonomy of cognitive learning outcomes.

Level	Typical descriptors of capabilities
1. Knowledge	Memorize, recall, repeat, label, list, define, duplicate, order, name, reproduce, state, recognize
2. Comprehension	Explain, restate, describe, classify, identify, select, translate, discuss, express, interpret, review
3. Application	Demonstrate, apply, interpret, illustrate, dramatize, solve, sketch, operate, use, schedule
4. Analysis	Analyse, compare, contrast, test, categorize, criticize, subdivide, differentiate, examine, question
5. Synthesis	Arrange, compose, construct, create, organize, develop, design, plan, propose, combine, assemble
6. Evaluation	Argue, defend, estimate, evaluate, support, predict, judge, rate, resolve, value, appraise, assess

4.4 Modelling Cognition: A Perception-Reasoning Pipeline

To develop cognitive solutions for network automation, we need to break the cognitive challenge into components that can technically be implemented. In other words, we need a technical model for cognition. Based on our understanding of the human cognitive system and in contrast to self-organization, we hypothesized [27] that cognition is about getting a piece of data, hereinafter called a data element (DE), and running it through a data processing engine that then generates understanding and action. Correspondingly, these two outcomes imply two internal sequential processes of the data processing engine, i.e. perception and reasoning. We hypothesize, as illustrated in Figure 4.3, that:

1) The two broad outcomes are realized from four quasi-orthogonal processes of this engine, namely conceptualization, contextualization, organization, and inference as illustrated by Figure 4.3. Quasi-orthogonality here means that it is hard to clearly demarcate the boundary between any two consecutive processes.
2) There are feedback loops between any of the later processes to each of its former processes, albeit to a different data element. Specifically, this is because knowledge or understanding derived from a given data element may lead to adjustment of knowledge and/or understanding about other previously processed data elements.

In principle, perception prescribes the ability to make sense of an incoming DE both on its own and in relation to data elements about its context. It leads

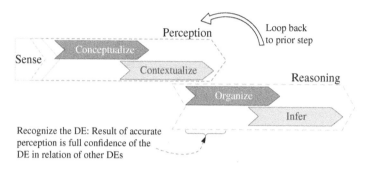

Figure 4.3 Cognition: A data processing pipeline.

to statements such as recognizing the element represented by the data or knowing the object. On the other hand, reasoning implies understanding the object and its implications. Correspondingly, this understanding leads to the selection of the most appropriate actions for the data element and its context.

4.4.1 Conceptualization

Given a DE captured through the sensory system, the first step in perceiving information related to the DE is conceptualization of the DE. Thereby, the cognitive agent makes a hypothesis about what the DE is or may be. Through such a conceptualization, the agent will, for example, decide if an animal it sees could be a fat cat, a small lioness or her cab. However, since such an initial conception of the DE may be inaccurate, the DE must be further processed for accuracy. The agent draws on its memory to crystalize the created concept. For example, it matches what it has seen with its in-memory features of different animals (e.g. length of whiskers relative to body size) to confirm that what it sees is, in fact, a fat cat.

In principle, a conceptualizer is a detector that compares the features of the incoming data element with features of multiple models that are already known to the conceptualizer to select the best matching hypothesis. Where the DE does not match any of the known models, a new model is created and stored in memory.

4.4.2 Contextualization

The conceptualization step identifies the isolated information that the DE delivers, but to choose the right course of action, the DE must be put in the right context. The contextualization step evaluates the situational, environmental and other circumstances related to the DE being processed to concretize the knowledge on that DE. Besides being used for further processing, the context may also partly help to confirm the accuracy of the conceptualization or to at least increase the confidence level thereof. For example, if the observed big cat or small lioness is found in the African wild, this context increases the possibility that it is a lioness. And, where the conceptualization was accurate, contextualization guides the decision making to select the most appropriate actions for the DE and its context. In the lioness case, a different decision will be made if the lioness is encountered in the wild African Savannah than if it is found a zoo.

Similar to conceptualization, a contextualization is also a detector but this detector evaluates supplementary features, or the contexts such as environmental information, to either confirm or adjust the conceptualized hypothesis. Correspondingly, it is not always the case that contextualization follows sequentially after conceptualization. The outcome of the two processes may sometimes be created through an oscillatory interchange between the two processes – i.e. hypothesize then add context, then re-hypothesize then add more context and repeat the process.

4.4.3 Organization

Given accurate conceptualization and contextualization of DE, the organization step defines relations amongst DEs. It continuously makes connections amongst DEs with descriptive words like 'is', 'is not', 'can', 'cannot', 'may', etc. In the case of the cat vs lioness, the organization step could create or fetch statements like: 'the African Savanah has Lions roaming around'; 'A lioness is dangerous'; 'A cat can be dangerous'; etc. These are partly previously inferred conclusions that have been stored in memory although the organization step can also create new connections amongst DEs, for example, such as when one decides two items are related without necessarily knowing how they are related; such a realization is part of the DE organization step. It is logical to consider that the outcome of the organization step is information organized in such a way that each connection is an Information Element (IE).

4.4.4 Inference

The inference step logically and arithmetically combines multiple DEs and IEs to create new DEs and relations. It undertakes the logical analysis of DEs and their relations through the logical operations (AND, OR, NOT) as well as the linguistic information conjunctions (such as BUT, WHILE, NOTWITHSTANDING, etc.). In the lioness example, consider three IEs that could be evaluated alongside the 'concept of a lioness' and the 'African savannah context' – A: 'I am watching TV' or B: 'I am out hunting' or C: 'I am seated inside a vehicle on a tour excursion'. The inference step evaluates the truthfulness of the different combinations of the lioness DE and IEs alongside these IEs to guide the appropriate decision. Correspondingly, different decisions will be made if watching TV as opposed to when one is out hunting.

4.4.5 Memory Operations

Each stage of the data processing cycle has access to the memory operations cycle, as illustrated by Figure 4.4. This cycle involves the four steps of Fetch(F), Read(R), Label(L), and store(S). At the conceptualization state, it is mainly the label and store functions that are executed. The correctly conceptualized, the DE, is appropriately labelled (e.g. it is a cat or a lioness) and then stored in memory if necessary, for example, if it is a new DE that is being encountered for the first time. Fetching and reading is mainly done after the conceptualization step. Thereby, the agent needs to check what it has in memory either to confirm its perception of the object; or to create new and/or edit existing relationships for the DE; or to even derive actions amongst the many possible actions known to the agent. In respect of this, a previously stored DE may also have to be updated, in which case, the agent fetches and reads the DE before it relabels it and stores it again.

4.4.6 Concurrent Processing and Actioning

Throughout all four processes, actions are triggered in response to the stimuli, knowledge and or understanding created at the respective process. We expect that the selection of actions is not deterministic but probabilistic over the hypotheses-action space. We can consider the processes as estimating the probability distribution over hypotheses for a given DE, i.e. estimating the probability that a given hypothesis is true given the DE. Similarly, actioning may

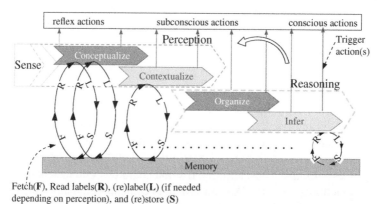

Figure 4.4 Cognitive data processing with memory operations and actioning.

also be modelled as the probability distribution over actions at each step, i.e. the actioning estimates the probability of a successful outcome if action **a** is taken given hypothesis **h**.

The actions may be distinguished between automatic and conscious actions, respectively related to automatic and controlled processes [28]. 'Automatic processes are inevitably engaged by the presentation of specific stimuli inputs, regardless of the subjects' intention.' Correspondingly, '... an automatic process is modelled after the reflexes, taxes, and instincts from physiology...'. In such processes, sensory inputs and perceptions are directly related to actions which are executed as soon as the DE is captured, for example, after understanding (perception and subsequent inference of) the effect of getting in contact with an elephant (either through experience or learning), the appropriate action (running) gets automated by matching it to the data element. Subsequently, when an elephant is encountered, for example, the default action of running is triggered through what may be called instinct. In such cases, the action may be considered a reflex action with a default way of behaving for a given event.

Although actions can be triggered on conceptualization, it is mainly more appropriate to trigger them after contextualization. This explains the case of actions which are appropriate in one context being taken in a wrong context, for example, one does not run on seeing a lion in a zoo because we contextualize that this is a zoo where the animal is caged. In other words, the zoo context does not allow for the reflex action to be triggered. It may, however, be triggered in cases where the contextualization is faulty or inappropriately developed, e.g. in young children.

Note that the automation is not always innate but can also be achieved through extensive practice. Meanwhile, automatic action can also 'consume attentional resources once invoked by appropriate stimuli conditions' [28]. This explains the presence of subconscious processes and actions as distinct from reflex processes and actions.

Action automation also exists at organization and inference, albeit to a much lesser degree, especially at inference. Here, the agent activates its cognitive functions to process the captured data and information and generate the most optimal action.

4.4.7 Attention and the Higher Processes

Attention has been defined as the allocation of scarce cognitive resources. Although it begins with conceptualization, its degree of significance increases as the processes go from left to right. It is co-dependent on the processes, in that, it feeds them even as they feed it and each of them then acts on these cross signals. In a way then, attention is less of a separate process but a measure of significance of DEs, IEs, and processes. Where a process observes significance in a DE or IE (be it sensory or from memory), the process triggers for more attention onto that DE/IE which, in turn, avails more resources for the process to be executed to completion.

Attention may, however, also be triggered externally, especially in response to intended higher processes. Since higher processes need to engage multiple fundamental processes (as illustrated by Figure 4.5), they may call for higher attention at specific sub-processes, DEs and IEs as they pursue their outcomes. For example, problem solving may put more emphasis on the data organization task to ensure that the relevant data for the problem at hand is obtained and organized in a way that allows for effective and efficient inference.

4.4.8 Comparing Models of Cognition

There are multiple activities that intend to apply cognition in communications networks, most focusing on dynamic and opportunistic spectrum exploitation in cognitive radio. Correspondingly, multiple models have been used for describing the developed cognitive solutions, with perhaps the most widely being Mitola's observe-orient-plan-decide-act cycle [29]. Other widely used models are the observe-decide-act cycle and the conjecture-test-modify-repeat cycle, both of which are in fact generalizations of the Mitola cycle. This section discusses how the Perception-Reasoning Pipeline (PRP) model compares with the existing models for cognitive decision making, specifically with the Mitola model.

As with all other models, the Mitola model starts with the outside world which provides stimuli that triggers the cognitive cycle. Then, as illustrated in Figure 4.6, the cycle executes the processes of 'observe', 'orient', 'plan', and

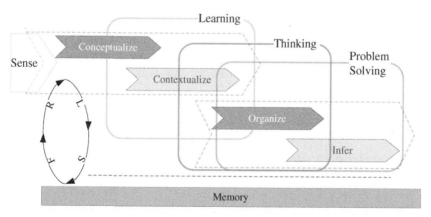

Figure 4.5 Higher cognitive processes (including thinking, learning, and problem solving) as combinations of multiple basic processes.

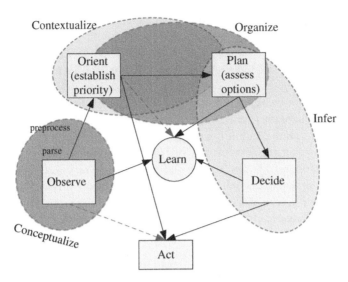

Figure 4.6 Comparing the perception-reasoning pipeline model and Mitola's cognitive cycle.

'decide' to generate an action that is then executed to affect the outside world whilst the agent learns from the cycle.

The model was developed in the context of cognitive radio which refers to radio systems that can adjust their behaviour in response to conditions, and changes thereof, of their operating environments, e.g. according to spectrum occupancy. Correspondingly, the steps in this model are heavily influenced by the necessary radio-related challenges that need to be addressed. Specifically, in the 'observe' stage, the cognitive radio parses the stimuli to extract contextual cues necessary for the tasks, for example, it analyses GPS coordinates, and light and temperature to determine if it is inside or outside a building. At the 'orient' stage, the agent parses incoming and outgoing messages for content to yield contextual cues necessary to infer urgency of tasks and to decide the next steps. Urgent tasks bypass the 'plan' stage, triggering quick decisions and subsequent actions. Normal events, however, which might not require time-sensitive responses, go through the full plan-decide-act cycle.

We find that the Mitola model is not appropriate as a model for cognitive decision-making. Firstly, it is incomplete in that some steps which are expected in a cognitive system (here represented by the dashed lines) were left out. Specifically, it should be possible to act based on an observation without any processing, the case for reflex actions. Similarly, there should also be something to learn from the orient stage, e.g. that a scenario which was initially assumed to be low priority should be raised in priority or that even a higher priority scenario should be processed through the planning step to ensure an appropriate response.

It is also evident that besides the model being too specific to the dynamic radio spectrum access challenge, it is more of a model for a higher-level task of problem solving than a model for low-level cognitive decision making. The model relies a lot on aggregate functions and provides few cues on how these functions can be built from smaller simpler functions. The involved processes are themselves not basic, but higher processes that are different manifestations of the problem-solving process and that require multiple basic processes. For example, planning is a higher process that involves both organization of data and inference according to that data. It requires knowledge on different DEs that are important for the case at hand, requires relationships amongst DEs and IEs as well as the inferences on the effects of different combinations of DEs, IEs, and actions. It is only then that it can generate the set of possible alternatives amongst which to choose the best action. Similarly, orientation is a higher process, which after contextualizing the DEs, combines the DEs, contexts and relations thereof to infer the implications of such combinations and thus to form the directions of interest.

Besides, the processes are not completely orthogonal. Both planning and orientation involve contextualization in generating their outcomes, i.e. both the urgency/priority and the alternatives must be evaluated according to the prevailing contexts. Similarly, planning is not different from deciding but is a subset thereof; deciding involves assessing options and choosing one of them. And both planning and deciding are heavily reliant on the organization of, and inference over, DEs and IEs.

Moreover, the processes in the model are non-consequential whilst others are simply too fuzzy. The observation process is too passive and does not prescribe an outcome whilst the pre-process sub step does not state what is undertaken therein. In principle, as also assumed in the PRP model, observation is a subpart of both the conceptualization and contextualization steps, i.e. both these two need to observe before they can form the respective hypotheses about the DEs and their contexts.

Meanwhile, observe-decide-act cycle is the generic control loop which would be true even for non-cognitive decision making. On the other hand, the conjecture-test-modify is more of a general learning process, specifically, the learning by doing model. Conjecture in this case is equivalent to the whole PRP since it must combine all the processes in deriving an action that must then be tested. Test then is equivalent to taking actions whilst modify represents the learning process through which the agent edits its understanding of the environment and the effects of different actions on that environment.

In general, this PRP model is the best attempt at breaking down cognitive decision making into a set of basic (quasi-)orthogonal sub processes.

4.5   Implications for Network Management Automation

The previous sections have developed the model for cognitive decision making, i.e. the so-called Perception-Reasoning Pipeline or PRP model. It remains open as to what extent this model is useful for NMA. This section exemplifies the value of the PRP model for NMA. Firstly, it answers questions on the complexity of the model especially for implementation in NMA. This informs the characterization of SON and the degree to which SON achieves the expectations of cognitive decision making when measured by the template of the PRP model. Then, contrasting the expectations from a Cognitive Autonomous Network (CAN), the last section defines the mechanism through which cognitive solutions should be characterized when evaluated according to the PRP model.

4.5.1   Complexity of the PRP Processes

In the PRP model, complexity reduces from left to right. Forming perceptions from data is the most complex whilst making inferences is the simplest of the four processes. The complexity of the different processes indicates the degree to which they will be easy to implement for NMA, i.e. inference will and is much simpler to implement compared to its prior processes.

Firstly, the accuracy of perception is very critical since accurate perception results in knowledge whilst the reverse results in delusion. However, it is very hard to objectively confirm the accuracy as there is no internal measure for the accuracy of knowledge, only proxy measures are possible, for example, by evaluating the appropriateness of the actions derived from that knowledge. Thus, to ensure accuracy, the cognitive agent must test multiple hypotheses and sub-hypotheses about a given data element and its features ensuring that there is consistence in the conclusions drawn from the diverse hypotheses. Even then, the agent may also have to test the actions derived from the conclusions before it confirms the captured knowledge as being true and not a delusion.

Data organization is simpler than perception but is still more complicated compared to inference. The availability of perceptive models – clear answers to what/who/where ... etc. questions regarding the objects captured by the sensory inputs – implies that objectivity is simpler when compared to the perception steps. However, it is still the case that unless through a proxy mechanism, it not easy to objectively measure the accuracy of the data relationship and implication model – the formed relationships and implications amongst data elements. In principle, the greatest challenge is that an accurate data relationship and implication model (as is also the case for the perceptive model) is not formed in one instance but must be learned and tuned with more sensory data.

Inference is simple because it assumes the availability of accurate perceptive and data relationship and implication models from which inferences are made and yet its derivatives (the computed decision or actions) can be evaluated for accuracy. The assumption is that the relationship model accurately describes the relationships amongst data elements and the implications of the different data elements and their derived actions in different context.

In line with this complexity, the development of NMA functionality will (and has been) from right to left.

4.5.2 How Cognitive Is SON?

Self-organization (and specifically, SON in communication systems) has mainly focused on inference – i.e. selecting the right actions that lead to consistent, predictable outcomes. In the case of SON, the outcomes are constrained to ensure that they fulfil the desired goal(s) of the SON function(s). Therein, the underlying data relationship and implication models that are applied by SON's inference engines are manually developed by the SON system design engineers. Correspondingly, SON cannot be considered a cognitive system, since the human designers derive the decision logic and only leave the inference part to the SON functions.

In general, it is expected that machine learning will help address the complex parts. Specifically, a combination of supervised and unsupervised learning will enable the development of classification engines to accurately hypothesize nature sensory inputs and their contexts. With accurate perceptive models, supervised learning will be able to learn relations amongst data elements which relations can then be used for inference and decision making either using supervised learning or reinforcement learning. The degree to which the existing technologies and algorithms can achieve the requirement of a cognitive decision maker are discussed in Chapters 5 and 6.

4.5.3 Expectations from Cognitive Autonomous Networks

Given the limitations of self-organization (i.e. its capability being limited to only inference), the expectation for CAN is that even the more complex components of the PRP model will be taken from the humans and be undertaken by the software module. In the end, the system should be able to perceive the sensed data in a way that is specific to every context as well, as be able to reason the relationships amongst the data elements, in order to develop the data relationship and implication model to be used for inference.

The PRP model presents an outline for cognitive decision making, suggesting that any cognitive/artificial intelligence (AI) solution should be measured against the PRP to characterize the degree to which the solution is cognitive. Correspondingly, the degree to which the NMA decision-making functions are cognitive will be evaluated along the presented quasi-orthogonal processes of the PRP model. In that context, a fully cognitive system should be able to combine different processes each of which achieves some functionality of the PRP model, and most importantly, in a way that the complete system achieves all the processes of the PRP model. For a given use case, the solution should prove that it achieves the outcomes of the sub-processes – not only for a single data element but in a general enough way, that the achieved cognitive outcome can be stored in memory for further use.

4.6 Conclusions

This chapter has evaluated self-organization in contrast to cognition showing that self-organization is evaluated from an outsider's view that focuses on the degree to which a system achieves consistent, and possibly predictable, outcomes without any control or trigger from an external entity. It shows that some self-organizing systems are in fact cognitive, in that decisions are not based on prescribed rules but involve a data processing system which takes sensory inputs and generates decisions.

To understand the nature of cognition as a basis for developing a model for cognitive decision making, the discussion contrasted the nature of human cognitive capabilities identifying that, at the core of human cognition, are two processes – perception and reasoning. Correspondingly, the discussion develops a PRP model of cognition that identifies two subprocesses in each of the two major processes – i.e. respectively conceptualization and contextualization as well as organization and inference. The discussion also contrasts the different actions that humans take, identifying that actions occur on the continuum over all four processes on the PRP model. The implication of this are that: (i) a cognitive agent should be able to select different kinds of action depending stimulus and context; and (ii) not all actions are derived from an inference processes but that some are reflexive to specific incoming stimuli.

The PRP model is contrasted against the current widely accepted model for decision making – the observe-orient-plan-decide model or simply the Mitola model. It was shown that the Mitola model has major limitations. Amongst these is the lack of orthogonality (or even quasi-orthogonality) amongst the subprocess of the model and the failure to address some data processing steps. Moreover, the Mitola model fails to account for the different kinds of actions that can be drawn by the cognitive agent. And this is no better for the other closely related models including the observe-decide-act cycle and the conjecture-test-modify cycle, both of which are found to be simpler models for generic decision making and not models for cognitive decision making.

Finally, the chapter evaluates the implications of the PRP model on NMA and highlights the expected conclusion that the current SON framework does not constitute a cognitive NMA paradigm. It has been shown, however, that applying machine learning will address the cognitive requirements for NMA, the details of which are left to subsequent chapters and future work. Specifically, the subsequent chapters will (i) summarize the capabilities of AI/machine learning algorithms and (ii) describe some existing transitional solutions that apply AI/machine learning (ML) algorithms for NMA as a precursor to CAN.

References

1 Next Generation Mobile Networks Alliance (2007). Use cases related to self organising network, overall description. www.ngmn.org (accessed 27 January 2020).

2 Hamalainen, S., Sanneck, H., and Sartori, C. (eds.) (2011). *LTE Self-Organising Networks (SON) Network Management Automation for Operational Efficiency*. Wiley.

3 Gierer, A. and Meinhardt, H. (1972). A theory of biological pattern formation. *Kybernetik* 12: 30–39.

4 Leibnitz, K., Wakamiya, N., and Murata, M. (2007). Biologically inspired networking. In: *Cognitive Networks – Towards Self-Aware Networks* (ed. Q.H. Mahmoud), –1, 19. Wiley.

5 Richmond, C.A. (1930). Fireflies flashing in unison. *Science* 71 (1847): 537–538.

6 Janeway, C.A., Travers, P., Walport, M., and Shlomchik, M.J. (2005). *Immunobiology*, 6e. Garland Science. ISBN: 0-8153-4101-6.

7 Kaneko, K. (2006). *Life: An Introduction to Complex Systems Biology*. Berlin: Springer-Verlag.

8 Heylighen, F. (1997). Self-organization. Principia Cybernetica Web. http://pespmc1.vub.ac.be/SELFORG.html (accessed 17 December 2018).

9 Yates, F.E. (ed.) (1987). *Self-Organizing Systems*. New York: Plenum.

10 Bonabeau, E., Doringo, M., and Theraulaz, G. (1999). *Swarm Intelligence: From Nature to Artificial Systems*. New York: Oxford University Press.

11 Camazine, S. (2003). *Self-Organization in Biological Systems*. Princeton University Press.

12 Diehl-Fleig, E. and De Araujo, A.M. (1996). Haplometrosis and pleometrosis in the ant *Acromyrmex striatus* (Hymenoptera: Formicidae). *Insectes Sociaux* 43 (1): 47–51.

13 Theraulaz, G., Gautrais, J., Camazine, S., and Deneubourg, J.-L. (2003). The formation of spatial patterns in social insects: from simple behaviours to complex structures. *Philosophical Transactions of the Royal Society of London, Series A: Mathematical, Physical and Engineering Sciences* 361 (1807): 1263–1282.

14 Dorigo, M., Di Caro, G., and Gambardella, L.M. (1999). *Ant Algorithms for Discrete Optimization*. Artificial Life: MIT Press.

15 Newton, I. (2008). *The Migration Ecology of Birds*. London: Academic Press.

16 Kern, K., Niehus, H., Schatz, A. et al. (1991). Long-range spatial self-organization in the adsorbate-induced restructuring of surfaces: Cu{100}-(2×1)O. *Physical Review Letters* 67 (7): 855.

17 Petrov, V., Gáspár, V., Masere, J., and Showalter, K. (1993). Controlling chaos in the Belousov–Zhabotinsky Reaction. *Nature* 361 (6409): 240.

18 US Central Intelligence Agency (2000). Major Narco Trafficking Routes And Crop Areas. CIA. https://www.pbs.org/wgbh/pages/frontline/shows/drugs/business/map.html (accessed 18 December 2018).

19 Schadschneider, A. (1999). The Nagel-Schreckenberg model revisited. *The European Physical Journal B: Condensed Matter and Complex Systems* 10 (3): 573–582.

20 Dijkstra, J., Jessurun, A.J., and Timmermans, H.J.P. (2001). A multi-agent cellular automata model of pedestrian movement. *Pedestrian and Evacuation Dynamics*: 173–181.

21 Steels, L. (2003). Intelligence with representation. *Philosophical Transactions of the Royal Society of London, Series A: Mathematical, Physical and Engineering Sciences* 361 (1811): 2381–2395.

22 Li, L., Peng, H., Kurths, J. et al. (2014). Chaos–order transition in foraging behavior of ants. *Proceedings of the National Academy of Sciences of the United States of America* 111 (23): 8392–8397.

23 Reed, S.K. (1982). *Cognitive Theory and Application*, 6e.

24 vocabulary.com. (2018). Dictionary. www.vocabulary.com. (accessed 18 December 2018).

25 Salovey, P. and Mayer, J.D. (1990). Emotional intelligence. *Imagination, Cognition and Personality* 9 (3): 185–211.

26 Adams, N.E. (2015). Bloom's taxonomy of cognitive learning objectives. *Journal of the Medical Library Association: JMLA* 103 (3): 152.

27 Mwanje Stephen S and Mannweiler, Christian (2018). Towards Cognitive Autonomous Networks in 5G. Proceedings of ITU Kaleidoscope: Machine learning for a 5G future (K-2018), Santa Fe.

28 Kihlstrom, J.F. (1990). The psychological unconscious. In: *Handbook of Personality: Theory and Research* (ed. L.A. Pervin), 424–442. New York: Guildford Publications.

29 Mitola, J. (1999). Cognitive radio for flexible mobile multimedia communications. IEEE International Workshop on Mobile Multimedia Communications, Sandiego.

5

Classic Artificial Intelligence: Tools for Autonomous Reasoning

Stephen Mwanje[1], Marton Kajo[2], Benedek Schultz[3], Kimmo Hatonen[4] and Ilaria Malanchini[5]

[1] *Nokia Bell Labs, Munich, Germany*
[2] *Technical University of Munich, Munich, Germany*
[3] *Nokia Bell Labs, Budapest, Hungary*
[4] *Nokia Bell Labs, Espoo, Finland*
[5] *Nokia Bell Labs, Stuttgart, Germany*

Network management is an information processing task, which, owing to the complexity of the networks, has traditionally been undertaken by human experts. However, as demonstrated by 3G and 4G Self-Organizing Networks (SONs), a proper application of cognitive techniques can do a lot in automating network management. Cognition, as defined in Chapter 1, refers to the capability to perceive a signal into a data element and subsequently reasoning over the data element to select an action. As such, cognitive techniques refer to the broad grouping of Artificial Intelligence (AI) related methods and technologies developed for information processing.

Autonomous reasoning requires that the cognitive entity perceives relationships amongst data elements and makes inferences about these elements and their relations to subsequently select the appropriate action. Classic AI techniques were developed with inference engines intended to map human reason and draw humanly relatable decisions from data. Consequently, these classic AI techniques can be very useful in designing automated reasoning engines, also applicable to networks and network management.

This chapter summarizes some of the available techniques for achieving autonomous reasoning. Essentially, it presents a toolbox of classic AI techniques that could be useful for automating inference tasks in networks. For each tool or technique, the presentation includes an evaluation of the accorded degree of cognition based on the cognitive decision-making model developed in Chapter 4. It also presents the basis on which the specific tool may be selected for application

Towards Cognitive Autonomous Networks: Network Management Automation for 5G and Beyond,
First Edition. Edited by Stephen S. Mwanje and Christian Mannweiler.
© 2021 John Wiley & Sons Ltd. Published 2021 by John Wiley & Sons Ltd.

towards specific network challenges. The chapter does not claim to exhaustively discuss the field but instead focusses on the most common methods that are also likely to have wider usage in network management. Specifically, the following methods are presented: expert systems, closed-loop control systems, case-based reasoning (CBR), and fuzzy inference systems. They are presented in an order that highlights ever more cognitive capability beyond simple inference.

5.1 Classical AI: Expectations and Limitations

The principles underlying intelligence must be separate from any implementation mechanisms and true for all forms of reasoning, be it logical, probabilistic, strategic, diagnostic reasoning or otherwise. Symbolic (or Classical) AI is the branch of artificial intelligence research that attempts to implement intelligence by explicitly representing human knowledge in a declarative form (i.e. through facts and rules). As the earliest form of AI implementations, symbolic AI has had some impressive successes, one of the earliest being the expert level performance in diagnosis of blood infections (see [1]).

Symbolic AI has, however, encountered deep and possibly unresolvable difficulties related to dealing with implicit knowledge. Implicit and procedural knowledge is knowledge and skills, which are not readily accessible to conscious awareness. Implicit knowledge is hard to encode into symbols, and so symbolic AI has forfeited the fields which rely on implicit knowledge to neural network architectures (discussed in Chapter 6) which are more suitable for such tasks. This challenge of dealing with implicit knowledge is exemplified by what is known as the common-sense knowledge problem.

5.1.1 Caveat: The Common-Sense Knowledge Problem

In early AI research, it was commonly accepted that knowledge would have to be explicitly represented. Researchers, however, did not anticipate the vast amount of implicit knowledge that all humans share about the world and humanity [2]. AI systems designers did not consider implicit rules about relationships, e.g. body parts, family relations, behaviours, etc. The following scenarios are, for example, obvious to humans and a human AI designer would not produce rules for them:

- 'If President Clinton is in Washington, then his left foot is also in Washington' [2]
- 'If a father has a son, then the son is younger than the father and remains younger for his entire life' [2]

The implicit nature of this knowledge in humans means that all humans take it for granted, and never have to state it or consider it explicitly be it for decision making or in communication.

A central part of the common-sense knowledge problem is that symbolic AI needs to translate this implicit knowledge into an explicit form using symbols and rules for their manipulation. However, the best approach on how to represent knowledge in artificial systems remains an open question. For example, is knowledge better represented as dictionary- or encyclopaedia-like entries, as a series of if-then rules or should multiple forms of representation be used?

It has become clear that not all human knowledge is represented in such an explicit or declarative form. Moreover, the implicit nature of knowledge applies not only to common sense knowledge, but also to a wide variety of human expertise and skills. Such domain-specific knowledge is often represented as procedures, rather than facts and rules. As such, areas which rely on procedural knowledge such as sensory/motor processes, are much more difficult to handle within the Symbolic AI framework. Correspondingly, classic AI methods are not likely to be successful for the development of general-purpose intelligent systems although they remain useful for some specific aspect of decision making and control.

5.1.2 Search and Planning for Intelligent Decision Making

The implicit knowledge challenge notwithstanding, today there are thousands of systems that employ symbolic AI techniques ranging from diagnosing cancer to designing dentures, where they often outperform doctors in clinical trials. These justify the need to study the symbolic AI concepts and consider the extent to which these concepts may be usable in solving network management problems.

For any given problem, there can be many ways of arriving at a solution, even when there is only one solution. In mathematical computation, for example, where there is typically only one solution to the problem, there can still be many algorithms possible to find the solution. In other cases, there may be multiple candidate solutions, each differently fulfilling the requirements of the problem, and this candidate space must be searched through to find the single exact solution. In effect, searching is one of the foundations for intelligence.

Whilst searching focuses on identifying a solution that fulfils certain requirements, planning is concerned with identifying sequences of actions that, when taken as a sequence, achieve specific goals of the intelligent system. Planning takes as input, the representation of actions and models of the world and undertakes reasoning about the effects of different actions and sequences of actions to derive the best sequence(s). It prescribes the techniques for efficiently searching the space of possible plans. The output of the planning process is the sequence of actions that convert a given initial state of world process into another state that satisfies particular goal(s) and constraints. The models of the world taken as input to the planning process are the descriptions of:

1) the set of all possible states of the world
2) the initial state of world for the planning process
3) the goals and constraints of the process
4) the set of available actions
5) the quality or implications (e.g. costs) of taking the different actions in the different states of the world actions.

Search and planning techniques underlie the capabilities of intelligent agents and have been applied in a variety of tasks including robotics, process planning, and autonomous control.

5.1.3 The Symbolic AI Framework

The general structure of the symbolic AI framework of the intelligent agent is illustrated by Figure 5.1. Given the input world and action models, for a specific problem input, the intelligent agent applies some internal methods to search and/or plan for the appropriate solution. In control systems, the solution is directly applied to the controlled process and the outcome of the controlled process may also be reused as part of the problem input that is subsequently taken by the intelligent agent.

The world model encapsulates the symbolically specified human knowledge. The different methods of symbolic AI are suitable for different problems and fields, but also differ in the amount and specialty of the knowledge they require. However, three following fundamental beliefs are common to these methods [3]:

1) a model for intelligent decision making can be defined in an explicit way
2) knowledge in such a model is represented in a symbolic way
3) cognitive operations can be described as formal operations over the symbolic expressions and structures of a knowledge model.

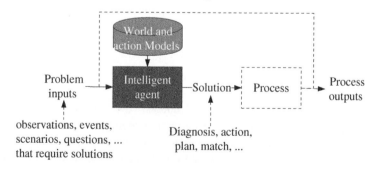

Figure 5.1 Intelligent control in a symbolic AI framework.

The subsequent sections in this chapter summarize the principles and utility of the most widely used symbolic AI methods and systems.

5.2 Expert Systems

An expert system can be defined as a computer system that emulates the decision-making ability of a human expert [4]. Expert systems are the simplest automation systems, in principle being the means of encoding and applying human expert's knowledge. The simplest expert systems do not hold any intuition or even structure about the knowledge, only realizing a way to find out what the human would do in certain circumstances. However, there are also more complex expert systems which encode structure and dependencies within the domain and, derive the used decision logic from that model [5].

In the simplest set-up, expert systems are generalized as rule-based systems, in which case, rules are expressed as a set of if-then statements (called IF-THEN rules or production rules) with some interpreter controlling the application of the rules, given particular facts [6].

5.2.1 System Components

For the system to be able to infer the appropriate solution, it requires two sub-systems – the knowledge base and an inference engine illustrated by Figure 5.2. The expert knowledge base or expert designed rule-base contains human expert's knowledge about the underlying problem, which may include the likely causes, the interrelations amongst different problems and the effectiveness of different responses. The inference engine takes the responsibility for matching observed problems to the right deductions/inferences. It evaluates the observed state or problem and applies the relevant rules, although it can, in advanced cases, also enhance the knowledge base with new knowledge. The solution derived by such a system could either give a recommendation of an action to be taken or a deduction of the cause or reason behind the observed state.

5.2.2 Cognitive Capabilities and Application of Expert Systems

Expert systems are, in principle, the simplest automaton tools, since all the knowledge structuring is still done by humans, and the system left only to select a course of action from known options. In this respect, expert systems only provide the very last cognitive sub-process – inference. Note that the subsequent action execution may not necessarily be automated but may, for example, require a human to check the recommended action or inference before executing it.

The most cited examples for expert systems are in medical treatment recommendations, where the knowledge base is populated with the opinions of expert doctors regarding the expected diagnosis for certain patient symptoms [6]. Expert systems have also been proposed for diagnostics in manufacturing processes, where again, the opinions of expert technicians are recorded in the knowledge base and if-else rules used to match observed faults to the best course of action.

In a similar way, therefore, expert systems can be extensively applied in communication networks, especially for fault detection and diagnosis, e.g. for alarm correlation. The typical solution registers the knowledge of communication systems expert into pairs of observations and implications and/or actions. Example knowledge could be the following pairs: network events and their causes; event sequences and their implications, or the event sequences and the most appropriate actions or responses thereto. The solution then adds an inference engine that evaluates observations to be matched to the known events and/or event sequences to eventually select the outcome, be it the cause, implication, or appropriate response.

5.2.3 Rule-Based Handover-Events Root Cause Analysis

The best example of an expert system in network operations is the identification of Mobility Robustness Optimization (MRO) problems as described in [7]. Therein radio system experts analysed the timing relations amongst a set Radio Resource Management (RRM) events in LTE to determine if and when handover problems are likely to happen. Such events include handover reports, radio link failures (RLFs), re-establishment requests and RLF indications, for which different combinations indicate different handover problems.

The subsequent solution, illustrated by Figure 5.3, implements a knowledge base of IF-THEN-ELSE rules which encodes the human-expert-known handover–related events, their sequences of occurrences as well as the related possible network states and configurations. The solution then implements an inference engine that evaluates the network state and event sequence to infer

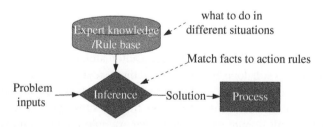

Figure 5.2 Rule-based expert system.

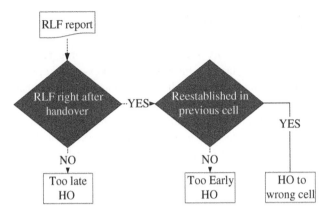

Figure 5.3 MRO Root cause analysis of Radio Link Failures (RLFs).

what the likely cause of such an observed state is and, in some cases, what the appropriate reconfiguration response should be. These kinds of solutions have been the focus on most SON approaches but improvements thereof have also been proposed as described in Chapters 7–9.

5.2.4 Limitations of Expert Systems

Despite their success in many sectors, the use of rule-based expert systems has several critical problems, amongst them being the time required to build knowledge, the complexity of maintaining the knowledge, the need for explicit definition of solutions and excessive dependence on the programmer [5, 8].

The construction of the knowledge base is difficult and time-consuming, specifically due to the complex and time-consuming expert knowledge elicitation. The level of complexity increases with the breadth and degree of unstructured nature of the knowledge. Moreover, changes in the network or in the environment may invalidate certain rules, necessitating them to be updated. A large rule base could even become a burden in the sense that all changes in the network become more expensive, since the rule base needs to be verified and possibly updated after each change.

The system is generally incapable of dealing with problems that are not explicitly covered by the utilized rule base, i.e. expert systems are only useful if the built-in knowledge is well formalized, circumscribed, established, and stable. Any additions to the system typically require a programmer's intervention, unless a more intelligent learning module is built into the system to either extend the rules, correlate/combine them or derive new ones.

5.3 Closed-Loop Control Systems

Closed-loop control systems are a mechanism for autonomic response to stimuli in which the system (as a state machine) selects responses appropriate for specific stimuli but in a way that it can also adjust its response using feedback from previous output. The simplest system, as illustrated in Figure 5.4, uses a controller that evaluates the input to select an appropriate response but that also evaluates the feedback from the effects of its actions to adjust the solution. This kind of system iteratively acts on the controlled process to continuously improve the solution for a given input until it attains a steady-state that is the best amongst all those achievable through the available/possible solutions.

5.3.1 The Controller

The controller adjusts its control output (the input to the controlled process or system) to reduce the error between the output and the desired goal. Closed-loop control was mainly developed for motor drives where several types of controllers have been developed: A proportional controller generates an output that is proportional to the present error which ensures an immediate response to the observed error. The drawback, however, is the error can be eliminated because when the error becomes zero – the output also becomes zero which creates a new error.

An integral controller generates an output that is the integral of the error so both the present and past errors will affect the output. This removes the error completely but will only do so slowly owing to the accumulated errors in the integral.

These error response strategies can also be employed in network optimization. Although the output cannot be directly compared with the input (they are not necessarily signals or functions with equal measure), the means of responding to the error can still be applied. For each input, the output is compared with the goal or desired output to determine the error. The input is then adjusted using one the above strategies in accordance with the observed error.

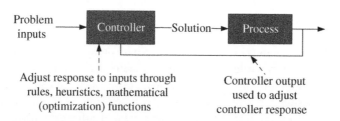

Figure 5.4 Closed-loop control system.

5.3.2 Cognitive Capabilities and Application of Closed-Loop Control

As closed-loop control is the simplest advancement of expert systems, their capabilities are limited to inference. In effect, they are mainly tools for automation with the ability to infer solutions for specific observed conditions or system states. This inference capability is built into the design logic through using rules, heuristics, and mathematical functions that often select actions through optimization.

Closed-loop control systems have been extensively used in electronic control systems where they are referred to as feedback control systems. Therein, the system applies the concept of an open-loop system as its forward path. Then, one or more feedback paths take some portion of the output back to the input, where it forms part of the system's excitation. In so doing, the system can automatically achieve or maintain a desired output state by generating an error signal as the difference between the output and the reference input.

Closed-loop control has been the bedrock of SON, having been used for different solutions. The problems where it has been applied, range from self-configuration of a newly installed base stations, to auto diagnostics and predictive maintenance as a solution for self-healing [9]. The closed-loop control approach has been successful in SON owing to this ability to instrument the output and take corrective action in case of an observed deviation when compared to the desired output. In principle, for a given observation, one or more actions may be taken to cause a certain desired outcome. This outcome is then instrumented with decision control that triggers a feedback in case the desired outcome is not achieved.

5.3.3 Example: Handover Optimization Loop

Closed-loop control can be used to improve the MRO Root cause analysis solutions to be able not only to determine what the candidate handover problem is, but also to decide the appropriate corrective action. The subsequent handover optimization control-loop decides the likely problem, chooses an action to adjust handover parameters and then evaluates handover events to re-trigger the execution if the desired outcome is not achieved.

As illustrated in Figure 5.5, given some initialization, the loop:

1. Evaluates handover performance-related indicators, including handover failures, handover ping-pongs, and Radio Link Failures.
2. Determines the improvement or degradation in handover performance (HP).
3. Adjusts the handover point (HOP) (changes the handover control parameters handover margin and/or Time to trigger) either in direction of optimization in case of a previous improvement or in the opposite direction in the case of a previous degradation.

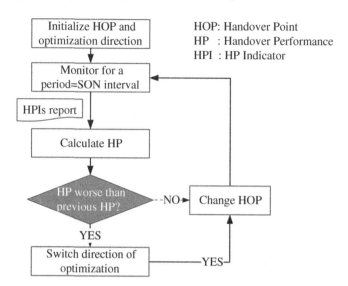

Figure 5.5 The handover optimization closed-loop control system.

4. Re-executes the process from Step 1 until a termination condition is reached that generally indicates that no further improvement in handover performance can be achieved.

5.4 Case-Based Reasoning

CBR is a reasoning mechanism for solving problems based on solutions of similar past problems. The cases therein are the problem-solution pairs for which the particular solution is the best for related problem. For a given observed sample data, instead of finding a closed form model of the data, CBR learns the matching between specific problems and solutions, both of which need to be represented in data. The learned matching is then later used to resolve similar new problems.

CBR was developed as an advancement of rule-based expert system, where rather than following a set of rules, CBR attempts to reapply previously successful solutions to newly encountered problems that are either exactly the same or even only related in context to the old problems. in this view, CBRs are seen as memory-based expert systems [10], in which:

1. The knowledge base is the enumeration of specific cases or experiences, which should be simpler than designing rules as is the case for rule-based expert systems.

2. For new problems with no matching case in the knowledge base, the system can reason from similarities between the new case and existing cases. This is built around the generalization skills in human reasoning, i.e. that a person can be reminded of one thing or experience through its similarity or differences with other experiences.
3. The learning process is built into the system, in that the memory of experiences in the knowledge base are changed and augmented by each additional case that is experienced.

5.4.1 The CBR Execution Cycle

The basis for CBR are the four characteristics of regularity, typicality, consistency, and adaptability which are observable in the natural world [11]. According to [11] these characteristics can be described as follows:

Regularity implies that the same actions executed under the same conditions will tend to have the same or similar outcomes whilst typicality indicates that experiences tend to repeat themselves. Meanwhile, consistency implies that small changes in the situation merely require small changes in the interpretation and in the solution whilst adaptability implies that when things are repeated, the differences tend to be small, and such small differences are easy to compensate for.

The implication of these characteristics is that, for as long as the current problem can be described in terms of the previously solved problems, an appropriate solution can be found. With such a description, the solution to the most similar of the solved problems can be applied to the new problem either directly or with some adaptation based on the differences between the current problem and the most often identified similar problem from the past. Once the identified solution is verified as the best for the new problem, a new case is created as an association of the problem-solution pair which can be used to resolve similar new problems in the future. Thus, this addition of new cases will improve results of a CBR system by filling the problem space more densely.

Correspondingly, the CBR working cycle involves four processing stages illustrated in Figure 5.6:

1. **Case retrieval**: After the problem situation has been assessed, the best matching case is searched for in the case base and an approximate solution is retrieved.
2. **Case adaptation**: The retrieved solution is adapted to better fit the new problem.
3. **Solution evaluation**: The adapted solution can be evaluated either before the solution is applied to the problem or after the solution has been applied. In any case, if the accomplished result is unsatisfactory, the retrieved solution must be adapted again or more cases should be retrieved.

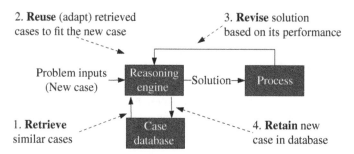

Figure 5.6 Case-based reasoning system.

4. **Case-base updating**: If the solution was verified as correct, the new case can be added to the case base.

5.4.2 Cognitive Capabilities and Applications of CBR Systems

CBR achieves higher cognitive capability, specifically some level of reasoning in the form of matching and deterministic inference. Whereas closed-loop control always needs to derive the solution each time the input changes, CBR minimizes this computation through its ability to match an input state to previously observed inputs. Thereby, it simply applies the previously computed best solution. However, this is still limited reasoning as the input state must match the previously observed states, which is not always the case in all systems. In communication systems, for example, most observations are continuously valued variables and so it is rarely the case that two observations will completely match. Consequently, the degree to which CBR may be useful is highly dependent on the degree to which the input states can be matched given the applied 'Measure of similarity'. Nevertheless, CBR already raises the level of cognitive capability compared to the famous closed-loop control.

The matching nature of CBR-based solutions implies that CBRs can be applied across a very wide range of applications. It has been applied [12] in interpretation applications to evaluate situations/problems in some context; in design challenges as a process of satisfying a number of posed constraints; in planning applications, for example, to arrange a sequence of actions in time; in classification applications as a means of explaining a number of encountered symptoms; and even in advising as a process of resolving diagnosed problems.

In networks, CBR systems can be developed for diagnostics [13] and classification, e.g. to classify network cell areas as rural, urban, or otherwise; or to classify users as static, slow, fast. However, CBR has not been extensively applied in SON solutions although it can be a tool for improving the existing closed-loop control SON as described in the example in the next section.

5.4.3 CBR Example for RAN Energy Savings Management

Consider then, an energy savings management solution has been designed (e.g. using closed-loop control mechanism) to evaluate the combination of load in a network, the cell utility over the last hour and other load-related characteristics like the load in neighbour cells. Then, assume that for a given cell S, the solution computes an estimate of the expected load profile over the next hour to determine if the cell S should be retained or switched to one of two states – normal operating state or an energy saving state.

Without a CBR mechanism, this solution will always undertake this computation for each hour in each cell. However, a CBR mechanism can be added to the solution such that each time a solution is computed (i.e. either remain in current state or switch energy saving states), the observation-to-solution pair is saved to be referenced later. Correspondingly, when the same conditions are observed at a later stage, the previously computed solution can be applied without recomputing the load and utility predictions. Therein, for the cell S at a certain time of day t, the CBR compares the observed conditions to those that have been observed before and applies the same solutions that were applied in those conditions. Corresponding entries in such a CBR database could, for example, show the pairs 'urban pico cell at 1900 hours → deactivate' or 'rural macro cell at 2100 hours → retain active'.

However, advanced variations of the CBR could also be considered. Consider, for example, that multiple actions have been tried under similar or related conditions with different results (e.g. how users in the neighbour cell areas were affected by those actions) and all stored in the CBR database. In this case, the CBR re-applies the best of those stored candidate solutions. Alternatively, the database may be centralized to be shared by multiple cells, which would indicate that a solution computed by one cell can be applied to another cell provided conditions are similar.

5.4.4 Limitations of CBR Systems

The major benefits of the CBR system are their relative simplicity in elucidating knowledge, expressing, or explaining solutions and even the maintenance of the system. This makes it applicable for sequential problems as well as complex and not-fully formalized solution spaces. Besides, the built learning process allows for the system to be initialized with a small set of cases but still to be beneficial in the long term.

These benefits notwithstanding, CBR systems suffer from memory limitations and time-consuming execution owing to the required processing in describing the problem and matching the cases to identify a solution. These directly imply that:

1. **Handling large case bases**: High memory/storage requirements and time-consuming retrieval accompany CBR systems utilizing large case bases. Although the order of both is at most linear with the number of cases, these problems usually lead to increased construction costs and reduced system performance. Yet, these problems are less and less significant as the hardware components become faster and cheaper.
2. **Dynamic problem domains**: CBR systems may have difficulties in handling dynamic problem domains, where they may be unable to follow a shift in the way problems are solved, since they are usually strongly biassed$$$ towards what has already worked. This may result in an outdated case base.
3. **Handling noisy data**: Parts of the problem situation may be irrelevant to the problem itself. Unsuccessful assessment of such noise present in a problem situation currently imposed on a CBR system may result in the same problem being unnecessarily stored numerous times in the case base because of the difference due to the noise. In turn, this implies inefficient storage and retrieval of cases.
4. **Fully automatic operation**: In a typical CBR system, the problem domain is not usually fully covered. Hence, some problem situations can occur for which the system has no solution. In such situations, CBR systems commonly expect input from the user.

5.5 Fuzzy Inference Systems

In discrete logic, specifically in Boolean logic, the truth values of a variable may only be integer values 0 or 1. In fuzzy logic, such truth values are continuous over the range [0,1], i.e. the truth ranges between being completely false and being completely true. Correspondingly, fuzzy sets describe vague concepts (fast runner, hot weather, weekend days). Fuzzy inference systems use fuzzy set theory in mapping from input to output i.e. although variables are defined independently of one another, Fuzzy operators (AND, OR, etc.) determine the outcomes of the variable combinations.

Owing to its multidisciplinary nature, Fuzzy Inference Systems (FIS) are associated with several names, including fuzzy-rule-based systems, fuzzy expert systems, fuzzy modelling, fuzzy associative memory, fuzzy logic controllers, or simply fuzzy systems. The terms used to describe the various parts of the fuzzy inference process are as such not standard, but the descriptions here attempt to define and be as clear as possible about the different terms introduced in this section.

5.5.1 Fuzzy Sets and Membership Functions

A fuzzy set is a set without a crisp, clearly defined boundary and may contain elements with only a partial degree of membership [14]. Correspondingly, the truth

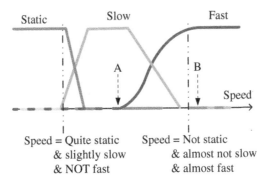

Figure 5.7 An example graphic description of fuzzy membership sets for three linguistic variables (static, slow, and fast).

of any statement becomes a matter of degree, which is the cornerstone of fuzzy logic. Fuzzy logic is multivalued logic since the membership to a fuzzy set can take any value between the bivalent logic values of yes or no, e.g. (Friday is sort of a weekend day, the weather is rather hot).

The degree of membership to a set, i.e. the degree of truthfulness is denoted by a membership value between 0 and 1. (Friday is a weekend day to the degree 0.8). A membership function (MF) associated with a given fuzzy set maps an input value to its appropriate membership value, i.e. the MF designated by μ is a curve that defines the membership value (between the limits 0 and 1) for a variable x. A fuzzy set A in X is defined as a set of ordered pairs.

$$A = \{x, \mu_A(x) \mid x \,\varepsilon X\}$$

As an example, Figure 5.7 shows three fuzzy sets of velocity and the corresponding MFs, - a linear function for the static-speed set, a truncated triangle function for the slow-speed set and a sigmoid function for the fast-speed-speed set. Consider that one is asked whether a velocity v belongs to the set 'fast'. The response thereto may either be one of the two binary values yes and no: (say yes for velocities below A and no for velocities above B) or it may be any one of the infinite values in between A and B.

Note that memberships are independently defined for each set so that the sum of membership values do not have to add to 1.

5.5.2 Fuzzy Logic and Fuzzy Rules

Fuzzy logical reasoning is a superset of standard Boolean logic. Standard logical operations hold if the fuzzy values are held at their extremes of 1 (completely true), and 0 (completely false). Thus, Boolean logic is the special case of fuzzy logic for which the values can only hold the extreme values. Considering two variables A and B, the outcome of an and operation can be achieved by a min operation, min (A, B) as seen in Figure 5.8a. Similarly, as illustrated in Figure 5.8b and c, the OR

A	B	Min(A,B)
0.2	0.1	0.1
0.1	0.9	0.1
0.9	0.1	0.1
0.9	0.9	0.9

(a) and (A, B)

A	B	Max(A,B)
0.2	0.1	0.2
0.1	0.9	0.9
0.9	0.1	0.9
0.9	0.9	0.9

(b) OR (A, B)

A	1-A
0.1	0.9
0.9	0.1

(c) NOT(A)

Figure 5.8 Fuzzy implementation of AND, OR, and NOT operations.

operation between the two variables A and B is achieved by the max(A, B) whilst NOT(A) the NOT operation on a variable A, can be implemented as $1-A$.

The above definitions are the classical implementation of fuzzy logical operators for fuzzy intersection or conjunction (AND), fuzzy union or disjunction (OR), and fuzzy complement (NOT). These definitions are not unique, and generalization has been proposed as binary mapping T for any two sets A, B, with T defined to fulfil the specific requirements for the respective operation (conjunction, disjunction, or complement) [14].

A single fuzzy if-then rule of the form {if x is A then y is B} evaluates an '*antecedent*' or 'premise' (x is A) and follows a *consequent* or conclusion (y is B) in case the antecedent is true. In the rule, A and B are linguistic values defined by fuzzy sets on the respective ranges (universes of discourse) X and Y. An example of such a rule could be stated as:

if velocity is high then handover delay is low

The input to an if-then rule (in this case *high*) is the current value for the input variable (in this case, *velocity*), represented as a number between 0 and 1, and so the antecedent is an interpretation that returns a single number between 0 and 1. The output (in this case, *low*) is an entire fuzzy set, and so the consequent is an assignment that assigns the entire fuzzy set B to the output variable y.

Note that the word 'is' gets used in two entirely different ways, i.e. as a test condition in the antecedent (equivalent to *if velocity* $== high$) and as an assignment in the consequent (equivalent to *then handover delay* $= low$)

5.5.3 Fuzzy Interference System Components

Fuzzy inference is the process of formulating the mapping from a given input to an output using fuzzy logic. The mapping then provides a basis from which decisions can be made, or patterns discerned. The fuzzy inference process involves four major components shown in Figure 5.9: the fuzzifier, the fuzzy rule base, the inference engine and the defuzzifier. These implement the five steps of the process: fuzzification of the input variables, application of the fuzzy operator (AND or OR) in the antecedent, implication from the antecedent to the consequent, aggregation of the consequents across the rules, and defuzzification.

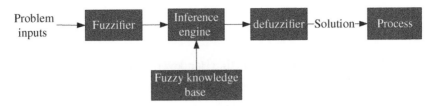

Figure 5.9 A fuzzy inference system.

The fuzzy knowledge base is the database that holds the if-then rules and the membership functions of the different fuzzy sets used in the fuzzy rules. The fuzzifier compares the inputs with membership functions to obtain the membership values of each linguistic label. The inference engine then combines membership values to get firing strength of each rule and thereafter weights the rules to obtain their fuzzy or crisp consequents based on their respective firing strengths. Finally, the defuzzifier aggregates the qualified consequents and defuzzifies the aggregate to produce a crisp output.

5.5.4 Cognitive Capabilities and Applications of FIS

Fuzzy inference systems can fulfil the requirements of the full reasoning engine of the data processing pipeline model of Cognition as described in Section 3.5. In other words, FIS can provide a full matching and inference engine for both discrete and continuous state spaces. Note, however, that fuzzy inference does not provide full automation capability, i.e. it is the only mechanism for decision making without perceptive capability or action execution. Subsequent action is typically left to the underlying managed system otherwise extra components need to be added to afford autonomous action subsequent the derived decisions.

Fuzzy inference systems have been successfully applied in fields such as automatic control, data classification, decision analysis, expert systems, and computer vision. The earliest applications were for speed controllers in steam engines (see [15]) and eventually for automotive systems. FIS have, however, also been extensively applied in production systems e.g. in scheduling, requirements planning as well as in capacity planning applications.

FIS can be very useful in network management owing to the continuous nature of the variables used thereof in decision making. For example, many network management applications need to classify variables to select the most appropriate actions. FIS' can enable more granular decisions to be taken based on a few discrete sets e.g. for coverage areas with the sets rural, urban, suburban, etc., or for user speeds classified in the sets {static, slow, or fast}. For learning-based network management applications, FIS' can help to manage the state space which is a challenge in network management. Instead of a learning solution for each possible state, the application learns only for a few discrete states which then act as the fuzzy sets to be combined in deriving solutions for the unknown states.

5.5.5 Example Application: Selecting Handover Margins

For an end-to-end flow of the decision-making process, consider a contrived challenge of selecting the handover margin for a user, based on the user's velocity and the cell's spectral characteristics. In particular, we want to use the fact that a cell operating at a high frequency is more prone to handover (HO)-related failures as the signal can easily be blocked by obstacles. Such a cell needs to execute HOs as soon as is practically possible. Correspondingly, the fuzzy inference problem is a two-input one-output three-rule handover-margin setting problem illustrated by Figure 5.10.

During the left-to-right processing, rules are evaluated in parallel which avoid sharp switching between modes based on breakpoints and instead allows the system to glide smoothly from regions where the system's behaviour is dominated by either one rule or another. The corresponding inference processing for the handover margins problem at a velocity of 68 kmph and spectrum of 2 GHz is implemented by the following five steps as also illustrated by Figure 5.11.

5.5.5.1 Step 1: Fuzzification

Fuzzification is the table lookup or a function evaluation process that takes the inputs and determines the degree to which they belong to each of the appropriate fuzzy sets via membership functions. The inputs are limited to the universe of

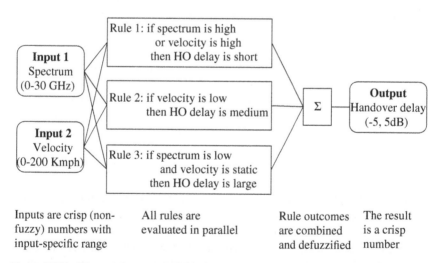

Figure 5.10 A fuzzy inference problem.

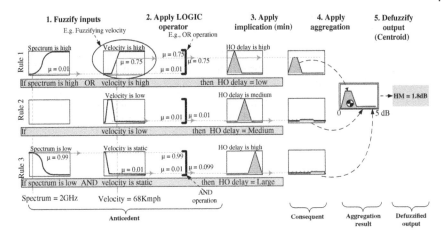

Figure 5.11 End-to-end process of the 3-rule Fuzzy inference computation showing the aggregation of outputs and their eventual defuzzification.

discourse of the respective input variable (here [0,30] for the spectrum and [0,200] for the velocity). The output of the Fuzzification process is a fuzzy degree of membership in the qualifying linguistic set (always over the interval [0,1]).

In the Handover configuration problem above, the inputs must be resolved into the different fuzzy linguistic sets: spectrum is low, spectrum is high, velocity is low, velocity is high, etc. In Figure 5.11 for example, it is shown how the membership function is used to qualify velocity for the linguistic variable 'high'. Here, the velocity at 68 kmph corresponds to $\mu = 0.75$ for the 'high' membership function. Each input is similarly qualified over the respective membership functions.

5.5.5.2 Step 2: Apply Fuzzy Operator(s)

The outcome of the fuzzification process is the degree to which each part of the antecedent has been satisfied for each rule. Then, for each rule that has more than one part in the antecedent, the respective fuzzy operator is applied to obtain a single number for the result of the antecedent for that rule. The applicable logic operator (as is described in Section 5.4.1) takes the two or more membership values from fuzzified input variables and computes the firing strength of the rule which is the result of the antecedent and describes the degree to which the rule matches the inputs. Figure 5.11 highlights the computation of the OR operation for the first rule using the max function and the and operation for the third rule using the multiplication function. The fuzzy OR operator simply selects the maximum of the two values, i.e. the firing strength is computed as 0.75 in this case.

5.5.5.3 Step 3: Apply Weighted Implication

After the firing strength is computed for each rule, the implication process computes the consequent of the rule. A consequent is a fuzzy set represented by a membership function, which appropriately weights the linguistic characteristics that are attributed to it. Each rule has a weight (a number between 0 and 1), which allows the rules to be weighted relative to each other. Generally, this weight is 1 (as it is for this example) and so it has no effect at all on the implication process. Implication combines the weight and the firing strength to compute the consequent of the rule, which is a fuzzy set. This combination can be implemented using the and fuzzy operator which, in that case, truncates the output fuzzy set according to the weights (see Figure 5.11).

5.5.5.4 Step 4: Aggregate All Outputs

To make the final decision after consequents are computed for all the rules, the rules must be combined through the aggregation process. Thereby, the output fuzzy sets (the list of truncated output functions returned by the implication process) for each rule are combined into a single fuzzy set for each output variable. Several methods can be used for the aggregate, the simplest of which are the *max* (which takes maximum of the functions) and *sum* (which takes the sum of the rules' output sets). The critical requirement, however, is that the aggregation method should be commutative to ensure that the order in which the rules are executed is unimportant.

In Figure 5.11, all three rules have been placed together to show how the output of each rule is combined, or aggregated, into a single fuzzy set whose membership function assigns a weighting for every output (HO-margin) value.

5.5.5.5 Step 5: Defuzzify

The final step is the defuzzification process which takes the aggregate fuzzy set and generates the final output as a single number. The process defuzzifies the range of output values (in the fuzzy set) in order to resolve a single output value from the set. The most popular defuzzification method is the centroid calculation, which as shown in Figure 5.11, returns the centre of area under the curve. However, there are other methods including the other four implemented in the MATLAB fuzzy logic toolbox [16]: bisector, middle of maximum (the average of the maximum value of the output set), largest of maximum, and smallest of maximum.

5.6 Bayesian Networks

The modelling of relationships between symptoms and diagnosis, or more generally, the modelling of a set of variables and the conditional dependencies between them is an important part in any diagnosis process, including mobile network diagnosis. One of the main methods for this are Bayesian Networks (BNs).

5.6.1 Definitions

A Bayesian Network is a Directed Acyclic Graph (DAG), where nodes correspond to random variables, and edges correspond to conditional dependencies. Formally a BN is defined as a pair $B = \langle G, \Theta \rangle$, with $G = \langle V, E \rangle$ is the DAG, and Θ is the set of parameters of the network. The nodes V represent random variables, and an edge exists in G between vertices $v, q \in V$, if the conditional probability $P(q \mid v)$ is a factor in the joint probability distribution. The set Θ contains these conditional probabilities, so $\Theta_{v_i, \pi_i} = P(v_i \mid \pi_i)$, where π_i is the parent set of $v_i \in V$ in G. BNs satisfy the local Markov property, so a vertex given its parents is conditionally independent of its non-descendants. As such, the Joint Probability Distribution (JPD) over V can be written as in Eq. (5.1):

$$P(v_1, v_2, v_3, \ldots, v_n) = \prod_{i=1}^{n} P(v_i \mid \pi_i) = \prod_{i=1}^{n} \Theta_{v_i, \pi_i} \tag{5.1}$$

The simplest BN is statistical (informational), where the BN does not contain any causal assumptions, i.e. there is no knowledge of the causal order between the variables and so the network can only be interpreted statistically (for informational purpose only). On the other extreme, a dynamic BN keeps track of the dynamic variables for entities in changing environments by representing multiple copies of the state variables with a copy each time step. A general BN, however, is a causal BN that describes the static causal relationships amongst the variables.

5.6.2 Example Application: Diagnosis in Mobile Networks

A Bayesian Network can be constructed to diagnose problems in mobile networks, using a model that was proposed in [17] (as illustrated in Figure 5.12). The model distinguishes between three kinds of nodes – causes, symptoms, and factors. Causes represent the possible faults that may be sources for the problems observed in the network, including operationally independent issues as hardware problems, interference, etc. Symptoms represent the manifestations of poor performance in the network, including issues such as degradation in signal level and increases in the number of handovers (HOs). Conditions represent the factors that can have an impact on the causes and symptoms. For example, since load problems are more likely to be realized in densely populated areas where cell density is also high, cell density becomes a critical condition to be evaluated for diagnosis. Similarly, the number of HO events increases with population and cell density.

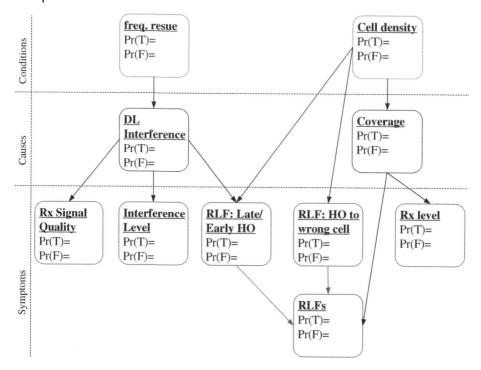

Figure 5.12 Example of Bayesian network for troubleshooting.

The relationships amongst these nodes are marked as the vertices in the BN graph – in this case, the Frequency reuse and Cell density (as conditions) are noted to have direct impact on DL Interference and Coverage (as causes) as well as on the symptoms of RLFs due to late/early HO and due to HO to a wrong cell. However, that the conditions also have indirect impact on these or other symptoms through the conditions, impact on causes that then directly impact on the symptoms.

Some symptoms may also be compounds of other symptoms as is the case of the symptom of 'RLFs' which is a union of the symptoms of 'RFL: Late/Early HO' and 'RFL:HO to a wrong cell'. Otherwise, given structure of the model, the states of the nodes are specified, and the probability tables should be completed as summarized in the next section.

5.6.3 Selecting and Training Bayesian Networks

In many cases, the BN is unknown and must be constructed through training from data, a challenge referred to as the BN learning problem. Formally, the learning problem requires one to estimate the graph topology (network structure) and the parameters of the JPD in the BN from training data and some prior information (e.g. expert knowledge, casual relationships) [18].

Using prior information to construct the BN simplifies the learning challenge. However, the computation of the probabilities is also a complex task that is highly influenced by the BN structure. Specifically, it is more stringent when nodes in the graph have multiple parents. Two types of models have usually been applied to construct the structure of BNs – naïve models and the causal independence models.

Naïve models, which have been extensively used in medical diagnostic systems, assumes a single parent for any child node fault for any given symptom, i.e. that only one cause is present at any time, and that the children (symptoms and conditions) are considered to be independent given that the cause is known. This may be the case in medical scenario where, a given time, each symptom (as the child node) may indeed have one cause or disease (as the parent node). However, when a node can have several parents at a given time, as can be the case in mobile networks, these assumptions become too restrictive, as probabilities for each node need to be set for each combination of the parent nodes.

Causal Independence models lift the restrictions imposed, by naïve models by assuming that multiple causes can have the same symptom, but that the causal mechanisms are mutually independent [19]. A common example model thereof is the Noisy-OR [7], which assumes that each binary cause C_i will bring about a related binary symptom S_i to happen unless an 'inhibitor' prevents it. Other modes with non-binary variables can, however, also be applicable, specifically the Noisy-MAX, Noisy-ADD allow any number of discrete states whilst the linear-Gaussian model allows for continuous cause and effect variables.

Usually, learning the structure of the DAG is a more difficult problem than learning the parameters. Missing data or hidden nodes also present the issue of partial observability, i.e. when there is insufficient information to explicitly build part of the network. Due to this, there are multiple cases of BN learning with different preferred learning methods. For unknown BN structures, the use of model-space search is usually recommended. For fully observable models, maximum-likelihood can be used for parameter learning, whilst for partially observable models, expectation maximization (EM) is a good candidate.

5.6.4 Cognitive Capabilities and Applications of Bayesian Networks

BNs provide an inference engine that reasons over the JPD of the set of random variables. Inference in BNs is most often considered in two ways; *Predictive support,* or otherwise known as top-down reasoning, is based on evidence nodes connected to a node through its parents, whilst *diagnostic support,* or bottom-up reasoning, is based on evidence nodes connected to a node through its children.

Bayesian networks have long been used to build medical diagnosis systems. These models are hand-built by experts and, after training, can be used to infer medical diagnosis from patient symptoms. The models provide the likelihood of different causes to the symptoms. Similar models are used in factories for diagnosis of faulty products.

A famous example of BNs is the Microsoft Office assistant Clippy. The Microsoft Research project Lumiere was built on Bayesian models, that captured the relationship between the goals and needs of users. It was built into Microsoft products first as an office assistant, with different characters to choose from. In the end, it turned out to be a distraction that users did not want, so eventually it was discontinued.

In practice, inference on Bayesian Networks is computationally intensive. It has been shown in [20] that exact inference in BNs is NP-hard and as later shown in fact #P-complete [21]. So, whilst BNs offer a rich and complex representation for machine learning use-cases, they are only realistically usable by adding restriction to ease the inference on the model. Usually, this is done by either restricting the topology, such as in naïve Bayesian Networks, or by restricting the conditional probabilities.

5.7 Time Series Forecasting

When data are collected over time, either on a regular or irregular time basis, it is natural to associate the entries with the corresponding time stamp, which results in a time series. In other words, a time series is a sequence of observations ordered according to the sequential time at which they were observed. There are many automation problems where the structure of a time series needs to be understood, i.e. the time series needs to be described through a process called time series analysis. In most applications, the understanding is also needed to make predictions on the future progress of the time series, in a process called time series forecasting.

5.7.1 Time Series Modelling

Time series analysis involves developing models that best capture or describe an observed time series in order to understand the underlying causes, i.e. time series analysis develops mathematical models that provide plausible descriptions from sample data. Generally, a time series $\{X(t), t = 0, 1, 2...\}$ may be assumed to follow a probability model which describes the joint distribution of a stochastic process that produces the random variable X_t, i.e. the sequence of observations of the series is actually a sample realization of the stochastic process that produced it.

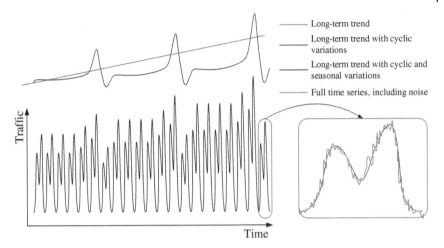

Figure 5.13 Example illustration of the constituent parts of a time series.

The analysis decomposes a time series X(t) into four constituent parts illustrated by Figure 5.13:

1. The trend T(t), which is the long-term monotonic change of the average level of the time series, i.e. either the increasing, decreasing, or stagnant behaviour of the series over time. Note also that a level L may also be distinguished as the baseline value for the series if it were a straight line.
2. The trend cycle C(t), which is the long-term variation of the time series.
3. Seasonality S(t), which describes the repeating patterns or cycles of behaviour over time. The cycles may be over very short-time periods, say a day or less, or over medium to longer periods like the seasons of the year or years.
4. Noise N(t), which describes the variability in the observations that are not regular, do not have any discernible pattern and cannot be explained by the model.

Two types of models are generally used for modelling a time series X(t), the multiplicative and additive models. The multiplicative model given by Eq. (5.1) assumes that the four components of a time series are not necessarily independent, and they can affect one another; whereas the additive model in Eq. (5.2) assumes that the four components are independent of each other.

$$\text{Multiplicative Model:} \quad X_t = L * T_t * S_t * N_t \tag{5.2}$$

$$\text{Additive Model:} \quad X_t = L + T_t + S_t + N_t \tag{5.3}$$

The concept of stationarity of a stochastic process is critical to the analysis of time series. Stationarity describes the degree of statistical equilibrium of a process, i.e. a stationary process has statistical properties such as mean and variance that do not depend upon time.

A process {X(t), t = 0, 1, 2...} is Strongly Stationary or Strictly Stationary if the joint probability distribution function of any possible set of random variables from the process $\{x_{t-s}, x_{t-s+1}, ...x_t, ...x_{t+s+1}, x_{t+s}\}$ is independent of t for all s. This is, however, a strong assumption which is not always needed and is, for most practical applications, replaced by a weaker form of Weak Stationary. A stochastic process is said to be of order k Weakly Stationary if its statistical moments up to k depend only on time differences and not the time of occurrences of the data used to estimate the moments. For example, a stochastic process {X(t), t = 0, 1, 2...} is second-order stationary if it has time independent mean and variance and the covariance values $Cov(x_t, x_{t-s})$ depend only on s.

5.7.2 Auto Regressive and Moving Average Models

The most widely used linear time series model is the Auto Regressive and Moving Average (ARMA). Namely, consider a univariate time series $\{X_t : t \in T\}$, where T denotes the set of time indices. The general ARMA model, denoted by ARMA (p, q), has p AR terms and q MA terms, and is given by

$$X_t = Z_t + \sum_{i=1}^{p} \alpha_i X_{t-i} + \sum_{j=1}^{q} \beta_j X_{t-j} \tag{5.4}$$

where Z_t is the white noise error process, and $\{\alpha_i\}_{i=1}^{p}$ and $\{\beta_j\}_{j=1}^{q}$ denote the learning parameters. Note that for $q = 0$ and $p = 0$ one can derive the simpler AR and MA models, respectively.

The ARMA model fitting process assumes *stationarity*. One possibility to be able to capture also *non-stationary* time series consists in *stationarizing* the series by differencing. in this case, we are dealing with ARIMA(p, d, q) model (a generalization of the ARMA model), where the parameter d indicates that after d differentiations the series reduces to an ARMA(p, q).

Finally, in case the time series presents seasonality, one can use SARIMA model (seasonal ARIMA), which is able to explicitly model the seasonal element of the univariate data. Those models are usually formalized as SARIMA$(p,d,q)(P,D,Q)m$, where the additional parameters P, D, Q, and m refer to the SAR (seasonal auto regressive) order, the seasonal difference order, the SMA (seasonal moving average) order and the length of the seasonal period, respectively.

5.7.3 Cognitive Capabilities and Applications of Time Series Models

As the core mechanism for studying the dynamic behaviour of process, or metrics thereof, time series models provide an engine for automated reasoning with data organization and inference capabilities. The understanding of the dynamic behaviour of a process implies that relationships can be developed for the process and its constituent and/or drivers. Then the capability to predict future behaviour of the process enables for inference and decision to be made about how to influence or take advantage of the future outcomes of the process.

Time series models have been applied in many areas, to e.g. forecast natural phenomena like weather, or to forecast economic and business processes like sales and inventory changes. In communication networks, time series models can be applied to predict future network behaviour and events. One can, for example, predict the load be it in one or multiple cells, the number of events like handover events or failure events or even the occurrences of a particular events like alarms. In general, the most usage is in troubleshooting and diagnostics as is discussed in Chapter 9.

5.8 Conclusion

This chapter has summarized the most critical classic AI techniques which can be applied as tools for autonomous reasoning. The topics presented were selected with a focus on those that are likely to have wider usage in Network Management Automation (NMA). Specifically, expert systems, closed-loop control systems and case-based reasoning allow for some sort of autonomic control through which experts develop and train the system to solve reasoning problems. These have also had extensive usage in NMA and, in particular, SON. Fuzzy inference systems have been presented as a specific capability for reasoning over continuous variables whilst Bayesian networks allow for probabilistic causal reasoning that can be highly useful in diagnostics. Finally, the last section presented time series models as the analytics, means to understand the dynamic behaviour of processes, a capability that would also be highly useful in predicting future states of a process or in this case a network.

Together, the techniques show that classic AI is indeed a system for autonomic reasoning that is applicable to problems of different contexts, like discrete and continuous variables or deterministic and probabilistic state values. However, even then, the capabilities of all these techniques are limited to reasoning, they do not provide any means for autonomic perception which a fundamental component of a fully cognitive system. Instead, for perceptive capabilities, one must turn to machine learning techniques which are the subject of discussion in the next chapter.

References

1 Fagan, L.M., Shortliffe, E.H., and Buchanan, B.G. (1980). Computer-based medical decision making: from MYCIN to VM. *Automedica* 3 (2): 97–108.

2 Reingold, E. and Nightingale, J. (1999). Artificial Intelligence Tutorial Review. http://psych.utoronto.ca/users/reingold/courses/ai/commonsense.html (accessed 13 February 2019).

3 Flasiński, M. (2016). Symbolic artificial intelligence. In: *Introduction to Artificial Intelligence*, 15–22. Cham: Springer.

4 Jackson, P. (1998). *Introduction To Expert Systems*, 3e, 2. Addison Wesley.

5 Nurminen, J., K, K., and Hätönen, K. (2003). What makes expert systems survive over 10 years—empirical evaluation of several engineering applications. *Expert Systems with Applications* 24 (2): 199–211.

6 Grosan, A. and Abraham, C. (2011). *Rule-Based Expert Systems*, Intelligent Systems Reference Library, vol. 17. Springer.

7 Laselva, D. et al. Self optimization. In: *LTE Self-Organising Networks (SON): Network Management Automation for Operational Efficiency* (eds. S. Hamalainen, H. Sanneck and C. Sartori), 135–235. Wiley.

8 Schank, R. (1984). Memory-based expert systems. Technical Report (# AFOSR. TR. 84- 0814).

9 Hamalainen, S., Sanneck, H., and Sartori, C. (eds.). *LTE Self-Organising Networks (SON): Network Management Automation for Operational Efficiency*. Wiley.

10 Riesbeck, C.K. and Schank, R.C. (2013). *Inside Case-Based Reasoning*. Psychology Press.

11 Kolodner, J. (1996). Making the implicit explicit: clarifying the principles of case-based reasoning. In: *Case-Based Reasoning: Experiences, Lessons & Future Directions*, 349–370. Menlo Park, USA: AAAI Press.

12 Pantic, M. (2001). *Facial Expression Analysis by Computational Intelligence Techniques*. Delft University of Technology.

13 Tang, H., Stenberg, K., Apajalahti, K. et al. (2016). Automatic Definition and Application of Similarity Measures for Self-Operation of Network. International Conference on Mobile Networks and Management, Abu Dhabi..

14 MathWorks (2018). Fuzzy Logic Toolbox TM: Getting Started with Fuzzy Logic Toolbox. https://de.mathworks.com/help/fuzzy/getting-started-with-fuzzy-logic-toolbox.html (accessed 29 August 2018).

15 Mamdani, E.H. and Sedrak, A. (1975). An experiment in linguistic synthesis with a fuzzy logic controller. *International Journal of Man-Machine Studies* 7 (1): 1–13.

16 MathWorks (2018). Fuzzy logic toolbox user's guide. https://www.mathworks.com/help/pdf_doc/fuzzy/fuzzy.pdf. (accessed 28 February 2019).

17 Barco, R., Guerrero, R., Hylander, G. et al. (2002). Automated troubleshooting of mobile networks using bayesian networks. IASTED International Conference on Communication Systems and Networks (CSN02), Malaga.

18 Ben-Gal, I. (2008). *Bayesian Networks*, Encyclopedia of Statistics in Quality and Reliability, vol. 1. Wiley.

19 Heckerman, D. and Breese, J.S. (1996). Causal independence for probability assessment and inference using Bayesian networks. *IEEE Transactions on Systems, Man, and Cybernetics-Part A: Systems and Humans* 26 (6): 826–831.

20 Cooper, G.F. (1990). The computational complexity of probabilistic inference using Bayesian belief networks. *Artificial Intelligence* 42 (2–3): 393–405.

21 Roth, D. (1996). On the hardness of approximate reasoning. *Artificial Intelligence* 82 (1–2): 273–302.

6

Machine Learning: Tools for End-to-End Cognition

Stephen Mwanje[1], Marton Kajo[2] and Benedek Schultz[3]

[1] *Nokia Bell Labs, Munich, Germany*
[2] *Technical University of Munich, Munich, Germany*
[3] *Nokia Bell Labs, Budapest, Hungary*

Network management (NM) is responsible for processing network and service-related information, to draw insights about the network and its related services and to derive network and service reconfigurations in order to optimize resource usage while minimizing cost. Traditionally, NM tasks were undertaken by human experts who interpreted network events and decided the best course of action. As networks became increasingly dense and complex, supporting a higher number of experts needed to manage the networks became rather expensive. The Self-Organizing Networks paradigm solved the challenge by automating the expert-like decision making through classic AI techniques, such as expert systems or rule-based closed control loops. However, as discussed in Chapter 5, classic AI techniques have major limitations – mainly, the inability to achieve end-to-end cognition owing to the need to explicitly describe the models of the world. Thereby, Machine Learning (ML) is the solution.

End-to-end cognition is expected to improve NM automation by reducing the need for explicit models underlying the NM decisions. For example, it should no longer be necessary for the operator to describe the full set of possible faults, the sequences in which they occur and the related solutions. Instead, the cognitive NM tool could learn these relations and their underlying symptoms, thus raising the operator's abstraction level. ML-based cognition overcomes the classic AI limitations, since it enables the ML-based functions to derive their own models from the network and service data, which model they subsequently use to derive network reconfigurations. Specifically, ML allows for development of perceptive and relational models through supervised, unsupervised and reinforcement learning methods.

Towards Cognitive Autonomous Networks: Network Management Automation for 5G and Beyond, First Edition. Edited by Stephen S. Mwanje and Christian Mannweiler.
© 2021 John Wiley & Sons Ltd. Published 2021 by John Wiley & Sons Ltd.

This chapter summarizes some of the ML concepts and how the related techniques can be employed as tools for end-to-end cognition in NM. The discussion starts with a generic conceptual review of what it takes to learn from data presenting the general underlying mathematical concepts and the specific use cases for learning with or without supervision. This is followed by a summary of neural network concepts as the foundation for recent successes in ML. A separate – albeit brief – discussion on deep learning is added to highlight the most successful neural network ML methods and how they have applied to different problem areas. Finally, a section on Reinforcement Learning (RL) discusses how RL can be used for autonomic control, highlighting concepts on the recently successful deep learning based RL models.

6.1 Learning from Data

Given a dataset, it is possible to learn and obtain insights from the data and to derive decisions and actions (see Figure 6.1). Besides the use case, the learned insights depend on the nature of the dataset and can generally be of two forms: unsupervised in the case of unlabelled data and supervised if the data is labelled to indicate a matching of observations and insights. In general, there are major commonalities between the two approaches, e.g. in terms of the concepts used in the learning process [1, 2]. This is the reason for discussing the underlying concepts in a single section.

Unsupervised learning algorithms can find structure within the dataset e.g. the principal or independent dimensions of the data, or how to cluster samples therein, without this specific information being explicitly attached to the observations. Conversely, for supervised learning algorithms, each example **x** in the training dataset has a label or target **y** with which the algorithm learns a function that maps **x** to the target.

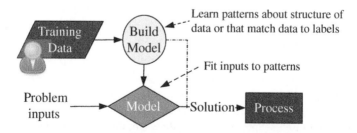

Figure 6.1 Learning from data: Supervised learning system learns to match data to labels while an unsupervised learning system learns patterns about the structure of the data.

In general, however, since the distinction between supervised and unsupervised learning is based on the nature of the data, the lines between them can be blurred, especially since the same learning technologies can be used to perform the same tasks. Correspondingly, some problems and algorithms do exist in the grey between supervised and unsupervised learning. These algorithms are classified as semi-supervised, training on a dataset where some examples do have a target and others do not. A fourth group of learning algorithms, grouped under reinforcement learning, is also discussed at the end of this chapter. These algorithms are arguably the most cognitive learning algorithms to date, generate their own training data by experimenting on a system implementing some form of autonomic control functionality.

6.1.1 Definitions

According to Tom Mitchell (1997): 'a computer program is said to learn from experience E with respect to some class of tasks T and performance measure P, if its performance at tasks T as measured by P improves with experience E' [3]. The learning is controlled by the nature of the experience E in which each sample is an example, and the complete set of examples is the dataset. Each example is a vector $\mathbf{x} \in \mathbb{R}^n$ that is a collection of features x_i. The characteristics of the experience E describes and distinguishes the type of learning that is undertaken as well as the algorithms required for that kind of learning.

Learning algorithms are typically most suited for use cases, where it is difficult to design the required model upfront, because the model is heavily dependent on the context and it is likely that a separate model would need to be designed for each deployment. In these cases, the required functionality is more easily described by humans through examples, showing not how the machine should process an observation explicitly, but what the desired outcome would be. It is the learning algorithm's task to generalize across the given examples, and create an internal ruleset that best implements the exemplified functioning for the tasks T. The application areas of machine learning are directly related to the task that can be accomplished. A list of non-orthogonal example tasks is given in Table 6.1 (see [1]).

The performance measure P, which is typically referred to as the cost function, is the quantitative description of how good the algorithm performs the specific task T. Besides telling how well the model performs, the performance measure also influences how and what the model learns from the dataset. In order to realize a given task, the learning algorithms usually tunes internal parameters, that are also called training parameters or weights, according to the observed performance. Typically, a learning machine's performance is measured through a continuous-valued score, which represents the deviation (loss/error/cost) from an envisioned perfect solution to the task. This perfect solution can either be

Table 6.1 A non-exhaustive characterization of machine learning tasks.

Task class	Task	Description	Example application
Classification	General	Find which of k categories an input x belongs to, i.e. learn function $f: \mathbb{R}^n \to \{1, \ldots, k\}$ that, for input vector x, computes a probability distribution over the categories. In the special case, f assigns x to a category y that has the highest probability.	Detecting network anomalies and/or events with specific required responses.
	Tasks with missing inputs	Learn a set of functions (instead of a single function,) each for classifying x with a different subset of its features missing.	
Regression	General	Learn a function $f: \mathbb{R}^n \to \mathbb{R}$ that predicts a numerical value for a given input vector x. Its difference from classification is only in the format of the output.	Predict how cell/network load will change over time.
	Synthesis and sampling	Generate new examples that are similar to those in the given training dataset. This is useful where manually doing the task would be expensive, boring, slow, etc.	Simulate network events that cannot be physically modelled.
	Imputation of missing values	For an example x with missing features x_I, the algorithm predicts the values of the missing features.	Impute lost or corrupted NM and other network data.
	Denoising	With some features x_i corrupted by some unknown process, the algorithm predicts the clean sample from its corrupted version	
Structured output	General	For a given input vector, generate an output vector with important relationships among its elements.	Reduce dimensionality of network KPIs.
	Transcription	Observe an unstructured data representation and transcribe the information into discrete textual form.	Google street view (for address numbers); Speech recognition for audio waveforms.
	Machine translation	Convert a sequence of symbols in one language into a sequence of symbols in another language.	Get descriptions of network events in one language from another language.
Density or mass function estimation	General	For a dataset, learn a function $P: \mathbb{R}^n \to \mathbb{R}$ where P(x) can be taken as the Probability Density or Mass Function (PDF or PMF) on the space from which x is drawn.	Used by the other tasks above that need to capture the structure of the probability distribution.

realistically achievable with the specific learning machine, but could also be impossible to reach due to the model complexity, computational constraints, etc. The main point is to be able to measure a (limited) numerical value, with which the training can be converted into a numerical optimization task. The number of, and the specific interactions between the internal parameters, as well as the way in which they influence the output, is unique to the learning algorithm and the chosen model (topology or complexity). These choices are called hyper-parameters, and can be numerical choices, such as the number of neurons in a layer, or non-numerical choices, such as the activation function used in a layer. These choices are usually user-made before training and cannot be easily changed once the model has been trained.

6.1.2 Training Using Numerical Optimization

Most machine learning methods use some form of numerical optimization during training, both for model learning and sometimes also for hyper-parameter optimization. The simplest method for tuning the parameters is grid search, but expectation maximization has had the widest usage. This section summarizes the related concepts and algorithms.

Despite its simplicity, *grid search* is widely used for hyper-parameter tuning. It is a simple exhaustive search on a manually specified set of parameters. It greatly suffers from the curse of dimensionality, i.e. as the dimensions of the feature space increase, the volume of the space increases so fast that the available data become sparse. This has the effect that the amount of data needed to support the result often grows exponentially with the dimensionality. However, grid search is highly parallelizable, which makes it a go-to algorithm for tuning hyper-parameters of neural networks and other machine learning methods.

Expectation maximization (EM) is a framework of two alternating steps – expectation and maximization. In the expectation step, the algorithm estimates the model parameters to be optimized. Then, in the maximization step, it updates the parameters based on the estimates in the first step. The EM algorithm is used in a wide range of applications. Most well-known is the k-means algorithm for quantization and clustering, or the fitting of mixture of Gaussians.

Simulated- annealing is a heuristic for optimization in large state-spaces, minimizing system energy. It is inspired by a metallurgic technique called annealing, which involves the heating and controlled cooling of a metal to reduce its defects. The algorithm considers a new point from the neighbourhood at every iteration, fully accepting it if it is a lower energy state and in a higher energy state accepting it with a probability. This probabilistic acceptance of some high energy points avoids local minima. The size of the neighbourhood is dependent on the 'temperature', which is slowly decreased at every iteration. Correspondingly, the success of the algorithm is highly dependent on the temperature schedule since as the temperature decreases the algorithm reduces the extent of its search.

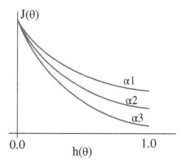

Figure 6.2 Variation of the cost function J(θ) with the learning rate α.

Currently one of the most (if not the most) frequently used numerical optimization algorithm is *gradient descent* and its various extensions. The basic idea behind gradient descent is to roll down (descend the gradient of) the profile of a cost function J(θ) (which calculates the previously mentioned performance measure P) in Figure 6.2 towards its lowest point. Consider a set of parameters $\theta := [b, W]^T$, resulting in an output $\hat{y}$ estimated by the hypothesis and the corresponding error. The gradient of the cost function is computed at the point of the applied parameters. Subsequently, the weights are adjusted in the direction of reduced cost using a learning parameter α. This learning rate parameter controls the extent to which the parameters are adjusted in the direction of the new gradient. This is repeated for multiple iterations until convergence (until there is no further reduction in the estimation error). This process represented in Algorithm 6.1 finds the least point of the cost function J(θ) and can be applied both for regression or classification.

Algorithm 6.1   Gradient Descent Algorithm

Given an estimate $\hat{y}$

1. Compute cost J(W, b, x, y) e.g. using Mean Square Error (MSE) as

$$J(W, b, x, y) = \frac{1}{n} \sum_{i=1}^{n} [y_i - \hat{y}_i]^2$$

2. Compute gradients $\frac{\partial}{\partial \theta} J(W, b) = g'(\theta)$. The special case of logistic regression has $g'(\theta) = g(z)[1 - g(z)]$

3. Update the parameters $W_i = W_i - \alpha \frac{\partial}{\partial W_i} J(W, b)$ and $b = b - \alpha \frac{\partial}{\partial b} J(W, b)$

Batch gradient descent (also called vanilla gradient descent and shown in Algorithm 6.1) computes the gradient for all batches in the data until all training points have been evaluated, and then updates the model using the computed gradients. By contrast *stochastic gradient descent* (SGD) updates the model for each training example. While batch gradient descent is computationally efficient and produces a stable gradient, this same stability can cause it to have problems with local minima. In comparison, SGD is more expensive due to the number of updates, but usually produces faster convergence. The go-to method, which is a combination of these two is called *mini-batch gradient descent*. This method slices up the data into batches, then performs the gradient computation and update for each batch. Usually, the term SGD is also used to refer to the mini batch method and, in many cases, where SGD is stated as used, it in effect refers to this model.

The usual problem plaguing gradient descent is that it can get stuck in local minima. While SGD minimizes this challenge, it does not solve the problem. *SGD with momentum* [4] attempts to fix this by adding 'inertia' to the system, in the form of a moving average of the gradients. A version of this, called *Nesterov momentum* [4] modifies the regular momentum by computing the gradient at the 'lookahead' position, where momentum is taking the system. In practice, Nesterov momentum is slightly better than regular momentum and also has some string theoretical guarantee of convergence in case of convex functions.

Gradient descent methods are highly dependent on the learning rate α and usually require expensive tuning of α. Adaptive learning rate methods have been devised to automatically and adaptively tune the learning rate, even for individual parameters. Whilst some of these methods still have hyperparameters to be optimized, usually these behave better than the raw learning rate. *Adagrad* [4] keeps a variable cache, that tracks the sum of the squared gradients per-parameter. Each parameter's learning rate is then normalized with this in every update step. This, in effect, decreases the learning rate of high-gradient parameters while increasing it for the ones with smaller gradients. Then, *RMSprop* modifies Adagrad by introducing a decay rate to the cache variable. This takes away some of the aggressiveness of the adaptation.

Adam extends RMSprop with a momentum, combining the best features of the previously mentioned algorithms. It is currently the most recommended learning rate adaptation method, usually working slightly better than RMSprop. It is important to mention that while currently gradient descent and its various extensions started getting more and more attention due to their role in training neural networks, they are not constrained to that application, and can be used for general purpose optimization.

6.1.3 Over- and Underfitting, Regularization

One of the biggest concerns while training learning machines is overfitting the training data. Overfitting occurs during training, when the machine starts to learn individual examples instead of general rules. Overfitting means that the network loses generalization power, because the fitted model trades off global accuracy for individual observation-local accuracy. Moreover, this also implies that the network starts to learn noise or irrelevant parts of the data. The main reason thereof is an excess modelling power in the learning algorithm, that makes it learn individual examples. The simplistic solution to overfitting then is to ensure to use an adequate amount of training samples for the model, i.e. to get more training data so that each parameter of the model is trained by multiple examples, or in other words, there aren't enough parameters to individually learn the examples. On the reverse, where the training data cannot be increased, this necessitates the use of a simpler model to achieve the same effect.

Fortunately, overfitting is measurable during training, by measuring the loss on a validation dataset, which is a wholly separate set of examples from the training data. The idea behind the validation dataset is that since it was never used to update the model parameters, the model could not have learned the individual examples in it, thus, the loss measured on the validation set should represent only the precision of the generalization. Overfitting is detectable when the validation loss – the loss calculated from the validation set – starts to increase, while the training loss still decreases (Figure 6.3).

The major concern for machine learning is overfitting the learned model, the condition where a statistical model begins to describe the non-valuable parts of the function, like the random error in the data, instead of describing the relationships between variables. The biggest cause of overfitting is using a very complex model. For example, an overfit regression model has too many terms for the number of observations, that the regression coefficients represent more the noise than the relationships in the data. Correspondingly, the model starts to memorize the training data observations and loses the capability to generalize a model based on them.

As one could guess, underfitting is the opposite of overfitting; the learning machine does not fit the model precisely, overlooking nuances in the data

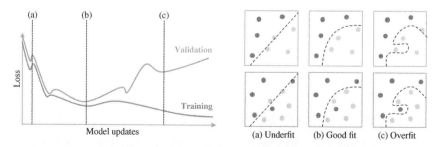

Figure 6.3 Example of under- and overfitting on a simple classification task.

which could be made into general rules. Underfitting can happen from an overtly short training, unresponsive cost functions, or the most common pitfall: overemphasizing a regularization method in the training.

Regularization methods are techniques that try to combat overfitting. They function by restricting the modelling power of the machine through additional costs on weights, or other means, such as adding noise to internal representations. Good regularization methods are sought-after, as they remove a lot of guesswork from machine learning design; the user no longer has to concern himself with model complexity and modelling power, the regularization will take care of overfitting. Regularization methods are unique to each learning algorithm and, as such, we will discuss them at their respective sections later.

6.1.4 Supervised Learning in Practice – Regression

Regression seeks to develop a tuned predictor function, the 'hypothesis' $h(x)$, on input multi-dimensional domain data x. The task of this model is to predict a scalar value $y \in \mathbb{R}$ for every input $x \in \mathbb{R}^m$. The simplest form of regression is linear regression, where the output is a linear combination of the input features:

$$h(x) = y = W^T x + b \tag{6.1}$$

where $W \in \mathbb{R}^m$ is a vector of parameters or weights that are multiplied with the features x_i before summing up the features' contributions, and b is a scalar parameter. Note, however, that many problems require more complex regression models. *Generalized linear models* or *general regression models* allow for linear combinations of higher-ordered predictors $W^T x$, e.g. where h(x) is a polynomial function of x. An even more generalized form is non-linear regression where h(x) is a non-linear function of both W and x, $h(x) = f(W, x) + b$.

The performance of the model is measured by the loss function $J(w)$. Regression algorithms usually use the linear Mean Squares Error or the (one-half) squared-error loss function of Eq. (6.2). The division by 2 is only added to simplify the derivative of the loss.

$$MSE: \quad J(w) = \frac{1}{2n} \sum_{i=1}^{n} [y_i - \hat{y}_i]^2 \tag{6.2}$$

For the algorithm to learn how to predict y, it iteratively tunes the parameters w to minimize the cost function J(w), over the multiple iterations. Note that one may be tempted to replace the iterative learning by simply deriving the normal equation, i.e. computing the weight parameters $\theta := [b, W]^T$ as $\theta = (X^T X)^{-1} X^T y$ using the inputs matrix $X = [x_1, ..., x_n]$ and the outputs vector $y = [y_1, ..., y_n]^T$. However, it is quite possible that the right 'hypothesis' h(x) is a polynomial construct of x, e.g. $h(x) = \theta^T(x * x)$, where it is complex to compute the inverse of a very large matrix. Such is the case, for example, if one attempt to predict the traffic distribution among thousands of cells over a long-time span. The iterative learning process manages the complexity that would otherwise be necessary in computing the matrix inverse.

6.1.5 Supervised Learning in Practice – Classification

Classification seeks to find one of k categories to which a given input x x belongs, so the algorithm learns a function $f : \mathbb{R}^n \rightarrow \{1, ..., k\}$ which computes, for the given input vector x, a probability distribution over the k categories. For each of the categories y, the hypothesis $h(x)$ returns a guess between 0 (No confidence) and 1 (full confidence) that states the degree to which the classifier is confident that x belongs to category y.

Here, the hypothesis is an activation function $g(z)$ that computes the probability for $z(x)$ that is a polynomial of the input vector x. There are different activation functions for classification of which the simplest is the sigmoid function of Eq. (6.3) as used for logistic regression.

$$h(x) = g(z) = \frac{1}{1 + e^{-z}}; \quad z = W^T x + b \tag{6.3}$$

Correspondingly, the cost function can still be the MSE between the predicted confidence and the true confidence value (0 or 1) for each category. Another widely used Cost Function $J(w)$ is the cross-entropy loss of Eq. (6.4) but others including the Mean Absolute Error, Hinge, Huber and Kullback–Leibler are also used for different kinds of problems.

$$\text{Cross Entropy Loss:} \quad J(w) = - \sum_{i=1}^{n} y_i \log(\hat{y}_i) + (1 - y_i) \log(1 - \hat{y}_i) \tag{6.4}$$

Besides overfitting, the biggest challenge with supervised learning is the strict dependency on labelled data. Unlike regression where the data may come from a process or population, training data for classification requires a human effort of labelling the data. This is a tedious error prone process, yet there are no simple answers for the challenge as the data must be suitably labelled for it to be used as training. However, concepts on Generative Models [5] may, in some cases, be used to reduce the burden by artificially generating new labelled data given a small set of correctly labelled data. A Generative Model is a way of learning the true data distribution of a dataset so as to generate new data points with some variations. This has proven to be useful in many cases where adequate labelled data is hard to get.

6.1.6 Unsupervised Learning in Practice – Dimensionality Reduction

Every distribution can be modelled as a combination of probabilistic generators which, if realized, can be used to understand and manipulate the distribution. Correspondingly, the cornerstones of unsupervised learning are probabilistic models defined in terms of latent or hidden variables. This information can then be used to do dimensionality reduction the training data. The generic version of these models is Factor Analysis (FA) from which specialized models including Principal Component Analysis (PCA) and Independent Components Analysis (ICA) can be derived [6].

The foregoing analysis considers a dataset D of N-dimensional real-valued vectors, $y = \{y_1, \ldots, y_N\}$; $R^y = N$ to describe the differences in the three models. As an example application of the models, Figure 6.4 illustrates the outcomes of two executions of the three models for a dataset of 10 000 points computed as a mix of a normal and an exponential distribution.

6.1.6.1 Factor Analysis
Here, each vector $y \in \mathbb{R}^N$ has latent model with model parameters $\theta = (\Psi, \Lambda)$ defined as [6].

$$y = \Lambda x + \varepsilon \tag{6.5}$$

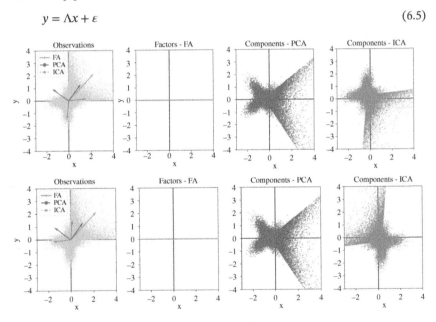

Figure 6.4 Differentiation between FA, PCA, and ICA.

where $x \in \mathbb{R}^K$ is a zero-mean unit-variance multivariate Gaussian vector whose elements are the hidden (or latent) factors, Λ is a N x K matrix of parameters, known as the factor loading matrix, and ε is a N-dimensional zero-mean multivariate Gaussian noise vector with diagonal covariance matrix Ψ. By choosing $K < N$, FA makes it possible to model a Gaussian density for high-dimensional data. For any data point, the posterior over the hidden factors, $p(x|y, \theta)$ provides a low-dimensional representation of the data (e.g. one could pick the mean of $p(x|y, \theta)$ as the representation for y).

The greatest challenge with FA is that it almost always finds factors, i.e. even when given random data FA will find factors. For example, a FA may still find apparent structure in the data generated from random numbers and it is difficult to tell if the factors that emerge reflect the data or whether they are simply the result of FAs ability to find any kind of patterns. Moreover, besides the risk of finding unnecessary factors, interpretation of the meaning of the factors is subjective. FA can identify the variables in the dataset that 'go together' but in ways that aren't always obvious. As such, interpreting what those sets of variables actually represent is up to the analyst.

6.1.6.2 Principal Components Analysis

PCA is the limited case of FA with two constraints: Firstly, it assumes isotropic noise in what is termed probabilistic PCA, i.e. each element of ε has equal variance $\Psi = \sigma^2 I$ where I is a DxD identity matrix. Secondly, taking the limit of $\sigma \to 0$ gives PCA, which is also called Singular Value Decomposition (SVD). If the columns of Λ are orthogonal, y will be projected onto the principal components $x = \Lambda^T y$.

PCA transforms the data into a new space, where each dimension is orthogonal to each other, and the components are sorted starting with those that explain the highest variance. The assumption is that the principal components with highest variance will be the most useful for solving the machine learning problems, such as predicting if a datapoint belongs to one of two classes. For example, with two classes in a classification dataset, the principal component with the highest variance would be the best feature that would allow the data to be separated. In other words, the assumption is that the interclass variance is larger than the intraclass variance. However, although logical, this assumption may not always be true. For datasets where their separation is not on the highest variance, the use of the most important components of PCA will not work and this becomes a limitation for using PCA.

6.1.6.3 Independent Components Analysis

ICA is another variant of FA, one where the factors are non-Gaussian [6]. Factors could, for example, have quasi-sparse distributions such as $p(x) = \frac{\lambda}{2} \exp\{-\lambda|x|\}$, which has a higher peak at zero and heavier tails than a Gaussian with corresponding mean and variance. Although sparsity requires non-zero probability mass at 0 and zero mass otherwise, such sparse models can be used to model real-world data sets (including images and auditory signals). This is because even although these datasets are not sparse, their structure can be modelled as combinations of sparse sources.

In general, for ICA to estimate independent sources, the sources must be statistically independent with non-Gaussian distributions, although ICA can still estimate the sources with small degree of non-Gaussianity. Moreover, the number of available mixtures N must be at least the same as the number of the independent components M and the mixtures must be (or can at least be assumed to be) linear combinations of the independent sources. Besides, there should be little or no noise and delay in the recordings. Even when all these constraints are met, ICA still suffers from two unavoidable ambiguities: (i) the order of the independent components cannot be determined (i.e. the order may change each time the estimation is executed) and (ii) the exact amplitude and sign of the independent components cannot be determined. Both these challenges are illustrated in Figure 6.4 where on the second execution, ICA returns different components from the first execution.

6.1.6.4 Implementations

There are many algorithms for performing FA, PCA, or ICA. In the case of ICA, for example, one very efficient one is the (fixed-point) *FastICA* algorithm [7] (see Algorithm 6.2), which finds directions with weight vectors $w_1, \ldots, w_n$, such that for each vector w_i, the projection $w_i^T y$ maximizes non-gaussianity. Thereby, the variance of $w_i^T y$ here must be constrained to unity which, for whitened data, is equivalent to constraining the norm of w to be unity. FastICA is based on a fixed-point iteration scheme for finding a maximum of the non-gaussianity of $W^T y$, which can be derived as an approximate Newton iteration. This can be computed using an activation function g and its derivative g' ($g = \tanh(au)$ and $g' = u \exp\left(-\frac{u^2}{2}\right)$, where $1 \leq a \leq 2$ is some suitable constant, often as $a = 1$). The basic form of the algorithm, shown in Algorithm 6.2, prevents different vectors from converging to the same maxima by *decorrelating* the outputs $w_i^T y$ after every iteration [8] (see step 4).

6.1.7 Unsupervised Learning in Practice – Clustering Using K-Means

Clustering attempts to assign each of the examples in a dataset to one of a number of groups or clusters, in a way that maximizes intra-class similarity and minimizes inter-class similarity. k-means clustering assigns the examples to one of K groups, using the groups' means as the prototypes of the respective groups. An example is assigned to a particular cluster if it is closer to that cluster's centroid than any other.

Algorithm 6.2   FastICA Algorithm Using Iterative Decorrelation

---

1. Choose an initial (e.g. random) weight matrix $\mathbf{W}$.

   Repeat until convergence:

2. Let $W^+ = E\{Yg(W^TY)\} - E\{g'(W^TY)\}W$; where $E\{.\}$ is the expectation

3. Let $= \dfrac{W^+}{\|W^+\|}$; where $\| \cdot \|$ is the norm e.g. the second norm

4. a. Let $W = \dfrac{W}{\sqrt{\|WW^T\|}}$
   Repeat until convergence
   b. Let $W = 1.5\,W - 0.5\,WW^TW$

---

k-Means finds the best centroids using the Expectation-Maximization optimization framework, by iteratively alternating between (i, expectation) assigning data examples to clusters based on the current centroids (ii, maximization) choosing new centroids as the examples (or sets of features) which are the centre of a cluster based on the current assignment of data examples to clusters.

A major concern of *k*-means and other similarity-based algorithms is that 'similarity' translates to 'closeness', which is subjective to the distance measure used. The typical measure is the Euclidean distance between points for which given a centroid k_i for a cluster *i*, the optimization objective simplifies to minimizing the squared distance to k_i. Given the required number of clusters k, the algorithm for a dataset $X = \{x_1, x_2, ..., x_m\}$ simplifies to the steps listed in and illustrated by Figure 6.5.

The main advantage of *k*-Means is speed, it requires only a few and simple computations of the distances between points and group centres, which results in linear complexity scaling O(n). The downside, however, is that one must manually select the number of classes *k*. Ideally, for proper insight into the data, the clustering algorithm should also have determined the number of groups. Nevertheless, this parameter is often required for other clustering algorithms too. Another disadvantage is that *k*-means starts with a random choice of cluster centres, and so could potentially yield very different cluster memberships on different runs of the algorithm. This lack of consistency implies that obtained results may not be reproducible.

6.1.8   Cognitive Capabilities and Limitations of Machine Learning

Combining supervised and unsupervised learning algorithms can provide a full cognitive data processing pipeline. Supervised learning offers a reasoning engine

---

Algorithm 6.3   *K*-means algorithm

1. Initialize centroids k_i (e.g. randomly sample the training data)

2. Loop on the following steps:
   a. Expectation: Calculate the distance between examples and centroids k_i and decide class memberships – assign example to the cluster with the nearest centroid.
   b. Maximization: For each cluster, re-estimate centroids as the centre of the cluster's memberships e.g. as $k_i = \frac{1}{n_i}\sum_{i=1}^{n_i} x_i$.

3. Exit if no object changes membership in the last iteration

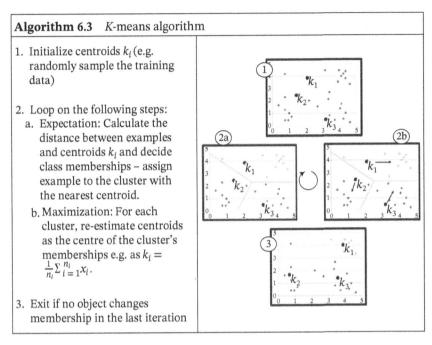

Figure 6.5 The K-means clustering algorithm and its example execution.

through its ability to organize data and make inferences, for example, regression models provide a means to answer questions related to quantity (how much or how many) and to mapping ('when' or 'where'), while classification can answer questions on relationships (i.e. question about 'which'). Conversely, unsupervised machine learning offers a strong perceptive engine (although with weak reasoning capability). Unsupervised learning can create insights by answering questions on 'what' and to some extent on 'where' and 'which'. This ability to create insights implies capability for means of making conceptualization and for improving such concepts through contextualization. Note that supervised learning also has a perceptive engine (albeit a latent one), since the learning process involves creating insights about the data e.g. by constructing probability density and mass functions of the data.

For any formed hypothesis (both for perception and/or reasoning), the training examples iteratively adjust the model parameters to improve the hypothesis by improving its accuracy. A single algorithm may not be able to offer a complete step of the cognitive pipeline, it may require a combination of multiple algorithms. There are multiple alternative algorithms for both supervised and unsupervised learning with different cognitive capabilities and mechanisms for minimizing or addressing the above challenges. Examples of these alternatives are illustrated in

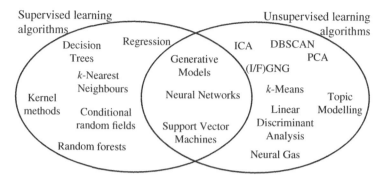

Figure 6.6 Supervised and unsupervised learning methods and algorithms.

Figure 6.6 with distinction of their respective usage as supervised or unsupervised learning solutions. The differences and similarities amongst these algorithms are discussed in many reference machine learning texts including [9].

6.1.9 Example Application: Temporal-Spatial Load Profiling

The possible range of applications for machine learning is wider than those listed in Table 6.1. They can be applied to classification and clustering related problems such as anomaly detection; regression problems such as cell grouping and prediction e.g. for cell load or user velocity as well as for general estimation problems which may also encompass classification and regression besides other challenges like compression and filtering. Detailed descriptions of such applications are the subjects of discussion for Chapters 6–9. Here, the discussion only shows a simplified case that exemplifies the line of reasoning behind the application of machine learning techniques for network automation challenges.

Figure 6.7 exemplifies a simplified structure of a Mobility Pattern Profiles (MPPs) use case. It describes a supervised learning solution for predicting future cell load profiles given data of past cell load statistics, often referred to as Mobility

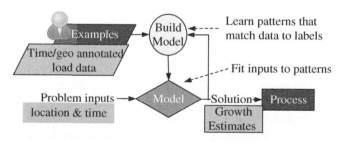

Figure 6.7 Supervised learning for temporal-spatial load profiling.

Pattern Prediction. The dataset examples are the timeseries with entries of time- (and preferable geo-) annotated cell load data, which are used to build the model of cell load variation. The problems of interest are time and location for which the load needs to be estimated. The solution takes the location-time coordinate as the input which it applies to the model to predict the likely cell load, which it then uses as input to respective process like a network resource optimization application.

6.2 Neural Networks

A neural network is a computational model that uses several simple units (artificial neurons, nodes), each connected to others and transmitting a signal that is a function of a combination of its multiple incoming signals. A single neuron, shown in Figure 6.8a, takes an input array $x = [x_1, x_2, ..., x_n]$ (plus a bias term b) and outputs $h_{W,b}(x) = f(W^T) = f(W_i x_i + b)$. The function $f : \mathbb{R} \to \mathbb{R}$ is the activation function.

The neuron learns by changing the parameters W and b. In the above model, a single neuron, without a nonlinearity, realizes the input–output mapping of a linear regressor. The neural network is the combination of many such single units set up in a way that the output of one unit serves as the input to others. Usually, neurons are organized into groups called layers, so that connections run between neurons in neighbouring layers, but not between neurons in the same layer. The neural network can be imagined as a computational graph, where neurons are the vertices, and values (activations) propagate along the directed graph edges. A network is called feed-forward, if there are no cycles in the graph.

An example feed-forward neural network is shown on Figure 6.8. Therein, each neuron is represented by a circle, whilst directed one-way connections are shown as arrows. This network has four layers: an input layer at which the input values

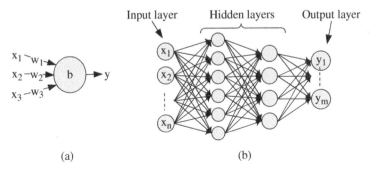

(a) (b)

Figure 6.8 A neural unit and an example neural network with two hidden layers.

are applied, an output layer where the network's computed output is observed and two hidden layers which compute intermediate values.

6.2.1 Neurons and Activation Functions

The simplest and oldest neuron type is the perceptron, which takes a binary input vector x and using a threshold b, computes a binary output y according to Eq. (6.6). The neuron learns by changing the values in the parameter vectors w & b to minimize the error made in computing the estimate y for each input x.

$$y(x) = \begin{cases} 0 \ if \ \sum_j w_j x_j \leq threshold, b \\ 1 \qquad\qquad otherwise \end{cases}$$

$$vectorized \ as \ y(x) = \begin{cases} 0 \ if \ w.x + b \leq 0 \\ 1 \ if \ w.x + b > 0 \end{cases} \tag{6.6}$$

The challenge with the perceptron is that small changes in w or b can lead to very large changes in y – according to $\delta y = f(\delta w) + f(\delta(b))$. At the discontinuity $w.x = -b$, any small change in w or b can lead to y switching from 0 to 1 or back. This is especially problematic as perceptrons are not differentiable on the whole activation range, which makes them unusable with gradient-descent-type optimization. Correspondingly, modern neural networks use other nonlinearities for the neurons, such as the sigmoid neuron and the hyperbolic tangent (tanh) functions shown in Figure 6.9 and Table 6.2 .

Given the polynomial $z = W \cdot X + b = \sum_j w_j x_j + b$, the sigmoid neuron computes y as a continuous valued function on the range [0,1] as shown in Figure 6.9b. The tanh neuron uses the hyperbolic tangent (tanh[z]) function of Figure 6.9c, which is a rescaled version of the sigmoid with an output range of [−1,1] instead of [0,1]. A more recent activation function is the rectified linear function (ReLu), which often works better in practice for deep neural networks. The ReLu activation function is different from sigmoid and tanh neurons (see Figure 6.9) in that it

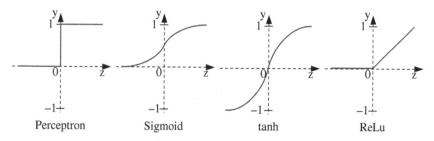

| Perceptron | Sigmoid | tanh | ReLu |

Figure 6.9 The perceptron, sigmoid, tanh, and ReLu activation functions for the polynomial $z = w. x + b$.

Table 6.2   The sigmoid, tanh, and ReLu activation functions and their derivatives.

Function	Definition, f	Derivative, f'
Sigmoid	$f(z) = \sigma(z) = \dfrac{1}{1 + e^{-z}}$	$f'(z) = f(z)[1 - f(z)]$
Tanh	$f(z) = tanh(z) = \dfrac{e^z - e^{-z}}{e^z + e^{-z}}$	$f'(z) = 1 - [f(z)]^2$
ReLu	$f(z) = max(0, z)$	$f'(z) = \begin{cases} 0 & z < 0 \\ 1 & otherwise \end{cases}$

is not bounded or continuously differentiable. Instead, it is piece-wise linear which saturates at exactly 0 whenever the input z is less than 0.

Note that differentiability has been an important criterion for selecting functions for neurons, since the training process involves evaluating the gradient of the activation functions to compute the effects of different parameters on the output. The gradients of the sigmoid and tanh functions are given in Table 6.2. Although the ReLu gradient is undefined at $z = 0$ from a mathematical standpoint, this doesn't cause problems in practice because the gradient can be forced to be either 0 or 1 there.

6.2.2   Neural Network Computational Model

Consider a 3-input 1-ouput neural network with a single 3-unit hidden layer as shown in Figure 6.10. For each layer, the number of neurons accorded to the layer does not include the bias unit. The number of layers in the network is denoted as n_L ($n_L = 3$ in Figure 6.10), the layers as L_l starting with the input layer as $l = 1$ (so that the output layer is L_{n_L}) and the number of neurons in layer l (excluding the bias unit) as s_l.

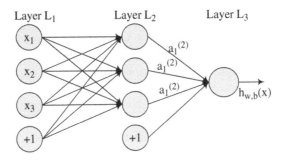

Figure 6.10   A 3-unit input, 3-unit single hidden layer, 1-unit output NN.

The neural network has parameters $[W, b] = [W^{(1)}, b^{(1)}, W^{(2)}, b^{(2)}]$, where weight $W_{ij}^{(l)} \in W^{(l)}$ and bias $b_i^{(l)} \in b^{(1)}$ are the respective parameters associated with the connection between unit j in layer l and unit i in layer $l+1$. In the example, $W^{(1)} \in \mathbb{R}^{3 \times 3}$ and $W^{(2)} \in \mathbb{R}^{1 \times 3}$. The bias units don't have inputs or input connections, as they always output a value +1.

The value $a_i^{(l)}$ denotes the activation (or simply the output value) of unit i in layer l, which for $l = 1$, implies that $a_i^{(1)} = x_i$ the ith input. $z_i^{(l)}$ denotes the total weighted sum of inputs to unit i in layer l, including the bias term. Expressed in terms of the outputs of the previous layer, for each unit i in layer $l+1$,

$$z_i^{l+1} = \sum_{j=1}^{s_l} w_{ij}^l a_j^l + b_i = w_i^l.a^l + b_i^l \tag{6.7}$$

where the vectorized product is an element-wise inner product of the vectors w_i^l and a^l. Correspondingly, the activation for z_i^{l+1} will be $a_i^{l+1} = f(z_i^{l+1})$. This computation can be vectorized such that with $a^1 = X$, for each layer $l+1$, $l = 1, 2, \dots, n_L - 1$,

$$z^{l+1} = W^l a^l + b^l \text{ and } a^{l+1} = f(z^{l+1}) \tag{6.8}$$

For a fixed setting of the parameters W, b, this process of computing polynomials and the corresponding activations for each unit is called *forward propagation*. It computes the hypothesis $h_{w,b}(x)$ of the network as a real number given by $h_{W,b}(x) = a^{n_L} = f(z^{n_L})$. For the example network in Figure 6.10, the hypothesis will be Eq. (6.9):

$$h_{W,b}(x) = a^3 = f(W^2 a^2 + b^2) \tag{6.9}$$

Note that organizing the parameters in matrices and using matrix-vector operations enables the use of fast linear algebra routines and hardware accelerators to quickly perform calculations in the network.

6.2.3 Training Through Gradient Descent and Backpropagation

Assuming a supervised learning problem, the usual neural network is trained using a set of training data $\{(x^1, y^1), (x^2, y^2), \dots, (x^m, y^m)\}$ and utilizing gradient descent with an appropriate cost function e.g. from amongst those in Section 6.1.4. For gradient-descent optimization, given the computation of the activations for each node in a forward pass, it is necessary to compute a 'loss term' $\delta_i^{(l)}$ that measures how much that node was 'responsible' for any errors in the output. In other words, one needs to determine how much the cost function, $J(W, b)$, changes for each change in the parameter (w or b) at each given node in the neural network, which is the partial derivative $\frac{\partial}{\partial [W_i^{(l)} \text{or } b_i^{(l)}]} J(W, b)$ (where $J(W, b)$ is the loss function).

Table 6.3 Implementations of backpropagation algorithm.

Step	Action	Non-vectorized form	Vectorized form
1.	Compute loss term for the output layer:	$\delta_i^{nl} = \dfrac{\partial}{\partial z_i^{(nl)}} \dfrac{1}{2} \|h_{W,b}(x) - y\|^2$ $= -(y_i - a_i^{(nl)}) \cdot f'(z_i^{(nl)})$	$\delta^{nl} = -(y - a^{nl}) \cdot f'(z^{nl})$
2.	Compute loss terms for the previous layers:	$\delta_i^l = \left(\sum_{j=1}^{s_{l+1}} W_{ji}^{(l)} \delta_j^{l+1} \right) f'(z_i^{(l)})$	$\delta^l = ((W^{(l)})^T \delta^{(l+1)}) \cdot f'(z^l)$
3.	Compute the desired partial derivatives as:	$\dfrac{\partial}{\partial W_{ij}^{(l)}} J(W, b) = a_j^{(l)} \delta_i^{l+1}$ $\dfrac{\partial}{\partial b_i^{(l)}} J(W, b) = \delta_i^{l+1}$	$\nabla_{W^{(l)}} J(W, b) = \delta^{(l+1)} (a^{(l)})^T$ $\nabla_{b^{(l)}} J(W, b) = \delta^{(l+1)}$
4.	Update the parameters:	$W_{ij}^{(l)} = W_{ij}^{(l)} - \alpha \dfrac{\partial}{\partial W_{ij}^{(l)}} J(W, b)$ $b_i^{(l)} = b_i^{(l)} - \alpha \dfrac{\partial}{\partial b_i^{(l)}} J(W, b)$	$W^{(l)} = W^{(l)} - \alpha \nabla_{W^{(l)}} J(W, b)$ $b^{(l)} = b^{(l)} - \alpha \nabla_{b^{(l)}} J(W, b)$

For nodes in the last layer (output layer nl), the 'loss term' $\delta_i^{(nl)}$ can be directly measured by the difference between the network's activation and the true target value. The measured difference is then used to define $\delta_i^{(nl)}$. For units in the previous layers, $\delta_i^{(l)}$ can be computed, based on a weighted average of the loss terms of the nodes in the next layer. By progressing through the network from the last layer to the first, all loss terms can be computed. This is called backpropagation, i.e. the loss is propagated backwards through the neural network.

Assuming a sigmoid function for the activation function (as defined in Table 6.2) and taking the (one-half) squared-error cost function, the backpropagation algorithm is as shown in Table 6.3. This is executed after performing a feedforward pass to compute the activations for layers L_2, L_3, and so on up to the output layer L_{n_l}. An interesting thing to note when using sigmoid activations: In steps 2 and 3, $f'(z_i^l)$ needs to be computed for each value of i. Since a_i^l can be stored during forward pass through the network, using the expression for $f'(z_i^l)$ in Table 6.2, it is possible to compute this as $f'(z_i^l) = a_i^l(1 - a_i^l)$.

The parameter $W_{ji}^{(l)}$ and $b_i^{(l)}$ are initialized to random values, say according to a Normal distribution over $(0, \varepsilon)$ for some small ε, say 0.01. Otherwise, when initialized with identical values (e.g. to all 0's), all the hidden layer units will learn the same function of the input (i.e. $W_{ji}^{(l)}$ will be the same for all values of i, so that $a_1^{(2)} = a_2^{(2)} = a_3^{(2)} = \ldots$ for any input x). The proper initialization of weights is especially critical in deeper networks and has seen much research in recent years, with more sophisticated initialization algorithms developed [10].

To train the neural network, gradient descent optimization is applied, usually, SGD as previously discussed in Section 6.1.2. Given a dataset, the set of training examples is divided into multiple subsets (minibatches that are usually random sampled) so that the parameter weights are updated after each epoch/minibatch. Assuming that dW and db respectively are the loss matrix and vector for, and with the same dimensions as W and b, a single iteration of SGD with epoch size p is implemented as in Algorithm 6.4.

Algorithm 6.4 Stochastic Gradient Descent Algorithm

Given an estimate $\hat{y}$

1. Set $dW^l := 0$, $db^l := 0$ (i.e. a matrix/vector of zeros) for all l.

For i in 1 to number of epochs,

2. For l in 1 to p,
   2.1. Backpropagate and compute $\nabla_{W^l}J(W, b; x, y)$ and $\nabla_{b^l}J(W, b; x, y)$
   2.2. Set $dW^l := dW^l + \nabla_{W^l}J(W, b; x, y)$
   2.3. Set $db^l := db^l + \nabla_{b^l}J(W, b; x, y)$
3. Update the parameters using the average loss across all training examples:
   $W^l = W^l - \alpha \left[\frac{1}{m}dW^l + \lambda dW^l \right]$ and $b^l = b^l - \alpha \left[\frac{1}{m}db^l \right]$

This is repeated iteratively to reduce the cost function, $J(W, b)$ and train the neural network. The main reason to use SGD instead of regular gradient descent is to regulate memory imprint. Specifically, hardware accelerators usually have a quite constrained working memory, that storing the gradients for the whole training dataset would not fit into memory. Splitting the input set into smaller batches and propagating one chunk at a time limits the required memory, regardless of the size of the training dataset.

6.2.4 Overfitting and Regularization

As stated earlier, the main cause of overfitting is excess modelling power in the learning algorithm. For neural networks, this may be due to too many layers, or number of nodes within the layers. The biggest problem is that it is very hard to guess the complexity of the learning task and, correspondingly, the optimal network topology. As such, instead of focusing on rules for optimal

Table 6.4 Computing partial derivatives with regularization.

Step	Non-vectorized form	Vectorized form
3. Compute partial derivatives as:	$\dfrac{\partial}{\partial W_{ij}^{(l)}} J(W,b) = a_j^{(l)} \delta_i^{l+1} + 2\lambda W_{ij}^{(l)}$ $\dfrac{\partial}{\partial b_i^{(l)}} J(W,b) = \delta_i^{l+1}$	$\nabla_{W^{(l)}} J(W,b) = \delta^{(l+1)} (a^{(l)})^T + 2\lambda W^{(l)}$ $\nabla_{b^{(l)}} J(W,b) = \delta^{(l+1)}$

topologies, research has focused on explicit methods to force regularization in networks. This has the benefit that 'oversized' networks become valid for simpler tasks, which removes the need for fine-tuning network topologies for different applications.

Regularization can be done through vastly different methods. One of the earliest and simpler methods is regularization of the network through regularization of individual weights, by attaching a loss to the amplitude of the weight, thereby forcing it to stay close to 0. This approach is called *weight-decay* and can be done through the adjustment of the cost function. Taking the (one-half) squared-error cost function for example, the overall cost function for the training set of m examples is:

$$J(W,b) = \left[\frac{1}{m} \sum_{i=1}^{m} J(W,b;x^{(i)},y^{(i)}) \right] + \frac{\lambda}{2} \sum_{l=1}^{n_l-1} \sum_{i=1}^{s_{l+1}} \sum_{j=1}^{s_{l+1}} (w_{ij}^{(l)})^2 \qquad (6.10)$$

where the objective function $J(W,b;x^{(i)},y^{(i)}) = \frac{1}{2} \|h_{W,b}(x^{(i)}) - y^{(i)}\|^2$ will remain convex as before. The total cost includes the average sum-of-squares error (the first term) and a regularization term (also called a weight decay term) that prevents overfitting by enforcing small magnitudes for the weights. The relative importance of the two terms is controlled using the weight decay parameter λ.

With weight decay, the partial derivatives in step 3 changes as shown in Table 6.4. The interpretation of the gradient remains the same as before, i.e. a weighted average of votes by the data points. However, the regularization term also exerts a force (that is proportional to the magnitude of the coefficients) that pushes the coefficients closer to 0. The parameters can correspondingly be updated as in step 4 in Table 6.3.

Dropout [11] is a recently discovered technique for regularization in neural networks. It functions by randomly masking neurons in each batch, restricting them from contributing to the output or getting updated in that batch. By randomly turning off parts of the network, dropout makes it hard for the network to form strong inter-dependencies between weights, resulting in a regularization effect. Dropout is preferred for deeper networks; however, weight-decay has not lost its relevancy, and it is still just as widely used.

6.2.5 Cognitive Capabilities of Neural Networks

Neural networks can be trained to achieve general purpose cognition, to accomplish any of the four tasks of cognition, e.g. they can be trained:

i) to conceptualize sensory data e.g. in computer vision to detect objects
ii) to contextualize data, e.g. in computer vision to recognize and differentiate objects
iii) to organize and relate data e.g. to learn sequences of words, and
iv) to makes inferences from data e.g. for diagnostics

Although neural networks are general computation mechanisms, particular architectures thereof have been found to perform better than others on certain problems. For the specific architecture, the degree to which a neural network can accomplish these tasks highly depends on its design, i.e. on the selection of its hyperparameters, such as topology, or activation functions used. Consequently, the use of neural networks to undertake a cognitive task is more of an art, and study of neural network architectures for the different cognitive tasks is an active field that is still evolving. The most successful results have used deep neural networks, which apply a large number of hidden layers (as discussed in Section 6.3).

6.2.6 Application Areas in Communication Networks

In communication systems, neural networks can be used both for supervised and unsupervised learning for different network related challenges. They could, for example, be used to make perceptions of network data to evaluate states of network devices and behaviour in order to detect and predict particular events. They can help organize and relate network data, for example, to distinguish network scenarios. Or, they can be used to make inferences based on network data, for example, to determine when and where a new network node may be required to be added to the network depending on the statistics of growth of network utilization in a region.

The applicable networks-related tasks can be grouped among:

– Identification problems e.g. for network profiling of traffic, user behaviour, resource utilizations, etc. to identify network-related events;

- Pattern recognition e.g. for classification of coverage areas as rural, urban, or otherwise; or for classification of users as static, slow, or fast.
- Estimation problems, which through regression determine/estimate network behaviour and events at particular points in time and space. This may be used, for example for requirements planning and capacity planning applications.

6.3 A Dip into Deep Neural Networks

Deep Learning (DL) may be defined as the art of learning multiple levels of representation and abstraction that together make sense of the data. Today, DL almost exclusively involves deep neural networks – so much so that the two expressions became almost interchangeable. This section will mostly discuss deep neural networks for a primary audience of researchers and engineers who are familiar with common machine learning techniques – including neural networks – but have not yet worked with deep learning. This section will not discuss the detailed mechanics of machine learning and neural networks. For these details, such as discussions on backpropagation and gradient descent, the reader is referred to Section 6.2.3.

6.3.1 Deep Learning

The quick spread of use and acknowledgment of deep learning cannot be tied to a single invention, but rather to multiple smaller factors that, together, enabled the training and use of complex learning machines that are today referred to as deep learning. Correspondingly, there is a lot of misuse of the term, stemming from misconceptions about what does and does not constitute as deep learning. The origins of the term are inherently tied to Deep Neural Networks (DNNs), the topology of complex neural networks that have many hidden layers. However, there is no breaking point, no set number of hidden layers above which one can call a network deep. For some deep neural network structures, the number of hidden layers keep on increasing, yet, as is shown later, there are some the deep network architectures in use today do not have more than a few hidden layers in the traditional sense. So, instead of defining deep learning by the applied number of hidden layers, it is necessary to find a definition that is also applicable outside the realm of neural networks and instead focusses on what is achieved by deep learning.

The first deep neural networks performed better than their non-deep predecessors because they could learn complex, hierarchical structures of rules or features present in the training data. This deep understanding is what defines whether a system achieves deep learning. Therefore, deep learning implies that the machine:

- Uses data to autonomously form a comprehensive model of a system with:
- rules-upon-rules, a hierarchical problem where the solutions (outputs) of simpler, smaller recognition tasks are used in higher levels to recognize more complex structures, culminating in:
- models that achieve close-to-human or better performance.

In other words, deep learning involves learning multiple levels of representation and abstraction of data and combining the different abstractions to make sense of the systems hidden in the data.

Deep learning architectures have been developed for both supervised and unsupervised learning. Many of these have also been extended for use in reinforcement learning, which is discussed Section 6.4. The most utilized deep learning architecture types for supervised learning are convolutional and recurrent neural networks. Both types expand on the basic neural network concept not only in the depth of cognition they can achieve, but also in an added aspect in which they perceive data. Convolutional Neural Networks (CNNs) account for the spatial properties of the information and can recognize the same structure in multiple positions in the input. Long-Short Term Memory Recurrent Neural Networks (LSTM RNNs) on the other hand pay attention to the temporal context of the information and are able to recognize types and events based on a history of observations. Autoencoders are the most widely used unsupervised deep learning architecture and implement true unsupervised learning, realizing a mix between feature extraction and clustering. Autoencoders are the simplest of the DNN architectures, having no real additional techniques compared to regular neural networks besides having many hidden layers. However, there is a reason why deep autoencoders were not used earlier; the vanishing gradients problem.

6.3.2 The Vanishing Gradients Problem

When adding hidden layers to the traditional neural network that utilizes sigmoid activations, and perhaps random initialization, gradient descent starts to break down. The network becomes slow to train, requiring more epochs to converge, or stops converging altogether. This phenomenon, called the vanishing gradients problem, is a prevalent constraint on the achievable depth of neural networks.

As the name implies, the problem comes from the amplitude of gradients in deep networks. When using sigmoid, or any other activation function that is continuously differentiable and limited to a range, a neuron with activations close to the limits, and all other previous neurons that receive the backpropagated gradient from this neuron, will have small gradients during backpropagation. Sigmoid activation functions are especially prone to this, as the derivative of the standard sigmoid function never exceeds 0.25. Backpropagating through multiples of these activation functions means that the gradient lowers exponentially, which in turn implies that the neurons in the earlier layers will not converge quickly (or at all). Non-learning early layers disrupt the functioning of the whole deep network, thus achieving less accuracy in these cases than their shallower counterparts. The opposite of this effect, gradient explosion, can also occur in a similar manner, however, this phenomenon is more likely with unconstrained polynomial or exponential activations, or extremely large initial weights, both of which are seldom used in neural networks. Gradient vanishing from the nonlinearities can be combated by the Rectified Linear Unit (and similar) activation functions, where the derivative of the function is fixed, and at least on one half of the non-limited activation range is also equal to 1, thus it is less likely that the gradients are reduced when backpropagated through it.

A second reason of experiencing gradient vanishing (or explosion) can stem from the initial value of weights. Considering a simple gaussian random initialization, the expected aggregated activation value in the first epoch will be the sum of the expected values of many gaussians, and as such, will be dependent on the size of the layer and the exact parameters of the random generation process. To combat this and move the expected activations in every layer to around the same value, several initialization methods were proposed. One such method is the Xavier-initialization [10], which, after randomly generating initial weights, normalizes the weights so that the probable output of each fully connected layer is around a 0 mean and has 1 standard deviation.

A more general and arguably the most effective method to combat gradient vanishing is to use weight-specific adaptive learning rates. RMSprop, ADAM, and other optimization methods utilize per-weight adaptive learning rates, trying to speed up training by equalizing the amount by which each weight is updated with each batch. This process directly cancels vanishing gradients on those parameters that would otherwise be affected, regardless of initial values or activation functions.

Although the above techniques can mitigate the vanishing gradients problem, one cannot just freely stack layers upon layers in a network; deep neural networks have large computational requirements, require large amounts of data, and necessitate state-of-the art hardware to be used effectively. These constraints and enablers will be discussed in the next section.

6.3.3 Drivers, Enablers, and Computational Constraints

The increased cognitive capabilities of deep neural nets come at cost: with great cognitive powers come great computational requirements. For deep neural networks, the drivers, and constraints are intertwined, creating a complex landscape through which one must navigate when aiming to develop these algorithms.

6.3.3.1 Computational Power

Parallel to the development of better ways to train neural network algorithms, machine learning's history is also closely tied to computational power. As the basic steps to train neural networks through backpropagation and gradient descent are very atomic algebraic tasks, there is little room to algorithmically optimize these calculations. However, as these tasks are highly parallelizable, there are significant opportunities to speed up the computation through implementations using hardware capable of massively parallel computations. This has created a unique two-step cycle of progression for DNNs:

- Neural networks are improved, allowing for deeper, more complex networks to be trained but with major increases in computational requirements.
- Hardware computational and storage capacity is improved, allowing for the complex DNNs to be trained and tested in humanly acceptable timeframes, e.g. purpose-built hardware and new implementations emerge to further speed up computations.

6.3.3.2 Timing Constraints

Current improvements sometimes achieve only a few 1/10 of a percent increase in accuracy at the cost of a tenfold increase in training time. However, these seemingly small increments in accuracy amount to a large forward-step in cognitional power, and a proportionally large decrease in error rates. Deep neural networks can have two types of constraints for processing time;

- A hard limit at inference runtime, so that the network can process the constant stream of information at the rate it is arriving, such as the case with object recognition in video processing.
- A softer limit at training time, so that the network can be trained in a reasonable timeframe.

The time it takes to go from concept to testing prototypes, mainly influenced by the models' training time, is critical in deep learning research. Although most common architectures can nowadays be trained on desktop computers, the need for speed and simplicity implies that buying expensive hardware dedicated to deep learning is not a waste. This is the reason why leading institutions in the field utilize huge supercomputers; being able to validate a concept quickly without spending much effort on optimization is invaluable in the deep learning race.

6.3.3.3 Quantity of Data

The companies most invested in deep learning are those that handle large amounts of data, usually in the form of images, videos, or text, where the automation of tasks previously solved by humans would mean a huge decrease in cost and a large increase in capabilities. This large amount of information, however, is not only a burden; the quantity and quality of available data is what enables the training of the learning machines more than any hardware or algorithmic improvement could.

6.3.4 Convolutional Networks for Image Recognition

Automated image recognition, such as image classification or object localization, is one of the most sought-after machine learning tasks today, stemming from the fact that a disproportionately large part of the hosted data on the internet today consists of images or videos. These tasks are unique in the deep learning world because they require the machine to recognize features in any part of the image and rotated/skewed in unusual ways. This can be achieved by applying convolution on the image.

Convolution is originally defined as a mathematical operation on two functions, which computes the amount of overlap of one function as it is shifted over the other function. In two-dimensions, convolution may be visualized as sliding one function, g, on top of another, f, while, at each shift, point-wise multiplying the functions and adding the products as illustrated by Figure 6.11. The function g may be referred to as a filter that detects the presence of specific features or patterns in the function f.

The correct recognition of objects in images also requires the detection of certain structures on multiple abstraction levels, the exact property that validates the use of deep learning for these tasks. By processing the previous layer's filtered output, each convolutional layer adds an additional level of abstraction to the network. This structure mimics the structure of the mammalian visual cortex, which also

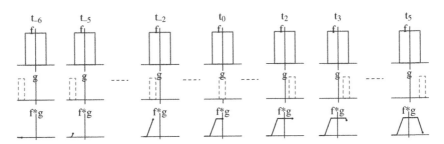

Figure 6.11 Convolution of two functions f and g.

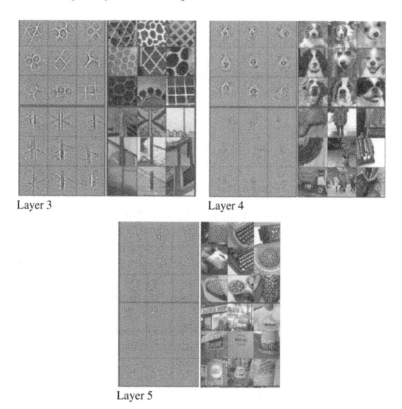

Layer 3 Layer 4

Layer 5

Figure 6.12 Recognition of increasingly complex structures in images [22].

shows similar composition of neurons activating on simple, and others on more complex features in what the eye sees (Figure 6.12).

To realize image recognition, the machine needs to exert a combination of both sensitivity and insensitivity to two-dimensional features, depending on what the detection of the object requires. CNNs implement this spatial awareness with two unique layer types; convolutional layers (containing filters) and pooling layers, arranged in an alternating sequence, as shown on Figure 6.13. Optionally, there can be a few fully connected layers attached at the end of the network running a classification on the internal representation of the convolutional layers. Going from the front layers close to the input towards the deeper layers, the filters learn to recognize more and more complex, higher-level structures.

A logical assumption is that the more layers that are stacked, the more the network will be capable of an in-depth understanding of what it sees. This trend leads to the creation of some of the deepest neural networks known today, topological monsters containing more than 100 layers of convolution and pooling. These

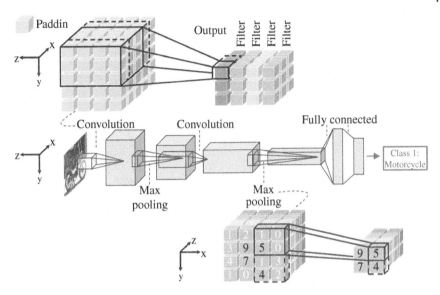

Figure 6.13 An example Convolution Neural Network highlighting the different layers.

networks would not be trainable with only the basic backpropagation techniques, succumbing to vanishing or exploding gradients quickly. To combat these effects, different techniques are employed in many aspects of the network and the training itself.

CNN layers capture the spatial and temporal dependencies in images through the application of learned convolutional and static pooling filters. The layers operate on three-dimensional tensors as both inputs and outputs, where the first two dimensions (x and y, or width and height) correspond to the plane of the original image and define the location of activations for the same filter output, while the third dimensions (z, or depth) is the differentiator between the outputs of the different filters, as seen in Figure 6.13.

6.3.4.1 Convolution Layers

Convolution layers are made up of small filters (such as a $3 \times 3 \times Z$ grids), each reacting when detecting a unique structure. This detection is done through the usual matrix multiplication also used in fully connected layers; the filters are learned weights, which are multiplied by a part of the input tensor and then summed to form a single output value. The filters receptive area extends through the whole depth (z) of the previous output tensor but does not cover the other two dimensions fully. The top of Figure 6.13, for example, highlights convolution on a $3 \times 3 \times 4$ tensor with a padding of 1, using four different $3 \times 3 \times 4$ filters with a stride of 1.

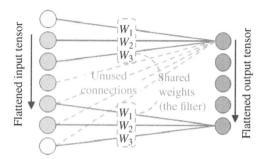

Figure 6.14 Convolutional layer viewed as a fully connected layer.

To process the whole input through a single filter, the filters scan the output of previous layer in the x and y dimensions and propagate their localized activations for each scanned position to the next layer. This notion of scanning the input through a filter is the realization of the convolution operation, hence the name of the layer. The size of the output of these layers is governed by multiple parameters: the actual filter size, the number of filters, the stride (the step size of the scan) and the amount of zero padding on the sides of the previous output.

Note that the convolution operation could also be realized with a simple fully connected layer. In fact, convolutional layers can be viewed as fully connected layers that have connections which are turned off, and others that share weights across the layer, as depicted on Figure 6.14. This view highlights the original aim of the convolutional layer; to reduce the number of weights in the network by leaving out often unused parts, so that deeper, more complex networks (specifically for image processing) could be fit into state-of-the-art hardware.

6.3.4.2 Max Pooling

The convolutional filters are receptive to spatial structures, realizing a sensitivity to position in their inputs, and encode the position of the sensed structures in their output. Pooling layers undertake a conceptually opposing task; their job is to mitigate the strong sensitivity of the convolutional layers with regard to exact position or orientation. They do this with a scanning compression method similar to the filters in the convolution layer, but only propagating the largest activation value from a small area of the input (max-pooling). The compression areas are restricted to a depth of 1, and are usually small, 2×2 or 3×3 in the x–y plane. The stride of the scan can either be set to create overlapping or non-overlapping scans, however, usually non-overlapping pooling is used. Pooling enforces sparsity in the inner representation of the network, acting as a regularization, and compresses information. Overall, this regularization manifests as an insensitivity to the exact orientation, skew or position of the objects in the input image, which is a sought-after property for robust and precise recognition. An example of max pooling is highlighted on the bottom of Figure 6.13 with a 2×2 compression area and no overlap.

Max pooling also combats the vanishing gradient problem by only routing the gradient to the most activated position in the previous layer. This avoids splitting the gradient into multiple parts which, in turn, avoids the repeated division present in fully connected layers. Besides, the nonlinear activation layers use ReLUs, which, as discussed previously, also mitigate the vanishing gradient problem. However, both components are chosen not only for these effects; their functions are very simple to calculate for both forward and backward propagation, speeding up the training of the whole network. More advanced techniques directly aimed at helping deep networks train are also deployed, such as the cross-cutting connections that 'jump' over multiple layers in Microsoft's ResNet [12].

6.3.5 Recurrent Neural Networks for Sequence Processing

Recurrent networks extend on the basic neural network concept by incorporating the temporal context of the information in sequences, such as written text or time-series. Traditional neural networks have a one-to-one assignment between input and output observations, while recurrent networks can have multiple mappings:

- one-to-many: as in image captioning, to add a descriptive sentence to images
- many-to-one: as in sentiment analysis, to categorize opinions expressed in a text.
- many-to-many: either un-synced as in machine translation to translate one sentence to another; or synced as in object localization to frame different objects in a video.

Some versions can work with variable length sequences, on the input or output side, which dispels the need for artificially generating equal length sequences for processing.

RNNs utilize a hidden state h_t, which is also the output of the RNN cell, that feeds back to the input at every step of the sequence, retaining a memory of all previous states. An RNN cell, illustrated by Figure 6.15a may be seen as a whole layer in the traditional sense, incorporating multiple neurons. In this view, h_t includes all atomic states from every neuron and is also fed back to every one of them, so that each neuron gains access to all neuron's hidden states in the same cell. This hidden state can be described by Eq. (6.11):

$$h_t = \phi(Wx_t + Uh_{t-1}) \tag{6.11}$$

where h_t is the hidden state at time t, x_t is the input vector, W and U are weight matrices and ϕ is the activation function (usually tanh, sigmoid, or ReLU). At every

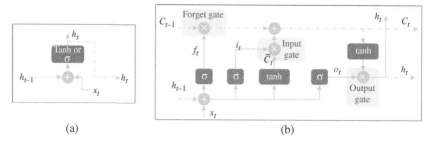

(a) (b)

Figure 6.15 RNN cells: (a) the Simple RNN cell and (b) the LSTM cell.

timestep, the recurrent cell not only receives the input of the next step (x_t), but also gets the previous output concatenated to it. This creates a feedback loop, enabling the network to make decisions based not only on the current input, but also on past states.

Typically, RNNs have a relatively low number of layers (cells), compared to modern convolutional networks. Why then are RNNs categorized as deep networks? The trick is to imagine an RNN as a network with the same cell repeated over and over for each step of the input sequence. The depth of recurrent networks comes from the fact that this 'unfolded' network can be very large, depending on the length of the input sequence. Due to this, in most cases, it is wise to limit the number of stacked (the actual) recurrent layers in a neural net, since the sequential nature of the network already introduces vast complexity to the system.

RNNs have lost favour mainly because of exploding and vanishing gradients. Exploding gradients imply that the algorithm assigns a stupidly high importance to the weights, without much reason. Conversely, vanishing gradients are due to the functional depth of the unfolded networks which makes the values of a gradient to be too small that the model stops learning or takes far too long to learn. The exploding gradients can be solved by truncating or squashing the gradients, but the vanishing gradients problem has proven to be much harder to solve, which is the reason why RNNs have in practice been replaced by Long Short-Term Memory architectures.

6.3.5.1 Long Short-Term Memory

A successful architecture that overcomes the vanishing gradients problem is the Long Short-Term Memory (LSTM) network. LSTM cells deal with vanishing gradients by introducing an internal memory that is not the same as the cell's output. This internal memory is governed by input-, output-, and forget-gates in the cell. Assuming, that W_i, W_o and W_f are the weight matrices, while b_i, b_o and b_f are the bias terms associated with the input, output and forget gates, the inner workings of an LSTM cell can be described by the equations in (6.12):

$$\left.\begin{array}{r} f_t = \sigma(W_f \cdot [h_{t-1}, x_t] + b_f) \\ i_t = \sigma(W_i \cdot [h_{t-1}, x_t] + b_i) \\ \tilde{C}_t = \tanh(W_C \cdot [h_{t-1}, x_t] + b_C) \\ C_t = f_t * C_{t-1} + i_t * \tilde{C}_t \\ o_t = \sigma(W_o \cdot [h_{t-1}, x_t] + b_o) \\ h_t = o_t * \tanh(C_t) \end{array}\right\} \qquad (6.12)$$

As shown in Figure 6.15b, in addition to the hidden state h_t, an LSTM cell also has an internal state C_t, which works as the long-term memory of the cell, in contrast to h_t, which is more akin to an external, short-term memory. C_t is controlled by the *input* and *forget* gates, while the output is regulated by the *output* gate. The forget gate keeps or clears the internal state based on the input and the previous hidden state. The input gate regulates how much of the new input is added to the internal state, while the output is a version of the new internal state filtered by the output gate. Intuitively, the LSTM's various gates regulate how much the RNN takes the past and new information into account, only considering the most useful inputs instead of using all the past timesteps with decreasing weight as in simple RNNs. Essentially, an LSTM resembles max-pooling in time; only routing information from specific temporal locations, instead of trying to utilize everything at once.

Today's RNNs for natural language modelling tasks mostly use LSTMs or Gated Recurrent Units (GRUs), which work on a similar logic as LSTMs. Natural languages are composed of a hierarchical structure of sentences, expressions, words, and finally letters or sounds. This multi-level representation can be learned using RNNs, by stacking multiple LSTM layers, similar to the way CNNs build up their feature representations of images; earlier layers learn lower level features (in this case words), while later layers learn high level representations (sentences). This makes it possible to apply RNNs to a variety of applications. For machine translation, the machine learns a context from a sentence based on the sequence of letters and words, and thereafter, generates the sentence in a different language. RNNs can also be used to generate text styled by a training set: a generative RNN taught using Shakespeare sonnets will generate eerily Shakespeare-like (albeit mostly nonsense) texts.

6.3.6 Combining LSTMs with Convolutional Networks

Temporal structures in the data are well modelled by recurrent neural networks, but they are not easily used for data that also have spatial variations, such as

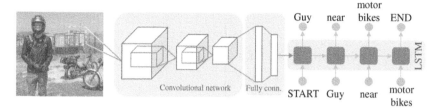

Figure 6.16 Image captioning with convolutional and LSTM networks.

images and videos. The combination of convolutional neural networks with RNNs, called CNN LSTM, allows one to solve sequential problems. There are three usual applications of CNN LSTMs: image captioning, activity recognition and video description [13].

Image captioning is an extension of the simple image classification problem: the network must look at an image and provide a short summary or caption for it, describing what the picture contains. Contrary to plain classification, the images can contain multiple objects in different relations, so the network also has the task of recognizing the logical connections between them, in addition, to correctly recognizing the targets themselves. This is done using a convolutional network as a detector sub-network of the system at the start, and a recurrent network (usually using stacked LSTMs) as a caption generator sub-network at the end. An illustration of this technology can be seen on Figure 6.16.

By contrast, activity recognition has an image sequence input (usually a video) and has to output a label denoting the activity. In this case, a CNN is used as a feature extractor sub-network on each image or video-frame. A recurrent neural network, usually an LSTM, is fed the extracted features and outputs a label. Video description is a similar problem to activity recognition, but it outputs a short summary of the video, instead of an activity label. In both activity recognition and video description the network recognizes logical connection both in time and space. In this sense, the main difference between the problems of image classification, image captioning, activity recognition, and video description is whether the input and output are sequential or not, which determines in which dimension (time and space) is the network looking for logical continuity.

6.3.7 Autoencoders for Data Compression and Cleaning

Autoencoders are the earliest deep neural network architectures, preceding the deep learning boom that started with convolutional networks, and the accompanying technological and algorithmic advancements. The reason why autoencoders were still possible to train – avoiding vanishing gradients and not requiring enormous computational power – is because of the greedy, layer-by-layer

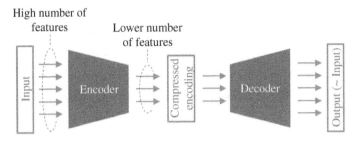

Figure 6.17 Autoencoders in data compression tasks.

training method they implemented. These so-called stacked autoencoders start with only a single hidden layer, and add layers one-by-one, only after the previous network has reached convergence. This method limited the calculations to a single layer at any time, allowing for deep networks to be trained. At the end of the training, the weights could be fine-tuned with end-to-end backpropagation; at this point vanishing gradients do not pose a big threat to the network. Nowadays, this training method has fallen out of favour, as gradient vanishing mitigation methods and GPU accelerated computations made training deeper autoencoders in an end-to-end fashion possible from the start.

An autoencoder's primary use is to compress data, all the while minimizing the loss of relevant information, making use of the fewer but more abstract internal features formed in the middle layers. The measure of goodness for an autoencoder is how well it can restore the original data from its internal representation, which is calculated as a distance between the input and the output of the network. The goal of compression is immediately visible in the usual topology of an autoencoder; at training, the networks narrow up to the middle layers, then widen again to regain the same width as the input layer (can be seen on Figure 6.17). This topology forces the network to form higher abstractions in the middle layers, causing irrelevant or redundant information not to be propagated. De-noising autoencoders make great use of the removal of irrelevant information; by running data through them, the middle layers separate noise from the learned features, only propagating what is considered relevant to the network. This causes the output of the autoencoder to have less noise.

Traditionally, autoencoders relied on the number of hidden units in the middle layer being small enough to achieve compression. Nowadays, by imposing sparsity constraints in the activations, a larger number of representation units can be used to discover interesting features in the data. These sparse autoencoders try to only allow a few neurons to be active at any time in the middle layer, either directly as in k-sparse autoencoders [14], or by using a regularization term. This type of topology emphasizes the feature selector aspect of autoencoders. In extreme cases, with sparsity strictly enforced, the autoencoder can even function as a clustering method, structuring the internal representation as finite clusters in the middle layer.

6.3.8 Cognitive Capabilities and Application of Deep Neural Networks

Deep learning structures have the capability to offer true cognition especially when multiple architectures can be combined. The ability to learn low-level features and aggregate them into higher more complex features maps directly to observations on the sequential computation of insight in the Perception-Reasoning pipeline model developed in Chapter 4. Specifically, unsupervised and supervised learning architectures like autoencoders can be used to develop a perception engine while supervised leaning structures can be useful in developing data organization and inference engines. As such, combining multiple structures can enable a combined perception and reasoning engine.

The specialized structures can, however, also be individual useful albeit to different extents. Autoencoders are probably the structures with the widest areas of usage in communication given the amount of data generated by communication systems which may need to be reduced in dimensionality in order to derive appropriate insight. This could be possible for data on a single element as well network-wide data. However, while deep autoencoders are straightforward to use in Communication Networks (CNs) data processing tasks, CNN, and RNN networks are highly specialized algorithms that will need considerable modifications to be applicable on CN data. Nevertheless, a number of uses may be imagined to be good candidates for applying these very specialized deep neural networks:

- Convolutional Neural Networks could be used to work on spatially encoded network data, that accounts for neighbour-relations between network elements. For this to work, however, the network data would probably need to be formed to represent network topology, on the other hand, the CNN would need to be able to understand and do convolutions on graph-like data structures.
- Recurrent Neural Networks could be used to process CN data in a sequential manner that also considers the temporal features in the data. This functionality could be utilized for anomaly detection, or even prediction purposes, to create more robust and resilient networks.
- Combining CNNs and LSTMs can be used to concurrently learn dynamics spatial and temporal

These uses would benefit from the specific strengths of each DNN type, bringing advanced cognition to CN management tasks.

6.4 Reinforcement Learning

Reinforcement learning is the body of theory and techniques for optimal sequential decision making [15, 16]. It encompasses the challenge of an agent learning by interacting with its environment through trial and error, and evaluative, sequential delayed feedback [2]. As illustrated by Figure 6.18, the agent learns to adjust its response to specific input states of the environment based on rewards and/or penalties that it receives for its actions.

6.4.1 Learning Through Exploration

RL agents explore the environment to learn how best to behave. For example, let us consider that the agent wishes to learn the right time to switch off a capacity enhancing radio cell with the aim of minimizing consumed energy during low traffic periods. The agent can learn the appropriate time-point through exploration, for which the following definitions hold:

- **Agent:** The agent is the entity that takes actions and hosts the learning algorithm that seeks to learn from these actions. This, in the energy saving example, is the energy saving algorithm tasked with learning the cell switch off time.
- **Environment:** The environment is the unknown, nonlinear, stochastic, and possibly complex world in which the agent acts. The environment takes the agent's action as input in given state and based on that action transitions to a new state and emits an output as the reward towards the agent. For the cell switch problem, the environment is the network with all its static and dynamic world including the cell of interest, the neighbour cells, the active users, etc.
- **State (S):** A state is a concrete situation, relative to the agents' objectives, in which the agent finds itself. It is an instance of the space of interest for the agent. For the cell switch problem, the states will be the active or inactive states of the cell at a specific time.
- **Actions (A):** Actions A is the set of all possible moves the agent can make. These could be as simple as a single number e.g. for a parameter configuration, a change in such a configuration or as complex as combinations of changes in different parameters and statuses. For the cell-switch off example, there are only two actions – either to switch off the cell or to leave it active.

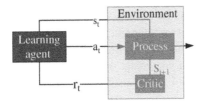

Figure 6.18 Reinforcement learning process.

- **Reward (R):** A reward is the feedback that measures the quality of the agent's actions. In any given state, when the agent takes an action, the environment transitions to a new state (resulting from that action). Then, the environment, or a critic thereof, sends a reward, if there are any, that quantifies the quality of the agent's action. The reward function could be a simple discrete function e.g. a binary function, (1, −1), that simply states if the action is acceptable (1) or not (−1). It could also be a continuous function, e.g. between +1 and −1, that characterizes the degree to which the action is acceptable e.g. with +1 indicating the best possible action and −1 indicating a totally unacceptable action. For the cell switch off example, a discrete reward function could for example assign −1 to any action that results in an unwanted event like call drop or the congestion of a neighbour cell and assign +1 otherwise. A continuous function on the other hand could assign +1 where no unwanted event is observed and reduce this reward in response to the observed number of unwanted events.
- **Policy (π):** The policy is the strategy of taking actions that is used by the agent to determine the next action based on the current state, i.e. it is the mapping from states to actions intended to maximize the agent's cumulative rewards in the long run. Policies can either be Deterministic, i.e. where the states are directly mapped to the actions: $a = \pi(s)$, or they may be Stochastic, in which case, the states are mapped a distribution over actions: $\pi(a\,|\,s) = P(a\,|\,s)$. The distribution from which the actions are drawn could, for example, be the Normal distribution with mean μ and deviation σ for which the policy predicts μ and σ. Note that the deterministic policy is the special case of the stochastic policy for which the variance of the distribution is 0.

The intention of RL is for the agent to learn a policy of how to act in an environment based on feedback about quality of different states visited or the quality of the actions taken. Given a state, a policy (control strategy) π returns an action to perform. The required policy (called the optimal policy) is the one that maximizes the expected return (cumulative, discounted reward) in the environment.

6.4.2 RL Challenges and Framework

Solving RL problems breaks down to solving two major concerns: the exploration-exploitation challenge and the credit assignment problem. An optimal policy must be inferred by trial-and-error interaction with the environment using the reward as the only learning signal. The agent needs to explore the environment before it can exploit the learned knowledge and this exploration-exploitation dilemma is a non-trivial problem for which a solution is guaranteed only if each state and action are visited (and respectively tried) infinitely many times. The best way to explore is randomly, although some heuristics, such as curiosity [17], can be added to improve the outcome but they are not guaranteed to work well in all cases. However, other solutions have been shown to work well in many scenarios including the ε-greedy algorithm [18].

The second problem is that the agent's observations can depend on strong temporal correlations of its actions. Agents must deal with long-range time dependencies in what is known as the (temporal) credit assignment problem: i.e. the consequences of an action only materialize after many transitions of the environment. Moreover, in many real-world tasks, the agent is incapable of observing the complete environment. As such, deciding about state- and action-space tiling is difficult although it is critical for the convergence of RL-methods. This has led to the use of continuous-valued RL solutions e.g. using function appropriation.

Formally, RL can be described as a Markov decision process (MDP), consisting of:

- A set of states S, plus a distribution of starting states $p(s_0)$.
- A set of actions A.
- Transition dynamics $T(s_{t+1} \mid s_t; a_t)$ that map a state-action pair at time t onto a distribution of states at time $t+1$.
- An immediate/instantaneous reward function $R(s_t; a_t; s_{t+1})$
- A discount factor $\gamma \in [0, 1]$, where lower values emphasize immediate rewards.

In general, the policy π is a mapping from states to a probability distribution over actions, i.e. $\pi : S = P(A = a \mid S)$. If the MDP is *episodic*, i.e. the state is reset after each episode of length T, the sequence of states, actions, and rewards in an episode constitutes a *trajectory* or *rollout* of the policy. Every rollout accumulates rewards from the environment, resulting in the return

$$R = \sum_{t=0}^{T-1} \gamma^t r_{t+1} \tag{6.13}$$

In the case of non-episodic MDPs, where $T \equiv 1$, $\gamma < 1$ prevents an infinite sum of rewards from being accumulated and only methods that use a finite set of transitions are applicable, i.e. methods that rely on complete trajectories are no longer applicable.

There are two main approaches to solving RL problems: methods based on *value functions* and methods based on *policy search*. There is also a hybrid, *actor-critic* approach, which employs both value functions and policy search. These approaches and other useful concepts for solving RL problems are discussed here.

6.4.3 Value Functions

Value function methods are based on estimating the value (expected return) of being in a given state. The *state-value function* $V^\pi(s)$ is the expected return when starting in state **s** and following π as in (6.14). The optimal policy, π^*, has a corresponding optimal state-value function $V^*(s)$, which can be defined as in (6.15).

$$V^\pi(s) = E[R|s; \pi] \tag{6.14}$$

$$V^{\pi*}(s) = \max_\pi V^\pi(s) \ \forall s \in S \tag{6.15}$$

With $V^*(s)$ available, the optimal policy is obtained by choosing amongst the actions available at s_t and picking the action a that maximizes $E_{s_{t+1}} \sim T(s_{t+1}|s_t, a)[V^*(s_{t+1})]$. However, since the transition dynamics T are unavailable in the RL setting, the *state-action value* or *quality function* $Q^\pi(s; a)$ is used. $Q_\pi(s; a)$ is similar to V_π except that the initial action a is provided, and π is followed from the succeeding state onwards:

$$Q^\pi(s; a) = E[R|s; a; \pi] \tag{6.16}$$

Given $Q^\pi(s; a)$, the best policy can be found by greedily choosing a at every state, i.e. choosing such that $a = arg\ \max_a Q^\pi(s, a)$ and $V_\pi^*(s) = \max_a Q^\pi(s, a)$.

6.4.4 Model-Based Learning Through Value and Policy Iteration

Value-iteration and policy-iteration are methods for solving MDPs based on the assumption that the agent knows the MDP model of the world. The agent knows the state-transition and reward probability functions, which it uses to plan its actions offline (i.e. before interacting with environment). Value iteration computes the optimal state value function by iteratively improving the estimate of **V(s)**. The algorithm (see Algorithm 6.5) initializes **V(s)** to arbitrary random values and then repeatedly updates $Q(s, a)$ and $V(s)$ values until they converge.

For policy iteration in Algorithm 6.6, which instead of repeatedly improving the value-function estimate, will re-define the policy at each step and compute the value according to this new policy until the policy converges. Policy iteration may even be preferred to value-iteration since the agent only cares about the finding the optimal policy and yet sometimes the optimal policy will converge before the value function.

The core of policy iteration is the Policy improvement theorem, i.e. given the value function $V^\pi(s)$ for any policy π, $V^\pi(s)$ can always be greedified to obtain a better policy $\pi'(s) = arg\ \max_a Q^{\pi'}(s, a)$ with $Q^{\pi'}(s, a) \geq Q^\pi(s, a)$. This implies that:

1. Any policy evaluates to a unique value function which can be greedified to produce a better policy – which then evaluates to another function which can be greedified.
2. Each policy is strictly better than the previous until eventually both are optimal.

Algorithm 6.5 Value Iteration

Initialize V(s) arbitrarily
loop until policy good enough
    for s in S:
        for a in A:

$$Q(s, a) = r(s, a) + \gamma \sum_{s' \in S} T(s, a, s') V(s')$$

$$V(s) := \max_a Q_\pi(s, a)$$

Algorithm 6.6 Policy Iteration

Chose a policy π arbitrarily

1. loop until policy good enough (e.g. until $\pi := \pi'$)
2. $\pi := \pi'$
3. Compute the value function of policy π by solving the linear equations:

$$V^\pi(s) = R(s, \pi(s)) + \gamma \sum_{s' \in S} T(s, \pi(s), s') V^\pi(s')$$

4. Improve the policy at each stage:

$$\pi'(s) := \arg\max_a \left[R(s, a) + \gamma \sum_{s' \in S} T(s, a, s') V^\pi(s') \right]$$

Both Value iteration and Policy iteration are guaranteed to converge to the optimal value and policy respectively, but Policy iteration often takes fewer iterations to converge. Both algorithms are robust to initial conditions, randomization and noise as well as the update criteria. Such criteria include delayed and asynchronous updating as in parallel and distributed implementations; incomplete evaluation and greedification, e.g. by updating only a subset of states, or even when a single state is updated at a time by a random amount that is only correct in expectation.

6.4.5 Q-Learning Through Dynamic Programming

Value iteration and Policy iteration methods require the environmental model, i.e. a description of the transition dynamics through the transition probability $T(s' \,|\, (s, a))$. This then requires a prior step of learning this probability through a simulation of the environment. This makes the algorithms impractical as the state-action space grows. A model-free learning method would as such be better placed to learn the policy of selecting actions by learning the Q-value function $Q^\pi(s; a)$,

To learn Q^π, the Markov property is exploited to define the function as a Bellman equation, which has the following recursive form [18]:

$$Q^\pi(s_t, a_t) = E_{s_{t+1}}[r_t + \gamma Q^\pi(s_{t+1}; \pi(s_{t+1}))] \tag{6.17}$$

This means that Q^π can be improved by *bootstrapping*, i.e. the current values of the estimate of Q^π can be used to improve the estimate, i.e.

$$Q^\pi(s_t, a_t) = Q^\pi(s_t, a_t) + \alpha\delta \tag{6.18}$$

where α is the learning rate and δ is the Temporal Difference (TD) error given by $\delta = r_t + \gamma\max_a Q^\pi(s_t, a) - Q^\pi(s_t, a)$. The rewards are discounted by γ which could, for example, be reduced over time ($\gamma \propto \frac{1}{t}$) or average over the number of episodes k ($\gamma = \frac{1}{k}$).

Q-learning is *off-policy learning*, as Q^π is updated by transitions that are not necessarily generated by the derived policy, Q^*. To find Q^* from an arbitrary Q^π, *generalized policy iteration in* Algorithm 6.7 is applied, with the two processes of *policy evaluation* and *policy improvement*. Policy evaluation improves the estimate of the value function by minimizing TD errors from trajectories derived by a policy. Then, as the estimate improves, the policy is improved by choosing actions greedily based on the updated value function. Instead of performing these steps separately to convergence (as in policy iteration), generalized policy iteration allows for interleaving the steps, such that progress can be made more rapidly.

Algorithm 6.7   Q-learning Algorithm

Initialize $Q(s, a)$ arbitrarily
loop until policy good enough (i.e. until $Q(s, a)$ converges)

– Select and apply an action a_t in state s_t
– Using obtained reward r_t^π, update the quality function

$$Q^\pi(s_t, a_t) = Q^\pi(s_t, a_t) + \alpha[r_t + \gamma\max_a Q^\pi(s_t, a) - Q^\pi(s_t, a_t)]$$

Eventually, choose a policy π^* that maximizes reward $r^{\pi^*}(s)$

6.4.6 Linear Function Approximation

Real-world challenges e.g. in network operations have very large and complex state spaces, so much so that value functions and policies can no longer be captured in tables and arrays. Function approximation can be used to represent the action-value function by a parameterized approximator with parameter θ, such that $\hat{Q}(s, a, \theta) \approx Q^*(s, a)$. The approximator could be any differentiable function, such as a linear function or a deep neural network.

Consider that an observation is the state for which the Q-learning algorithm needs to learn the quality of the state or the state-action pairs. Then, assume a well selected n-dimension observation state-space, where well selected implies that: (i) it is sample efficient on training yet generalizes well on unseen states; and (ii) it offers opportunity to encode expert knowledge. Assume also, that each sample in observation state can be described by a feature vector $\phi(s, a)$. Instead of learning the state(-action) values, function approximation wishes to learn a differentiable function that will generate the state(-action) values given $\phi(s, a)$. This is useful where the high-dimensional observations, for example, have a low-dimensional underlying state. Such a function can be learned either through generalized approximation, as represented by a neural network, but linear approximation gives a simpler albeit useful solution.

The simplest approximator is a linear function where the state(-action) value is parametrized by a weight vector θ, i.e. $Q(s_t) = \theta^T \phi(s_t, a)$. From the value function $v^\pi(s_t) = E[R(s_t) + \gamma v^\pi(s_{t+1})]$, given samples s_t, a_t, r_t, s_{t+1} and the feature vector $\phi(s)$, the temporal difference error will be (6.19). For Q-learning, the policy evaluation step becomes (6.20) with the equivalent squared loss given by (6.21)

$$\delta \leftarrow r_t + \gamma \theta^T \phi(s_{t+1}) - \theta^T \phi(s_t) \tag{6.19}$$

$$\delta \leftarrow r_t + \gamma max_{a \in A}[\theta^T \phi(s_{t+1}, a)] - \theta^T \phi(s_t, a) \tag{6.20}$$

$$J(\theta) = \|r_t + \gamma max_{a \in A}[\theta^T \phi(s_{t+1}, a)] - \theta^T \phi(s_t, a)\|^2 \tag{6.21}$$

The loss function is minimized by updating the parameters θ, as in (6.22) for a linear function. The Q-learning algorithm can then be completed through the policy improvement step to derive policy $\pi(s)$ as (6.23)

$$\theta \leftarrow \theta - \alpha \nabla_\theta J(\theta) = \theta + \alpha \delta \nabla_\theta [\theta^T \phi(s_t, a)] = \theta + \alpha \delta \phi(s_t, a) \tag{6.22}$$

$$\pi(s) \leftarrow max_{a \in A}[\theta^T \phi(s, a)] \tag{6.23}$$

6.4.7 Generalized Approximators and Deep Q-Learning

In the general case, θ is any differentiable function θ_t that allows to learn a parameterized Q-value function $Q(s_t, a, \theta_t)$ [19]. Such a function can be a neural network as is the case for a Deep Q-Network (DQN). However, instead of learning using a single neural network, the DQN uses a target network, with parameters θ_t^-. This network is the same as the online network except that its parameters are only copied from the online network every τ steps but kept fixed on all other steps.

Consequently, the policy evaluation step in the DQN gives δ as in (6.24) for which the parameters can be updated using SGD with the samples drawn from set D as (6.25) and (6.26)

$$\delta_t \leftarrow r_t + \gamma \max_{a \in A} Q[\phi(s_{t+1}, a); \theta_t^-] - Q[\phi(s_t, a); \theta_t] \qquad (6.24)$$

$$\nabla_{\theta_t} J(\theta_t) = E_{s_t, a_t, r_t, s_{t+1} \sim D}[r_t + \gamma \max_{a \in A} Q[\phi(s_{t+1}, a); \theta_t^-] - Q[\phi(s_t, a); \theta_t]].$$

$$\nabla_{\theta_t} Q[\phi(s_t, a); \theta_t] \qquad (6.25)$$

$$\theta_{t+1} = \theta_t + \alpha \delta_t \nabla_{\theta_t} Q[\phi(s_t, a); \theta_t] \qquad (6.26)$$

Subsequently, the policy improvement step gives

$$\pi_t(s) \leftarrow \max_{a \in A} Q[\phi(s, a); \theta_t] \qquad (6.27)$$

In practice, standard tools from supervised learning (deep-learning frameworks and optimizers) can be used to compute gradients and apply parameter updates. Stochastic gradient methods are key to using powerful function approximators. In particular, while older methods used typically linear function approximators to estimate the Q function, the idea in deep Q-learning is to use a neural network as a non-linear function approximator for this task. This allows the algorithm to directly work on the sensory input, without need for hand-engineered features, since the neural network learns such features automatically.

A challenge in reinforcement learning tasks using deep Q-learning is that, compared to unsupervised learning, the training samples are not truly independent; subsequent steps are highly correlated, which can skew the training. Also, the data distribution is non-stationary since the data itself is changing as the agent learns new behaviours. To combat this [20] introduces the technique of experience replay, where the agent experience is stored and pooled into a replay memory, introducing a hybrid approach, instead of the traditional fully online reinforcement learning. This causes the training to be more efficient and eliminates the problem of correlated, non-stationary data.

6.4.8 Policy Gradient and Actor-Critic Methods

Function approximation can also be used to directly learn the optimal policy. Given:

a) a policy π of choosing an action a_t in state s_t to be parameterized by the function θ, i.e. $P(a_t | s_t) = \pi_\theta(a_t | s_t)$,
b) a trajectory τ as the sequence of states, actions and instantaneous rewards starting at t = 0 and terminating $t = T$, i.e. $\tau = \{s_0, a_0, r_0, s_1, a_1, r_1, \ldots, a_T, r_T, s_T\}$
c) the return for the trajectory τ to be $R(\tau) = \sum_{t=0}^{T} r_t$,

the expected return for the policy is

$$J(\theta) = E\left[\sum_{t=0}^{T} r_t; \pi_\theta\right] = \sum_\tau P(\tau, \theta)R(\tau) \tag{6.28}$$

To obtain the optimal policy, it is necessary to find θ that maximizes the expected return, i.e. $\theta = \text{argmax}J(\theta) = \text{argmax}\sum_\tau P(\tau, \theta)R(\tau)$. This can be achieved by iteratively improving the policy using its gradient, i.e. $\theta = \theta + \alpha\nabla_\theta J(\theta)$ where α is the learning rate.

Given that $\nabla_\theta \log f(x) = \frac{\nabla_\theta f(x)}{f(x)}$, the gradient, which increases the likelihood of trajectories with high return and decreases it otherwise, can be obtained as:

$$\nabla_\theta J(\theta) = \nabla_\theta \sum_\tau P(\tau, \theta)R(\tau)$$

$$= \frac{\nabla_\theta P(\tau, \theta)}{P(\tau, \theta)} \sum_\tau P(\tau, \theta)R(\tau) \tag{6.29}$$

$$= \sum_\tau P(\tau, \theta)\nabla_\theta \log P(\tau, \theta)R(\tau) \tag{6.30}$$

The gradient can be approximated using a set of m trajectories sampled from π_θ,

$$\nabla_\theta J(\theta) = \frac{1}{m} \sum_{i=1}^{m} \nabla_\theta \log P(\tau^i, \theta)R(\tau^i) \tag{6.31}$$

$$\text{Since: } \nabla_\theta \log P(\tau^i, \theta) = \nabla_\theta \log\left[\prod_{t=0}^{T} P(s_{t+1}|s_t, a_t)\pi_\theta(a_t|s_t)\right] \tag{6.32}$$

$$= \nabla_\theta \sum_{t=0}^{T} P(s_{t+1}|s_t, a_t) + \nabla_\theta \sum_{t=0}^{T} \log \pi_\theta(a_t|s_t) \tag{6.33}$$

Since the transitions are independent of the policy ($\frac{dP}{d\theta} = 0$), $\nabla_\theta \sum_{t=0}^{T} P(s_{t+1}|s_t, a_t) = 0$ and

$$\nabla_\theta J(\theta) = \frac{1}{m} \sum_{i=1}^{m} \sum_{t=0}^{T} \nabla_\theta \log \pi_\theta(a_t|s_t)R(\tau^i) \tag{6.34}$$

From Eq. (6.34), no model of the transition dynamics is required, so it is possible to learn the optimal actions without a model of the underlying behaviour of the system. The gradient is an unbiased estimate of the policy gradient, i.e. it will in the limit (as m $\rightarrow \infty$) converge to the true gradient of J.

6.4.8.1 Reinforce Algorithm

The earliest policy-gradient based method is the Reinforce Algorithm which estimates the gradient for each trajectory and updates the policy parameters. The corresponding algorithm is illustrated by Algorithm 6.8.

Algorithm 6.8   The Reinforce Algorithm

Initialize policy π_θ

Repeat until convergence

1. Act in the environment $a_t \sim \pi(s_t)$ for an episode storing states, action, and rewards: $s_0, a_0, r_0, s_1, a_1, r_1, \ldots, a_T, r_T, s_T$
2. *compute discounted return* $R = \sum_{t=0}^{T} r_t$
3. Compute policy *gradient* $\nabla_\theta J(\theta) = \sum_{t=0}^{T} \nabla_\theta \log \pi_\theta(a_t|s_t) R$
4. *Take a step in the direction of the gradient:* $\theta = \theta + \alpha \nabla_\theta J(\theta)$

Ideally, the algorithms wish to learn a low bias and low variance estimator. The reinforce algorithm has a low bias (it is unbiased) but the variance in return is quite high. $R(\tau)$ encourages/discourages full trajectories instead of the individual actions and so can result in high variance, e.g. with $R(\tau) = 0$, policy gradient learns nothing, yet it should still be able to learn from individual actions.

6.4.8.2   Reducing Variance

The variance can be lowered using discounted step-wise returns and/or baselines. For discounted step-wise returns, the variance can be reduced by replacing the episodic return $R(\tau)$ with the return at each time step, which encourages good actions and discourages bad actions. With the RL discount factor γ and a step-wise discounted return $R_t = \sum_{t'=0}^{T} \gamma^{t'} R_{t'}$, the gradient then becomes

$$\nabla_\theta J(\theta) = \sum_{t=0}^{T} \nabla_\theta \log \pi_\theta(a_t|s_t) R_t \tag{6.35}$$

The Baselines approach applies the estimate of the expected performance of the policy as a baseline $b_t = E(R_t, \theta)$. This discourages trajectories below the baseline (even with positive returns) and encourages those above the baseline. The gradient then becomes

$$\nabla_\theta J(\theta) = \sum_{t=0}^{T} \nabla_\theta \log \pi_\theta(a_t|s_t)(R_t - b_t) \tag{6.36}$$

The baseline could be a constant – the average of returns of individual trajectories as in Eq. (6.37) or time dependent as in Eq. (6.38). The base could, however, also be State varying as in Eq. (6.39) in which case it is simply the value function $V^\pi(s_t) = E_\pi[r_t + r_{t+1} + \ldots + r_T | s_t]$

$$b_t = b = \frac{1}{m} \sum_{i=1}^{m} R(\tau^i) \tag{6.37}$$

$$b_t = \frac{1}{m} \sum_{i=1}^{m} \sum_{t'=t}^{T} r_{t'}^i \tag{6.38}$$

$$b_t = E_\pi \left[\sum_{t'=t}^{T} r_{t'}^i \mid s_t \right] \tag{6.39}$$

6.4.8.3 Policy Gradient Algorithm

By applying the state-varying baseline of Eq. (6.39), the gradient becomes Eq. (6.40). The value $V_\emptyset$ can be estimated as in Algorithm 6.9 by: (i) collecting a set of m trajectories $\tau_1, \tau_2, \ldots \tau_m$; and (ii) regressing $V_\emptyset$ towards the empirical return as in Eq. (6.41).

$$\nabla_\theta J(\theta) = \frac{1}{m} \sum_{i=1}^{m} \sum_{t=0}^{T} \nabla_\theta \log \pi_\theta(a_t|s_t)(R_t - V_\emptyset(s_t)) \tag{6.40}$$

$$\emptyset_{i+1} = \mathrm{argmin}_\emptyset \frac{1}{m} \sum_{i=1}^{m} \sum_{t=0}^{T} \left[V_\emptyset^\pi(s_{t'}^i) - \sum_{t'=t}^{T} r_{t'}^i \right]^2 \tag{6.41}$$

Algorithm 6.9 Policy Gradient Algorithm

Initialize policy π_θ and value function $V_\emptyset$
For each episode

1. *Collect trajectory τ by running current policy*
2. *For each step in τ, compute discounted return $R_t = \sum_{t'=t}^{T} r_{t'}$*
   Re-fit baseline minimizing $\|V_\emptyset(s_t) - R_t\|^2$ over all steps
   Update π_θ using gradient $\sum_{t=0}^{T} \nabla_\theta \log \pi_\theta(a_t|s_t)(R_t - V_\emptyset(s_t))$

6.4.8.4 Actor-Critic

Actor-Critic algorithms consist of two parts – the actor which is the policy responsible for selecting and executing actions in the environment and the critic which is responsible for evaluating the quality of the selected actions and suggesting direction for improvement. The likelihood ratio policy gradient can be considered an actor-critic algorithm since it uses a separate part for the policy and value function. In Eq. (6.40), the actor is the $\log \pi_\theta(a_t \mid s_t)$ part while the reward part $(R_t - V_\emptyset(s_t))$ is the critic that quantifies the quality of acting.

This policy Gradient algorithm can further be improved since there are still high variance parts of the critic that are not learned – specifically, R_t. Such improvements are discussed in advanced Actor-Critic algorithm like the Asynchronous Advantage Actor-Critic (A3C) algorithm. Details on this algorithm, as discussed in [21] are left to the reader as are others reflected in the algorithm cloud of Figure 6.19.

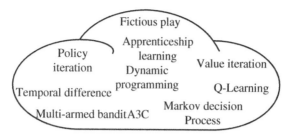

Figure 6.19 Reinforcement learning algorithms cloud.

6.4.9 Cognitive Capabilities and Application of Reinforcement Learning

Reinforcement learning is capable of general-purpose cognition as it can be trained to perceive different sensor data and to reason over different data elements. It can be trained to combine both perception and reasoning, i.e. to observe some event, characterize it and, in combination with its context, to reason and infer the best course of action. In general, reinforcement learning could provide the best candidate for general purpose artificial intelligence, although it is also the hardest to train.

With the ability to learn how to perceive and reason, RL can be trained to accomplish the core challenges in network management automation – to characterize observed events and to decide how to behave in the different scenarios. For example, an RL solution can be trained to diagnose network faults and decide the best response to the different faults. Similarly, as will be discussed in Chapter 7, an RL solution can learn how to optimize network processes e.g. where to set network performance baselines or how to optimize user mobility robustness.

Although RL is a great promise for general intelligence, it also has the challenge that it requires time to learn and in online form. This is a great challenge in communication network and other mission critical systems where there is no room for failure. Any solution is only applied when the operator is sure how it works and that it will not cause any degradation. RL needs to be accorded the ability to fail for it to learn the best solution but this need to fail makes RL non-applicable to communication networks.

The current perspective is that RL can be used for optimization solution over and above the manual or model-based expert-systems solutions. Therein, the operator sets a controlled optimization space in which he is sure that the worst performance is still acceptable. Alternatively, the RL solution can first be trained in a simulation environment to learn the problem structure and the applicable range of good control parameters. This range then becomes the parameter-space which is applied in the controlled online optimization.

6.5 Conclusions

This chapter has summarized the state-of-the-art machine learning concepts and highlighted how these concepts may be of use as the cognitive techniques to address network management automation challenges. The sections summarize the general concepts on learning from data through supervised and unsupervised learning; the concepts on the use of neural networks and the related deep learning as a computation mechanism for implementing machine learning as well as RL as a means for autonomic control.

In all sections, the discussion is sure to highlight the cognitive capabilities of the different concepts as measured against the perception-reasoning model presented in Chapter 4. This evaluation of capabilities is then justified with a summary discussion of the application where the different techniques have been or could be applied. Where it was deemed critical, specific applications have been described with a little more detail to enable the reader to better relate the concepts to the candidate challenges.

The concepts presented in this chapter have, however, already been variably applied to different network automation challenges. The details on these applications describing the design decisions and summarizing the achieved results are presented in Chapters 7–10. Note, however, that although these cognitive techniques provide opportunities for solutions, they also raise a number of challenges, specifically with regard to the control and management of a system of independently learning agents. These challenges and the candidate solutions thereof are also discussed in Chapter 11.

References

1 Goodfellow, I., Bengio, Y., and Courville, A. (2016). *Deep Learning*. MIT Press.

2 Hastie, T., Tibshirani, R., and Friedman, J.H. (2009). *The Elements of Statistical Learning: Data Mining, Inference, and Prediction*. New York, NY, USA: Springer.

3 Mitchell, T.M. (1997). *Machine Learning*, 870–877. Burr Ridge, IL: McGraw Hill.

4 Ruder, S. (2016). An overview of gradient descent optimisation algorithms. arXiv:1609.04747.

5 Goodfellow, I., Pouget-Abadie, J., Mirza, M. et al. (2014). Generative adversarial nets. Advances in neural information processing systems, Montréal.

6 Ghahramani, Z. (2004). *Unsupervised Learning*. London, UK: Gatsby Computational Neuroscience Unit, University College London.

7 Aalto University (2018). School of Science, Department of Computer Science, Independent Component Analysis (ICA) and Blind Source Separation (BSS). http://research.ics.aalto.fi/ica/fastica (accessed 02 July 2018).

8 Hyvärinen, A. (1999). Fast and robust fixed-point algorithms for independent component analysis. *IEEE Transactions on Neural Networks* 10 (3): 626–634.

9 Smola, A. and Vishwanathan, S.V.N. (2008). *Introduction to Machine Learning*. Cambridge, UK: Cambridge University.

10 Glorot, X. and Bengio, Y. (2010). Understanding the difficulty of training deep feedforward neural networks. International Conference on Artificial Intelligence and Statistics, Sardinia.

11 Srivastava, N., Hinton, G., Krizhevsky, A. et al. (2014). Dropout: a simple way to prevent neural networks from overfitting. *Journal of Machine Learning Research* 15: 1929–1958.

12 Kaiming, H., Zhang, X., Ren, S. and Jian, S. (2016). Deep Residual Learning for Image Recognition. IEEE Conference on Computer Vision and Pattern Recognition (CVPR), Las Vegas.

13 Donahue, J., Hendricks, L.A., Guadarrama, S. et al. (2015). Long-term recurrent convolutional networks for visual recognition and description. IEEE conference on computer vision and pattern recognition, Boston.

14 Alireza Makhzani, B.J.F. (2014). k-sparse autoencoders. arXiv:1312.5663.

15 Gosavi, A. (2009). Reinforcement learning: a tutorial survey and recent advances. *INFORMS Journal on Computing* 21 (2): 178–192.

16 Kaelbling, L.P., Littman, M.L., and Moore, A.W. (1996). Reinforcement learning: a survey. *Journal of Artificial Intelligence Research* 4: 237–285.

17 Schmidhuber, J. (2010). Formal theory of creativity, fun, and intrinsic motivation (1990–2010). *IEEE Transactions on Autonomous Mental Development* 2 (3): 230–247.

18 Sutton, R. (1998). *Reinforcement Learning: An Introduction*. MIT Press.

19 Van Hasselt, H., Guez, A. and Silver, D. (2016). Deep reinforcement learning with double q-learning. International Conference on Artificial Intelligence and Statistics, Sardinia.

20 Mnih, V., Kavukcuoglu, K., Silver, D. et al. (2013). Playing atari with deep reinforcement learning. arXiv:1312.5602.

21 Nielsen, M.A. (2015). *Neural Networks and Deep Learning*. Determination Press.

22 Zeiler, M.D. and Fergus, R. (2014). Visualizing and Understanding Convolutional Networks. European Conference on Computer Vision, Zurich.

7

Cognitive Autonomy for Network Configuration

Stephen S. Mwanje[1], Rashid Mijumbi[2] and Lars Christoph Schmelz[1]

[1] *Nokia Bell Labs, Munich, Germany*
[2] *MSD, Dublin, Ireland*

Network configuration is typically used as the general term to describe the processes of making changes to one or more network parameters, be it to one or multiple devices, to achieve some objective. Although procedures thereof have been described in the fault, configuration, accounting, performance, and security (FCAPS) configuration management framework [1], the automation of network configuration remains an open field of work within the wider Network Management Automation (NMA) research. Specifically, there remains opportunities for cognitive decision-making mechanisms aimed at raising network's autonomy at configuration management.

Configuration, and specifically the term self-configuration, has been used within the Self-Organizing Networks (SONs) context in a restrictive way, to refer to the activities and processes that are needed to put network devices into operation. This chapter takes the general view of network configuration as the processes for managing the network configuration parameters both on commissioning as well as in operation. Herein, four use cases are discussed here as an attempt to answer the question on how cognition and autonomy could be realized in network configuration. The considered use cases apply cognitive concepts to different degrees, but all propose means for increasing autonomy of configuration management.

The first section presents concepts on context aware auto-configuration showing how objective models can be combined with context models to allow for autonomic derivation of network and function configuration policies. The section presents general ideas that can be applied to different kinds of network automation problems to derive configuration policies for both the automation functions and the network's configuration parameter values.

Towards Cognitive Autonomous Networks: Network Management Automation for 5G and Beyond,
First Edition. Edited by Stephen S. Mwanje and Christian Mannweiler.
© 2021 John Wiley & Sons Ltd. Published 2021 by John Wiley & Sons Ltd.

Thereafter, three sections present use cases on network configuration where cognitive and model-based techniques are applied. A section on 'multi-layer PCI auto configuration' presents a simple learning mechanism employed to learn cell Neighbour relationships that are then applied to (re)configure the cells' Physical Cell Identities (PCIs). The next section then presents concepts on how a model-based approach can be applied to select the optimal cell deactivation and reactivation sequences to address challenges of multi-layer energy saving management. The last section then shows how machine learning can be used to automate the adaptation of network baselines.

7.1 Context Awareness for Auto-Configuration

Auto-configuration involves the use of automation functions to manage the configuration of network parameters or to adjust the behaviour automation functions. Each Network Automation Function (NAF) is characterized by what may be called Function Configuration Parameters (FCPs), which are the parameters that control the behaviour of the NAF. For SON-style control-loop type NAFs, the parameters are designed into the algorithm and include parameters like the thresholds that determine the points at which different solution options are taken by the algorithm. On the other hand, for Cognitive Functions (CFs) such parameters are the hyperparameters that control the learning, such as a neural network's number of hidden layers or the exploration-exploitation control parameters for reinforcement learning (RL).

A NAF's behaviour can be changed by adjusting its Function Configuration Parameters Values (FCVs), i.e. depending on the FCVs, the NAF will differently adjust the network's behaviour. Typically, in SON systems, one default FCV set is used, often the one supplied by the SON Function vendor. However, networks have infinitely many contexts, which cannot be addressed by a single function setting. Ideally, changes to a Key Performance Indicator (KPI) target or to the context require adjusting the behaviour of the NAF in such a way that the NAF can positively contribute to the changed KPI target.

For SON Functions, the change of behaviour can be achieved by adjusting the FCV Sets using a CF that manages the SON Functions. This has been studied under the construct of Objective-driven SON management in [2, 3], where the SON objective manager, as the CF in that case, manages the SON Functions (the NAF) and is responsible for setting the Functions' Configuration Parameter Values as illustrated in Figure 7.1. Another option is that the CF is directly responsible for adjusting the Network Configuration Parameter Values (NPVs). In that case, the CF may have to adjust its internal structure in such a way that it responds to the different contexts differently.

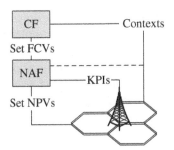

Figure 7.1 Context-aware adjustment of Function and Network Configuration Parameter Values (FCVs, NPVs).

The discussion here generalizes the concepts developed in [2, 3] in such way that they can be applied for configuring both NAFs and the NPVs. First, the distinctions between environment, network, service, and functional contexts are presented, followed by a discussion of the concepts on how context-aware configuration can be implemented to configure NAFs. Then, the next sections present the objective and context models and the discussion on how the context model can be derived from the objective model. Finally, the last section discusses how the two models can be combined to derive network and function configuration policies.

7.1.1 Environment, Network, and Function Contexts

The expectations from the network or the NAFs may not be the same at all time-space points within a mobile network but depend on a certain context. The different contexts represent changes in at least one of the dimensions of interest. Examples for such dimensions are environmental information such as the season of the year or time of day, or network information such as the cell locality which can be urban, rural, or semi-urban or service-related information like traffic patterns and user speed.

For each of the dimensions of interest the context describes a subspace of that dimension, the most widely used environmental and network contexts being:

- **Time of the day:** Differentiation of KPI targets and their importance is necessary for peak traffic periods, e.g. the time from $08:00$ to $12:59$, in comparison to periods of normal traffic, e.g. the time from $13:00$ to $22:59$, or periods with very low traffic, e.g. the time from $23:00$ to $07:59$.
- **Cell location:** The KPI targets and their importance may be different in urban, suburban, and rural areas, due to differences in the number of users, behavioural patterns of the users, and the nature of the network in terms of the available coverage and capacity.

- **Cell type:** Different cell types and sizes require different configurations (e.g. for coverage, capacity, or user behaviour). These differences must be contextualized, e.g. as macro, micro, or outdoor/indoor cell.
- **Network states:** The network may be in different performance or fault states, each of which, may require a different kind of response from the automation function. These states may consider the nature of the alarm status, e.g. major failure, critical failure, etc. or by the performance states such as congested, outage, or according to the KPI values.

Amongst the most considered traffic and service contexts are:

- **Traffic patterns:** Network traffic fluctuates heavily both in time and space. As such, contexts may be considered on the different statistics or patterns of the traffic. Typical contexts include the peak traffic, the lowest traffic and the variance of traffic between different time points (e.g. a weekday vs a weekend). For example, it has been found that the main determinant for weekend traffic patterns is day of the week, yet this has insignificant effect on weekdays where the traffic patterns are instead mostly specific to groups of networks elements [4].
- **User speed:** Typically, users will have different speeds in any environment. The context may as such consider different values on a statistic of the velocity. The most widely used statistic is the simple average velocity although others such as the range or the moving average are used. Moreover, for the considered statistic, the contexts may also be considered related to either time, location or both.
- **Service types:** As the network typically concurrently services different kinds of services, the service type may also be another dimension of the context-space. Service context may be in broad categories such as telecommunications, entertainment and media, and Internet/Web services, but typical contextualization considers the specific service types purchased by users and for which characteristics can be enumerated. These mainly include voice communication, messaging, on demand or streaming video and connectivity e.g. leased lines. Internet connectivity may also be considered a single context but as more and more over the top (OTT) services are developed over the internet connectivity as a platform, for example, WhatsApp, YouTube, etc. sub contexts are being designed so that a context can either be a specific OTT service or a set of related OTT services.

Contexts may also be function specific which then requires the functions to be configured differently as per the contexts. Example differences include the function type or its scope. The type differentiates functions according to their nature, for example, between: physical NAFs for functions that change the physical characteristics like antenna tilts and power; and logical NAFs for those that change logical parameters like handover settings. The scope defines the area of control of the function, in which case, functions may be differentiated amongst intra-cell

NAFs like congestion control; inter-cell NAFs like load balancing or Automatic Neighbour Relations (ANRs); or multi-cell NAFs which may include interreference management or PCI assignment.

7.1.2 NAF Context-Aware Configuration

Given a NAF, for example, a SON Function that controls network parameters, will ensure that the behaviour of this NAF accounts for changes in context. Moreover, the different contexts may have different performance targets and the different KPI targets may even be competing against each other, implying that they cannot be achieved together. All this justifies the need for context-specific behaviour which, in case of SON, requires a context-specific configuration of the SON Functions.

The KPI objective is the expected optimization action for the specific KPI, i.e. either minimizing/maximizing over, or fitting to a value range. The KPI targets are specifications of the objectives and the values or value ranges that must be achieved by the NAFs either as individual entities or as a group. Typical examples for KPI objectives and targets are given in Table 7.1. The KPIs may be prioritized amongst each other depending on their expected effects on the offered services in the different contexts. For example, since energy consumption KPIs focus on the optimization of operational expenses while dropped calls directly translate into poor service quality, the dropped-calls KPI targets are likely to be prioritized higher than energy saving KPI targets in scenarios where premium services are offered.

To account for different contexts and priority KPI targets, a Cognitive Autonomous Network (CAN) applying a context-aware auto-configuration mechanism selects FCV according to the objective model and context class. As illustrated in Figure 7.2, the context classes are the combinations of different cell and environment attributes while the objective model describes how different rules on KPI targets are weighted depending on contexts.

Table 7.1 Typical examples for KPI targets.

KPI	Objective	KPI target	Description
Dropped call rate	Minimize	<2.5%	The percentage of dropped voice calls due to, e.g. failed handovers or bad radio conditions
Cell load	Minimize	<90%	The radio resources usage per cell
Handover success rate	Maximize	>99.5%	The percentage of successful handovers between cells
Energy consumption	Minimize	<80%	The average consumed energy by the base station compared to the maximum energy consumption

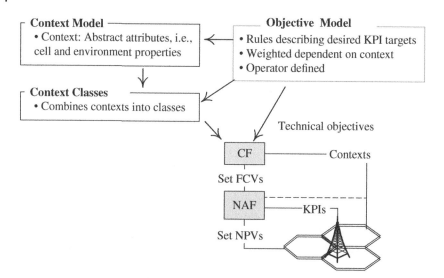

Figure 7.2 The cognitive function applies a context-weighted objective model to derive context-specific configuration values for the network automation function.

7.1.3 Objective Model

The objective model describes the targets that should be achieved by the NMA system and their relative prioritizations. The operator needs to define the importance of the KPI targets in a way that allows them to be traded off against each other. This can be expressed by allocating a priority measure to the individual KPI targets that indicates the precedence of the specific KPI targets. Prioritizing the KPI targets in a context-specific way then creates the technical objectives, i.e. a technical objective is a specific combination of context-specific KPI targets with a prioritization measure. Therefore, the objective model is the description of these technical objectives.

The objective model formalizes the operator's desired outcomes of the automation processes. Each outcome must specify the prioritization of a target for a given KPI in one or more specified contexts. Examples of such required outcomes highlighting the priority, the objective, the KPI, and the contexts are listed in Table 7.2.

To formalize these expectations, the objective model is implemented as a set of three-part 'IF.. THEN... WITH' rules of the form:

IF condition **THEN** KPI target **WITH** priority measure

The *condition part* is a logical formula over predicates, which evaluates context properties to determine the specific contexts to which the rule applies. This allows specification under which conditions, e.g. time periods or cell locations, a given

Table 7.2 Examples of desired outcomes highlighting the contexts, objectives, KPI targets, and prioritization.

No.	Priority	Objective	KPI target	Context
1	With very high priority	Minimize	Cell load to less than 80%	In an urban location during peak hours
2	With high priority	Minimize	Dropped call rate to less than 2%	In an urban location
3	With moderately high priority	Maximize	Handover success rate to more than 98%	During peak hours
4	With moderate priority	Minimize	Energy consumption to less than 40%	In non-urban locations
5	With very low priority	Minimize	Energy consumption to more than 50%	During periods with low traffic

prioritization of KPI target is active. The case of a general objective that is always true can also be represented in the form of such a rule by using an empty condition which will always evaluate to a logical true.

The *KPI target*, as defined in Section 7.1.2, specifies the KPI that the system should optimize and the desired value or value range to be achieved.

The *priority measure* encodes the relative importance of the KPI target to the operator in the specific context. The simplest solution is to prioritize the rules using values with non-bounded limits, i.e. one objective each with priority values 1, 2, 3, ..., etc. However, this raises the challenge that managing the prioritization requires a strict ordering of the rules, which then makes the objective management very manual and tedious given the number of potential rules. Instead, a better solution is to use weights with bounded limits which ensures that where the targets may be conflicting in a given condition, the weights can be used to prioritize the rule that sets precedence. In such a prioritization, for example, each rule is encoded with a weight in the range [0,1] that indicates the degree to which the rule may be prioritized, i.e. the degree to which the rule is important. Therein, a value of 0 indicates that the achievement of the target is unimportant while 1 indicates that the target must be achieved regardless of other constraints. With the five discrete priority levels very low, low, moderate, high, and very high as the weight ranges {[0.0–0.2), [0.2–0.4), [0.4–0.6), [0.6–0.8), [0.8–1.0]}, examples of such a machine-readable form may be as illustrated by the formal objective rules in Table 7.3.

Table 7.3 Examples of objectives and their respective formal technical objectives defined using prioritization weights.

No.	Desired outcome	Formal objective rule
1	With a very high priority, minimize the cell load to less than 80% in an urban location during peak hours	**IF** time in [08 : 00, 17 : 59] **AND** location = urban **THEN** cell load < 80% **WITH** 0.9
2	With a high priority, minimize the dropped call to less than 2% rate in an urban location	**IF** location = urban **THEN** Drop Call Rate < 2% WITH 0.7
3	With a moderately high priority, maximize the handover success rate to more than 98% during peak hours	**IF** location = urban **THEN** Drop Call Rate < 2% WITH 0.7
4	With a moderate priority, minimize energy consumption to less than 40% in non-urban locations	**IF** *NOT* location = urban **THEN** energy consumed < 40% WITH 0.5
5	With a very low priority, minimize energy consumption to more than 50% during periods with low traffic	**IF** *NOT* time in [08 : 00, 17 : 59] **THEN** energy consumed < 50% WITH 0.1

The operator may also impose an extra constraint that, for a given context, the weights of the KPI rules must always sum up to 1. This ensures that the degree of importance of each objective is stated, relative to the other objectives that apply to that context, as opposed to a generic statement of how important the objective is. For example, one could consider the call-drop rate, energy saving, and cell load in a given context to have the targets 0.3, 0.4, and 0.3, respectively, i.e. that saving energy is more important compared to minimizing overload and call drops. This might be a possible objective in multi-layer scenarios where the cell overlays assure coverage which makes call drops unlikely, and where adequate capacity is also available that congestion events are minimized. In that case, a configuration that fulfils the energy saving objective will be favoured, but only if it does not affect the other targets.

The objective model does not need to be complete, i.e. not all KPI targets need to be defined in all contexts. Since the KPI targets are defined globally, it follows that for those contexts where no specific prioritized objectives are defined, the system considers the default global objective defined in the KPI target and considers all KPIs in such a context to be equally important. In that case, the KPI objectives are only added on top of the global KPI target to allow for the expected outcomes to be varied for a subset of the context space.

The prioritization of KPI targets does not need to be unambiguous in some specific context. It is acceptable for one KPI target to have two different assigned priorities or weights, which can be the case if two objective rules with overlapping conditions and the same KPI targets are triggered. An overlap therefore means that at least one specific context exists in which both conditions are true. In such cases, this conflict is resolved by solely considering the higher priority. For instance, if both energy consumption rules in Table 7.3 (rules 4 and 5) are included in the objective model, the two will conflict (i.e. will both be true) for non-urban areas outside the time period [08 : 00, 17 : 59]. In this case, rule 4 will take precedence since it has a higher priority (higher weight) even if its desired target is actually lower than rule 5.

Rules are not the only way of modelling the technical objectives. An alternative solution for modelling the technical objectives is to define a utility function which maps the contexts to utilities for the KPI targets [2]. Whereas priorities only allow to rank the KPI targets according to their importance, these utilities would allow a trade-off between the satisfaction of different KPI targets. This is especially useful if there are conflicting KPI targets like the minimization of the energy consumption and the minimization of the cell load. However, the elicitation of the utility function requires far greater effort than the writing of objective rules.

7.1.4 Context Model – Context Regions and Classes

The context model provides a description of the context properties, i.e. it defines the domain or possible values of the context properties. For example, for the time and location contexts respectively defined over the ranges [00 : 00, 23 : 59] and {wilderness, rural, semi-urban, urban}, the context model is:

time : [00 : 00, 23 : 59]
location : {wilderness, rural, semi-urban, urban}

Owing to the possibly infinitely large context-space in networks, it is inefficient to derive configurations that are specific to each possible context combination. Instead, it is desirable to base the configuration decisions of the CF on combinations of the contexts. Correspondingly, the different contexts can be combined into context classes each of which may require a different action. In the case of the time and location example above, the example context classes are the four classes A, B, C, D illustrated in Figure 7.3. Therein, the context classes may be interpreted under normal conditions as four levels of traffic corresponding to very low, low, moderate, and high traffic.

To derive the context classes, the system needs to build up the complete context-space, i.e. a listing of all possible expected contexts which are derived

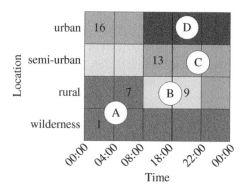

Figure 7.3 Context classes for the 'Time' and 'Location' contexts.

from the context model. Each context property in the context model represents a dimension in the context-space which can be continuous (like time with the scale [00:00, 23:59]) or discrete (like the location with the scale over the values wilderness, rural, semi-urban, and urban).

For manageability, the context-space needs to be divided into a finite number of context-space regions, or simply context regions. Each context region is a specific overlap of all the objective rules and as a coordinate in the context-space, represents a region in the context-space with the same KPI objectives and priorities, i.e. a sub-space of the context-space in which the NMA system should behave similarly. Some examples of such regions are shown in Figure 7.3, i.e. the regions 1, 7, 9, 13, and 16, where Region 16 represents the tuple (time in [00:00, 03:59], location = urban) and context class A has seven context regions including regions 1 and 7.

The number of regions can become very large and may not be useful for meaningfully grouping function or network configurations according to contexts. For instance, a Context model with 10 parameters and one threshold for each parameter results in 2^{10} regions. For this reason, regions with the same expected behaviour for a specific KPI can be combined. This results in the definition of a *context class type* as the combination of dimensions and the *context class* as the group of regions with identical targets and priority for a given context class type. Note that the regions do not need to be congruent in any or all dimensions. Figure 7.3, for instance, shows four context classes (A, B, C, and D) for the context class type 'time and location'.

The context model may be defined by the operator but, owing to the complexity of defining and delineating the complete context space, it should preferably be derived automatically. The next section describes a candidate method of deriving a context-space from the objective model, albeit without any guarantee that the derived space is complete.

7.1.5 Deriving the Context Model

The context-class types are ideally derived from the automation functions which must define the different dimensions and regions of interest. An ESM function may, for example, define three regions of interest that indicate aggressive, active, or laissez-faire energy saving based on the time of the day and location as the dimensions of interest. However, even where such functional-based definitions are possible, the context-class types must be initialized by the operator as part of the objective setting process, i.e. the operator must specify the contexts of interest and then set the desired targets in those contexts. Note, however, that the context space defined in this form by the operator does not need to be complete, it may be defined only for the contexts for which the operator has specific KPI targets. Then, the complete delineation of the context space is left to the automated context definition which derives the context classes and their types from the objective model and, where applicable, also using the functions. In the following, an example is described of how contexts can be derived from the objective model.

Computing the context classes first requires computation of the context regions. The regions can be computed by first partitioning the dimensions of the context-space according to the conditions of the objective rules: for each predicate p, the dimension of the context property in p is partitioned according to the value in p. The partitioning is further refined with each rule that has a predicate with the same context. For instance, for the predicate of '*time in [08 : 00, 17 : 59]*', the dimension for the context property *time* is split into three partitions: [00 : 00, 07 : 59], [08 : 00, 17 : 59], and [18 : 00, 23 : 59]. Then, if another rule has the predicate '*time in [22 : 00, 03 : 59]*', the first and third portions of the dimension for the context property *time* are split into two partitions each to give the full set of time contexts as {[00 : 00, 03 : 59], [04 : 00, 07 : 59], [08 : 00, 17 : 59], [18 : 00, 21 : 59], and [22 : 00, 23 : 59]}, see Figure 7.3. After partitioning the dimensions for all objective rules, the regions are defined as the elements of the cross product of the partitions of all dimensions. In Figure 7.3 for example, the combination of the contexts '*time in [00 : 00, 03 : 59]*' and location = wilderness result in context region 1.

After creating the regions, the context classes are created in consideration of the KPIs in the objective rules. First, a context-class type is created for each KPI by combining the context-space's dimensions of interest for the KPI that are represented in the Objective rules. For example, for energy saving with the two rules 4 and 5, the important dimensions are location and time and so the two are combined to create a context-class type.

To derive the context classes for the specific KPI, the objective rules are mapped to the regions to mark each region with the expected target and priority. As stated earlier, where there is a conflict amongst the prioritized targets, the higher priority

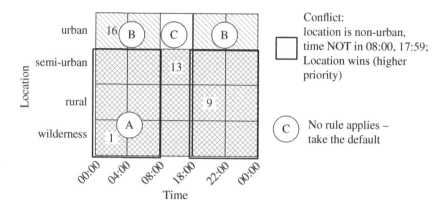

Figure 7.4 The 'Location + Time' context classes for Energy savings for the objectives {**IF NOT** location = urban **THEN** EC < 40% **WITH** 0.5 and **IF NOT** time in [08 : 00, 17 : 59] **THEN** EC < 50% **WITH** 0.1}.

rule is taken. For example, rule 4 will take priority in the non-urban areas in the time outside [*08 : 00, 17 : 59*]. The context classes are the groups of context regions with the same targets and priority. For the energy consumption KPI, these as in Figure 7.4 are context-class A for the target *EC < 40% with weight 0.5* (all time, non-urban areas) and context-class B for target *EC < 50% with weight 0.1* (urban locations, non-working hours).

Any context regions that remain unmarked after applying the rules are also classified as a single context class to apply a default global KPI target. Two options are possible here for setting the default global KPI target and can be configured by the operator. If the governing assumption is that all KPI targets are moderate priority unless specifically stated, then the default global KPI target will have median priority (i.e. weight = 0.5). Otherwise, however, if it is interpreted that all rules take the lowest priority unless otherwise stated, then default global KPI target will have the lowest possible weight.

7.1.6 Deriving Network and Function Configuration Policies

After computing the context classes, the final step of the context-specific configuration is the derivation of configuration policies, i.e. the function configurations or function behaviour models for each context class.

In the case of SON, a SON Function is typically designed and implemented with a pre-set number of operating modes, each of which indicates a different kind of behaviour. The ESM SON Function, earlier discussed for example, may have the three operating modes – aggressive, active, and laissez faire. For such SON Functions, this step involves matching the possible modes with the computed context classes, which for ESM implies matching the context classes (in Figure 7.4) to the three defined modes. This is implemented by updating the policy framework of the NMA system with the configurations indicating this matching, i.e. aggressive ESM for class A, active ESM for class B, and laissez faire ESM for class C. The corresponding SON Function will then activate the respective FCV set for the specific mode in each of the context classes.

Where the SON Function does not define the operating modes, but where more than one context-class has been identified for that SON Function's KPI(s), the CF that manages the SON Function configuration (referred to in [2, 3] as the SON objective manager) learns the appropriate FCV sets for each of the context-classes. A good candidate for such learning would a RL agent that tests different sets of FCVs for a specific SON Function to identify the best set in a specific context class. Note however, that a model-based FCV configuration is also feasible (as is proposed in [2, 3]), albeit with great effort in modelling the SON Functions to derive the FCV-set to network performance relationships (referred to in [2, 3] as the SON Functions' effect models).

The most interesting case is the case of cognitive network automation where the CF directly controls the network configuration values. Therein, there is no need for the intermediate step of computing the FCVs and instead the CF needs to update its internal policy function to directly respond to the context changes. Such an internal change may imply either learning using different hyper parameters or using the same hyper parameters but learning different models or NPVs. An example where an RL function is used for ESM, and the three ESM context-classes are interpreted as three modes requiring different behaviours (i.e. the aggressive, active, and Laissez faire behavioural types). Here, the RL agent may have to learn using different learning hyper-parameters for each of the three context classes. For example, such an RL agent may use a larger neural network or Q-table for the aggressive class to ensure that it learns a large and detailed model for activating or deactivating cells at a very high granularity for the time between evaluations and the periods for which cells may be activated or deactivated. For the laissez-faire class on the other hand, the RL agent may only apply a simple regression model that looks only for the time periods in which there is too much resource redundancy and simply does nothing in all the other time periods.

7.2 Multi-Layer Co-Channel PCI Auto-Configuration

A critical network configuration use case is the assignment of cell identities and an automation solution thereof is an example for auto-configuration. Automation of cell identity assignment has typically been an expert-design based solution, i.e. the system experts write rules as to when and how cell identities may be configured or reconfigured. This has typically worked well in simpler networks, but more cognitive and inherently flexible solutions are needed in more complex deployments. This section discusses how a simple learning mechanism can be employed to learn cell Neighbour relationships that are then applied to configure or reconfigure the cells' PCIs.

7.2.1 Automating PCI Assignment in LTE and 5G Radio

Demands for higher user throughput for mobile broadband applications can be addressed by densification of the Radio Access Network (RAN), where the traditional macro cells are overlaid by one or more layers of smaller cells resulting in what is called a Heterogeneous Network (HetNet). The overlay is typically called the small cell layer or simply the pico layer even although it may also include micro cells which are typically larger than pico cells. This pico layer may use the same or different Radio Access Technology (RAT) as the macro cells. In such deployments, PCI assignment becomes a complex task when the layers share the same frequency.

The well-known problem in allocating PCI is that since there are only a limited number of PCI values available (504 in an Long-Term Evolution [LTE] network), the PCIs must be reused. At the same time, however, there are constraints for the PCI assignment. For example, no two Neighbouring cells, or two cells sharing a same Neighbour, must use the same PCI value. In multi-layer, co-channel deployments these constraints need to be satisfied not only within the layers, but also in interlayer cell Neighbour relationships.

A PCI is a combination of a cell's physical-layer cell-identity group $N_{ID}^{(1)}$ and its physical-layer cell identity $N_{ID}^{(2)}$, as given by Eq. (7.1). These IDs are derived from the cell's synchronization signals from which the User Equipment (UE) obtains the cell ID and derives time and frequency synchronization with the cell during the cell search procedure.

$$PCI = 3 \cdot N_{ID}^{(1)} + N_{ID}^{(2)} \tag{7.1}$$

In LTE, there are 168 physical-layer cell-identity groups each containing three unique physical-layer cell identities, leading to 504 available PCI values. Each PCI value is related to the cell's Primary Synchronization Signal (PSS) and to the Secondary Synchronization Signal (SSS) – see Section 3.3.3 for a detailed description of these relationships.

In 5G, the PCI space has been expanded to 1008 unique IDs also given by Eq. (7.1) but derived differently [5]. The PSS still carries the physical layer cell identity $N_{ID}^{(2)}$ in {0, 1, 2} but is constructed using frequency domain-based BPSK M-sequence with generator polynomial $g(x) = x7 + x4 + 1$ and three cyclic shifts (0, 43, 86) to generate the PSS signal. The SSS carries the physical layer cell identity group $N_{ID}^{(1)} \in \{0, 1, \ldots, 335\}$. The SSS sequence is constructed using two frequency-domain-based Binary Phase Shift Keying (BPSK) M-sequences with generator polynomials $g0(x) = x7 + x4 + 1$ and $g1(x) = x7 + x + 1$. As in LTE, 5G only has three PSS implying that every third PCI will have the same PSS value, i.e. PCI mod 3 will be the same after two PCIs. This can cause interference on the PSS and delay the synchronization. Also, 5G uses a DeModulation Reference Signal (DMRS) in Physical Broadcast Channel (PBCH) and these follow PCI mod-4 rule which means that every fourth PCI has the same location for PBCH DMRS in the frequency domain.

It was shown for co-channel scenarios [6] that, as networks get denser, existing state-of-the-art-methods are unable to effectively assign PCIs in each cell layer independently. Specifically, these solutions are likely to compromise the PCI integrity in inter-layer adjacencies. PCI confusions, especially, increase dramatically, as often two small cells with the same PCI (equi-PCI cells) are both Neighbours to the same macro cell.

7.2.2 PCI Assignment Objectives

PCI assignment objectives are comprehensively discussed in [6, 7] and in Section 4.3.3, but, for completeness, the most critical ones are briefly reviewed here especially for the case of 5G new radio. Although seemingly simple, PCI assignment is not trivial owing to the limited number of available values and the need to minimize conflicting PCI values between Neighbouring cells. With limited PCIs, values inevitably need to be reused. Consequently, a good assignment should at least ensure to:

1) Minimize the number applied PCIs.
2) **Avoid PCI collision** as illustrated by Figure 7.5a, to avoid ambiguity in cell measurements.
3) **Avoid PCI confusion** as illustrated by Figure 7.5b, to confusion at cell handovers and the subsequent handover failures.
4) **Avoid PCI mod 3 and PCI mod 4 conflicts.**
5) **Minimize the need for PCI reconfigurations** and the related cell restarts since each reconfiguration requires a restart.

PCI conflicts are typically avoided by defining a Safety Margin (SM), which is the minimum number of cells between two cells C1 and C2 sharing the same PCI value. For example, as shown in Figure 7.5c, if SM is set to zero, direct Neighbours may share the same PCI, while SM = 2, the minimum required for confusion-free assignment, leaves two cells between C1 and C2. Meanwhile, a larger SM (>2) is often necessary to allow for network extension. For example, if more cells will

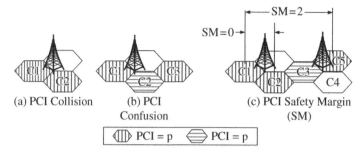

Figure 7.5 Critical PCI conflicts and the safety margin (SM).

be added to the network after initial deployment, a larger SM allows a PCI to be assigned to the new cells without changing PCIs of the existing cells.

7.2.3 Blind PCI Auto Configuration

The main complexity in co-channel multi-layer deployments comes from two contradicting requirements. Firstly, it is desirable to assign the PCIs separately for each layer, to remove the need to share Network Management (NM) information between them. This would require advanced features from the small cell layer, in case it is not managed by the same NM solution as the macros and can often mean multi-vendor integration. On the other hand, it is paramount to ensure optimal performance, i.e. that the PCI conflicts are, as much as possible, minimized.

The appropriate solution, as proposed in [7] as the PCI Assignment Function (PAF) shown in Figure 7.6, learns the relationships amongst the cells using its observations from radio measurements. Consider a network with a macro and a (typically independent) pico layer that:

1) the pico layer cells do not have access to macro layer neighbour relations (NRs). Such is the case, when pico cells have no X2 interface [8] and so cannot request the information from their Neighbours
2) pico layer Operation, Administration, and Management (OAM) systems do not know macro NRs, a typical case when the two layers are supplied by different vendors with incompatible OAM systems
3) the PCI range is separated into subranges, one for each layer.

The PAF automates the PCI assignment using four sub-functions, one for each of four sequential processes.

Firstly, the PAF blindly assigns (or configures) PCIs to the pico layer cells with each cell independently assigned. It then learns the actual NRs, beginning with the pico-macro NRs and then learning the macro-macro NRs using the learned

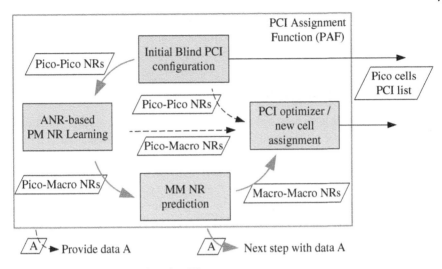

Figure 7.6 PCI assignment function [7].

pico-macro NRs. Lastly, it optimizes the PCIs assignments using the full graph of NRs as learned in the previous steps. The subsequent section discusses how the individual steps are accomplished.

7.2.4 Initial Blind Assignment

For the initial pico layer configuration, the PAF blindly assigns PCIs to all pico cells from the pico layer PCI range, ensuring that PCIs are re-used far enough apart that there are no PCI confusions. The problem here is on deciding the appropriate SM so that PCI confusions are avoided, but yet PCIs are assigned efficiently and not wasted. This is estimated utilizing the expected macro cell coverage radius/distance, or simply the macro cell radius and the pico cells Inter-Site Distance (ISD). For confusion-free assignment, the pico SM is set according to Eq. (7.2).

$$SM \geq \frac{MacroCell\ Radius}{SmallCell\ ISD} + 2 \tag{7.2}$$

Equation (7.2) fulfils the minimum requirement for confusion-free assignment, i.e. that no two equi-PCI pico cells should be Neighbours to the same macro cell. For pico islands, i.e. subnetworks of pico cells which are not connected to other cells in the pico layer with an adjacency, the reuse distance is determined using geography, i.e. a PCI is reused in two cells if the distance d between them is

$$d \geq MacroCell\ Radius + 2 * SmallCell\ ISD \tag{7.3}$$

The assignment uses a pico-pico NRs graph, in which NRs are obtained by applying geometry and radio prediction rules (similar to those used in network planning tools) to select the pico cells that are likely to be Neighbours.

7.2.5 Learning Pico-Macro NRs

After the pico PCIs have been blindly assigned, the PAF needs to learn the actual macro PCIs in order to optimize the initial assignments and make optimal new assignments, if needed. The macro PCIs are learned during operation from information provided by the ANR process [9], through which unknown NRs for a given cell are learned. As shown in Figure 7.7, each pico cell learns its macro NRs by requesting its associated UEs to read the Neighbour cells' PCI and Cell Global Identity (CGI) after which it updates OAM with the newly learned NR (steps 1–3). This is the standard ANR procedure to which a new step is added to achieve the desired objective. In this last step (step 4), the OAM updates the PAF with the new pico-macro NR. Note however, that the final updates can be combined by having each pico cell directly update the PAF with the new pico-macro NRs as soon as they are learned.

7.2.6 Predicting Macro-Macro NRs

In order to enable the later extension of the network with new cells, the used SM must be sufficiently large to incorporate cell additions. Furthermore, the SM must consider extensions in the pico as well as the macro layer, which requires complete

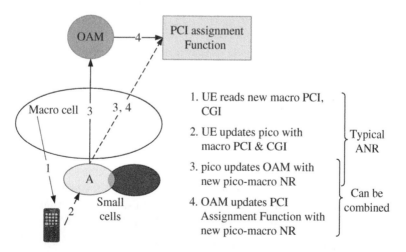

Figure 7.7 Learning macro cells' Neighbour relations.

knowledge of all NRs at the PAF. Thus, macro-macro NRs need to be derived for the pico layer.

Using the list of pico-macro NRs (as learned according to Section 7.2.5 above), the macro-macro NRs can be derived from the observation/rule that:

Rule: if two (2) macro cells are Neighbours to one pico cell, then the two macro cells are Neighbours to each other.

This rule is derived from the observation that pico cells are so small that they cannot span two macro cells that are not Neighbours to one another.

7.2.7 PCI Update/Optimization and New Cells Configuration

After the learning process, all three NR types (pico-pico, pico-macro, and macro-macro) are available to the PAF and a complete NR graph for the network can be constructed from the three NR types. Then, applying graph-colouring-based methods, this complete NR graph can be used to either optimize the initially blindly assigned PCIs or to assign PCIs to new cells that are added to the network. Note that a subset of the complete solution (without the learning function) is possible if pico-macro NRs and macro-macro NRs are available. These NRs can, for example, be derived from macro and pico planning data, if available. Similarly, macro-macro NRs could be retrieved from the macro layer OAM system. Either way, the separation of ranges and the availability of the complete NR graph ensures that PCIs are assigned to different layers separately and in a confusion-free manner.

7.2.8 Performance Expectations

The potential benefit of learning-based PCI assignment for LTE was evaluated in a simulation of a Ultra-Dense network (UDN) deployment as described in [7], where the evaluation also compares the ANR-based learning (called blind Small Cell layer) with simple range separation as well with blind assignment before learning the NRs (called blind assignment). The evaluation considers a stringent scenario where the PCI assignment is also tasked to minimize the mod30, m0, and m1 conflicts. It was observed that range separation is unusable even in low-density scenarios. It performs poorly in all scenarios and only manages to avoid PCI confusions when the SM is high. Blind assignment significantly minimizes PCI confusions and even eliminates them in low-density scenarios. It is, however, unable to reduce mod30, m0, and m1 conflicts, since that requires each cell to know the exact PCI assignments in its Neighbour cells. Correspondingly, by learning the macro NRs, the Blind Small Cell layer eliminates mod30, m0, and m1 conflicts in the PCI assignments, since it allows each cell to consider the assignments in all its Neighbour cells. This, however, comes at the cost of utilizing more PCIs than those used in the initial blind assignment. The result is that the PCI space is quickly exhausted in very dense network scenarios.

In practice, where the PCI space is quickly exhausted, the operator allows the softer conflicts (mod30, m0, and m1 constraints) to be violated instead to allow all cells to be assigned with PCIs. However, the knowledge of the cell relationships allows for the most minimal or least impactful violations. These results clearly show that combining an initial blind PCI assignment with the ANR-based learning of cell NRs allows the PAF to compute assignment based on the full network graph of the assigned PCIs. Subsequently, it achieves the desired compromise of enabling independent assignment in each layer, while minimizing the subsequent conflicts. Although the available results simulated for LTE, similar performance is also expected in 5G where the increased PCI space is expected to be confronted with even higher cell density.

The PCI assignment challenge has demonstrated an example of how simple cognition can be used to improve the outcomes of automated network configuration. It is an example of a use case that is applicable both at commissioning and during network operations. For example, if a new cell is being commissioned, the PCI assignment procedure can be triggered to do the initial assignment as well as to improve the cell's assignment after the NRs have been earned. On the other hand, in case of any changes in a given cell's propagation environment, the PCI assignment may be re-triggered to relearn the NRs based on the new propagation condition and to reoptimize the PCI assignment as necessary.

7.3 Energy Saving Management in Multi-Layer RANs

Cellular network traffic varies over the course of a day, as illustrated by the typical traffic profile in Figure 7.8. For example, the maximum busy hour load can be many-fold of the load in the quiet hours [10]. Yet, since networks are dimensioned to guarantee the desired Quality of Service (QoS) and avoid congestion during peak traffic, this results in over-provisioning and unnecessary energy consumption outside the busy hours. In Heterogeneous Networks (HetNets), additional layers – typically small cells – are deployed to enhance the capacity of the network, while the macro layer guarantees adequate coverage. Obviously, network energy consumption can be reduced if the capacity enhancing cells are deactivated whenever they are not needed.

The energy consumption of the RAN does not scale with the traffic, there's a large component that is constant regardless of the carried traffic. Note that there are attempts to make the energy consumption scale with traffic, but no successful results have been published. As such, significant Energy Saving (ES) can be achieved since the radio network makes up to 90% of the total energy consumption of a mobile network [11], and this holds even if only the power amplifier is switched off, given its share of up to 50–65% of the overall base station power consumption [12]. This section presents concepts and solutions on Energy Saving Management (ESM) for different network contexts. The presented solutions emphasize more the autonomy of the network, or elements thereof, in making the ESM decisions, although cognitive concepts can be added, e.g. to learn the critical trigger thresholds.

7.3.1 The HetNet Energy Saving Management Challenge

Many publications have justified the need for ESM (e.g. [13–16]). However, only a few [17] delivered concrete solutions for distributed HetNet environments and even fewer for the problem of optimal switch-on order. The selection of cells to be switched on or off is not trivial. For example, when the cells are off, their potential load is not known (without using a beacon feature to listen for UE within range). A simple solution is to activate all inactive cells. This is very inefficient, since unnecessary capacity is provisioned, as shown in Figure 7.8, even with a subsequent action to again deactivate any cells with low load. Alternative approaches include, for example, activating cells based on a fixed time sequence or using historical data and traffic profiling to determine the switch-on order. Although better than the first approach, they are also inefficient in a dynamic HetNet environment since the cell activation/deactivation is not matched to the inherent load in the cells. The best deactivation/reactivation decision should, as described here, be based on the aggregated traffic in the area and should aim to maximize the network's spectral efficiency.

A similar challenge is true in selecting a cell to switch off. The typical assumption is to deactivate a cell whose load has significantly decreased (preferably zero).

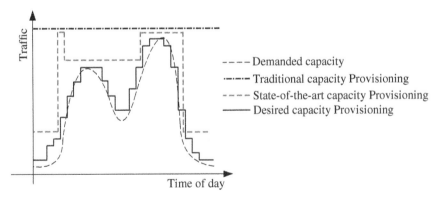

Figure 7.8 A typical diurnal load pattern contrasted against the traditional and the state-of-the-art and the desired capacity provisioning.

However, since the Neighbour cells serve a common region (or the exact same region in the HetNet scenario), the load will typically increase or decrease concurrently in all cells. Consequently, waiting for one cell's load to decrease implies waiting for the load in all cells, and so the ES objective will not be achieved. As such, the best alternative must consider the group of cells together but choose one of the cells to activate or reactivate even when the load in that particular cell is not zero for as long as the other cells will take that load.

For ESM in a Multi-layer UDN, three problems must be solved:

1) How to group the cells to monitor the load and the free capacity in the network and how to select the cell(s) that can be put to energy saving state?
2) How to ensure that the coverage layer has enough capacity to compensate for the cells that are switched off?
3) How to select the cells to activate if multiple cells are inactive?

These problems translate into three configuration challenges: (i) configuring the groups of cells which must be collectively evaluated for Energy saving (herein after called Power Saving Groups [PSGs]), (ii) Configuring the order in which the cells in the group should be (de)activated, and (iii) configuring the thresholds that control the cell switch off-switch on procedures.

7.3.2 Power Saving Groups

A PSG is a group of cells, which cover a given area and amongst which some cells can compensate for those cells that can be deactivated when there is low demand. Within a given area, cells need to be grouped in PSGs and within each PSG the cells are categorized into two types: (i) reference cells that ensure coverage and (ii) helper cells that provide extra capacity and throughput but can be deactivated and reactivated as needed.

The simplest solution for defining PSG is to consider a PSG as the set of overlapping cells. This is transferred from the historical deployments where capacity cells were installed at the same location as the original coverage cell, e.g. the case of deploying a capacity cell in the 1800 MHz bandwidth at the same location as a 900 MHz coverage cell. In HetNets, however, it is no longer the case that coverage and capacity cells are deployed at the same site, specifically, they will not have completely overlapping coverage. For example, it is the case that a single LTE cell may overlap with multiple LTE and 5G small cells each of which only covers a small part of the coverage cell's area. Moreover, some of the cells will provide capacity to more than one coverage cell, e.g. if such a small cell is deployed at the boundary of two coverage cells. And such a small cell may also contribute to the coverage, e.g. in the case where the coverage at that boarder area is as good as desired. In such cases, the simple rule of selecting co-located or overlapping cells as a PSG will not work. Instead, the selection of the PSG must concurrently select the reference cells as it selects the PSG memberships.

For the area of concern, the set of reference cells must together provide full coverage for the area. So, all the cells that together provide the coverage, including small cells that contribute thereto, must be added to the list of reference cells. Then, to simplify the PSG selection, a PSG can then be defined per reference cell to be the list of all the Neighbour cells to it that are not themselves reference cells. The other cells are considered as helpers to the reference cell and are candidates for deactivation (and reactivation). Note from this definition that a helper cell can, in practice, belong to more than one PSG with the implication that an accurate evaluation of its effects has to consider its contribution to all its PSGs. However, allocating each helper to only one PSG will simplify the ESM computations and if the thresholds are selected appropriately will not cause adverse overload effects.

7.3.3 Cell Switch-On Switch-Off Order

Cells that are candidates for deactivation and reactivation need to be selected based on their expected spectral efficiency, i.e. always ensuring to retain those helpers that result in the highest spectral efficiency for the network/area. This requires, for a given PSG, that activation starts with the small cells which are furthest, in radio terms, from the reference cell, i.e. the cells which are closest to the reference cell's edge. A user that is close to such a small cell would have the worst spectral efficiency at the reference cell. So, if that user is transferred to the small cell more resources are freed at the reference cell. Deactivation then goes in the reverse direction starting with the small cells that are closest to the reference.

This deactivation-activation mechanism requires that the ESM engine has knowledge of the radio distances of these cells from the reference cell. The radio distance here implies a measure of how much the coverage areas of the cells overlap, i.e. it must account for not only the physical separation of the transmission points but also the respective antenna radiation patterns, gains, azimuths, and tilts.

Given the above requirements, the cell selection becomes analogous to the triangulation of heat flow, as shown in Figure 7.9 [18]. The figure depicts a reference macro cell with seven helper small cells that overlap its coverage. The triangulation of heat flow considers the load at each reference cell j as an amount of heat induced in the cell due to the small cell secondary sources whose mobile devices are the primary distributed heat sources. According to the analogy, maximum load (heat) is induced if the primary source (the UE) is at the edge of the reference cell, and so, maximum load is transferred from cell j if a new small cell i is activated at or closer to the edge of cell j.

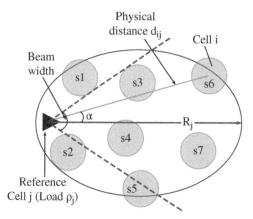

Figure 7.9 Selecting cell switch on/off candidates with one reference cell.

Consider a reference cell j with cell range R_j and having a set of helper cells $i \in I$ ($i = 1, 2, 3, \ldots$) as shown in Figure 7.9. For the helper cell i with unit load and a distance d_{ij} to cell j, the induced heat intensity (from a hotspot near and covered by i), is given by Eq. (7.4) [18].

$$h_{ij} = \begin{cases} [r_{ij}]^r & ; \quad r_{ij} < R_j \\ [r_{ij}]^r \cdot \left[1 - \left[\dfrac{r_{ij} - R}{R} \right]^{\frac{1}{r}} \right] & ; \quad r_{ij} > R_j \end{cases} \qquad (7.4)$$

$$\text{with } r_{ij} = \frac{d_{ij}}{cos^\tau(\alpha)}$$

In Eq. (7.1), α is the angle between the direction of cell j and the line between cells j and i, which for an Omni-directional cell would be $\alpha = 0$, since the line between the two cells lies along the path of maximum gain. Differences in propagation environments may be accounted for using the coefficient of heat floor r which is assumed to have a default value of 1. Since no study has been found to justify other values, the subsequent sections also assume a value $r = 1$. τ is the beamwidth factor that accounts for how much, for a given distance d, the received signal changes as a function of the antenna's beamwidth. Specifically, for the different antenna beam-widths of 60°, 90°, 120°, or 360° (omni), $\tau = 0.3; 0.45; 0.5; 1$, respectively. Note that the heat, i.e. the cell load, can be measured in terms of carried data, e.g. Mbps, or in terms of used cell resources, e.g. LTE Physical Resource Blocks.

7.3.4 PSG Load and ESM Triggering

The execution of cell activation and deactivation can be done either in a central location like the OAM or in a distributed form at each individual cell. However, a hybrid solution where the algorithm is executed at the reference cell may, in many cases, be the preferred alternative since it allows for fast response based on the observed PSG load and yet it avoids the complexities of evaluating the decision at multiple points, i.e. in the small cells.

A critical challenge of the execution is the evaluation of the PSG load, i.e. an optimal decision needs to ensure that the actual load in the PSG can be handled by the remaining cells before one is deactivated or that a cell is reactivated just in time to take any increase in load. A weighted average of the load that gives more weight to the reference cell can be considered. This is, in general, a complex decision and in general also needs to consider the weighted average of the load in the cells of the considered network scope. Ideally, the execution is triggered if this load average increases above a threshold ThH or reduces below a threshold ThL. However, if the UEs are always pushed to the small cell layer via Traffic Steering (TS), the activation decision based only on the reference cell load will be just as accurate. This would be a special case of the weighted average load scenario in which the reference cell weight is set to one while the helper cells weights are all set to zero. This is the assumed trigger mechanism here owing to its ease of implementation and its likelihood in practice since TS is default in many deployments.

Note also that: (i) the appropriate values of the thresholds ThL and ThH can be determined subjectively and operationally; and (ii) the ESM engine can optionally be periodically activated to calculate the induced heat at all active and inactive cells to decide if some cells could be activated or deactivated.

7.3.5 Static Cell Activation and Deactivation Sequence

The ideal cell activation and deactivation procedure must account for the load in the different small cells and how much load each small cell is likely to induce in the reference cell. This then requires continuous dynamic evaluation of the load to select the cells to be activated or deactivated at each time instant. A simpler static solution can, however, also be considered in which case the obtained ordering of the cells can be used as a fixed sequence for cell activation and deactivation.

Without considering load, the heat flow mechanism in Eq. (7.4) is used to compute the cell activation order for the given PSG. E.g. for the PSG in Figure 7.9, the deactivation order may be $s6 \rightarrow s5 \rightarrow s7 \rightarrow s1 \rightarrow s3 \rightarrow s4 \rightarrow s2$ with the activation order being the reverse. Note that although s7 is further than s5, it is possible that activating s5 results in higher spectral efficiency compared to s7, since owing to their angles from the reference cell's boresight, a user near s7 will have a higher SINR from the reference cell.

Such a (de)activation order can then be statically configured at network planning and can be used by different kinds of ESM solutions be it centralized or otherwise. For example, although not optimal, it can serve as a reasonable first approximation for the cell switch on/off order in a PSG. Correspondingly, a time-of-day-based ESM solution can be configured to have those cells deactivated at a specific time of day according to the sequence. On the other hand, a PSG-load triggered fixed sequence solution will follow the sequence each time the PSG load changes even if such load triggered solution does not consider the small cell load in its decision.

7.3.6 Reference-Cell-Based ESM

As stated above, the simplest PSG configuration is to consider one reference cell per PSG and reduce the challenge of finding the helper cells that best support that reference cell's coverage area. With such a PSG system, the PSG load can be effectively represented by the reference cell's load. Then, the activation/deactivation decisions can be taken at the reference cell which is expected to have visibility of all the helper cells [17].

Deactivation disables those helper cells according to the amount of their induced 'heat' at the reference cell starting with those cells inducing the lowest heat. When the reference cell's load reduces below the threshold ThL, the ESM engine deactivates small cells starting with cells with low load and/or closer to the reference cell centre. The threshold ThL must be selected in a way that when the helper cell is deactivated, the load it releases can be carried by the reference cell either on its own or in combination with the other small cells which would, in that case, take some of the load from the reference cell. Ideally, a different threshold is required for each small cell depending on the load expected to be abandoned by that cell when it is deactivated.

The helper cell users, who are moved to the reference cell through this process, will be served well, since they are close to the reference cell. Assuming the reference cell j and small cell i, the heat intensity hji will be as defined in Eq. (7.1). The ESM engine deactivates helpers with the lowest induced heat (see Eq. (7.5)), but for which the expected total load added to the reference cell is $\Delta \rho j < (TMhigh - TMlow)$.

$$Candidate = \underset{i}{\mathrm{argmin}}\{\rho_i \cdot h_{ij}\} \tag{7.5}$$

Activation of cells, although simpler than deactivation, is also not straightforward. It is simpler however, because the small cell is initially not carrying any traffic, so the decision only considers the induced heat from the reference cell j and towards the helper cells as computed in Eq. (7.6). However, since the reference cell j is the same for all considered cells, the decision reduces to a comparison of the induced heat and thus to the static sequence.

$$Candidate = \underset{i}{\mathrm{argmax}}\{\rho_j \cdot h_{ij}\} \tag{7.6}$$

7.3.7 ESM with Multiple Reference Cells

As stated earlier, there will be cases where some helper cells belong to several reference cells, as shown in Figure 7.10, and where it is desired that the activation/deactivation decisions be made in consideration of all the respective reference cells.

Cell deactivations in such a case require that the selected deactivation candidate is the one with the highest aggregate induced heat amongst the set of reference cells. For each helper cell i, the ESM engine aggregates *hij* the induced heat to each of the reference cells j. It then deactivates the helper cell with the lowest total induced heat *hij* as per Eq. (7.7), but which maintains the load in all reference cells to a value below the maximum threshold.

$$Candidate = \underset{i}{\text{argmin}} \left\{ \sum_{\forall j} (\rho_i \cdot h_{ij}) \right\}$$

$$\text{s.t. } \Delta \rho_j < (TMhigh - TMlow); \quad \forall j \tag{7.7}$$

In the case of a cell activation with more than one cell inactive, the ESM engine determines the helper cell that will assist as many reference cells as possible, i.e. it selects the helper cells that will take the most combined load from the reference cells. The ESM engine uses the ranking Eq. (7.4) and, for each helper cell i, it aggregates the induced heat *hij* from each of the reference cells j. It then activates the helper cell with the highest total induced heat *hi*, considering the different reference cell loads, i.e.

$$Candidate = \underset{i}{\text{argmax}} \left\{ \sum_{\forall j} (\rho_j \cdot h_{ij}) \right\} \tag{7.8}$$

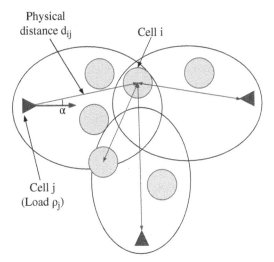

Figure 7.10 Selecting cell switch on/off candidates based on multiple reference cells.

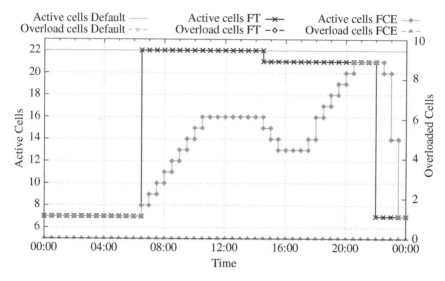

Figure 7.11 Comparing the number of active cells.

The potential performance of a reference-cell based ESM solution, called a Fluid Capacity Engine (FCE) [19], is shown in Figure 7.11 evaluated in a simulation of a UDN deployment described in [6], the figure compares FCE with two other solutions – the default where all cells are always active and a Fixed-Time (FT) cell deactivation solution. The FT represents the state-of-the-art ESM solutions in two respects: (i) it activates or deactivates small cells at fixed times respectively at $06:30$ and at $22:00$ hours, and (ii) it deactivates small cells when their traffic falls below a set threshold of $\rho i < 0.1$.

The observation is that the FCE reacts to the traffic variations in a more dynamic way. Without being manually pre-set, it still activates the first cells at the same time as FT, but only activates a few cells and then gradually activates more as the load increases. Its activations start with small cells that have the highest estimated spectral efficiency for the given traffic distribution which enables it to activate a much lower number of cells compared to FT. The effect of evaluating the spectral efficiency also enables the FCE to deactivate some of small cells when the load reduces, for example, in the early afternoon. It then reactivates more cells when the load increases again, tracking the load until the day's peak at about $22:00$ hours when aggressive deactivation is triggered.

Note, however, that even although the deactivation is aggressive after $22:00$ hours, it is still not as sudden as for the FT since it must account for low traffic that still exists in the network. And as with the activations, the deactivations also consider the cells' estimated spectral efficiency ensuring that the shutdown sequence is optimized to start with least spectrally efficient small cells, i.e. the ones that least offload the macro/reference cell.

The net benefit of this kind of ESM solutions has been noted to be up to 30% less small cells' energy consumption over the course of the day. And a major benefit of the solution is that it is technology agnostic, it can be applied even to the old RATs for as long as the cells have mechanisms for remote deactivation and activation.

7.3.8 Distributed Cell Activation and Deactivation

The centralized (or hybrid) solution like the FCE works well in many scenarios, but in some scenarios, a distributed solution may be required, For example, where the small cell layer is supplied and managed by a different vendor from the coverage cell's vendor, it may not be easy to integrate all the cells in order to have a common decision engine. In such cases, a distributed solution [19] allows small cells to independently determine when to deactivate or reactivate themselves with either or all the following constraints:

1) in consideration of network traffic
2) without a specific trigger from OAM or the macro cell
3) with minimized number of X2 links amongst the cells given the high number of small cells
4) without the need for new X2 messages (proprietary or otherwise).

The ESM module for such a solution, that must be implemented in each small cell, is initially configured, e.g. by OAM, with the cell switch-on/off ranking order. Such an order specifies the helper cell's rank in the switch-on/off process, e.g. with a lower rank indicating that the cell is deactivated before a cell with a higher rank, with the reverse true for activation. Then, each helper cell has a deactivation counter (dc) that is initially set to zero (0) with which to count-up/down the activation-deactivation sequence and to consequently track the value of the 'Highest-Ranking Deactivated Cell'.

For the Distributed ESM (DESM) mechanism to be able to base its decisions on the PSG load, means for exchanging the PSG load are needed. The simplest solution is to periodically (e.g. every 5/10/60 seconds) communicate the PSG load (here the reference cell load) to all helper cells. In this case, each helper cell requires only one X2 link (i.e. only to the reference cell) and no new X2 messages (proprietary or otherwise) are needed. For LTE, load exchange is achievable using the Composite Available Capacity Information Exchange included in the 3GPP specified X2 Resource Status Update message exchange [8] with an equivalent specification expected for 5G. To minimize the amount of X2 signalling, communication may be limited only to when the PSG load fulfils criteria for cell activation or deactivation.

With the PSG load availed to each helper cell, The ESM module in the helper cell compares the load to the low- and high-load thresholds to determine if one or more helper cells should either activate or deactivate. If so, to decide if a cell itself should be the one to activate or deactivate, the DESM in the cell:

1. Updates the deactivation counter to indicate the value of the Highest-Ranking Deactivated Cell and,
2. takes an action if the corresponding count is equal to the cell's own rank as originally configured by the OAM.

To deactivate itself, the helper cell's DESM module tracks the value of the *PSGLoad* and increases its *dc* whenever a low-load event is detected, i.e. when *PSGLoad* < *ThL*. If *dc* reaches the cell's rank, the helper cell enters the 'Deactivating' state.

For activation, even whilst inactive, the helper cell receives the PSG load at the fixed time interval (e.g. every 5/10/60 seconds), and tracks the overload and cell inactivity status. Since one cell is deactivated each time the low-load condition is fulfilled, the number of deactivated cells is equal to the deactivation events. Consequently, the 'Highest-Ranking-Deactivated-Cell' is the one whose rank is equivalent to the prevalent count of deactivation events. So, by incrementing the counter even after deactivation, the cell continues to track the highest-ranking deactivated cell at any time. When a high-load event occurs, i.e. when *PSGLoad* > *ThH*, (i) each cell compares its rank with the value of *dc* so that the 'Highest-Ranking-Deactivated-Cell' (one whose rank is equivalent to *dc*) enters the 'Activating status' to be reactivated; and (ii) all cells count down the value of *dc*. This is repeated for each subsequent high-load event either until all cells are reactivated or a low-load event occurs triggering new deactivations.

The end-to-end process can be summarized by Figure 7.12 which starts with deactivations and eventually reactivates the three cells in sequence. Each time the

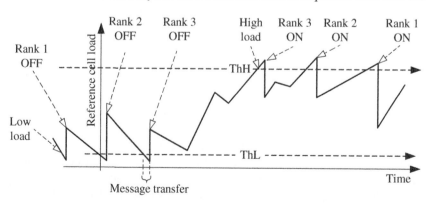

Figure 7.12 Deactivation-activation sequence for a PSG with three helper cells.

load falls below ThL, a cell is deactivated and later the cells are reactivated in the reverse order when the load rises above ThH.

7.3.9 Improving ESM Solutions Through Cognition

The solutions so far presented are all based on heuristic models and so there is a possibility that the activations and deactivations will not always follow the proposed sequences. Firstly, the heuristic models only consider generic propagation models and not the specific propagation of the specific cells in a given network. This difference in propagation implies that two cells may have different conditions from what was expected. Moreover, not all models consider the dynamic traffic characteristics in each cell. One result of these two variations may be that one cell may always be heavily loaded at the time when it is recommended to be deactivated.

Ideally, an operator will not want to deactivate a highly loaded cell. This, besides the PSG load constraint, there will typically be an extra constraint that a cell is only deactivated if its individual load is also below some threshold ThC. With such a constraint, the high-loaded cell will not always be deactivated at the time when it is due according to the sequence. Instead, either the next candidate will be deactivated or the deactivation stalls.

This conundrum can be alleviated by adding a learning solution on top of the selected ESM process. Thereby, the ESM learner will learn when and how often a given cell does not follow its prescribed position in the sequence. The learner may then propose a change in sequence to ensure that the local sequence accounts for those local disparities.

Another critical observation from the presented ESM solutions is that the achieved benefit is always highly dependent on the applied trigger values, e.g. the threshold ThL and ThH (and now also ThC if applicable as above described). It is highly unlikely that the operator can set these thresholds appropriately for each ESM scenario and context. The operator will typically set a global threshold to be applied at cells or, at best, the threshold may be set for each cell type like macro, pico, etc. However, a learning solution can be added to learn the appropriate thresholds. Thereby, an RL approach tests different threshold values for the specific local environment to determine which of them offers the best ESM benefits.

7.4 Dynamic Baselines for Real-Time Network Control

Monitoring ensures high reliability and availability of communication Network and its value will be even more pronounced in future, more complex

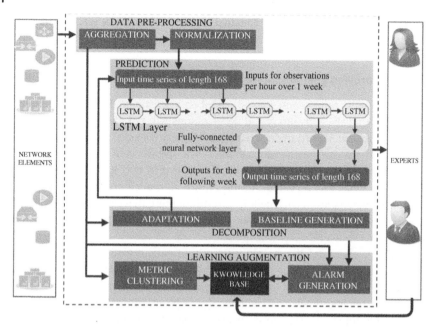

Figure 7.13 The system design for learning the dynamic baselines.

networks. Creation of network baselines is one of the major challenges faced by state-of-the-art monitoring tools. Baselines are the context-specific thresholds used to determine whether monitored values represent normal network operation or not. The size and complexity of current (and future) networks makes it unfeasible to manually determine and set baselines for each network operator and metric, let alone adapting the thresholds to changes in network conditions. Yet, the use of default baselines, or baselines that are not adapted over the lifetime of network elements, does not only cause inefficient operation but could also have implications for network reliability and availability. This section presents an example solution to this challenge, a dynamic mechanism for Real-time Network Control (DARN) [20]. DARN, shown by Figure 7.13, is a collection of analytics and machine learning-based algorithms aimed at automatically setting and adapting network baselines.

7.4.1 DARN System Design

As illustrated in Figure 7.13, DARN is composed of four major processes – data pre-processing, prediction, decomposition, and learning augmentation. DARN interacts with external entities such as the monitored network elements (on the left) from which the data is collected, and experts (on the right) who provide feedback to improve the learning process.

Consider a network made up from E elements and M metrics across all elements. If x_{me}^t is the value for metric $m \in M$ of element $e \in E$ at a time t over a period T, the collected data can be represented by the time series given in (7.9)

$$X = \{x_{me}^t \ \forall \ m \in M, e \in E, t \in T\} \tag{7.9}$$

The first problem is to determine the lower and upper thresholds L and U given in (7.10) and (7.11) respectively, for a time period $T' \geq T$. A specific example of T' that has been used for the evaluations here is 'starting at the end of T and lasting one week (168 hours)'. The motivation for the value of T' is the consideration that most metrics in communications systems vary on an hour and daily basis but are generally periodic on a weekly basis [21].

$$L = \{l_{me}^t \leq x_{me}^t \ \ \forall \ m \in M, \ e \in E, \ t \in T'\} \tag{7.10}$$

$$U = \{u_{me}^t \geq x_{me}^t \ \ \forall \ m \in M, \ e \in E, \ t \in T'\} \tag{7.11}$$

With U and L determined, any state-of-the-art monitoring application can be used to raise alarms when the baselines are breached. To optimize network operational efficiency, reliability, and availability, it is necessary to ensure that: (i) the values in U and L are adapted to changes in network conditions (such as a change in traffic due to user behaviour or due to network elements' failures or upgrades), and (ii) the number of false alarms is minimized.

To address those needs, DARN periodically collects data (metrics) from network elements (such as NodeB, radio network controller (RNC), -, etc.) using several protocols (such as Simple Network Management Protocol [SNMP]). The collected data is pre-processed before being stored in a database. The pre-processed data for each monitored element is then fed into a Long Short-Term Memory (LSTM) network (see Section 6.3.4), which creates a prediction for the element under consideration over a given future period. Time Series Analysis (TSA) [22] is then used to determine some characteristics of the monitored metric such as the trend and noise. The noise is used, together with the predicted values, to determine the baseline for each metric, while the changes in trend are used to detect a change in network conditions and hence trigger baseline adaptation. In addition, the performance of DARN is augmented by an expert, who, on violation of given baselines gives a 'YES or NO' feedback to the system in response to whether the present violation constitutes an anomaly or not. This feedback is used to build a knowledge base (KB) for the system, such that similar violations in the future are automatically handled by the system. Finally, DARN also groups different network elements into clusters based on their observed values and applies the knowledge learned for a given element to make decisions for other elements in the same cluster.

The subsequent sections describe each of the modules in detail and end with an evaluation of the system.

7.4.2 Data Pre-Processing

Data pre-processing involves the subprocesses of aggregation and normalization. At aggregation, observed values are averaged to hourly averages to minimize stochastic variations. In practice, different elements may be monitored at different frequencies within a given one hour-period, so the number of observations might be different for different metrics. As a result, depending on the observation times, the periodicity of some time series may be affected. It might, therefore, be necessary to use techniques such as frequency analysis to determine the subset of metrics for which such averaging is useful.

Since the monitored (and hence average) values may have varying scales, it is important to rescale the input data into a uniform range. This is especially important for LSTMs, which are sensitive to the scale of the input data when the (default) sigmoid or tanh activation functions are used. Therefore, all input values are normalized to time series that have approximately 0 mean with a standard deviation that is close to 1. This ensures that the prediction focuses on structural similarities and dissimilarities rather than on the amplitudes.

7.4.3 Prediction

The prediction module is an LSTM network that takes the one-week long sequence of observations from the pre-processing step for each metric of a given element, and outputs that metric's predicted observations for the following week. The number of layers and units (and consequently the required amount of historical data) is usually application-dependent and determined through experimentation. Considering the desire to have weekly predictions based on historical observations for the past week, the LSTM is found to require one layer of 600 units, and is designed to have 168 inputs, which is one input for each hour during the week. To ensure the output length of 168, i.e. one week's hourly data and allow the required number (600) of units in the LSTM layer, a fully-connected neural network layer is added at the output. This takes as input any number of outputs produced by the LSTM layer and only produces the desired (fixed) number of outputs (i.e. 168).

The prediction module runs iteratively to make predictions for each metric. Given the predictions, the output is determined by reversing the normalization step carried out before. Note that the system can be improved by using more historical data provided the LSTMs are designed with weights that emphasize the very last week's evolution. However, the conceptual considerations remain the same.

7.4.4 Decomposition

The decomposition module serves two objectives: (i) generating the network baseline for each metric, and (ii) continuously adapting these baselines to changes in network conditions. To achieve these objectives for a given metric, the module decomposes the observations over a given time period into its trend and noise using time series decomposition. Specifically, the observations for a metric x_{me}^t are decomposed into three additive components such that $x_{me}^t = r_{me}^t + s_{me}^t + n_{me}^t$ where r_{me}^t, s_{me}^t, and n_{me}^t are the trend, seasonal, and noise components respectively.

The trend r_{me}^t is determined as a centred, weighted moving average of order $k+1$ given by (7.12), with $k = 168$ (since weekly trends are needed).

$$r_{me}^t = \sum_{j=-n}^{k} c_j \times x_{me}^{t+j} \qquad \forall\ m \in M, e \in E,\ t \in T \qquad (7.12)$$

where $n = k/2$ and all observations take weight c_j given by $1/k$ except the first and last observations where the weights c_{-n} and c_n are given by $1/2k$. A weighted moving average is used (rather than a moving average) because the period/order (168) corresponding to hourly observations in a week is not symmetric (or centred), which is the usual requirement for using the moving average as a trend.

The seasonal component s_{me}^t can be determined by calculating the seasonal factors from the de-trended series $\widehat{x_{me}^t} = x_{me}^t - r_{me}^t$. Since the aggregated data is hourly, the seasonal factors s_{me}^t include 168 observations over one week, which are repeated over the desired time period. This is obtained by estimating the effect of each hour (within a week) by averaging the values in $\widehat{x_{me}^t}$ for each hour. For example, to get the seasonal value for observations at 8 a.m. on Mondays, the average s_{me}^t for values in $\widehat{x_{me}^t}$ is determined for observations at 8 a.m. for every Monday in the dataset. And the same is done for the other times and days. These seasonal factors are then normalized using (7.13) so that the normalized values in any given week sum to zero. The result of (7.13) is the seasonal component of x_{me}^t.

$$s_{me}^t = s_{me}^t - \sum_{t=1}^{168} \left(\frac{s_{me}^t}{168} \right) \qquad (7.13)$$

The noise component n_{me}^t can be determined as $n_{me}^t = x_{me}^t - r_{me}^t - s_{me}^t$. It is worth mentioning that DARN only uses the trend and noise components of x_{me}^t. The seasonal component is only presented here because it has to be determined in order to determine the noise component. The seasonal component can, however, also be used as weights within one week's LSTMs to ensure that the relatively high seasonal values do not affect the result of boundaries more than the low seasonal values.

7.4.4.1 Adaptation

To ensure that the prediction models remain accurate as the network evolves (i.e. to account for such evolutions as traffic changes, or network topology and module changes), the adaptation sub-module pro-actively assesses the change in the trend of observations for a given metric (e.g. from increasing to decreasing, constant to increasing, etc.) since the preceding prediction model. To this end, for every week, DARN determines the slope $\Delta r_{me}^t / \Delta t$ of the trend for each metric. A positive value for the slope represents an increasing trend, while a negative value represents a decreasing trend.

In order to determine changes in the trend for an observed metric, DARN keeps track of the slopes over time, and compares the current and previous slopes. In case of a change in the trend, from positive to negative slope or vice versa, the system triggers an adaptation of the prediction model by retraining the LSTM network with the latest data for the metric under consideration. However, an extra consideration is required before adapting a model if the detected change in trend leads to a breach of the currently set baselines: here the adaptation is only performed if the system confirms from an expert or its KB (see Section 7.4.5) that the said breach was a not a result of an anomaly. This is aimed at avoiding that the model is re-created from anomalous observations.

7.4.4.2 Baseline Generation

To translate the predictions obtained from the prediction module into network baselines, it is necessary to determine the variability of the observations for any given metric. To this end, the noise component n_{me}^t of its observations is used. The main idea is that since neural networks are trained to generalize how to map inputs to outputs, their predictions would generally try to eliminate noise. Therefore, the noise can be used as a measure of the metrics normal variation from the predicted values to create lower and upper limits. The required network operating baselines are derived by taking the standard deviation σ_{me} of n_{me}^t as the maximum allowed deviation above and below the predicted values for any given hour in a week. This results in the baselines as the predicted lower and upper boundaries l_{me}^t and u_{me}^t for each metric as given in (7.14, 7.15), which are then updated each time the prediction model for a given metric is updated by the adaptation module.

$$l_{me}^t = y_{me}^t - \sigma_{me} \qquad \forall \ m \in M, e \in E, \ t \in T \qquad (7.14)$$

$$l_{me}^t = y_{me}^t + \sigma_{me} \qquad \forall \ m \in M, e \in E, \ t \in T \qquad (7.15)$$

7.4.5 Learning Augmentation

Whenever the observed value for any metric is outside the upper and lower baselines set as in Section 7.4.4, the system detects this as a breach. The learning augmentation module enables an expert to augment the predicted baselines by providing feedback whenever an alarm is raised. Thereby, the system learns from all the alarms it raises and uses this information in deciding whether to raise future alarms or not. Moreover, the behaviour of many metrics in a given network is usually similar, for example, the CPU utilization of two nodes may have similar magnitudes, and peaks and troughs at the same times since traffic in the network would affect them in the same way. Correspondingly, alarms raised for a given metric may also be used as a basis to learn towards future decisions regarding 'similar' metrics.

To achieve these objectives, the learning augmentation is composed of three sub-modules: (i) knowledge base, (ii) alarm generation, and (iii) metric clustering, as described here:

7.4.5.1 Knowledge Base

The KB contains records of all breaches that happened in the past, and feedback from an expert, and is used as a basis of making decisions on whether alarms should be generated or not after a breach has been detected. Figure 7.14 gives an example of how breaches are converted into KB entries for a metric with four breaches A, B, C, and D. Each breach is modelled as a three-parameter tuple (α, β, γ) indicating its start time α, its duration β, and its maximum deviation γ from its value at α. An example of these parameters is illustrated by the small box for breach A. Since network dynamics can generally be grouped into different daily time periods based on user behaviour (e.g. peak and non-peak periods), the values of α are categorized into one of four daily time periods, $\{\alpha_0, \alpha_1, \alpha_2, \alpha_3\}=:\{(0000, 0600], (0600,1200], (1200,1800], (1800,000]\}$ hours. As an example, if the time of

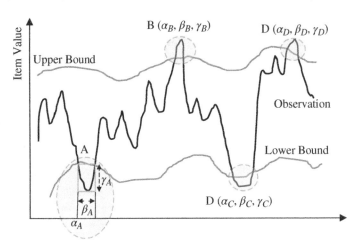

Figure 7.14 Knowledge-based modelling.

breach, α is 1330 hours, then $\alpha = \alpha_2$. These three parameters (α, β, γ) form part of a given entry in the KB that is determined solely from the breach.

7.4.5.2 Alarm Generation

This module continuously compares each metric's observations and baseline to establish whether there is a breach of the operating bounds, i.e. if the observed value is outside the lower and upper bounds. In case of a breach, the module then uses the KB to decides whether an alarm should be raised or not; not all breaches of the baselines result into raising alarms.

To decide on whether to raise the alarm, the module checks if the metric under consideration has an entry in the KB. If no entry is found, an alarm is generated, and a new entry created in the KB with the applicable available values (such as metric identification and α). Unavailable values (such as β and γ) are updated whenever they become available. A domain expert (e.g. network manager) receiving the alarm must then respond with a YES or NO answer on whether this breach represents an anomaly or not. This response is added to the KB as the fifth element of the KB entry – besides the metric identification, α, β, and γ. On the other hand, if, after a breach the metric is found to already have at least one KB entry, the current breach is evaluated against all existing entries to establish similarity. The parameters $(\alpha', \beta', \gamma')$ of the current breach are compared against the corresponding parameters $(\alpha'', \beta'', \gamma'')$ for each of the existing entries for the metric according to (7.16).

$$(\alpha\prime == \alpha\prime\prime)\, AND\, (\beta\prime \leq \beta\prime\prime)\ \ AND\, (\gamma\prime \leq \gamma\prime\prime) \tag{7.16}$$

If the condition in (7.16) is satisfied for any of the existing entries, and the anomaly column for the same is 'NO', then no alarm is generated. The logic behind this is that the current breach should also not be an anomaly, if an expert ruled in the past that a breach of the given metric: (i) within the same time of the day, and (ii) by at least the same deviation, and (iii) for at least the same amount of time as the current breach, does not constitute an anomaly.

A number of critical concerns are worth mentioning: Firstly, even when for a given metric a breach is detected and an entry (or entries) found in the KB, the decision to generate an alarm is not immediate, since determining whether β or γ is violated might necessitate continuous monitoring of the metric immediately after the breach. Secondly, the implementation of the logic can easily be improved if it is required to distinguish between a lower and upper bound breach. This is done by adding an extra column to the KB to store this binary value for each entry. Thirdly, cases with $\beta' > \beta''$ and/or $\gamma' > \gamma''$ would lead to new KB entries for each value of β' and γ', so to minimize KB size, entries with smaller limits are replaced with the new ones if the feedback from the expert is the same.

7.4.5.3 Metric Clustering

Experience shows that many elements in a given network are affected in a similar way by network dynamics. For example, Virtual Network Functions (VNFs) in a service function chain which process packets in a sequential manner would be expected to have ebbs and flows in the observations of their metrics such as CPU and bandwidth occurring at almost the same time [23]. Correspondingly, the CPU utilization of two VNFs hosted in adjacent nodes would have similar magnitudes, and peaks and troughs at similar times. When one of the nodes breaches a baseline for a given metric (e.g. CPU), it is very likely that a breach will also occur in an adjacent one. Thus, DARN groups metrics to generate alarms for metric clusters and not for individual metrics.

Each metric is attached to a cluster (a cluster may have one or more metrics), and the KB rules are stored based on cluster IDs rather than on individual metric IDs. When a breach occurs, KB entries are searched in the cluster in which the current metric is contained. Clustering metrics serves two purposes: (i) Since KB entries are grouped, it reduces the number of entries that must be kept in the KB, and (ii) it reduces the number of false alarms that can be presented to the expert since, for a given metric cluster, a given breach is responded to once.

The clustering of the metrics uses a variation of Dynamic Time Warping (DTW) [7], which is a well-known technique for finding an optimal alignment between a given number of (time-dependent) series under certain restrictions. In DARN, the DTW distances between all the metrics in a given time period are determined to generate a symmetric similarity matrix in which the value at the intersection of each row and column represents a similarity measure between the two respective metrics. The higher the value, the less similar the two metrics are and any two metrics m and f belong to the same cluster only if the 'normalized' distance $d(x_{me}^t, x_{fg}^t)$ between two metrics m and f (of elements e and g respectively) $d(x_{me}^t, x_{fg}^t)$ is less than a constant, d. The distances are 'normalized' by dividing each distance by the sum of the means of the absolute values of the two metrics using (7.17) where n is the number of observations.

$$d(x_{me}^t, x_{fg}^t) = \frac{d(x_{me}^t, x_{fg}^t)}{\frac{1}{n} \sum_{t-n}^{t} (|x_{me}^t| + |x_{fg}^t|)} \tag{7.17}$$

The normalization in (7.17) ensures that metrics that are grouped together have comparable magnitudes since the magnitude is also used in alarm generation decisions (using parameter γ in the KB). The appropriate value of d can be set through experimentation although the default values of d = 1 have been found to work satisfactorily. Note that clustering is also triggered by the adaptation sub-module (see Section 7.4.4) whenever a change in trend is observed in a metric, to re-generates new clusters using the most recent observations for each metric.

7.4.6 Evaluation

To obtain good performance, the LSTM needs to be appropriately configured, i.e. the optimum parameters (such as the necessary amount of training data, number of layers, number of units per layer, and number of epochs) for the LSTM network need to be determined. The experiments executed in [20] found the best parameters for the LSTM network to be: one layer each with 600 units, three weeks of training data, and 10 training epochs. These parameters are used for the rest of the evaluations.

The amount of training data means that every time a model for a given metric has to be obtained, three weeks of historic data is used to train LSTM. It was observed in this case that exactly three weeks of training data is needed and not more. This is contrary to what would generally be expected that 'the more the historic data, the more the accuracy'. The reasoning, however, is that with metrics' observations continuously changing, increasing the training data so far into the past increases the likelihood that the training data represents the past and not the current trend. Therefore, for each metric in Figure 7.15a–c, the first three weeks (504 hours) of data are used for training the LSTM network.

7.4.6.1 Accuracy of Generated Baselines and Clusters
Figure 7.16a shows a sample result of the prediction for the metric in Figure 7.15a while Figure 7.16b shows the corresponding baselines. It can be observed from these figures, that when there are no anomalies or outliers in the metric observations (e.g. between time 2000 and 2168), the generated metric baselines are accurate. The baselines are variable over the week's period to take into consideration periodic observations, indicating that what is accepted as a normal metric value at midday could raise an alarm if observed at midnight.

Figure 7.16c,d shows sample results from the clustering functionality with only two clusters shown for brevity. It can be observed that while all the metrics in the two clusters have a similar shape and periodicity, they are still clustered differently because of the relative differences in their magnitudes.

7.4.6.2 Effect of Baseline Adaptation
Figure 7.17a shows the trend (in red) of the metric in Figure 7.15b, and the corresponding changes in its weekly slope (blue line). It can be observed that whilst the long-term trend of the metric is upward until around 3000 hours, on a weekly (short-term) basis, the trend continuously changes as highlighted by the dotted vertical black lines which are drawn at the points where the slope crosses the zero value. These crossings are used to trigger system adaptation with the result in Figure 7.17b. It is observed that, even with this unchanged long-term trend, adapting the system to the short-term changes has a noticeable effect on the accuracy

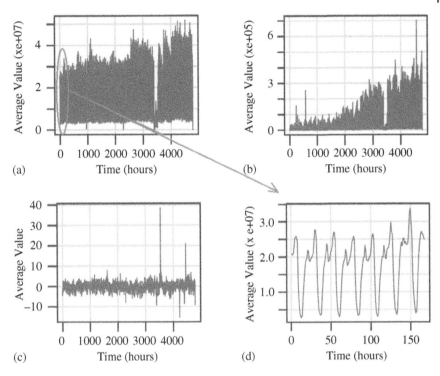

Figure 7.15 Metrics 1, 2, and 3 used for evaluation in figures (a, b, and c) and three weeks of Metric 1 in figure (d).

of the predictions for the following week. Specifically, there is a net improvement in prediction error of 22% over the entire period.

7.4.6.3 Effect of Learning Augmentation

Figure 7.18a shows the KB generated for the metric in Figure 7.15b within the time range 3000–4600 hours. Since the metric is generally stochastic, there are times when its value falls outside of the set baselines, specifically 99 baseline breaches in total, but not all are necessarily anomalies. Whilst this is an extreme case due to the fact that this metric is generally stochastic, it can be expected that even for seasonal metrics, some observations may sometimes fall out of the set ranges. There are, however, three breaches at 3500, 4200, and 4400 which would be of concern.

Figure 7.18b shows the effect of learning augmentation to this scenario. For brevity, the table shows 21 breaches which happen during the time period α_1 (0600–1200 hours), and the values of β and γ that are determined by the system are given, as well as the feedback from an expert regarding whether such a breach represents an alarm or not. As can be observed from Figure 7.18c, about

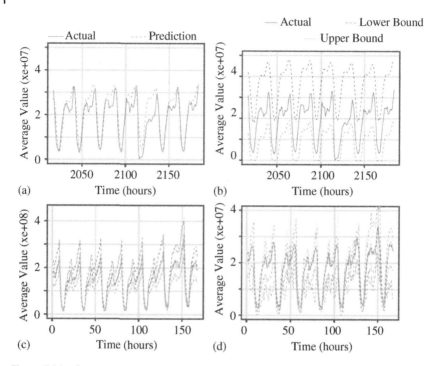

Figure 7.16 Generated baselines and clusters: (a) actual vs prediction, (b) operating bounds, (c) Cluster 1, (d) Cluster 2.

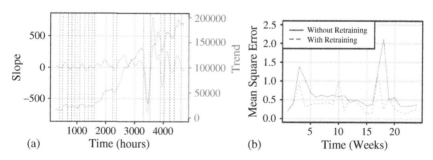

Figure 7.17 (a) Element trend and slope, (b) effect of adaptation.

(72%) of these breaches do not lead to an alarm being generated, i.e. many are found not to be anomalous based on a previous alarm and the related expert feedback. This allows for a generic system where some generated baselines may be breached without necessarily being anomalous, based on the system's learning from experience. The outcome is that this minimizes the false alarm load on a network administrator.

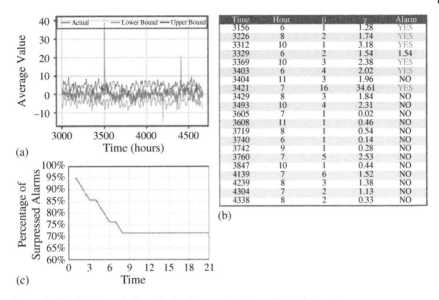

Figure 7.18 (a) Example knowledge base and (b) the effect of learning.

Besides the results presented here, the variation in computation time was tracked for each of the settings and found to be a low cost. In the evaluated implementation, model generation is only performed when a change in trend is detected, which means that most of the time, the required machine learning prediction time is comparable to that required by other processes of the system. Even when a model has to be updated, the combination of the chosen parameters takes about 10s to update a model. In general, these results show that a system can be developed to enable the automated adaptation of network baselines and improve the efficiency of network operations.

7.5 Conclusions

This chapter has presented concepts on the application of cognition and autonomy for network configuration. Configuration here takes the general perspective of setting network parameter values both at commissioning and during operations. The presented use cases differently apply cognitive concepts but are all targeted towards increased autonomy with minimal operator involvement even in configuring the network automation functions.

Firstly, concepts on context-aware auto-configuration are presented to show how context models can be derived from objective models and, subsequently, how the models can be combined to enable autonomic derivation of network and function configuration policies.

The considered use cases apply cognitive concepts to different degrees, but all propose a means for increasing autonomy of configuration management. The discussion starts with concepts on context-aware auto-configuration which show how objective models can be combined with context models to allow for autonomic derivation of network and function configuration policies. The ideas discussed therein are generic in a way that they can be applied to different kinds of network automation problems both to configure the automation functions or for the functions to derive the network's configuration parameter values.

Then, two specific use cases on network configuration are presented. Firstly, the challenge of 'multi-layer PCI auto configuration' is presented with a simple learning solution, through which cell Neighbour relations are learned amongst co-existing macro and small cells. The learned relationships are then used to evaluate the re-use distance between cell assigned with the same PCIs and to configure or reconfigure the cells' PCIs as and when needed.

In the second configuration use case, the challenges of multi-layer energy saving management (ESM) are discussed and heuristics-based models presented to show how a model-based approach can be applied for ESM to select the optimal cell deactivation and reactivation sequences. Concepts are then presented, albeit at a high level, on how the model-based approaches can be the improved through simple learning-based solutions.

References

1 CCITT (2000). *Recommendation M.3400. TMN management functions.* ITU.

2 Frenzel, C., Lohmuller, S. and Schmelz, L.C. (2014). Dynamic, context-specific SON management driven by operator objectives. IEEE Network Operations and Management Symposium (NOMS), Krakow.

3 Frenzel, C., Lohmüller, S. and Schmelz, L.C. (2014). SON management based on weighted objectives and combined SON Function models. IEEE International Symposium on Wireless Communications Systems (ISWCS), Barcelona.

4 Kumpulainen Pekka, H.K. (2012). Characterizing Mobile Network Daily Traffic Patterns by 1-Dimensional SOM and Clustering. International Conference on Engineering Applications of Neural Networks, London.

5 3GPP NR 38.211 (2017). *Technical Specification Group Radio Access Network.* Sophia Antipolis: 3GPP.

6 Mwanje, S.S., Ali-Tolppa, J. and Sanneck, H. (2015). On the limits of pci auto configuration and reuse in 4g/5g ultra dense networks. The 27th International Teletraffic Congress, Paris.

7 Mwanje, S.S. and Ali-Tolppa, J. (2017). Layer-Independent PCI Assignment Method for Ultra-Dense Multi-Layer Co-Channel Mobile Networks. International Symposium on Integrated Network Management (IM), Lisbon.

8 3GPP TS 36.300 (2009). Technical specification group radio access network; evolved universal terrestrial radio access network (e-utran). Sophia Antipolis: 3GPP.

9 3GPP TR 36.902 (2010). Evolved Universal Terrestrial Radio Access Network (EUTRAN); Self-configuring and Self-Optimizing Network (SON) use cases and solutions. Sophia Antipolis: 3GPP.

10 Blume, O., Eckhardt, H., Klein, S. et al. (2010). Energy savings in mobile networks based on adaptation to traffic statistics. *Bell Labs Technical Journal* 15 (2): 77–94.

11 Micallef, G., Saker, L., Elayoubi, S.E., and Sceck, H.-O. (2012). Realistic energy saving potential of sleep mode for existing and future mobile networks. *Journal of Communications* 7 (10): 740–748.

12 Dilupa, R., Withanage, R., and Arunatileka, D. (2011). Infrastructure sharing & renewable energy use in telecommunication industry for sustainable development. In: *Handbook of Research on Green ICT: Technology and Business and Social Perspectives* (ed. B. Unhelkar), 317–331. IGI Global.

13 Fang, J. (2014). Method and apparatus and system for realizing coverage compensation. China Patent WO2 014 110 910 A1/CN103 945 506A.

14 Nenner, K.H., Jacobsohn, D. and Polsterer, H. (2013). Method for saving energy in operating a first and second mobile communication networks and a mobile communication networks. Germany/USA Patent EP2 676 491 A1/US20 130 344 863.

15 Voltolina, E. and Kuningas, T. (2011). Energy-saving mechanisms in a heterogeneous radio communication network. USA Patent WO2 011 021 975 A1/US12 790 055.

16 Mwanje, S.M. and Ali-Tolppa, J. (2016). Fluid capacity for energy saving management in multi-layer ultra-dense 4g/5g cellular networks. IEEE International Conference on Network and Service Management (CNSM), Montreal.

17 Roth-Mandutz, E. and Mitschele-Thiel, A. (2013). LTE energy saving son using fingerprinting for identification of cells to be activated. Future Network & Mobile Summit, Lisbon.

18 Mwanje, S.S. and Ali-Tolppa, J. (2016). Fluid Capacity for Energy Saving Management in Multi-Layer Ultra-Dense 4G/5G Cellular Networks. International Conference on Network and Service Management (CNSM), Montreal.

19 Mwanje, S.S. and Ali-Tolppa, J. (2018). Distributed Energy Saving Management in Multi-Layer 4G/5G Ultra-Dense Networks. IEEE Wireless Communications and Networking Conference (WCNC) IWSON Workshop, Barcelona.

20 Mijumbi, R., Asthana, A., Koivunen, M. et al. (2018). DARN: Dynamic Baselines for Real-time Network Monitoring. IEEE Conference on Network Softwarization, Montreal.

21 Iversen, V.B. (2001).Teletraffic engineering handbook. ITU-T.

22 Hamilton, J. (1994). *Time Series Analysis*. Princeton, NJ: Princeton University Press.

23 Mijumbi, R., Hasija, S., Davy, S. et al. (2017). Topology-aware prediction of virtual network function resource requirements. *IEEE Transactions on Network and Service Management* 14 (1): 106–120.

8

Cognitive Autonomy for Network-Optimization

Stephen S. Mwanje[1], Mohammad Naseer Ul-Islam[1] and Qi Liao[2]

[1]*Nokia Bell Labs, Munich, Germany*
[2]*Nokia Bell Labs, Stuttgart, Germany*

Optimization is the act, process, or methodology of making something (such as a design, system, or decision) as fully perfect, functional, or as effective as possible [1]. In Mathematics, it refers to the techniques for finding a local or global maximum or minimum value of a function of several variables subject to a set of – constraints. Both these definitions apply to networks where, after the network (or an entity thereof) is configured and commissioned into operation, further actions may be needed to ensure the most effective use of that network's resources. Reasons for this may include:

- the availability of real operating conditions which may be different from the models used in planning and configuration
- changes in the operating points requiring reconfiguration of network parameters for the new conditions (including changes in traffic behaviour, propagation conditions due to new buildings, weather changes, or new deployments).

Self-optimization (SO) implies that the entity under optimization undertakes the optimization process or methodology by itself without being triggered or controlled by another external entity. In networks, this implies that the network triggers, runs and controls the optimization algorithms without any intervention from the human operator(s). Note that there is no strict boundary between optimization and the other two network operations use-case areas – configuration and healing. For example, 'self-configuration' functions like neighbour list update can be triggered in the operational state by changes in operating points whilst 'self-optimization' functions like coverage optimization can be run for preventive failure handling, i.e. to minimize failure risk.

Towards Cognitive Autonomous Networks: Network Management Automation for 5G and Beyond, First Edition. Edited by Stephen S. Mwanje and Christian Mannweiler.
© 2021 John Wiley & Sons Ltd. Published 2021 by John Wiley & Sons Ltd.

Owing to the diversity of SO challenges, different tools from the cognitive tool box can be employed. This chapter presents solutions that apply these tools towards solving SO related challenges in 4G and 5G networks. The chapter does not claim to answer all the challenges in self-optimization but raises examples of solutions that have been developed with the intent to provide a basis on which other solutions can be developed in future. The first section highlights how cognition and autonomy fit into the SO cycle. The next sections present a framework that promotes Q-learning (QL) as a good model for online learning in closed-loop optimization and show that it can even be more robust with the fuzzification of states and actions in the QL loop. It then discusses how these QL models have been applied towards the optimization of mobility robustness and antenna tilts. Finally, a solution using fixed-point iterations discusses how to address the interference-aware open-loop optimization of radio resource assignments in 5G.

8.1 Self-Optimization in Communication Networks

SO in communication networks encompasses the processes in which autonomic functions continuously interact with the network with the intention of improving the network's operating state. And cognition can be employed to achieve autonomy of the SO functionality. The functions themselves may be of different kinds and may have different capabilities. Correspondingly, even although the core functional behaviour is the same, SO functions can be seen from multiple perspectives (discussed in Sections 8.1.2–8.1.4).

8.1.1 Characterization of Self-Optimization

The core activity of any SO function is to determine if the network is in a suboptimal state and devise mechanisms to improve that state. The typical operation is a loop (the SO Loop of Figure 9.1) with the four functions – measure (evaluate state), trigger, compute changes, and activate.

The measure step characterizes the network state in the perspective of the respective function. This could be based on Key Performance Indicators (KPIs) like background block error statistics or call and handover failure rates for which the network needs to adjust configurations to minimize the failure and error rates. It could, however, also be based on changes in environmental characteristics like user speed or weather conditions to which the network may need to adapt its behaviour. In the Cognitive Self-optimization paradigm, this characterization of states based on KPIs and environmental conditions can also be made cognitive. Thereby, there is no fixed mapping hardcoded by human designers, but the required mapping is learned in a flexible way in that the association of observations to states can be reconfigured by the operator.

Given the network state, the trigger step determines the quality of such a network state and decides if there is a possibility of moving the network to a better operating state. The simplest trigger mechanism is a level-crossing device that initiates configuration if a certain threshold is crossed. More complex trigger mechanisms can however be developed, for example, to trigger based on different thresholds for different contexts. Moreover, cognitive techniques can also be applied here, e.g. by matching the states to performance such that the trigger evaluates not only the observed state but also the performance of such a state in comparison to other candidate eventual states.

The computation step determines the best configurations for the specific observed states. Here is where most of the optimization logic is implemented. Such logic is, in general, specific to the use case even when the underlying mechanism is the same. In the Self-Organising Network (SON) paradigm, the mechanism was almost always a rule-based decision engine or expert system, called a SON Function (SF), that evaluated the truthfulness of different rules for the specific state and then executed a specific command when such a given rule evaluated as true. Similarly, it is at this step that most of the higher cognitive technologies will be applied in the CNM/CAN paradigm. In the end, the compute step output a recommendation for new configurations to be activated.

Finally, the activate step implements the computed configuration on to the network. The range of possible actions extensively varies here from running a single command to change one parameter to running a series of scripts or sending a technical team to implement a physical change. As such, the aim here is less about cognition and more about autonomy, i.e. there should be as little human activity as possible – be in executing the scripts or in writing such scripts. This may, however, also require further advancement in the physics of networks, e.g. development of highly reconfigurable antennas for which the antenna beamwidth can be controlled to between 0° and 360° or the tilt changed by up to 90°.

The four steps will typically be executed iteratively many times before a steady (possibly optimal) state is reached. The optimization interval per iteration, the time between computing the configurations and observing the effect of such changes, will depend on the use case, influenced mainly by the measurement step (according to the required observation interval) and the length of the activation step.

Although cognitive automation can be done at each point in the SO loop (albeit to different degrees), most cognitive techniques will be applicable in the computation of the solution, whilst less is done in the measurement and triggering and possibly none in the activation of computed reconfigurations. The reverse is true for automation, i.e. more automation techniques are needed to activate computed reconfigurations on the network. However, the degree to which cognition and automation are applicable in the different steps will also vary depending on use case and proposed solution.

The subsequent sub-sections show that the solutions will also be at different points of the continua along the dimensions of openness vs closeness, reactive vs proactive nature, and model-based vs online optimization. For example, the applicable proactiveness of a solution, measured in terms of the horizon for which actions can be computed, varies depending on use case.

8.1.2 Open- and Closed-Loop Self-Optimization

The SO loop of Figure 8.1 is a closed-loop operation. For each iteration therein, the optimization function activates the computed solution on to the network, evaluates the effect and reinitiates optimization if needed. This is the typical desired behaviour since the function is guaranteed to eventually retain only the best of the known solutions.

There are many scenarios however, where the closed-loop mechanism is too expensive. One main challenge is when the monetary cost of activating the change is too high that it cannot be done iteratively many times. This is the case for example if the computed action is changing an antenna's mechanical tilts or azimuths, i.e. where a technician must drive to the site location. The other unacceptable cost is when the 'cost of failure' is so high that trial and error mechanisms cannot be applied. In that case, a computed solution for any network state needs to be checked for accuracy or acceptable performance before it is activated on the network.

These two high costs motivate the use of open-loop optimization of Figure 8.2, where, for a given network state, the action is computed from and possibly tested on a model before it is finally activated on to the network.

Correspondingly, owing to these costs, it may also be necessary for the solution computed in the open-loop mechanism to be verified by human operator

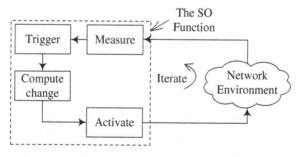

Figure 8.1 The self-organization loop.

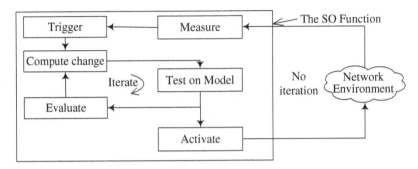

Figure 8.2 Open vs closed loop self-organization.

before activation. This is like self-planning where the autonomous function determines when and where a new solution is needed and proceeds to compute a recommendation for such a solution which is then verified and activated by the human operator. Such a mechanism would be the kind applied for the planning/optimization of antenna tilts and azimuths.

8.1.3 Reactive and Proactive Self-Optimization

Conventional SO strategies follow the *reactive* process as shown in Figure 9.1, where the system monitors system performance and network status, and activates optimization or compensation functions after detecting critical states. A growing trend for the next generation wireless networks is not just to react to changes but anticipate them and take *proactive* actions to guarantee reliable network performance. Such a trend is enabled by the growing information (data) availability as well as the emergence of the massively scalable data centres. Proactive network optimization opens an opportunity to take full advantage of good future conditions (e.g. good radio propagation conditions) and to mitigate the impact of negative events (e.g. coverage outage or ultra-high amount of data demand). A typical proactive SO solution is usually characterized by the following three attributes:

- **Context** defining the types of information considered to forecast the evolution of the network system.
- **Prediction** specifying how the system evolution is forecast from the current and past context.
- **Optimization** describing how prediction is exploited to meet the application objectives.

A thorough context-based analysis, two handbooks on the prediction and optimization techniques respectively, and an analysis of the applicability of proactive networking techniques are provided in a recent survey [2].

8.1.4   Model-Based and Statistical Learning Self-Optimization

When choosing the appropriate methods for SO functions, it is necessary to study the rationale for adopting the methods. In general, the methods are of two classes:

Model-based methods: When the physical properties of the system are well-known and thus a physical model with closed-form expression is available, such a model can be used to obtain an accurate prediction and robust optimization. They are usually preferable for mid- to long-term SO solutions and exhibit good resilience to poor data quality. One example is the long-term radio map inference with linear regression based on the path-loss model [3].

Statistical learning methods: If the physical model of system parameters is either unavailable or too complex to be used, probabilistic models based on stochastic processes and statistical properties offer more robust prediction based on the observation of sufficient data. In addition, such probabilistic models can quantify the uncertainty of the inference, prediction, and optimization, based on the probability distribution of the outcomes. For example, Markovian models and Bayesian models are amongst the most widely used models for statistical learning and optimization. A typical use case is to use Markovian models to predict the mobility of users in a combined space-time domain and proactively optimize the resource allocation based on the predicted channel condition and throughput [4].

The degree to which these methods are applicable will, in general, depend on the problem. Specifically, for both computing and/or testing the solutions, models are applied to different degrees depending on use case. Similarly, these different dimensions, e.g. of cognition vs automation and the degree of proactivity or reactivity of the solution, will also be differently applicable to the different problems. This will also be reflected in the different problems discussed in the subsequent sections.

8.2   Q-Learning Framework for Self-Optimization

It has been shown in Section 8.1.1. (Figure 9.1) that the typical SO Loop involves four steps of measuring the environment parameters to evaluate state, trigger action where necessary depending on the observed state, compute changes befitting the state and activate those changes in the environment. The measurement step and action execution steps are visible to the external environment with the measurement being seen in two respects. On one hand, the SO agent must establish the state of the environment before the action, but it also needs to establish the state after the action as well as measure the effectiveness of that action.

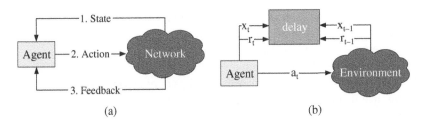

Figure 8.3 (a) The abstract optimization function and (b) the QL process.

So, from an external observer, the SO process may be interpreted as consisting of the three steps similar to those in Figure 8.3a: (i) establishing the state of the environment, (ii) selecting and executing action fit for the observed state and (iii) receiving and using the feedback about the executed action. These are also the core steps for a learning agent, so this section describes how a generalized learning framework can be applied to SO.

8.2.1 Self-Optimization as a Learning Loop

QL is one of the widely discussed Reinforcement Learning (RL) algorithms which allows for model-free online learning. As summarized in Section 6.4.4, QL uses Temporal Difference (TD) to solve learning problems without using models. For each state action pair (x, a), QL maintains a value function (the Q-value, $Q^\pi(x, a)$) defined as the expected total discounted reward received when starting with action a in state x and following the policy π thereafter. The Q-values denoted $\hat{Q}$, can be estimated iteratively as derived in Section 6.4.4. For the optimal policy π^* [5], the estimate $\hat{Q}_{t+1}(x, a)$ at $t + 1$ is updated by adding a small portion (i.e. α) of the error to the current estimate according to Eq. (8.1).

$$\hat{Q}_{t+1}(x, a) = (1 - \alpha_t)\hat{Q}_t(x, a) + \alpha_t . \{r_t(x, a) + \gamma . \max_{a_{t+1}} Q_{t+1}(x, a)\} \qquad (8.1)$$

In (8.1), r is the received instantaneous reward, $\gamma \in (0, 1)$ is the discount factor that balances between immediate and future rewards, wherein $\gamma = 0$ only considers the instantaneous rewards whilst $\gamma = 1$ equally weighs the immediate and the future rewards. Similarly, $\alpha \in (0, 1]$ is the learning rate that balances new information against previous knowledge. $\alpha = 0$ implies no learning whilst $\alpha = 1$ means that only the latest information is applied, and the old knowledge completely ignored.

In principle, QL represents the iterative closed loop of Figure 8.3b where an agent evaluates the state of its environment, computes and activates an action and eventually receives a reward that describes how good the action was to the environment. Thus, the agent is able to learn how best to act in its environment over multiple interactions through this loop. This loop is similar to the SO loop of Figure 8.3a for the basic optimization function characterized by a trigger that initiates the execution of an algorithm to configure a set of network parameters and optimize a particular network metric.

On the network, the optimization function (in Figure 8.3a) is a control agent that: (i) observes the network to evaluate trigger conditions, (ii) takes an action to optimize its metrics and (iii) gets feedback on the effect of that action on the network. Given the similarity of the QL loop and the optimization function, it has been proposed that optimization functions be designed as QL agents that act and, using network's feedback, learn from the effects of their actions. This advances network optimization functions beyond the rule-based control loops towards cognitive functions (CFs) that autonomously learn the desired optimal configurations. QL has been applied in some SON functions with positive results, e.g. in [6, 7], and this provides the basis to advance the concept towards a QL framework applicable to all optimization functions.

8.2.2   Homogeneous Multi-Agent Q-Learning

The previous section presents the QL algorithm as the implementation solution for a single instance of an automation function. However, network automation (be it with SON or otherwise) is a Multi-Agent System (MAS). The case of concurrent learning for different use case is typically addressed through coordination mechanisms as discussed in Chapter 11. For a single use case, however, optimization must be done for each cell individually so the concurrent learning of multiple instances of a single automation function also results in a MAS scenario.

The first challenge in MAS is how much the instances should cooperate or compete during their coexistence. Cooperative learning implies that the agents for common global objectives may be realized as learning based on the global rewards of all agents. On the other hand, competitive learning is where each agent is selfish and only considers its objectives and rewards during the learning process. In networks, since the entire network has common objectives, i.e. the objectives of the individual functions and instances are sub-objectives of the global network objectives, there is no need to compete. Instead, the functions must cooperate so that the learnings of one function or instance are not detrimental to another function or instance.

Even with cooperative learning, a decision must be made on the degree of shared learning amongst the automation function instances. Two forms of cooperation are possible, either shared or fully independent learning. For a given QL problem, if an observed state at one agent A will, at some other time, be observed by another agent B, then A and B should share their observations and learn a shared policy instead of learning independent policies. This results in Cooperative-shared QL where the agents independently select actions but they all update a single Q-table,

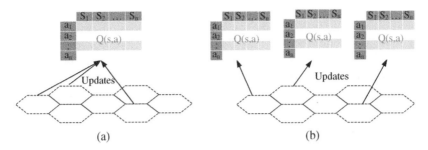

Figure 8.4 (a) Shared and (b) independent cooperative-learning strategies.

as shown in Figure 8.4a. The other alternative is independent QL where each agent learns and updates an independent Q table (as in Figure 8.4b) with, as expected, the reverse merits and demerits.

Besides the benefit of enabling cells to share experience, Cooperative-shared QL also speeds up the learning processes. Given the access to the common policy learned by all cells, each cell does not necessarily need to experience each state on its own for it to learn the best action in that state. The challenge here is that the single policy may not be optimal for all agents in all states. Nevertheless, where the QL states are defined in a way that they are consistent across agents (cells), Cooperative-shared QL should improve the convergence speed. Mobility Robustness Optimization (MRO) and Mobility Load balancing (MLB) are example SFs where similar states are observed in different cells. This makes MRO and MLB good candidates for cooperative QL, the challenges above notwithstanding.

8.2.3 The Heterogeneous Multi-Agent Q-Learning SO Framework

As stated earlier, RL has been applied in several SON related works, e.g. on Coverage and Capacity Optimization (CCO) [8, 9], Interference Management [10, 11], MLB [12, 13], and MRO [14]. In general, most (if not all) optimization functions can be considered as QL problems – and some of these are presented in the subsequent sections. For each use case, the QL-based learning functions, hereafter referred to as a CF, takes the network as the environment for which the required state(s) and the candidate actions thereof must be defined.

The states must be related to the observations that trigger for optimization, e.g. MLB states should consider the degree of cell overload since MLB is triggered by cell overload. Given the states, the action set is defined as the different possible combinations of parameter values that can be applied in such states. Then, owing to the need to quantify the quality of the actions, a feedback mechanism in the form of a reward system for the actions must also be added. For each action taken, a reward is derived for the CF using the defined reward function, from which the QL agent learns the best actions over multiple interactions with the network. Table 8.1 summarizes how some common SON Functions can be mapped to this QL framework.

Considering the different optimization problems together, the cellular environment becomes a Heterogeneous Multi-Agent learner environment, i.e. a different learner is needed for each problem although they all use the same learning framework. Note, however, that each problem may also be modelled as the Homogeneous Multi-Agent learner as described in Section 8.2.2 e.g. where a specific instance of the learner is required for each cell.

The biggest benefit of the framework is that each CF can be designed to learn based on all metrics influenced by its actions. As such, each CF will learn not only to optimize its metrics but also to minimize its effects on peers' metrics.

In Table 8.1, with knowledge of the dependence of MRO from MLB, the MLB QL agent can be designed to learn based on MLB's metrics (load; user dissatisfaction; ...) and on handover metrics (ping pongs; Radio Link failures [RLFs]; ...). However, details on the coordination of such functions using this framework are discussed in Chapter 11.

8.2.4 Fuzzy Q-Learning

The main disadvantage of QL is that it cannot handle problems where the state or action space is continuous. The solution that is typically applied is adding Fuzzy Logic (FL) to QL which results in what is referred to as Fuzzy Q-Learning (*FQL*). Fuzzy Logic is used to discretize the continuous state or action space so that *Q-values* can be learned for these discrete variables and then interpolated to get the *Q-value* for the continuous variable. FQL, as such, combines Fuzzy Logic with QL to optimize the consequents of the rules in fuzzy controllers. FQL uses fuzzy labels to discretize the continuous variables and implements a fuzzy rule-based inference system to calculate an action for these discretized states. The target of FQL is to find the best consequences for each fuzzy rule that maximizes the overall reward.

Note that FQL can also overcome the human expertise deficiency in the Fuzzy Inference System (FIS) design phase. With a precisely known rule, only one output action is enough to define that rule. However, where the knowledge is only partial or imprecise, a subset of possible actions can be integrated in the FIS. In the worst case, all the possible actions are included in the consequent part of that rule if no *a priori* knowledge is available.

As discussed in Section 5.4, the design of a fuzzy controller, involves the two tasks: (i) defining the fuzzy sets and membership functions of the input signals, and (ii) defining the rule base which determines the behaviour of the controller.

Table 8.1 Mapping the typical SO use cases to the QL framework.

SF	States	Solution: adjust...	Action to learn	Agent rewards...	Agent penalizes...
MRO	Mobility state, e.g. average UE speed in cell	Hys, TTT	Hys-TTT tuple in (dB,s), e.g. (2.5,0.64); ...	Reduction in no. of RLFs and ping-pong HOs	Increase in no. of RLFs and ping-pong HOs
MLB	Serving and neighbour cell load, user distribution	CIO by δ	Sizes of δ in dB, e.g. [δ = 0.5,1.0, ..]	Reduction in cell load	Increase in HO effects (RLFs and ping-pong HOs) and increase in cell load
CCO	Serving\neighbour cells' spectral efficiency, transmit power	Antenna tilt or transmit power by δ	Sizes of δ in dB, e.g. [δ = 0,1,2..]	Change in spectral efficiency	Increase in user dissatisfaction (e.g. reduction in throughput) increase in HO failures,
:	:	:	:	:	:
:	:	:	:	:	:

Legend: Hys, Hysteresis (Hys); TTT, Time to Trigger (TTT); Tx, Transmit; CIO, Cell Individual Offset; RLF, Radio Link Failure.

The execution of the controller is comprised of three steps: (i) fuzzification of the inputs, (ii) fuzzy reasoning, and (iii) defuzzification of the output. The fuzzifier projects the crisp data onto fuzzy information using membership functions whilst the fuzzy engine reasons on information based on a set of fuzzy rules and derives fuzzy actions. The defuzzifier reverts the results back to crisp mode and activates an adaptation action.

The implementation of the fuzzy controller with QL is summarized by Figure 8.5. Therein, the fuzzy controller is complemented with the Q-value update rule and the associated Q-value function (or Q-table) to create the FQL agent. The simplest FQL controller fuzzifies the states to be matched to each action. For a given continuous-valued input vector, multiple of the discrete states may be activated at a time but with only one action for each such state. This implies that multiple discrete actions, with one action for each discrete state, will be activated for the input vector. This framework can easily be extended to the case of multiple actions for each discrete state. In that case, the single action per state considered in the general case is replaced by a single compound action that is a composite of the multiple distinct actions.

Consider the following notation of the FIS for the vector of continuous valued variables $x = [x^1, x^2, \ldots x^m]$: The FQL has a set of fuzzy states S, and membership functions which define the truth values for the predicates that 'input vector x is a member of the fuzzy state S'. For the FIS composed of N rules (i.e. with N possible fuzzy states), the state $(S_i)_{i=1}^N$ with up to m dimensions is defined as

$$x \text{ is } S_i := \{(x^1 \text{ is } S_i^1) \text{ and } (x^2 \text{ is } S_i^2) \text{ and } \ldots \ldots .(x^m \text{ is } S_i^m)\} \qquad (8.2)$$

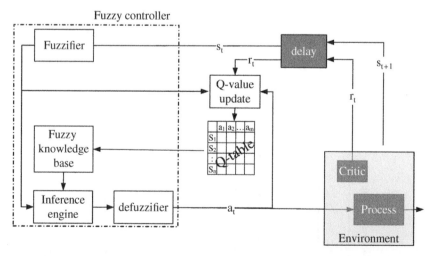

Figure 8.5 The Fuzzy Q-learning based control process.

where $(S_i^j)_{j=1}^m$ is the fuzzy label for subpart j of the state $(x \text{ is } S_i)$. The set of actions that can be chosen in state S_i is A. With one action $a^k \in A$ allowed per fuzzy state, each rule i will be of the form:

$$\textbf{\textit{If}} \quad x \text{ is } S_i \quad \textbf{\textit{then}} \quad \text{action } a_i = a^k \quad \textbf{\textit{with}} \quad q(i, a^k)$$

If for the input vector x, the truth value of rule i is the utility function $x \rightarrow u_i(x)$, the FQL algorithm is implemented as in Figure 8.6 that is adapted from [15].

As with QL, FQL can also apply cooperative learning if multiple FQL agents must learn within the same environment. Again, cooperation is enforced

---

Algorithm 1: Fuzzy Q-Learning

Require: γ, η

1. Initialize Q-values: $q[i, k] = 0$; $1 < i < N$, $1 < k < m$

Repeat for the state x_t at iteration t:

2. Select an action for each fired rule: e.g., using ϵ-greedy as

$$a_i = \begin{cases} \underset{k}{\arg\max}\, q(i, a^k) & \text{with probability } 1 - \epsilon \\ \text{random } \{a^k, k = 1, 2, \cdots, K\} & \text{with probability } \epsilon \end{cases}$$

3. Calculate the fuzzy control action using $\mu_i(x)$ the firing levels of each rule i:

$$a = \sum_{i=1}^{N} \mu_i(x)\, a_i$$

4. Approximate $Q(x_t, a)$ the Q-function estimate for the state x_t and the action a from the current Q-values and the rules' firing levels:

$$Q(x_t, a) = \sum_{i=1}^{N} \mu_i(x_t) . q(i, a_i)$$

5. Execute action a to transit the system to the next state x_{t+1} and then compute the value for the new state using the reinforcement signal, r_{t+1}:

$$V(x_{t+1}) = \sum_{k=1}^{n} \mu_i(x_{t+1}) . \underset{k}{\max}\, q(i, a_k)$$

6. Calculate the error signal using a discount factor γ

$$\Delta Q = r_{t+1} + \gamma . V(x_{t+1}) - Q(x_t, a)$$

7. Update Q-values with α as the learning rate:

$$q(i, a_i) = q(i, a_i) + \alpha . \mu_i(x_t) . \Delta Q$$

Figure 8.6 Fuzzy Q-learning algorithm. Source: Adapted from [15].

ensuring that each agent learns based on the global reward of all agents, as it was demonstrated in [7].

8.3 QL for Mobility Robustness Optimization

This section shows how the QL framework applies to a typical SON problem – MRO. MRO seeks to find the optimum handover (HO) settings, with the Hysteresis (Hys) – Time-To-Trigger (TTT) tuple (also called the Trigger point, TP) as the control parameter. For each HO, either a HO success, a Ping-Pong (PP), or an RLF occurs. MRO then optimizes radio-link robustness amidst the UE's mobility and subsequent HO, i.e. it minimizes RLFs and concurrently reduces PPs and unnecessary HO [16]. The HO problem is not new since it has been discussed, for example, in [8]. However, the existing solutions have mainly been ruled based loops designed by the experts. Yet it has been shown that HO optimization would be a good candidate for a learning-based solution. Such a solution is described here.

8.3.1 HO Performance and Parameters Sensitivity

The MRO goal is to dynamically select the optimum settings (called the Optimum Trigger point, OTP) for a network with a dynamic mobility profile. HO performance is evaluated in terms of the HO aggregate performance (HOAP) [3, 4], a weighted combination of the rates of RLFs and PPs. The individual sub-metrics are defined as:

Radio Link Failure Rate (F): RLFs occur if the UE Signal-to-Interference-plus-Noise Ratio (SINR) stays below a threshold for a duration equivalent to the critical time (T310) [16]. The RLF rate (denoted by F), due to either a too early HO (F_E) or a too late HO (F_L), is the per-second number of RLF events in the cell or the network.

Ping-Pong rate (P): A PP or HO oscillation arises if, for a user, an HO success from a cell B to another cell A occurs in a time less than the 'PP-Time' following a previous successful HO from A to B. The PP rate (P) is thus the rate of occurrence of PP per second in the cell or network. The *PP-Time* is not standardized and is set here to five seconds, which is approximately equal to the longest TTT.

Number of HO Candidates (NH): During learning, all rates are normalized to NH in the cell to ensure that all cells use comparable statistics in evaluating their actions. Here, an HO candidate is one who has attempted an HO or recorded an RLF within the time interval, ensuring that each user is counted only once even where a single user experiences multiple events within the interval.

Subsequently, the HOAP is defined as:

$$HOAP = w_1 P + w_2 F_E + w_3 F_L \; ; \sum w_i = 1 \qquad (8.3)$$

The weight vector is set to w = (0.2, 0.3, 0.5) to balance the effects of early HO (PP and F_E) against those of late HOs (F_L). It has been shown in [6, 10] that a cell's optimal control parameters are specific to the speed of the UEs. Sweeping a selected range of the parameter space for four speed scenarios (with UEs moving at constant speed in each case), it was observed that the OTP changes with speed as shown in Figure 8.7. The figure gives the linear variation of the HOAP with both Hys and TTT in Figure 8.7a and the detailed variation with TTT (using a TTT log scale) in Figure 8.7b.

According to Figure 8.7a, very high Hys values lead to high HOAP at all velocities and so are unacceptable, although a combination of moderately high Hys and low TTT could be acceptable. High TTTs are only acceptable at low velocities and when combined with low to medium Hys values. This shows that HOs can moderately be delayed without great penalty since the risk of RLF is low yet even the possibility of PPs is low owing to the low speed. This is evident in the 10 kmph environment where, for most of the TTTs, the performance is good at a Hys of 2 dB.

As the speed increases, the HO delay needs to reduce, especially using the TTT, since the performance significantly changes with TTT but is fairly constant relative to Hys. At high speed, even the Hys has a major effect and so both parameters should be low. This is evident in the 60 and 90 kmph environments in Figure 8.7a where the OTPs are restricted to the lower left corners of the grid, i.e. the part where TTT are within the range of 0–0.64 seconds. In general, however, although there is major variation in the HOAP with TTT for most Hys, this variation is blurred at points near the optimum point as seen within the small range of Figure 8.7b. In that case, adjacent TTTs will have a practically similar performance. The most obvious conclusion from Figure 8.7 is that the OTPs for different velocities lie in a large region and so any MRO algorithm must scan the entire parameter-space or at least more than half the space in order to determine the required TP.

8.3.2 Q-Learning Based MRO (QMRO)

Q-Learning based MRO (QMRO) [6, 11] aims to determine the best Hys-TTT actions that minimize the HOAP in any mobility state in a cell. Since the actions only change the cells' performance and not UE's mobility states, it is adequate to learn an action a in state x that maximizes the expected instantaneous reward r at time t. The constituents of QMRO are as follows:

QMRO State Space: The optimal settings in a cell depend on the mobility of the UEs in the cell. Thus, the states x characterize the degree of mobility in the

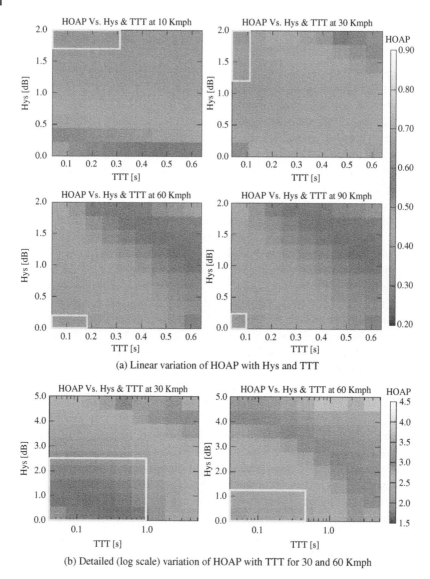

(a) Linear variation of HOAP with Hys and TTT

(b) Detailed (log scale) variation of HOAP with TTT for 30 and 60 Kmph

Figure 8.7 Sensitivity of handover control parameters.

cell and are defined as the average speed over the *SON interval*. Ideally, the states should represent different mobility environments, e.g. static, pedestrian, cyclist, residential area, city limit, suburban limit, rural highway, motorway, etc. This is modelled by discretizing the velocities into intervals, specifically the 12 states {0–4; 4–8; 8–12; 12–17; 17–22; 22–28; 28–34; 34–41; 41–48; 48–56; 56–65;

65–75; 75+}, for which the appropriate HO settings must be learned. Multiple techniques for estimating the velocities exist, and so with a good speed estimate, QMRO learns the best configuration for the given cell.

QMRO Action Space: Actions are the *Hys-TTT* tuples signalled by the cells to their associated UEs. Without SON, an operator configures a cell with default parameter settings obtained from network planning or through experience. Meanwhile, a local search SON solution finds and, at all times, applies usable fixed settings. With the observation that OTPs depend on speed, the settings need to be changed based on the instantaneous speed in the cell. Earlier results [5] showed that for all practical speeds, performance is almost always suboptimal at Hys > 5 dB. Thus, Hys values of only up to 6 dB are considered. Meanwhile, for some TTT settings, especially on the lower end, HOAP differences are unresolvable. So, only a subset of the TTT values is considered, i.e. the 11 values {0.04, 0.10, 0.128, 0.256, 0.32, 0.48, 0.512, 0.64, 1.02, 1.28, 2.56, 5.12} in s. The resulting action space for each state is the 143 possible combinations of the considered Hys and TTT.

QMRO Rewards: QMRO needs to concurrently minimize RLF, PP, and HO. Since the learner is a rewards-maximizing agent, the reward $r_{x,t}$ should be, and is, the negative HOAP over the SON interval. The weight vector applied during learning may need to be adjusted to enforce particular results, especially given the small SON interval. Particularly, in the case of no early RLFs during the SON interval, which could tilt the result to favour too many PPs, the vector is changed to w = (0.4, 0.0, 0.6).

8.3.3 Parameter Search Strategy

Even with cooperative learning, evaluating each action multiple times (with 143 possible actions for each speed state), would require a long time to converge to the desired solutions. To accelerate convergence, the actions are sub-grouped such that for any state, three learning regimes R1–R3 (Figure 8.8) are executed. R1 actions are selected from different regions of the grid to determine the area in which the desired action lies. It was shown in [11] that combinations of low Hys and high TTT are never optimal, so this region is excluded from the possible candidates. The outcome TP of R1 (TP_1), which specifies the region in which the optimum point lies, is then used to define the search space for the next regime R2.

During R2, actions along the diagonal through TP_1 are explored to obtain the approximate delay that is acceptable for the observed mobility state. Again, as seen in Figure 8.8, only a subset of the actions along this diagonal is evaluated. The obtained TP (TP_2) is then used to set the search space for regime R3, which refines the results by exploring points near TP_2. R3 compares TP_2 with its four neighbour

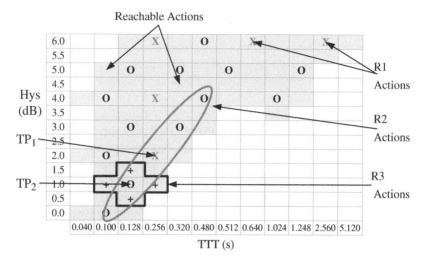

Figure 8.8 QMRO action space within the three learning regimes R1, R2, R3.

points to the left, right, top, or below. In Figure 8.8, 'R3 Actions' is the exploration region for R3 assuming $TP_2 = (1.0\,dB, 0.128\,s)$.

8.3.4 Optimization Algorithm

Given the QL elements as discussed above, the optimization algorithm can be summarized as follows: For each possible state, the action set is initialized with R1 actions and the Q-table entries initialized to 0. Learning is triggered to be executed after every SON interval t. Each cell c observes its environment over the interval, and at the end of t, it determines if an optimization is necessary, i.e. if the cell's speed state has changed. During the learning phase, c selects an action as described in the previous section, otherwise, it selects the best action that would have been learned. It then signals that action to all its associated UEs and starts collecting the necessary performance statistics for the next interval $(t + 1)$. At the end of interval $t + 1$, c evaluates its HOAP and derives the reward r_t for the action at t. In this case, the reward is based only on the resulting HOAP, otherwise, other KPIs that are affected by the actions could, in principle, be included in the reward calculation. Then, given the reward, the SF instance updates the learning agent (the Q-table) before repeating the process.

8.3.5 Evaluation

Performance was studied with an LTE system-level simulator (tool) based on libraries by Nokia Bell Labs and the University of Stuttgart's Institute of

Communication Networks and Computer Engineering. The tool simulates the LTE downlink [16] but was extended with classes defining the required SON functionality. For example, *RLagent,* a class that defines a generic QL agent, is added so that each SF, here MRO, instantiates its own learner as an object of *RLagent*. Then, to maintain local state of the optimization, each cell instantiates its own local objects of the SFs.

A network of seven tri-cell eNBs in a regular hexagonal grid of 500 m inter-site distance is considered, with wrap-around included for better interference estimates. Initially, users randomly placed in the coverage area move with a random walk mobility model, with the resulting deployment as in Figure 8.9. Details of the simulator have been described in [10] but the most important parameters are summarized in Table 8.2.

QMRO must adjust the HO settings congruent to the varying mobility states. To prove applicability to any network, performance is evaluated in the five different speed scenarios described in Table 8.2. These could, for example, represent three typical city districts – a city centre, city edge and residential suburb for the three 'normal' scenarios (10, 30, 60 km/h), and two extreme scenarios like an office park and a highway for the 3 and 120 km/h cases respectively. In each scenario, all UEs are allocated random velocities at the start but also independently randomly adjusted at the start of and during every batch, in each by up to 40%.

To evaluate QMRO, its performance in each speed scenario is compared against the reference network, referred to as 'Ref', which represents the case when all the

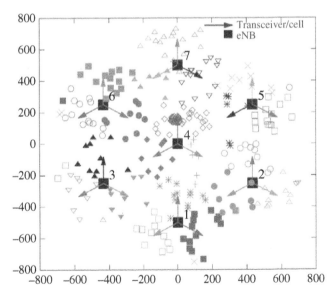

Figure 8.9 Initial network with deployed users coloured according to serving cell.

Table 8.2 Simulation parameters.

Parameter	Value		
System bandwidth	10 MHz		
Inter-site distance	500 m		
Users	240 mobile, 40 static		
Mobile users' velocity (kmph)	City area	Mean velocity	Velocity range
	Office park	3	2 – 4
	City centre	10	6 – 14
	City edge	30	18 – 42
	City suburb	60	36 – 84
	Highway	120	72 – 168
Mobility	Random walk model		
Pathloss	$A + B.\log_{10} [\max(d[km],0.035)]$; $A = 128.1$ and $B = 37.6$		
Shadowing	Standard deviation $= 6$ dB; Decorrelation distance $= 50$ m		
eNB Tx power	46 dBm		
eNB Tx antennas	1 per sector, gain 15 dBi, at height $= 32$ m		
UE antennas	1 Omni, gain 2 dBi, at height $= 1.5$ m		
Data rate	512 kbps		

cells in the network apply the best static settings. These are obtained by extensive search through the parameter space at different velocities [4]. The performance, shown in Figure 8.10, is evaluated in terms of averages of the metric(s) values throughout the network although cell-specific results would demonstrate the same trends.

Figure 8.10a describes the differences in performance between QMRO and *Ref* for all the five speed cases. The figure tracks the HOAP differences at different points in time during the simulation. It can be observed in the figure that QMRO initially performs poorly as it executes the first learning regime R1, but the performance improves as QMRO zooms into the OTP in regimes R2 and R3. In fact, QMRO eventually performs as well as the best static settings in *Ref*. This is evidenced by the fact that all charts converge towards 0 showing that the QMRO has learned settings that offer the same performance as those in *Ref*.

Where user velocities are widely spread, QMRO actually performs better by applying the right setting for each speed range as opposed to the default static setting for all velocities. This is evident for the 120 kmph case shown in Figure 8.10a where, after learning, QMRO is consistently better than *Ref*. Meanwhile, it has

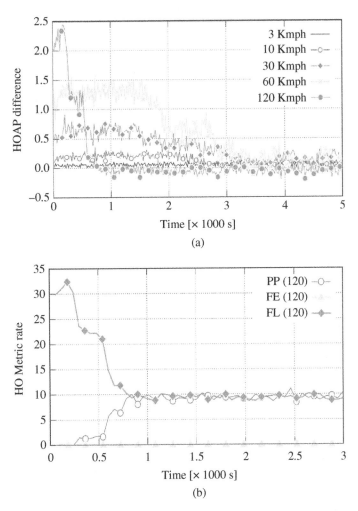

Figure 8.10 QMRO performance: (a) average network-wide HOAP gains for all velocities (b) variation of average rates for all three core-metrics in 120 kmph environment.

also been observed in Figure 8.10a that QMRO has faster convergence as the speed increases. This is because, as the speed increases, each user undertakes more HO with the effect that cells will have shorter SON intervals at higher velocities, i.e. the cells reach the minimum event count much faster.

Figure 8.10b considers the scenario at 120 kmph to evaluate QMRO's effect on the trends of the individual metrics (P, F_E, and F_L). It can be observed that the agent learns to minimize late RLFs (F_L) by trading them with PPs, which have less effect on the user's quality of experience. This is all whilst ensuring that early RLF

(F_E) remains low. QMRO suffers from many RLFs at the beginning as it evaluates settings across a large parameter space. Over time, however, the agent continuously reduces F_L by trading such reduction with increase in PP. This continues until it registers decreasing returns, i.e. when every extra reduction in F_L generates too many PPs or if, instead, it causes early RLFs to occur.

These results demonstrate that with an appropriate definition of states that capture UEs' mobility and adequate learning time, a CF, like QMRO, is able to learn the best Hys-TTT settings for a given mobility environment.

8.4 Fuzzy Q-Learning for Tilt Optimization

A critical challenge in network operations is the maximization of systems capacity without compromising the coverage. The antenna downtilt affects the shape and size of the antenna radiation pattern and subsequently, the received signal strength in the own cell and the interference to neighbouring cells. Too high values of antenna downtilt can result in coverage outage at the cell boundary, whereas, too low values can degrade the network capacity by introducing significant inter-cell interference (ICI). Therefore, antenna downtilts must be optimized in order to ensure that the targeted areas are covered by the network whilst maximizing the cell separation for acceptable cell edge performance. In particular FQL applied to tilt optimization can allow for concurrent optimization of the coverage-interference tradeoff amongst multiple cells. There are multiple learning strategies that can be applied as studied in [7, 17]. These approaches are summarized here.

8.4.1 Fuzzy Q-Learning Controller (FQLC) Components

QL needs to maintain a Q-Value for each state-action pair, which becomes complex in the case of continuous state and/or action spaces. FQL overcomes these shortcomings by applying FL to discretize the continuous variables using fuzzy membership functions. The components of the Fuzzy Q-Learning Controller (FQLC) for tilt optimization are the following:

8.4.1.1 State and Action Fuzzy Variables

Cell states can be measured in terms of several KPI statistics, which characterize the system's performance. The best KPI for the coverage-capacity trade-off is the spectral efficiency as it directly reflects how efficiently the available spectrum is utilized. As such, the input state is, is defined in (8.4), a vector of DT, SE^{center}, and SE^{edge} which are respectively the current downtilt value of the cell, and the mean

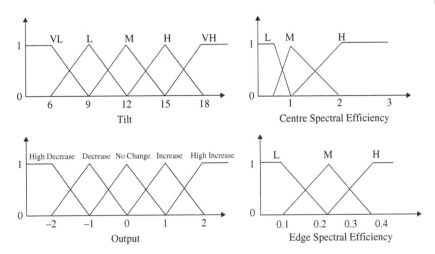

Figure 8.11 Fuzzy membership functions for FQL-based tilt optimization.

and edge spectral efficiency for the cell.

$$s = [DT \quad SE^{center} \quad SE^{edge}] \tag{8.4}$$

The edge spectral efficiency measured from the lower 5-percentile of the mobile SINR reports in the cell. Including SE^{edge} ensures improved performance not only on average but also at the cell borders.

With the focus on tilt optimization, the possible actions for a cell are the changes that can be applied to the cell's tilt. For fuzzification, a finite number of fuzzy labels is defined over the domain of each input and output variable as shown in Figure 8.11. Here, the downtilt and output have five labels, whereas, the SE^{center} and SE^{edge} have three labels each. Each label is assigned a membership function that maps the degree of membership of a particular fuzzy variable to each label within the real interval [0,1]. Strict triangular membership functions are used so that the sum of membership values of all the functions equals 1 at any point over the domain of a specific variable.

8.4.1.2 Rule-Based Fuzzy Inference System

The FIS that maps the input states to the output actions consists of rules defined by different combinations of the fuzzy variable membership functions. A typical FIS rule in FQL is initialized as [17]:

$$IF\ s^1\ is\ L_i\ and\ s^2\ is\ L_i^2\ and\ \dots s^n\ is\ L_i^n$$
$$THEN\ y = o_i^1\ with\ q(L_i, o_i^1) = 0$$
$$or\ y = o_i^2\ with\ q(L_i, o_i^2) = 0$$
$$or \dots$$
$$or\ y = o_i^k\ with\ q(L_i, o_i^k) = 0$$

where, L_j^i is a fuzzy label for a distinct fuzzy set defined in the domain of the nth component of s^n of the state vector $s = [s^1, s^2, ..., s^n]$ and o_k^i is the kth output action for rule i, O being the set of K possible actions. The vector $L_i = [L_i^1, L_i^2, ..., L_i^n]$ is known as the modal vector of rule i and represents one state of the controller. $q(L_i, o_k^i)$ is the Q-value for the fuzzy state Li and action o_k^i and is initialized to zero. The total number of rules depends on the number of variables in the state vector (here three) and the number of membership functions for each of variable (here five). Here no *a-priori* information is assumed to be available for the FQL design and so all actions are equally probable in all states at the start of the learning process. This results in 45 rules/states for the tilt FIS each with five possible actions, except that the states with downtilt in very low or very high state can only increase or respectively decrease the downtilt or keep it constant.

8.4.1.3 Instantaneous Reward

As expected, rewards provide the controller with feedback about its previous action. Let us assume a *State Quality* (SQ), as the measure which characterizes how good a particular state is. Then, for a cell c, the reward is, as in Eqs. (8.5) and (8.6), the difference in quality of two consecutive states.

$$SQ_c = SE_c^{center} + w\, SE_c^{edge} \tag{8.5}$$

$$r_{c,t+1} = SQ_{c,t+1} - SQ_{c,t} \tag{8.6}$$

w is a unitless factor intended to make the edge value comparable to the centre value. This can be set subjectively but a value of 2 has been found to be appropriately representative. $rt + 1$ is the instantaneous reward when the controller transits from state st to $st + 1$ with state qualities SQt and $SQt + 1$ respectively. Thus, actions that lead to better SQ are positively rewarded whilst those leading to poorer SQ are punished with negative rewards.

8.4.2 The FQLC Algorithm

The FQLC algorithm starts by identifying the current state of the learning agent. This is done by calculating the degree of truth of each FIS rule i which is the product of membership values of each input state label for that specific rule i as in (8.7).

$$\beta_i(s) = \prod_{n=1}^{N} \mu_{L_i^n}(s^n) \tag{8.7}$$

Let P be the set of activated rules with a non-zero degree of truth. The Q-value for the input state vector can be calculated in (8.8) as an interpolation of the current Q-values of the activated rules and their degree of truth.

$$Q(s, a(s)) = \sum_{p \in P_s} \beta_p(s) \cdot q(L_p, o_p^k) \tag{8.8}$$

Using the ε-greedy Exploration/Exploitation policy, an output action is selected for each activated rule according to (8.9). The action to be applied by the FQLC is then calculated from all the chosen actions of activated rules according to (8.10).

$$o_p = \begin{cases} \arg \max_{k \in K} q(L_p, o_p^k) & \text{with prob } \varepsilon \\ random (o_p^k) & \text{with prob } 1 - \varepsilon \end{cases} \tag{8.9}$$

$$a(s) = \sum_{p \in P_s} \beta_p(s) o_p \tag{8.10}$$

Applying the action transits environment to a new state $st + 1$ and the FQLC receives a (reinforcement) reward $rt + 1$. The FQLC then calculates the value of new state using (8.11) and subsequently computes the reward (the change in the value of new and old state) according to (8.12)

$$V_t(s_{t+1}) = \sum_{p \in P_{s_{t+1}}} \beta_p(s_{t+1}) \cdot \max_{k \in K} q(L_p, o_p^k) \tag{8.11}$$

$$\Delta Q = r_{t+1} + \gamma V_t(s_{t+1}) - Q(s_t, a(s_t)) \tag{8.12}$$

Finally using (8.1) the Q-values of the FIS rules are updated as in (8.13). As the final action was a combination of multiple actions of different rules, the update accounts for the relative contribution of each rule using the degree of truth $\beta_p(S_t)$ of each activated rule.

$$q(L_p, o_p^k) = q(L_p, o_p^k) + \alpha \beta_p(s_t) \Delta Q \tag{8.13}$$

8.4.3 Homogeneous Multi-Agent Learning Strategies

Owing to the interreference amongst cells, it is evident that the whilst learning the optimal configuration for one cell, there may be negative or positive effects on other cells. The question then is how the learning should be distributed amongst the cells, i.e. either with one cell, all the cells or a subset of the cells learning at a single snapshot. The relative performances of these strategies in terms of their learning speed and convergence properties has been studied using the network in Figure 8.12 that was implemented using the simulator in Section 8.3.5. In the network, with homogenous hexagonal cells, cells with solid boundaries form the simulated network, whereas, the cells with dashed lines are the wraparound cells.

With the considered scenario of a homogenous hexagonal cell structure with equal inter-site distance of 500 m and uniform traffic distribution, the best tilt configuration is an equal value at all cells, which was manually found to be 15°. Performance is evaluated in terms of the SQ at different snapshots of the network, where each snapshot is taken after 200 seconds of simulation time, at which point, each mobile reports its SINR to its serving cell from which the cell computes the SE^{center} and SE^{edge} KPIs. For comparison purposes, the performances of the three

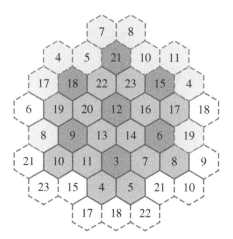

Figure 8.12 Network scenario with a cluster of cells (green) that can simultaneously update their tilts.

strategies can be compared to a reference scenario where all the cells have the best fixed antenna downtilt, i.e. 15° for all cells.

The default learning strategy is to have one cell to learn per snapshot, with the cell for example, randomly selected following a uniform distribution. The cell executes the FQLC algorithm and updates its downtilt whilst the other cells maintain fixed downtilts. This simplifies the learning problem as the environment is only affected by one agent's action making it easy to identify the effects of the agent. However, it, slows down the learning process as the network size increases.

To speed-up the learning process, the cells could update their tilts in every snapshot. This, however, makes the feedback from the environment more complicated since the change in the environment is a result of actions taken simultaneously by multiple agents. The ideal solution in such an environment would be to learn the best joint action of all the cells, which exponentially expands the joint action space instead and correspondingly the Q-table for state-action pairs. As such, the solution here relies on independent learning, where each cell maintains its own Q-table regardless of the actions of its neighbouring cells. This makes the reinforcement signal after each cell's action inaccurate as it also includes the effect of the actions of all the cells.

The best alternative then, is to allow a cluster of agents to update their downtilts per network snapshot with the cluster formed in a way that cells in the cluster are not direct neighbours to one another. This allows a relatively large number of agents to simultaneously update their tilts but with an easy characterization of the change in the environment. Such a cluster, identified by green cells in Figure 8.12, can easily be formed by graph colouring [18].

8.4.4 Coverage and Capacity Optimization

CCO is a network optimization task intended to autonomously provide the required capacity in target coverage areas, minimize the interference and maintain an acceptable quality of service. Optimization of antenna tilts using the FQL algorithm can thus be effective in achieving the CCO objectives and its performance thereof has been studied using the network in Figure 8.12.

The CCO challenge is to find the best configurations for all cells when applying each of the three candidate learning strategies. The corresponding comparison of the performances for the complete network is Figure 8.13a in which the horizontal and vertical axes respectively represent the snapshots and the average of SQ as defined in Eq. (8.5) for all cells. The 'Reference' curve represents the average SQ for the reference system with 15° tilts for all the antennas whilst the other three curves represent the results for the three learning strategies. For all three strategies, the network is initialized with 6° downtilt across all the cells, resulting in very low SQ due to very large coverage areas and high interference for all the cells. The system quickly learns the quasi-optimal antenna tilt settings and improves the overall SQ, albeit to different levels of success.

Allowing only one sector to update its tilt at each snapshot significantly increases the convergence time, yet the achieved steady-state performance is also the worst. As the only one taking action, each cell sees little interference and so it learns the most aggressive tilts which instead increases that cell's interreference to the other cells. On the other hand, with 'all sectors' concurrently updating their tilts, convergence is fast – requiring only about 100 snapshots to reach steady-state performance. But, as expected, the simultaneous actions of neighbouring cells limit the learning to the extent that the performance never exactly matches the performance of the reference system.

By allowing only non-adjacent cells to update their tilts at each snapshot, the clustering mechanism trades speed for performance. With only a subset of cells updating convergence is slower, requiring about 200 snapshots to reach steady-state performance. The achieved steady-state performance is, however, better than the other two strategies and is consistently maintained close to the optimal level.

8.4.5 Self-Healing and eNB Deployment

The FQL algorithm can also be used for self-healing in the event of a cell outage and self-configuration when a new eNB is deployed. For the self-healing study an outage is created by deactivating the centre eNB with cells 12, 13, and 14 in Figure 8.12. This creates a coverage hole and the neighbouring eNBs try to extend their coverage to compensate the coverage loss. Reactivating the centre

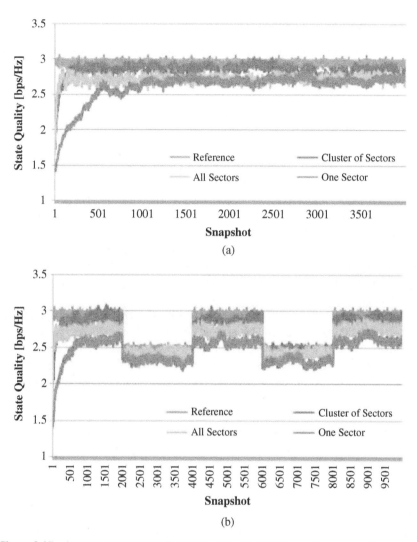

Figure 8.13 Average state quality for (a) the network and (b) at cell outage and recovery [17].

eNB is equivalent to deploying a new eNB and requires a (re)configuration of antenna tilts for all cells. Gradual deployment requires the eNB is first deployed with a very high downtilt value (in this case 22°) to ensure that the initial coverage of the deployed eNB is very small. The eNB can then extend its coverage to balance the coverage and capacity requirements of the whole network neighbourhood.

The SQ results for both the outage and deployment scenarios are presented in Figure 8.13b. The vertical-axis represents the average of SQs for all 21 cells even when the centre three cells are in outage. As such, the average SQ in outage is always less than that in normal operation, since the three cells in outage have zero contribution to the average. The network is initialized as in the previous section and results with the same performance observations.

The system suffers a performance degradation at the outage time at snapshot 2000. Neighbour cells react by adjusting downtilts to extend their coverage and compensate for the degradation. As a result, the SQ improves again for all three learning strategies. Finally, at snapshot number 4000, the centre eNB is reactivated but with a higher downtilt of 22°. This improves the SQ a little and the cells again adjust their downtilts using the proposed learning mechanisms. As a result, the gap between the reference system and the learning systems decreases after some snapshots.

8.5 Interference-Aware Flexible Resource Assignment in 5G

In the 5G era, the evolution of heterogeneous networks (HetNets) results in cell densification with cells of different sizes. Due to the time- and spatial-dependent service requirements and traffic patterns, it is expected to have time-varying asymmetric traffic load in both uplink (UL) and downlink (DL) in different cells. Flexible duplex is one of the key technologies in 5G to optimize the resource utilization depending on traffic demand. The main objective is to adapt to asymmetric UL and DL traffic with flexible resource allocation in the joint time-frequency domain, such that the distinction between time division duplex (TDD) and frequency division duplex (FDD) is blurred, or completely removed. It is considered as the next step of TDD/FDD convergence in 5G networks, as it allows UL and DL transmissions to share a joint set of resource and to dynamically split between the resource allocated to UL and DL. The split ratio is time-variant and cell-specific as shown in Figure 8.14.

Despite the advantage of adaptation to the dynamic traffic asymmetry, the challenge is the newly introduced ICI between duplexing mode DL and UL, hereinafter referred as inter-mode interference (IMI). The DL-to-UL interference plays a more important role due to the large difference between DL and UL transmission power. Promising approaches to address this challenge are the muting schemes. This section describes novel partial resource muting schemes [19] that have the objective of finding a good trade-off between resource reuse and interference avoidance in wireless networks and evaluate their applications in both 4G networks and flexible duplex-supported 5G networks.

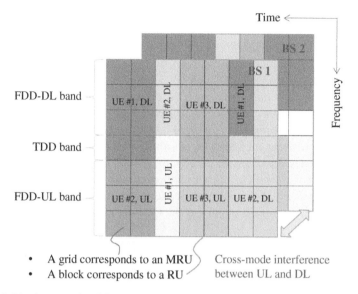

- A grid corresponds to an MRU
- A block corresponds to a RU

Cross-mode interference between UL and DL

Figure 8.14 An example of flexible resource allocation and inter-cell cross-mode interference between UL and DL.

8.5.1 Muting in Wireless Networks

Muting schemes are promising approaches to address the challenge of finding a good balance between interference avoidance and resource spectrum reuse, which include the almost blank subframe (ABS) approach in time domain [20] and the mutually exclusive resource block allocation approach in frequency domain [21]. However, in these approaches two questions remain largely unanswered:

- At which point does full resource reuse becomes inefficient? In other words, when is it appropriate to activate resource muting?
- How to develop efficient resource muting schemes that deal with complex interference patterns caused by disruptive architectures in future networks?

These questions can be answered with fundamental analysis on the resource efficiency region within a unified max-min utility fairness framework. Since Yates [22] introduced standard interference functions (SIFs) and formulated the constrained power control problem into an equivalent fixed-point framework, power control problems have been widely studied. Various max-min fairness problems with SINR related utilities have been proposed to optimize variables such as transmit power, beamformers, and base station (BS) assignment [9, 12]. Max-min fairness problems with rate-related utilities characterized by nonlinear load coupling have also attracted much attention recently [13].

Although many works have provided solutions to these problems, they have not studied the properties of the solution depending on the budget. A recent work [14] fills this gap, and it provides tight bounds for the performance of the solutions by exploiting the concept of asymptotic functions. The authors have also introduced a transition point that indicates whether a network operates in an energy-efficient region. This same technique is used in this work to examine the resource efficiency depending on the resource budget for resource allocation problems, with special focus on complex interference patterns in future networks.

The approach involves characterizing resource-efficient regions of the solutions to a general class of resource allocation problems and leveraging it to develop a partial resource muting scheme to improve resource efficiency. The developed scheme, which is based on the solutions to a series of subproblems using fixed point iterations, efficiently detects a set of bottleneck users assigned to the muting region and optimizes the service-centric resource allocation. This framework is then applied to flexible duplex enabled 5G networks with complex interference models using novel heuristics that are introduced to approximate successively the interference and to jointly optimize the uplink/downlink resource configuration.

The numerical results show a significant performance gain of the resource muting scheme, compared to the optimal solution to the max-min utility fairness problem with full frequency reuse. The gain is more significant as the network gets denser.

8.5.2 Notations, Definitions, and Preliminaries

Consider that the nonnegative and positive orthant in k dimensions are denoted by $\mathbb{R}_+^k$ and $\mathbb{R}_{++}^k$ respectively. If $\mathbf{x} \leq \mathbf{y}$ denotes the component-wise inequality between two vectors, a norm $\|\cdot\|$ on $\mathbb{R}_+^k$ is said to be monotone if $(\forall \mathbf{x} \in \mathbb{R}_+^k)(\forall \mathbf{y} \in \mathbb{R}_+^k) 0 \leq \mathbf{x} \leq \mathbf{y} \Rightarrow \|\mathbf{x}\| \leq \|\mathbf{y}\|$. A diagonal matrix with the elements of $\mathbf{x}$ on the main diagonal is denoted by $\mathrm{diag}(\mathbf{x})$ and the cardinality of set $\mathcal{A}$ by $|\mathcal{A}|$ whilst $[f(x)]^+ := \max\{0, f(x)\}$ defines the positive part of a real function. SIFs are defined as follows:

Definition 8.1 A function $f : \mathbb{R}^k \to \mathbb{R}_{++} \bigcup \{\infty\}$ is an SIF if the following axioms hold: (i) (Monotonicity) $(\forall \mathbf{x} \in \mathbb{R}_+^k)(\forall \mathbf{y} \in \mathbb{R}_+^k) \mathbf{x} \leq \mathbf{y} \Rightarrow f(\mathbf{x}) \leq f(\mathbf{y})$; (ii) (Scalability) $(\forall \mathbf{x} \in \mathbb{R}_+^k)(\forall \alpha > 1) \alpha f(\mathbf{x}) > f(\alpha \mathbf{x})$; and (iii) (Nonnegative effective domain) $\mathrm{dom} f := \{\mathbf{x} \in \mathbb{R}^k | f(\mathbf{x}) < \infty)\} = \mathbb{R}_+^k$. A vector function $\mathbf{f} : \mathbb{R}_+^k \to \mathbb{R}_{++}^k : \mathbf{x} \mapsto [f_1(\mathbf{x}), \dots, f_N(\mathbf{x})]$ is called an SIF if each of the component functions is an SIF.

Most utility maximization problems in wireless networks can be seen as particular instances of the following optimization problem [13, 14, 22]

$$\max_{\mathbf{x} \in \mathbb{R}_+^k} \min_{k \in \mathcal{K}} u_k(\mathbf{x}), \quad \text{subject to } \|\mathbf{x}\| \leq \theta \tag{8.14}$$

where $\mathcal{K}:=\{1,\dots,K\}$ is the set of network elements, $u_k : \mathbb{R}_+^k \to \mathbb{R}_+$ is the utility function of network element $k \in \mathcal{K}$ and $\|\cdot\|$ is a monotone norm used to constrain the resource utilization $\mathbf{x}:=[x_1,\dots,x_K]$ to a given budget $\theta > 0$. In the above problem formulation, the k-th component x_k of the optimization variable $\mathbf{x}$ is the resource utilization of network element $k \in \mathcal{K}$. Now, consider the following conditional eigenvalue problem (CEVP):

(The conditional eigenvalue problem:) Given a monotone norm $\|\cdot\|$, a budget $\theta > 0$ and a mapping $\mathbf{T} : \mathbb{R}_+^k \to \mathbb{R}_{++}^k : \mathbf{x} \mapsto [T^{(1)}(\mathbf{x}),\dots,T^{(K)}(\mathbf{x})]$, where $T^{(1)} : \mathbb{R}_+^k \to \mathbb{R}_{++}$ is an SIF for each $k \in \mathcal{K}$, the CEVP is stated as follows:

$$\text{Find } (\mathbf{x},c) \in \mathbb{R}_+^k \times \mathbb{R}_{++} \text{ such that } \mathbf{T}(\mathbf{x}) = \frac{1}{c}\,\mathbf{x} \text{ and } \|\mathbf{x}\| = \theta \qquad (8.15)$$

As an implication of the results, if the utility functions in (8.14) and the SIF $\mathbf{T}$ in (8.15) are related by $(\forall k \in \mathcal{K})(\forall \mathbf{x} \in \mathbb{R}^k)u_k(\mathbf{x}) = \frac{x_k}{T^{(k)}(\mathbf{x})}$, then $\mathbf{x}^*$ solves (8.14) if $(\mathbf{x}^*,c^*)$ solves (8.15). Furthermore, with $(\forall k \in \mathcal{K})u_k(\mathbf{x}^*) = c^*$, solving (8.14) simplifies to devising efficient algorithmic solutions to (8.15). To this end, [23] *has proved that* (8.15) *has a unique solution* $(\mathbf{x}^*,c^*) \in \mathbb{R}_{++}^k \times \mathbb{R}_{++}$ *and that* $\mathbf{x}^* \in \mathbb{R}_{++}^K$ *is the limit of the sequence* $(\mathbf{x}^{(n)})_{n \in \mathbb{N}}$ *generated by*

$$\mathbf{x}^{(n+1)} = \frac{\theta}{\|\mathbf{T}(\mathbf{x}^{(n)})\|}\mathbf{T}(\mathbf{x}^{(n)}), \quad \text{with } \mathbf{x}^{(0)} \in \mathbb{R}_+^k. \qquad (8.16)$$

With knowledge of $\mathbf{x}^*$, c^* is recovered from the equality $c^* = \frac{\theta}{\|\mathbf{T}(\mathbf{x}^*)\|}$.

The properties of the solution to Problem (8.14) can be obtained using the properties of the asymptotic functions [14]. Proposition 2 in [19] has provided the properties in terms of utility efficiency $U : \mathbb{R}_{++} \to \mathbb{R}_{++} : \theta \mapsto c_\theta$ and resource efficiency $E : \mathbb{R}_{++} \to \mathbb{R}_{++} : \theta \mapsto \frac{U(\theta)}{\|\mathbf{x}_\theta\|}$, where $(\mathbf{x}_\theta, c_\theta)$ denotes the solution to (8.14) for a given budget $\theta \in \mathbb{R}_{++}$. As an example, the properties of the utility function are shown in Figure 8.15. The transition point of θ serves as a coarse indicator of whether substantial gains in utility are achievable by increasing the budget θ. More precisely, if the given budget θ is greater than the transition point, then the network is likely operating in a regime where the performance is limited by interference, so increasing θ even by orders of magnitude typically brings only marginal gains in utility. In contrast, if θ is smaller than the transition point, then noticeable gains in utility can be obtained by increasing the budget θ. The observation is illustrated in Figure 8.15 and it is heavily exploited by the algorithms discussed in the next subsections.

8.5.3 System Model and Problem Formulation

Consider an orthogonal frequency division multiplexing (OFDM)-based wireless network system supporting flexible duplex, consisting of a set of BSs

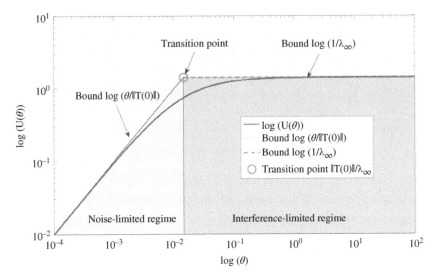

Figure 8.15 Utility as a function of the power budget θ.

$\mathcal{N} := \{1, \dots, N\}$ and a set of communication links $\mathcal{K} := \{1, \dots, K\}$ in DL. Hereafter, the terms 'communication links', 'services', and 'users' are used interchangeably, i.e. without loss of generality, one can assume that each user sets up one communication link for one service at a unit of time. Let $\overline{W}$ denote the total amount of resource in Hz $\cdot$ s (in both frequency and time domain) and, let $\mathbf{w} \in [0, 1]^K$ be a vector collecting the fraction of resource units allocated to the services. Also, assume that the transmit power spectral density (in Watts/Hz) of all services are given and collected in a vector $p \in \mathbb{R}^K_{++}$ and let the data rate demand of all services (in bit/s) be collected in $\overline{r} \in \mathbb{R}^K_{++}$. The matrix $\mathbf{A} \in \{0, 1\}^{N \times K}$ denotes the BS-to-service assignment matrix, where $a_{n,k} = 1$ if k is served by BS n, and 0 otherwise.

Let $v_{k,l}$ denote the channel gain between the transmitter of service l and the receiver of service k. $v_{k,k} > 0$ is the channel gain of link k whilst for $k \neq l$, $v_{k,l}$ is the interference channel gain, which is positive if l causes interference to k, otherwise $v_{k,l} = 0$. And let σ_k^2 denote the noise spectral density (in Watt/Hz) in k's receiver.

By enabling the flexible duplex, the IMI between UL and DL needs to be considered. To approximate such interference, the UL/DL resource overlapping factors $(c_{k,l})$ collected in $\mathbf{C}(\mathbf{w}) \in \mathbb{R}^{K \times K}_+$ are introduced to incorporate the probability that intra- and inter-mode interference appear for a given resource allocation $\mathbf{w}$ [19]. Let $\boldsymbol{\eta}^{(x)} := [\eta_1^{(x)}, \dots, \eta_K^{(x)}]$, $\mathrm{x} \in \{\mathrm{u}, \mathrm{d}\}$ denote the UL or DL load (i.e. the fraction of occupied resources) of all cells that can be computed with $\mathbf{w}$. For example, $\eta_n^{(u)} := \sum_{k \in \mathcal{K}_n^{(u)}} w_k$ denotes the load of BS n in UL, where $\mathcal{K}_n^{(u)} \subset \mathcal{K}_n$ is the set of UL services in BS n. The SINR of service k incorporating IMI is approximated by

$$\text{SINR}_k(\mathbf{w}) \approx \frac{p_k}{[(\mathbf{C}(\mathbf{w}) \cdot \widetilde{\mathbf{V}})\text{diag}(\mathbf{p})\mathbf{w} + \widetilde{\sigma}]_k} \tag{8.17}$$

with $\mathbf{C}(\mathbf{w}):=(c_{k,l}) \in \mathbb{R}_+^{K \times K}$,

$$c_{k,l} := \begin{cases} \left[\dfrac{(\eta_{n_l}^{(x_l)} + \eta_{n_k}^{(x_k)} - 1)}{\eta_{n_k}^{(x_k)}} \right]^+, & \text{if } x_l \neq x_k \\ \min\left\{ 1, \dfrac{\eta_{n_l}^{(x_l)}}{\eta_{n_k}^{(x_k)}} \right\}, & \text{if } x_l = x_k \end{cases} \tag{8.18}$$

where $\circ$ denotes the Hadamard product, $x_k \in \{u, d\}$ denotes the duplex mode of service k, and n_k denotes the serving BS of k. $\widetilde{\mathbf{V}} \in \mathbb{R}_+^{K \times K}$ denotes the interference coupling matrix with the (k, l)-th entry defined as $\frac{v_{k,l}}{v_{k,k}}$, and $\widetilde{\sigma} := \left[\frac{\sigma_1^2}{v_{1,1}}, \ldots, \frac{\sigma_K^2}{v_{k,k}} \right] \in \mathbb{R}_{++}^K$.

The multiplication of $c_{k,l}$ by w_l loosely approximates the probability that a resource unit allocated to l in duplex mode x_l in BS n_l causes interference to service k in duplex mode x_k in BS n_k. The achievable rate (in bit/s) of service k is computed by

$$r_k(\mathbf{w}) = w_k \overline{W} \log_2(1 + \text{SINR}_k(\mathbf{w})). \tag{8.19}$$

8.5.4 Optimal Resource Allocation and Performance Limits

The objective is to maximize the worst-case service-specific QoS satisfaction level, defined as the ratio of the achievable rate $r_k(\mathbf{w})$ to the rate demand $\overline{r}_k$, subject to a per-BS load constraint. Formally, the problem is stated as follows:

$$\max_{\mathbf{x} \in \mathbb{R}_+^k} \min_{k \in \mathcal{K}} \frac{r_k(\mathbf{w})}{\overline{r}_k}, \quad \text{subject to } \|\mathbf{A}\mathbf{w}\|_\infty \leq \theta \tag{8.20}$$

where $\|\cdot\|_\infty$ denotes the L_∞-norm. Note that with full load constraint, $\theta = 1$.

Besides the solution to Eq. (8.20), it is also necessary to answer the question: Is the load limit $\theta = 1$ with resource reuse factor 1 an efficient operation point? The resource efficiency of the solution to Problem (8.20) as a function of budget θ can be studied by first examining the properties of the mapping $\mathbf{T} := [T^{(1)}(\mathbf{w}), \ldots, T^{(K)}(\mathbf{w})]$, where $(\forall k \in \mathcal{K})T^{(k)} : \mathbb{R}_+^K \to \mathbb{R}_{++} : \mathbf{w} \mapsto \frac{r_k(w)}{(\overline{W}\log_2(1+SINR_k(\mathbf{w})))}$. Unfortunately, due to the introduction of $\mathbf{C}(\mathbf{w})$ in (8.18), $\mathbf{T}$ is nonlinear, nonconvex, and nonmonotonic, which leads to possibly more than one fixed point of the resulting CEVP (8.15).

However, if $\mathbf{C}$ is estimated as a matrix of fixed constants, then, $\mathbf{T}_{\widehat{\mathbf{C}}} : \mathbb{R}_+^K \to \mathbb{R}_{++}^K$, where $T_{\widehat{\mathbf{C}}}^{(k)} : \mathbb{R}_+^K \to \mathbb{R}_{++} : \mathbf{w} \mapsto \frac{r_k(w)}{(\overline{W}\log_2(1+SINR_k(\widehat{\mathbf{C}}, \mathbf{w})))}$ is an SIF (as proven in Lemma 1 in [19]). Assume that each BS serves at least one user and every user is served by a BS, which guarantees that each BS serves a nonempty and unique set of users, then all rows of the assignment matrix $\mathbf{A}$ are linearly independent. Therefore, $\mathbf{A}$ is a nonnegative full (row) rank matrix, so the function $g(\boldsymbol{w}) := \|\mathbf{A}\boldsymbol{w}\|_\infty$ is a monotone norm. Thus, with estimated $\hat{\mathbf{C}}$, the problem in (8.20) is an instance of that in (8.14), and the solution to (8.14) can easily be obtained with the iterations in (8.16). Furthermore, Proposition 2 in [19] can be used to compute the asymptotic mapping $\mathbf{T}_\infty$ associated with $\mathbf{T}$. Then, as shown below, $\boldsymbol{w}^*$ can be computed with the fixed-point iteration given in (8.14), and the performance properties as a function of θ become readily available.

8.5.5 Successive Approximation of Fixed Point (SAFP)

The developed novel Successive Approximation of Fixed Point (SAFP) to approximate the near-optimal fixed point of the CEVP, performs the following two steps iteratively: (i) fixed-point iteration with approximated $\hat{\mathbf{C}}$ to optimize $\boldsymbol{w}$, and (ii) estimating $\hat{\mathbf{C}}$ with $\hat{\mathbf{C}} := C(\boldsymbol{w}^{(t+1)})$. The proposed novel algorithm assisted with *random initialization* and *successive approximation* is summarized below:

- The algorithm runs for $Z^{(\max)}$ times, where at the i-th time different random initializations of $\mathbf{w}^{(0)} = \hat{\boldsymbol{w}}_i$ and the corresponding $\hat{\mathbf{C}} := \hat{\mathbf{C}}(\hat{\boldsymbol{w}}_i)$ are used.
- For each initialization, the following two steps are iteratively performed: (i) use the fixed-point iteration (8.14) with respect to the approximated $\hat{\mathbf{C}}$ and derive $\mathbf{w}^{(t)}$, and (ii) update $\hat{\mathbf{C}} := \hat{\mathbf{C}}(\boldsymbol{w}^{(t)})$ and increment t. The iteration stops if $\|\mathbf{w}^{(t)} - \mathbf{w}^{(t-1)}\| \leq \varepsilon$, where ε is a distance threshold.
- Each random initialization converges to a fixed point (not necessarily different from those derived from other initializations) and the solution with the maximum utility is chosen.

As shown in [24], the algorithm converges for each random initialization. With a limited number of random initializations, the algorithm can find a solution with low computational complexity.

8.5.6 Partial Resource Muting

Using Proposition 3 in [19] the asymptotic performance limits $U(\theta)$ and $E(\theta)$ as well as their solutions (w_0, λ_0) and $(w_\infty, \lambda_\infty)$ can be derived for the resource budgets $\theta \to 0$ and $\theta \to \infty$. Further using Proposition 2 in [19] the transition point (as shown in Figure 8.15) $\theta^{(\text{trans})}$ can be obtained. Due to the lack of space, the calculation is omitted here, and the reader is referred to [14].

The remaining problem is: If the transition point yields $\theta^{(\text{trans})} < 1$, full resource reuse (i.e. $\theta = 1$) may not be an efficient operation point because of the likelihood of operating in an interference limited region where the resource availability θ can be decreased without noticeable changes in utility – see Figure 8.15 A new challenge arises, i.e. How to improve the resource efficiency if the network is operating in an interference-limited inefficient region.

Since the bottleneck users usually consume most of the resources and impair the performance, it is adequate to consider muting partial resources in the neighbouring cells to mitigate the interference received in (and generated by) the bottleneck users. Based on $(w_\infty, \lambda_\infty)$ obtained with Proposition 3 in [19] and the transition point $\theta^{(\text{trans})}$ derived with Proposition 2 in [14], a resource muting scheme can be developed to consist of the following steps.

8.5.6.1 Triggering the Resource Muting Scheme

$\theta^{(\text{trans})} < 1$ implies inefficient usage of resources in the region $[\theta^{(\text{trans})}, 1]$ due to heavy interference. Instead of allocating all resources in at least one BS to achieve only a slight increase in utility, the network may better benefit from muting partial resources to reduce the interference of bottleneck users. Therefore, the resource muting scheme can be triggered if $\theta^{(\text{trans})} < 1$.

8.5.6.2 Detection of Bottleneck Users

The asymptotic behaviour $(\mathbf{w}_\infty, \lambda_\infty)$ provides a good guide to efficiently detect the bottleneck services, because it indicates the existing limits of the utility and the fraction of allocated resources as $\theta \to \infty$. More precisely, let the k-th entry of $\mathbf{w}_\infty$, be denoted by $w_\infty^{(k)}$, The larger the value of $w_\infty^{(k)}$, the higher the chance that the corresponding user impairs the system performance owing to the large amount of occupied resources, and, consequently, the higher the possibility that this user causes heavy interference.

To show how $\mathbf{w}_\infty$ can be useful in the development of efficient heuristics for selecting the set of bottleneck users, *successive selection* is applied. The approach is suboptimal, but its computational complexity is substantially smaller than that of exhaustive search. By sequentially selecting users with the highest values $\mathbf{w}_\infty$, the resulting optimized utility has a general trend of first increasing and then decreasing, as shown in Figure 8.16. The users with the next highest value of $\mathbf{w}_\infty$ are successively added until the utility does not increase.

Suppose that a set of bottleneck users, denoted by $\mathcal{K}^{(b)}$, is selected. The motivation is to allocate $\mathcal{K}^{(b)}$ to a muting region, such that each user $k \in \mathcal{K}^{(b)}$ neither receives interference from the neighbouring cells nor generates interference to those cells, as shown in Figure 8.17.

Since the received/generated interference of the bottleneck users from/to the neighbouring cells is cancelled, the interference coupling matrix for the set $\mathcal{N}_k$ of neighbouring cells of cell k is updated as follows

$$(\forall k \in \mathcal{K}^{(b)})(\forall m \in \mathcal{N}_k)(\forall l \in \mathcal{K}_m)\, v_{k,l} = v_{l,k} = 0 \tag{8.21}$$

On the other hand, to incorporate the muting region in the constraint, with $\mathcal{K}_m^{(b)}$ as the set of bottleneck users in cell m, the load constraint in (8.20) is

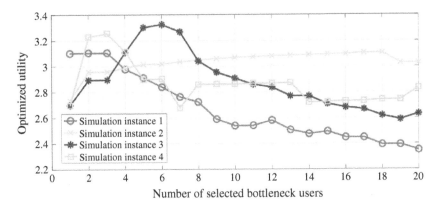

Figure 8.16 Example of successive selection of bottleneck users.

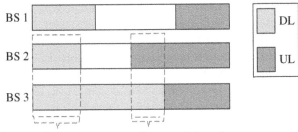

Figure 8.17 Resource muting region where inter-mode interference appears.

modified as:

$$g(\mathbf{w}) \le 1, g : \mathbb{R}^K \to \mathbb{R}_+ : \mathbf{w} \mapsto \max_{n \in \mathcal{N}} \left(\sum_{k \in \mathcal{K}_n} |w_k| + \sum_{m \in \mathcal{N}_n} \sum_{l \in \mathcal{K}_m^{(b)}} |w_l| \right) \qquad (8.22)$$

Constraint (8.22) implies that, for each cell n, the sum of resources allocated to its serving users $\mathcal{K}_n$ (including both the bottleneck and non-bottleneck users) and the resources allocated to all the bottleneck users served by its neighbouring cells $m \in \mathcal{N}_n$ is limited to the total amount of resources. Note that the modified function g is a monotone norm. Thus, the problem with the modified constraint (8.22) is still an instance of that in (8.14).

8.5.7 Evaluation

A real-world scenario with 15 three-sector macro BSs and 10 Pico BSs in the city centre of Berlin, Germany is considered. The locations of the macro BS are given by the real dataset [25], whilst the Pico BSs are placed uniformly at random in the playground. The macro BSs are equipped with directional antennae with transmit power of 43 dBm, whilst the pico BSs have omnidirectional antennae with transmit power of 30 dBm. The total bandwidth is 10 MHz. The macro-cell pathloss is obtained from the real data set [25], and the picocell pathloss uses the 3GPP LTE model in [26]. Uncorrelated fast fading characterized by Rayleigh distribution is implemented on top of the pathloss. A fixed number of users are randomly and uniformly distributed on the playground. The traffic demand per user is uniformly distributed between [0, 10] MBit/s with an average value of 5 MBit/s. Flexible duplex is enabled such that resources can be dynamically allocated to UL or DL services.

Figure 8.18 illustrates how the transition point of the utility and the resource efficiency change when the resource muting scheme is activated. The left side shows that the resource efficient region increases (or, equivalently, the interference-limited region decreases) when applying resource muting. The achieved utility $U(\theta)$ and resource efficiency $E(\theta)$ with the muting scheme are significantly higher at $\theta = 1$.

Owing to the space limitation, only one exemplary numerical result is presented in Figure 8.19. The performance was compared for the following four protocols: (i) 'FIX' for fixed ratio between UL and DL resources; (ii) 'SAFP' for dynamic UL/DL resource configuration with SAFP without muting; (iii) 'Resource muting based on I' for the resource muting scheme based on interference indicator I, and (iv) 'Resource muting based on $\mathbf{w}_\infty$' for the resource muting scheme based on the asymptotic behaviour $\mathbf{w}_\infty$. The performance is obtained by averaging 1000 Monte Carlo simulations with random distribution of user locations and traffic demands conditioned on the given traffic asymmetry for $K = 200$. A measure called *inter-cell traffic distance* is defined to reflect both the inter-cell traffic asymmetry and the intra-cell UL/DL traffic asymmetry between a pair of cells m, n. This is computed as $D_{m,n} := \|\mathbf{d}_n - \mathbf{d}_m\|_2$, where $\|\cdot\|_2$ denotes the L^2-norm, and the per-cell UL/DL traffic demands $\mathbf{d}_n := [d_n^{(u)} - d_n^{(d)}]$ are normalized. Figure 8.19 shows that dynamic UL/DL configuration achieves a twofold increase in the average utility compared to fixed UL/DL ratio. Resource muting brings further improvement, varying from 20 % − 100%, by adapting to the traffic asymmetry. Note that the per-formance gains increase with the traffic asymmetry.

In general, these results show that an efficient resource utilization region can be characterized using the asymptotic behaviour of solutions to max-min utility optimization problems with interference models based on the load imposed by services. This then enables a partial resource muting scheme that is suitable for flexible duplex enabled 5G networks. And the results measured by the

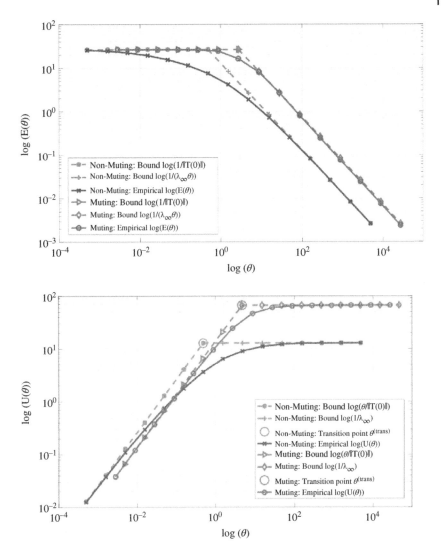

Figure 8.18 Comparison between the performance limits of non-muting scheme and muting scheme, $K = 200$.

worst-case QoS satisfaction level show that, compared to the optimal solution to the max-min optimization problem without resource muting mechanisms, significant performance gains are possible in both scenarios and increases with traffic asymmetry.

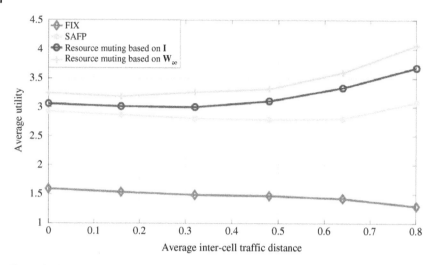

Figure 8.19 Average utility with different inter-cell traffic distances.

8.6 Summary and Open Challenges

This chapter has shown how cognitive techniques can be applied to SO tasks, i.e. for processes in which autonomic functions continuously interact with the network with the intention of improving the network's operating state. The nature of SO tasks is characterized, describing application of specific use-cases that demonstrated the proposed advancement of network optimization.

An RL framework is presented as an effective mechanism for implementing CF for network optimization. It can be used for online learning in cases where the cost of the trial and error activations during the learning period is low enough. Otherwise, it can also be used to learn the optimal configuration in an offline manner and transfer the learned policy functions onto the real network. Moreover, the framework is flexible enough to allow for incorporation of the operator's knowledge and desires mainly through the design of the reward function. The discussion also showed how such a framework can even be used for problems with continuous-valued state or action parameters using Fuzzy Inference.

The following sections described how the QL framework can be applied to two specific challenges – the optimization of user mobility robustness and cell downtilts. In the first case, the RL agent learned to optimize handover control parameters eventually learning configurations as good as the manually confirmed optimal settings. The tilt optimization agents on the one hand is capable of learning the optimal tilts to optimize the capacity-coverage trade off. Moreover, the same agent can be used to manage cell outages by learning how to adjust the downtilts of all cells that are neighbours to the failing cell to compensate for the lost coverage. And, in the event of a new cell deployment, the same agent was able to gradually readjust the tilts in each of the existing cells as well as in the new cell to allow for a graceful addition of the new cell's capacity without affecting the coverage and interreference state of the existing cells.

The reader should note, however, that presented results are only indicative and multiple improvements are possible in actual implementations both for MRO or for any other function. For example, using a neural network in place of the Q-table would enable the agents to learn more generic functions that combine the multiple control values with a high resolution without having to be concerned about the explosion of the state-action space. In the case of MRO, besides considering the Hys and TTT, this may translate in including the RSRP filter coefficient k and multiple traffic states in combination with user speed as part of either the state vector or the control parameter space. It would also be possible to learn based on the observations of individual UEs (i.e. train learner for each event observed by each UE) as opposed to learning based on aggregated statistics. The results also show the benefit of incorporating expert knowledge into the learning – understanding the effects of multiple cells learning together, allowed the design of learning exploration structure that minimized the cross interference during learning yet allowed for fast convergence. It is expected that this will still be useful going forward, i.e. cognitive algorithms should be designed to account for the known physics of the networks if they are going to be useful in network automation.

References

1 merriam-webster.com (2019). Optimization. merriam-webster.com https://www.merriam-webster.com/dictionary/optimization (accessed 7 Feb 2019).

2 Bui, N., Cesana, M., Hosseini, A. et al. (2017). A survey of anticipatory mobile networking: context-based classification, prediction methodologies, and optimization techniques. *IEEE Communication Surveys and Tutorials* 19 (3): 1790–1821.

3 G. R. MacCartney, J. Zhang, S. Nie and T. S. Rappaport, "Path loss models for 5G millimeter wave propagation channels in urban microcells," in *IEEE GLOBECOM*, 2013.

4 Y. Sun, X. Yin, J. Jiang, V. Sekar, F. Lin, N. Wang, T. Liu and B. Sinopoli, "CS2P: Improving video bitrate selection and adaptation with data-driven throughput prediction," in *ACM SIGCOMM*, 2016.

5 Sutton, R. (1998). *Reinforcement Learning: An Introduction*. MIT Press.

6 Mwanje, S.S., Schmelz, L.C., and Mitschele-Thiel, A. (2016). Cognitive cellular networks: a Q-learning framework for self-organizing networks. *IEEE Transactions on Network and Service Management* 13 (1): 85–98.

7 M.N. ul Islam. and A. Mitschele-Thiel, "Cooperative fuzzy Q-Learning for self-organized coverage and capacity optimization," in *Personal Indoor and Mobile Radio Communications (PIMRC), 2012 IEEE 23rd International Symposium on*, 2012.

8 Hamalainen, S., Sanneck, H., and Sartori, C. (eds.). *LTE Self-Organising Networks (SON): Network Management Automation for Operational Efficiency.* Wiley.

9 Yates, R.D. and Huang, C.-Y. (1995). Integrated power control and base station assignment. *IEEE Transactions on Vehicular Technology* 44 (3): 638–644.

10 Mwanje, S.S. (2015). *Coordinating Coupled Self-Organized Network Functions in Cellular Radio Networks.* TU Ilmenau.

11 S. S. Mwanje and A. Mitschele-Thiel, "Distributed cooperative Q-learning for mobility-sensitive handover optimization in LTE SON," in *Computers and Communication (ISCC), 2014 IEEE Symposium on. IEEE*, Funchal, Portugal, 2014.

12 N. Vucic and M. Schubert, "Fixed point iteration for max-min SIR balancing with general interference functions," in *IEEE ICASSP*, 2011.

13 Ho, C.K., Yuan, D., Lei, Y., and Sun, S. (2015). Power and load coupling in cellular networks for energy optimization. *IEEE Transactions on Wireless Communications* 14 (1): 509–519.

14 R. L. G. Cavacante and S. Stanczak, "Performance limits of solutions to network utility maximization problems," arXiv preprint: http://arxiv.org/abs/1701 .06491, 2017.

15 P. Jamshidi, A. Sharifloo, C. Pahl, A. Metzger and G. Estrada, "Self-learning cloud controllers: Fuzzy Q-learning for knowledge evolution," in *arXiv preprint arXiv:1507.00567*, 2015.

16 3GPP, "E-UTRA Radio Resource Control (RRC) Protocol specification (Release 8)," 3GPP, 2011.

17 M. N. ul Islam and A. Mitschele-Thiel., "Reinforcement learning strategies for self-organized coverage and capacity optimization," in *Wireless Communications and Networking Conference (WCNC), IEEE*, Paris, France, 2012.

18 Rosen, K.H. (2006). *Discrete Mathematics and Its Applications*, 6e, 103. Boston, MA: McGraw-Hill.

19 Q. Liao and R. L. G. Cavalcante., "Improving Resource Efficiency with Partial Resource Muting for Future Wireless Networks," in *IEEE Wireless and Mobile Computing, Network and Communication (WiMob)*, Rome, Italy, 2017.

20 3GPP, "TR 36.133, Requirements for support of radio resource management, Rel-14," 2016.

21 G. Hegde, O. D. Ramos-Cantor and M. Pesavento, "Optimal resource block allocation and muting in heterogeneous networks," in *IEEE ICASSP*, 2016.

22 Yates, R.D. (1995). A framework for uplink power control in cellular radio. *IEEE Journal on Selected Areas in Communications* 13 (7): 1341–1347.

23 C. J. Nuzman, "Contraction approach to power control, with nonmonotonic," in *IEEE GLOBECOM*, 2007.

24 Q. Liao, "Dynamic uplink/downlink resource management in flexible duplex-enabled wireless networks," in *IEEE ICC Workshop*, 2017.

25 MOMENTUM, "Models and simulations for network planning and control of UMTS," 2004. [Online]. Available: http://momentum.zib.de.

26 3GPP, "TR 36.814, Further advancements for E-UTRA physical layer," www .3gpp.org, 2010.

9

Cognitive Autonomy for Network Self-Healing

Janne Ali-Tolppa[1], Marton Kajo[2], Borislava Gajic[1], Ilaria Malanchini[3], Benedek Schultz[4] and Qi Liao[3]

[1]*Nokia Bell Labs, Munich, Germany*
[2]*Technical University of Munich, Munich, Germany*
[3]*Nokia Bell Labs, Stuttgart, Germany*
[4]*Bosch, Budapest, Hungary*

While self-optimization functions optimize a set of configuration parameters to improve the network performance from a given network state, the self-healing functions focus on ensuring that the network can fulfil its purpose and serve its customers even in case of failures, unexpected changes, or events that risk a degradation in the network performance [1]. In other words, their objective is to make the network more resilient [2–5].

This chapter discusses the methods for improving resiliency of future cognitive autonomous networks. In particular, concepts for cognitive self-healing functions that are designed to detect or even predict network performance degradations and apply corrective actions are presented. Unlike in optimization, the self-healing functions can only monitor the symptoms and the root causes are not known *a priori*. Due to this and the less-constrained problem space, the self-healing process is often more complex than the optimization control loops. This complexity is accentuated by the distributed and heterogeneous nature of mobile networks. Finally, diverse fault states often occur only in very rare cases, which makes it difficult to collect statistically meaningful data. The lack of statistical samples is a problem especially in diagnosing the detected anomalies. As such, mobile network self-healing functions typically require more sophisticated machine learning and augmented intelligence methods, as well as knowledge-sharing between domains.

The chapter begins with a generic discussion of resilience as regards self-healing highlighting how resilience is preferred to robustness. It then gives a general overview of the self-healing process and where cognition can be applied in this process. This is followed by a discussion of anomaly detection as applies to Radio

Towards Cognitive Autonomous Networks: Network Management Automation for 5G and Beyond,
First Edition. Edited by Stephen S. Mwanje and Christian Mannweiler.

Access Networks (RANs), highlighting the challenges of profiling a dynamic non-deterministic environment and how to characterize anomalies in such a system. Subsequent sections then treat the challenge of deriving auto diagnostic actions and the concept of transferring knowledge amongst network and automation instances to ease the self-healing process. The chapter then concludes with a discussion of future challenges and ideas on how they may be addressed.

9.1 Resilience and Self-Healing

Resiliency is the capability of a system to recover to a stable, functioning state after failure or adverse events [2]. By definition, it is not the same as robustness. A robust system is strongly designed to withstand any foreseen problems or failures but may be too rigid and fail to survive and adapt in case of unforeseen circumstances, which are inevitably bound to happen in complex systems. For example, a farmer may prepare his crop against fire, flooding, and local pests, but the crop can be destroyed by a foreign plant virus introduced in the environment. Paradoxically, a very robust system can sometimes be more susceptible to failure due to its increased rigidness and complexity [2]. Modern mobile telephone networks are often said to be amongst the largest and most complex human-created systems and their distributed nature makes them even more complex to manage and predict. So, simple robust design principles (redundancy etc.) are not sufficient to ensure ultra-reliable network performance required for many 5G use cases, such as remote surgery [6].

Making networks more robust and redundant makes them more complex which, in turn, may create new possibilities for failures. It can also make the problems harder to diagnose, when they occur. This also applies to the self-healing functions. The functions may not always be able to diagnose the complex causes and effects that lead to the degradation and the applied corrections can sometimes be rather work-arounds than really addressing the root cause. Such work-arounds remedy the immediate symptoms, but since the root cause has not been resolved, the problem may still appear later. A risk is that at a later, more escalated stage, the problem may be more severe and harder to troubleshoot and correct [6].

It is often necessary to monitor the corrective actions taken by self-healing functions. A self-healing function that triggers an erroneous corrective action, e.g. when a change in the context is detected as an anomaly, may itself cause a fault, analogous to an allergic reaction caused by the human immune system. Additional verification layers may be placed to prevent and mitigate such false positive triggers, but this, in turn, adds further complexity to the system. As such, it is important to find the right balance between robustness and complexity, also when applying self-healing.

9.1.1 Resilience by Design

Resilient systems follow principles that allow them to recover even in case of completely unforeseen disastrous events. For example, by diversifying the crop, a farmer can ensure that a new plant virus will not be able to wipe out all the production. Typically, resilient systems follow a number of design principles [2, 6]:

Monitoring and adaptation: Resilient systems must be responsive to change, and so they need to monitor the system to detect changes early. In network management, this is typically done by the Fault Management (FM) and Performance Management (PM) systems that monitor for fault events or alarms, reported by the managed network functions, and the network Key Performance Indicators (KPIs), respectively. Alarms are correlated in a Network Management System (NMS) to reduce their number and to build a more comprehensive picture of the context of the fault.

Ideally, the network should even be able to predict any degradations before they result in an alarm reported by a network function, at which time, it may have already led to a service degradation. An anomaly detection based self-healing system can profile and learn normal behaviour at runtime and detect deviations from it even before it is causing an alarm to be reported, giving an early warning even in case of unforeseen circumstances. If connected to a diagnosis function, it can also trigger automatic corrective or mitigating actions. Self-healing functions based on anomaly detection and diagnosis are discussed in the following sub-sections.

Redundancy, decoupling, and modularity: In addition to duplicating capacity for redundancy, resilient systems often have a decoupled and decentralized structure, so that the failures are localized and do not easily propagate through the system, avoiding a single point of failure. In mobile networks, the radio access has often relatively more limited redundancy. In 5G RAN, one solution is the RAN multi-connectivity. It is often utilized to increase the throughput but can also be used to exploit the inherent macro-diversity effect of multiple simultaneous connections, such that the probability that at least one connection is sufficiently strong is increased [7].

Focus: When changes are detected, resilient systems may focus on the problematic area to respond to a problem or a change. In network management, excess resources can be deployed where unexpected events are detected. Cloud-based deployments especially offer more flexibility on the resource distribution. For example, the telco clouds enable more flexible placement and allocation of FM functions depending on the criticality of the monitored resources and the network slices or services they are used in. For very critical Ultra-Reliable Low-Latency Communications (URLLCs), more resources can be assigned for the FM than, for example, for the less critical Mobile Broadband (MBB) services, where the FM function can be more centralized and shared by several network functions [8].

9.1.2 Holistic Self-Healing

A related recurring principle in resilient systems is holism. In a complex system, improving the resilience of only one part or level of the system can sometimes (unintentionally) introduce fragility in another. To improve the resilience, it is often necessary to work in more than one domain, scale, and time granularity at a time [2].

In mobile network self-healing, this means that the different management domains, levels, and areas cannot be considered in isolation [6]. The different management areas, although operating on different time scales and on different managed objects, need to be able to share knowledge and create a holistic picture of the whole system:

- **Quality of Experience (QoE) driven management:** Optimizing the end-to-end customer experience per subscriber and application.
- **Network Management (NM):** Management automation aggregated on a (Virtual) Network Function (VNF) level.
- **VNF and Service Orchestration:** Orchestration of cloud resources, CPU cores, memory, storage, links etc.

To give an example, consider the corrective actions done by a QoE-driven self-healing function in a transport network automation system [9]. Such a system monitors the QoE Key Quality Indicators (KQIs) of a user or service and can very quickly reroute traffic around problematic links. On the other hand, it does not know the context of the network functions that are implementing the services or the infrastructure they are running on. There may be underlying issues that create the need for rerouting and if these issues are not corrected, eventually it may be that the QoE management system is no longer able to fulfil the customer expectation with the available resources. The network will fail and, if this happens at a later and potentially more escalated stage of the problem, the failure will be more catastrophic and more difficult to troubleshoot. If you are failing, it is often better to fail quickly [6].

Information and knowledge sharing on the management and optimization actions executed on different levels is important. A more holistic view can enable early detection of problems. In the QoE example for instance, the NM-level self-healing functions may monitor the corrective actions performed by the QoE-driven, application, and user-centric backhaul Self-Organizing Networks (SONs) system [9]. The actions are modelled and aggregated as additional KPIs that are used as input features for the NM self-healing, in addition to the normal

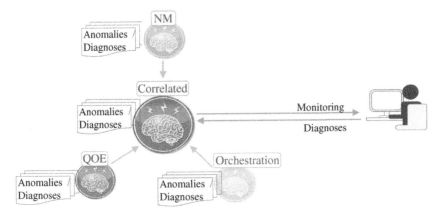

Figure 9.1 Holistic self-healing over different management areas.

NM-level alarms and performance KPIs. Several corrective QoE-driven actions may be an indication of an emerging problem, especially if they occur together with other network performance degradation indicators [6]. This concept may be extended to include VNF orchestration, for example. The different management areas can collaboratively monitor each other's corrective actions or determine the best course of action in a cooperative manner. As illustrated by Figure 9.1, the monitoring also enables coordination of the actions taken by management agents in different management areas.

9.2 Overview on Cognitive Self-Healing

Self-healing can be seen as the ultimate goal of automation: a system that is capable of recovering from any failure is a system conceptually bigger than *itself*. Ironically, it is easy to imagine how unimaginably complex such a perfect self-healing system could be. As with most other complex problems, the way to tackle self-healing is to break down the structure into smaller, simpler pieces that perhaps only realize a limited functionality, and start adding complexity once the simpler, limited pieces have crystallized.

In this section, an overview of a cognitive self-healing system is given, by discussing a few, simple building blocks. It is entirely possible that in the future, self-healing systems will be designed in a wholly different way. However, these building blocks serve as an easy-to-grasp starting point for the discussion throughout the chapter. This section only discusses the general aspects of a cognitive self-healing system, leaving details, and example implementations until later sections, specifically, Sections 9.3 and 9.4.

9.2.1 The Basic Building Blocks of Self-Healing

The simplest self-healing solutions are rule-based systems, that trigger a specified automated corrective action when a given condition is fulfilled, e.g. by a specified alarm. Such systems only work reliably on anticipated problems and typically under-perform in unforeseen circumstances, or in changing environments. Furthermore, the creation and maintenance of the rule base is expensive and laborious. Rule-based self-healing functions are very rigid and, because of this, rigidity can even make the system less resilient against unexpected changes, as discussed previously in 9.1.2 [6].

The rules governing which corrective actions to trigger and when, can be formulated as a classification problem and learned through machine learning. In this context, the network or network element can reside in one of many states, where each state describes either normal or degraded operation. Degraded states can be connected to one or more corrective workflows. Since the anomalous or degraded states are, by definition, rare the rules are learned on a skewed training dataset. If the system learns to detect degraded states from examples, it may also fail to recognize new, unforeseen problematic states. An additional complication can be the scarcity of such labelled training datasets, a topic that will be discussed later [6].

Self-healing functions are often implemented as a four-step process: profiling the normal states of the system, detecting deviations from the normal (anomalies), diagnosis and performing corrective actions. This process is depicted in Figure 9.2. The advantage of learning the normal behaviour is that any deviations from it, even unforeseen ones, can be detected. On the other hand, not all deviations are degradations and so a diagnosis function is required to diagnose the detected anomalies and connect them to possible corrective actions. Additionally, to adapt to trend-like and seasonal changes in the normal network behaviour, for example, to the evolution in the network traffic characteristics, the profiles for the normal states need to be able to adapt to these changes.

The four steps are not self-contained, each stage relies on the results of the previous one, thus making the reliability of the earlier stages more important for the overall performance of the entire self-healing process. On the other hand, the first steps of profiling and anomaly detection are also the relatively simpler ones, as the complexity of the steps keeps increasing for both training and runtime decisions,

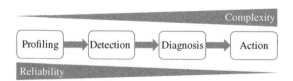

Figure 9.2 Four-step self-healing process.

with the most complex step being the decision for the right corrective action to take. This complexity mostly arises from the increasingly large problem space and the lack of repetition, which calls for highly sophisticated ability of generalizing and applying knowledge between similar but different domains or tasks, an ability that is notoriously hard to achieve with machine learning and artificial intelligence. As such, the further to the right of Figure 9.2, the more human operator involvement is likely to be required.

9.2.2 Profiling and Anomaly Detection

By definition, anomalies are deviations from normal states or behaviour patterns of the system. As such, an anomaly does not necessarily signal a fault or a degradation, or any other potentially significant event. On the other hand, some of the KPIs used as input for the anomaly detection are also usually direct fault indicators, for example, failure counters. Using these KPIs, the anomaly detection inherently becomes fault detection – part of the diagnosis may already be inherently built into the anomaly detection step. There are both pros and cons in combining or completely separating both the detection and the diagnosis steps. Here, these steps are discussed separately, to highlight the possible differences between them.

Typically, the network management information that can be used for training an anomaly detection system does not include labelling information, which would indicate which system states are considered normal and which are anomalous. To overcome this, an assumption is made that the network is operating normally *most of the time*. This is a strong assumption, but it holds if the profiling period is long enough that any anomalies that might have occurred during the period will average out and that the average system behaviour overwhelmingly represents the normal states.

The choice of features to be profiled for the detection depends on the kind of anomalies that are possible to detect. The selection process often requires domain knowledge but can be aided by methods to estimate the information content of the different features, such as the Principal Component Analysis (PCA). However, care must be taken when selecting the features, as anomaly detection performance is largely dependent on the quality and relevance of the input data. Noisy data containing irrelevant information can disturb the detector which, in turn, can produce false detections, or allow important anomalies to remain undetected, undermining the functioning of the whole self-healing process. For this reason, anomaly detectors should use only a few of the strongest anomaly indicators (i.e. features), whenever possible. These indicator features might not be able to tell the specific nature of the anomaly but are sensitive and strong indicators that something is wrong. This is analogous to a doctor measuring the patient's fever, as fever is a strong indicator that something is wrong, although it may be caused by many different factors. Feature selection is discussed more in detail in Section 9.2.6.

Apart from the detection features, additional context information is required to suitably localize an anomaly. If this context information is not explicitly included in the dataset, the dataset needs to be split according to the different context groups. For example, to profile the normal daily traffic pattern of a mobile network cell, one might choose to profile the traffic for each hour of the day, perhaps separately for working days and weekends. Full localization can be achieved in a way similar to how a doctor diagnoses a patient, by identifying where the anomaly happened (spatial localization, where the patient experiences pain on the body), how long it has been going on (temporal localization), which features it involves (dimensional localization, such as whether there is fever), and how strong it is (numerical quantification of the severity of the anomaly). If all this information is included in the dataset in some form, profiles can be created through a learning algorithm.

From the created profiles, an *anomaly level* can be calculated for the observations of each feature. It describes the anomalousness of the given feature at a given point in time with a numerical value, similar to how a doctor might ask the patient to describe the level of pain he might be experiencing. However, the general goal is to detect distinct anomaly events, which represent the aggregates of anomalous behaviour that share the same root cause. This is done through a process called *anomaly event aggregation*. A single root cause, e.g. a single fault, can manifest in several symptoms, effecting several features in one or more managed objects (e.g. cells) and over varying time periods. Keeping the analogy of the patient and the doctor, anomalies can affect multiple elements at once, the same way illnesses can affect multiple organs, or cause a plethora of symptoms. Figure 9.3 depicts these different dimensions of an anomaly event, where the same root cause is affecting both KPI_1 and KPI_2, but at subsequent timeframes and in both $cell_1$ and $cell_3$.

Overfitting (described in Section 6.2.4) can be an issue in anomaly detection, by using an overtly complex model with excess modelling capacity. Overfitting can

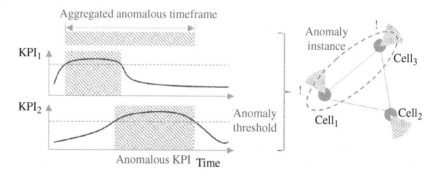

Figure 9.3 Temporal, dimensional, and spatial anomaly aggregation.

lead to high numbers of false positive detections, as the model starts to detect any observation not included in the training set as an outlier. This problem should not be tackled by regularization in the model only, but rather by supplying the system with more training observations. One way to improve this is to consider sharing profile data between several managed objects. For this purpose, the objects may be clustered based on their behaviour and similar objects may share a common profile to increase the amount of input data available for the generation of the profile.

9.2.3 Diagnosis

The detection process indicates that an unusual event has occurred, whilst the anomaly event aggregation process provides the context and the dimensions of the anomaly, as well as the anomaly levels of the indicator features. Then, it is the responsibility of the diagnosis function to describe the anomaly in depth, determine the impact it may have on the system and, if necessary, find out its root cause.

Compared to detection, the diagnosis phase poses different requirements for the used features. Whereas detection prefers a small set of features that are strong indicators for unusual network behaviour (to reduce the noise in the process), to determine the impact and root causes of the anomaly, typically a considerably larger set of features is required. Using this extended feature-set, the first task of the diagnosis function is the description of the anomaly, the generation of the *anomaly pattern.*

The anomaly pattern should aggregate the most important features for the diagnosis from the anomaly event scope. Which features are required again depends on the kind of anomalies to be diagnosed. For example, for some features the temporal context is important for the diagnosis, in which case, the feature cannot be aggregated in time for the duration of the event, because critical information for the diagnosis would be lost. In many cases, most of the detection features would be included in the set, as they also often provide important information for the diagnosis. An example of anomaly event aggregation and an anomaly pattern is shown in Figure 9.4. In this case, each KPI is aggregated to a single value to describe it in the anomaly pattern.

Once the anomaly event is described by the diagnosis function, it should find the best matching diagnosis for it, or if one cannot be made with sufficient confidence, it could escalate the anomaly to a human operator. Learning the diagnoses from the data is difficult, because the anomalies are, by definition, rare. So, data and knowledge sharing from a wider scope is often required for a reliable diagnosis.

The acquisition of correct labelling for the stored anomaly observations is also a challenging task; a comprehensive manual labelling of thousands of anomalies is not usually possible. In a way, the anomaly pattern dataset is often simultaneously both too small and too large; it might contain an insufficient

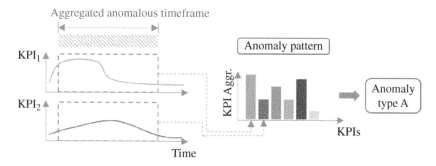

Figure 9.4 Anomaly patterns.

number of observations for machines to learn from, but an overwhelming amount for humans to process. One approach to tackle this problem is to create active learning systems, so that the user is only asked to label specific anomalies that are the most valuable for the classifier, reducing human labour.

9.2.4 Remediation Action

Even after the anomaly is detected and the root-cause pinpointed precisely, the action to be taken may not necessarily be directly derived from the outcome of the diagnosis. When deciding how to act, contextual information needs to be considered, such as the action's effect on neighbouring network elements, other actions that might still be in effect from a previous time, or actions undertaken on neighbours that might affect the anomalous element. The planning, execution, and validation of corrective actions is a complex task, that requires high cognitive capabilities from the automating system.

Also, there is no general way of automating the healing of previously unseen faults, since determining the right corrective actions in these cases is an extremely complex problem that requires deep reasoning ability. Established self-healing functions usually focus on specific anomalies with pre-defined actions to be taken. In the case of unforeseen anomalies, certain actions may be tried before delegating the task to a human supervisor. Typically, these actions only include less intrusive operations, which can be executed with little risk, such as a restart of the system. Alternatively, more intrusive healing actions could be first tried in an environment, where they can be evaluated without risking a Quality of Service (QoS) degradation, for example, in a simulation.

9.2.5 Advanced Self-Healing Concepts

Self-healing systems in mobile networks share their operating environment with other control functions. The actions of these functions can influence the network in a way that triggers anomaly detection. Additionally, the functions often operate in an iterative manner and require some time to arrive at a new 'stable' state. If the duration of the intermediate steps is on the same timescale that the anomaly detector operates on, the temporary states could be registered as anomalies and self-healing triggered. If the self-healing should include the supervision of the cognitive functions in the network, training data needs to be obtained from the network with the cognitive functions already in place. Otherwise, the self-healing needs to disconnect from these effects, either through an explicit suppression signal coming from the other functions, a kind of 'maintenance mode' mechanism, or by labelling these anomalies so that no further action is taken.

Networks also typically have gradual, trend-like changes, such as the constant growth of users or the volume of transmitted data. If the anomaly detection models are not robust against such changes, or if they are not able to adapt, this may result in anomalies that are considered false. However, a clear distinguishing line between a trend-like change and a long-lasting anomaly might be hard to find. Instead of relying on predefined values, trends should be learned from training data, and incorporated into the models. By being able to extrapolate the expected behaviour on a longer timeframe, models gain predictive power that leads to a more precise and robust self-healing system. Trends can, of course, only be learned from training data that covers a sufficiently large timeframe.

Although the training data for model building should be extensive so that it covers as many aspects, network elements, and as long a timeframe as possible, during inference, there are multiple ways to reduce the amount of collected and stored information. One such possibility is to only collect an extended KPI set when the smaller detection set signals an anomaly. This on-demand data collection can greatly reduce the transmitted and stored data volume, but it also entails a downside; later refinements/changes in the detection system cannot retrospectively use this information as training data, since any required extra information is not available. Thus, on-demand data collection is best used on more mature self-healing systems.

In an idealistically simple self-healing case, the three-step procedure of detection, diagnosis, and action suffices. Of course, most real-world self-healing scenarios play out in a far more complicated way. Returning to the analogy of a real-world doctor, it is not atypical that the first diagnosis is false, and the doctor must be visited a second time. Such iterative self-healing systems are also possible in cognitive autonomous networks, where the first action after a diagnosis can be to collect more data for a subsequent, more refined diagnosis. In fact, any of the three steps could be repeated several times before the self-healing procedure is completed, as depicted in Figure 9.5. However, self-referring systems can be very complex.

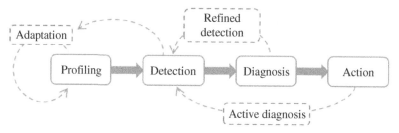

Figure 9.5 Recursive self-healing possibilities.

9.2.6 Feature Reduction and Context Selection for Anomaly Detection

This section still shortly looks more closely at feature and context selection. Both concepts have been mentioned heavily in both Sections 9.2.2 and 9.2.3, as they provide the essential data for anomaly detection and diagnosis. A badly selected feature set can break the functioning of even the most robust self-healing systems.

9.2.6.1 Feature Reduction

Feature reduction is mostly used simply for reducing the data volume to speed up processing, or even just to make a process computationally feasible at all. Many algorithms scale badly with the number of input features in terms of both processing time or runtime memory requirements. In the worst cases, this can cause a process to run out of hardware resources even with a moderate (~100) number of features. The reduction of computational requirements is not the only intended purpose of most feature reduction techniques, however. Feature reduction removes redundant and irrelevant information from the input, in effect removing noise from the data. By removing the duplicated or irrelevant features, models fitted in later processing stages are less prone to fit irrelevant parts of the data.

It is important to notice, however, that many anomaly detection methods rely on modelling the correlations between different features and detecting any unusual deviations from the profiles. Two features that are highly correlated in the profiles and may as such seem redundant, can offer a reliable indicator for an anomaly in the rare cases that this correlation is broken. Typical unsupervised dimensionality reduction methods, for example the PCA described in Section 6.1.5, do not work well with anomaly detection for this reason, because these remove highly correlated features from the dataset. However, the removal of irrelevant information or noise is still as valuable an undertaking for anomaly detection methods as well, if one can distinguish between highly correlated and irrelevant features.

Feature reduction can be done through either feature selection or feature extraction. In feature selection, a set of features is simply selected from a larger, original set. In feature extraction, on the other hand, an entirely new set of features is generated from the original set through a transformation process, so that the output is often not easily derivable from the original feature set, or even intuitively understandable. Feature selection can be preferred in anomaly detection cases, as its output is easier to present to human users. Conversely, feature extraction methods are usually able to remove more noise from and keep more valuable information in the original data.

If labelled training data is available, a value metric for each selected feature set can be calculated based on how well the algorithm performs with it. However, in most cases, such labelling is not available, and the anomaly detection process is used as a data-exploration tool. In these cases, only unsupervised methods are usable and as stated before, they can work sub-optimally in anomaly detection use cases; as strange observations in very predictable features can indicate more significant anomalies than in less predictable ones. This means that the more interesting features are usually the ones that do not show a large variance or have a high correlation with others in the training data, making the usual information estimation techniques not applicable here.

In such exploratory anomaly detection cases, rather than trying to estimate information content, it is better to directly estimate anomalousness. Consequently, separating feature reduction from the rest of the detection process is non-intuitive; any measure of anomalousness that would help evaluate the value of each feature will already carry out some form of anomaly detection. By moving the feature reduction into a separate step, the whole process will undertake anomaly detection multiple times. Instead, it is probably better to implement feature reduction as a wrapper around the actual detection process. Wrapping is an iterative process, where the selected or extracted features are evaluated in the detector, and the resolution is fed back to the feature reducer to be used in the next iteration. Wrapper feature selection can be implemented with many types of general-purpose search algorithms, such as a grid searches or genetic algorithms.

9.2.6.2 Context Selection

A context in the general sense means the external information that is not directly accessible from the supplied data. The goal of finding and selecting contexts, especially for anomaly detection, is to split the models into multiple parts to better follow individual archetypes, thereby producing less unexpected behaviour. This directly translates to less anomalies being detected, but also gives the remaining detected anomalies more significance. In this formulation, contexts are groupings or types occurring in the data. Contexts can be found in the:

- Temporal domain, such as separating weekdays from weekends, holidays from normal days, or lunch hours from the rest of the day.

- Spatial or element-wise domains, such as differentiating between busy cells at airports from cells with less traffic in rural areas or separating cells with mainly pedestrian users from cells close to highways.

These contexts can automatically be defined using unsupervised learning algorithms, such as clustering algorithms, by defining user-set parameters that influence the likelihood of group formations. However, the parameters need to be set carefully and conservatively for anomaly detection; too many contexts can overfit the data, causing some contexts to incorporate anomalous behaviour, hiding it as another type of norm.

9.2.6.3 Feature Reduction and Context Selection in the Future
Feature and context selection are important basic building blocks of a machine learning model, also of anomaly detection. The performance of the model is closely connected to the selection process and the selection typically happens in an iterative manner, by testing the model with different input features and models. Modern machine-learning algorithms, especially deep neural networks, can function as feature extractors as well as automatically detect contexts. However, they also require large quantities of labelled training data, which is often not available for anomaly detection. Anomaly detection is an explorative process, and as such, the algorithms that realize this process should also be explorative in nature.

9.3 Anomaly Detection in Radio Access Networks

The RAN is comprised of many network elements that are physically distributed in the serving area with relatively limited resources and less redundancy. Detecting and correcting, circumventing, or preventing potential faults in these elements is not straightforward. Simply rerouting around a broken link or replacing faulty equipment with a spare on standby is not always possible [1].

In many existing network deployments, RAN fault management is carried out by expert-designed KPI metrics for threshold-driven Fault Management (FM) process, generating alarms that are processed by operations personnel. This approach requires constant supervision, adjustments, and interpretation by experts [1].

Because radio network elements operate in diverse environments, static, rule-based methods may perform sub-optimally. The resulting false alarms unnecessarily increase the workload and can even lead to real alarms being missed because of the noise. At the same time, more complex or unanticipated problems may go undetected, or they are only detected at a very late stage, when the quality of service has already been compromised. Alarm correlation is complicated and determining any required corrective actions often requires deployment-specific expertise and manual analysis.

Utilizing machine learning based anomaly detection methods to augment the traditional FM provides a practicable solution: pre-configured KPI thresholds can still be used to detect well-known problems, while anomaly detection can reveal problems that are more complex or previously unforeseen. It can achieve this by learning the context of each managed network function and, in practice, adapting the detection thresholds to that context [10]. It may also monitor more complicated features, such as complex cross-correlations between several network KPIs [6]. In the following sections, the design considerations of a RAN anomaly detection method are discussed.

This section looks at self-healing for mobile network operations point of view, i.e. focusing on issues that would be typically monitored at a network operations and administration centre and on the corrections that can be deployed in operation. The self-healing use case, however, extends further, e.g. into technical support offered by the network function vendors and how to automate the diagnosis process there.

9.3.1 Use Cases

Before designing a self-healing solution, one should of course consider the kind of problems it is intended to detect and try to remedy. There are at least five types of faults in radio networks:

Software problems: A software fault may lead to any kind of erratic behaviour. In the worst case, a software problem can lead to an outage, e.g. to a sleeping cell, or to a loss of particular service, for example, to a failure of the packet-switched network while the circuit switched is okay. Such failures might be recurring. For example, a leak in resource usage, e.g. a memory leak, may require frequent resets, which leads to service interruptions and reduced availability. Special focus on detecting and diagnosing any software-related issues might be required after software version upgrades or deployment of new network functions.

Hardware problems: A radio hardware failure, such as a faulty power adapter, amplifier etc., can also lead to a service degradation or a coverage hole. Possible corrections include activating a spare unit, if one is available, or trying to route around the problem. In some hardware failures, it is possible for the self-healing anomaly detection to detect signs of the problem before the unit completely fails. For example, a power adapter may have jitter or reduced power before the failure.

Misconfiguration: Either a human operator or a misbehaving SON function can introduce a degradation by misconfiguring a network function. Examples of misconfigurations could be incorrect transmission power, Remote Electronic Tilt (RET) setting in an antenna, Physical Cell Identifier (PCI) or a physical misconfiguration, such as physical antenna setting or cabling. Depending on the misconfiguration, different symptoms may be observed: interference, radio link failures, dropped calls etc. Possible corrections include triggering of SON functions to optimize the network in order to mitigate the degradation or reverting to last known stable configuration, as is done in the automatic configuration change verification discussed in Chapter 11.

Environmental impacts: These include, for example, shadowing effects created by changes in the environment. The changes could be the construction of new buildings or be caused by weather and can influence the network coverage.

Unexpected exceptional traffic profiles: For example, unknown special events, that introduce high load in the network that it isn't prepared for, which can lead to overload conditions and service degradation. Load balancing may be triggered to try to better distribute the traffic.

9.3.2 An Overview of the RAN Anomaly Detection Process

Figure 9.6 depicts an overview of an anomaly detection and diagnosis function for RAN. Profiles of normal behaviour are created based on selected features in selected contexts, for example, per cell and using a chosen set of Performance Management (PM) KPIs, which may be aggregated over given time periods, for example, hourly.

Following the self-healing concept described in 9.2, once the profiles are created, an anomaly level can be calculated for each KPI in each context, based on

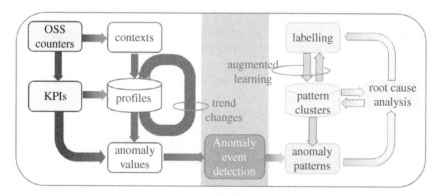

Figure 9.6 Anomaly detection and diagnosis process [6].

which distinct anomaly events are detected. The next sections will look at these detection steps.

9.3.3 Profiling the Normal Behaviour

The model that captures the learned normal behaviour is called a profile. Various types of profiles can be used to profile different features of a RAN: The profiles can be statistical models of normal distributions with a fitted set of parameters [10], or the profile may consist of cluster centroids in an encoded feature space [11]. The choice of profiling algorithm, features, and context is dependent on application-specific design choices that need to be considered. Some of these considerations are listed below:

- Where the algorithm is implemented: Distributed at the edge of the RAN, for example in a base station, or in a more centralized location, such as in an NMS?
- Is labelled training data available or do the profiles need to be learned in an unsupervised fashion?
- The context and granularity, in which granularity faults should be detected, i.e. are profiles created per selected Network Functions (NFs), Network Elements (NEs), a collection of NEs or even to profile the state of the whole network including all NEs in it? Should working days and holidays be profiled separately, e.g. a network for traffic-related KPIs?
- Which features are required for detection of the intended types of anomalies? For example, are PM KPIs profiled individually or are their cross-correlations important indicators for anomalies? Should they be profiled time series, e.g. for network traffic dependent KPIs, and in which granularity?
- What kind of resources (computational, memory, bandwidth for data collection) are available (e.g. distributed vs centralized implementation)?
- Should the profiles be intuitively understandable for a human operator to verify the algorithm's performance?

Figure 9.7 shows an example of two types of profiles for a RAN cell. The first one is a simple diurnal profile for modelling the behaviour of a KPI time series. Typically, such profiles are used to profile traffic-dependent KPIs, in this example, the number of Radio Resource Control (RRC) releases. It depends on the number of users in the network and follows the daily behaviour pattern of the network users, with clearly less traffic during the night-time and one or more peaks during the day or in the evening. In the figure, the y-axis depicts the number of the RRC releases and the x-axis is the time of day. The number of RRC releases is aggregated (resampled) per each full hour of day by summing them and then a normal distribution is learned for each hour from the observations in the training data. The lines are drawn for 1 and 2.5 standard deviations from the mean for each hour and the parameterizable boundary for anomaly detection.

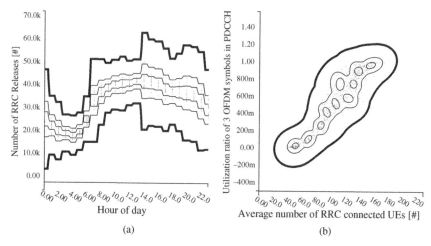

Figure 9.7 Diurnal (a) and cross-correlational (b) profiles [6].

Such profiles are only meaningful for KPIs and other features that exhibit a clear time dependency. A more generic way of profiling correlations in the network is shown on the right in Figure 9.7. It profiles the normal correlation of a pair of KPIs, in this example, the average number of RRC connected user equipment (UEs) and utilization ratio of three Orthogonal Frequency-Division Multiple Access (OFDM) symbols in Physical Downlink Control Channel (PDCCH). A nonlinear dependency pattern can be observed that is shown as the enclosed area of a hysteresis curve. The scatter plot points in the figures are the data points from the training data. The ellipse curves represent quanta in the profile to which bivariate normal distributions have been fitted respectively. The further a point is from a centroid of any cluster, the more anomalous it is. Note that since normal distributions are fitted only locally within a cluster, such a profile is also able to represent more complex distributions [6].

9.3.4 The New Normal – Adapting to Changes

A radio network does not work in a static environment. There are typically both seasonal and trend changes in the network traffic. A 61× to 115× increase in RAN data bearer traffic is predicted over the period of 2015–2025 [12]. If the absolute total amount of traffic is profiled on a network level, for example, and the profiles are not adapted after the training period, the detection function will soon indicate higher than normal traffic simply due to the trend of growing data demand. While this is true, the traffic is higher than the normal traffic at the time of profiling, it is still not considered as an anomaly. It is the norm that has changed.

And, it is not only the continuous trends of traffic growth, but there are often seasonal variations, which are considered normal. For example, a holiday destination in the Alps exhibits a clear increase in network traffic during the skiing season when all ski resorts are full or, in the summer, when people go to the mountains for hiking, mountain biking, paragliding etc. Between the seasons, traffic can be significantly lower. If normal traffic has been profiled during the high season, it would indicate an anomaly in the low season and vice versa. Such seasonal variations in traffic heavily depend on the location of the RAN site and so to profile these changes, individual profiles are needed for each cell or site and they also need to be adapted individually.

A third type of change to be considered are the disruptive, discontinuous changes. To adapt to the increase in network traffic, new cells may be deployed, and the existing radio equipment and software upgraded. However, these deployments imply changes in the context of profiled network functions and may trigger false anomalies or may invalidate the existing profiles altogether if the environment changes radically.

To remove the known trend and seasonal variations from data and make it stationary, simply moving average or exponential smoothing methods, such as the Holt-Winters double-exponential smoothing, Auto-Regressive Integrated Moving Average (ARIMA), or Seasonal ARIMA (SARIMA) (see Section 5.8.2), are often used [13]. Profiling can then be done on the more stationary data. This typically requires very long periods of training data, however, to reliably learn the seasonal variations, for example, considering the annual traffic variation. This can be a problem, especially since the profiles are very dependent on the environment of the profiled network function, e.g. a cell, and often need to be learned per function.

With less training data, a moving average can be used [10]. As shown in Figure 9.8, the re-learning of profiles is performed in the sliding window manner, which is controlled by two parameters: the *span* and the *slide interval*. The *span* controls the amount of data each profile is created from. The *slide interval*

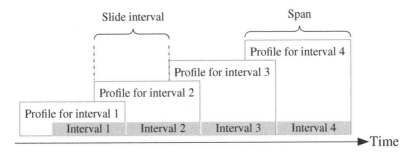

Figure 9.8 Continuous reprofiling of usual behaviour.

controls the time difference between the successive profiles. Inherently, the two parameters *together* define the amount of overlap between the data from which consecutive profiles are created, i.e. the smoothness of transition. Here, the trade-off decision of course is that the profiling window is long-enough to average out any anomalous behaviour, so that it is not learned in the profile as normal but, at the same time, the profiles are able to adapt to the seasonal changes and do not indicate them as anomalies.

The disruptive, discontinuous changes can be difficult to handle. Depending on the nature of changes and re-configurations, the effects on the validity of the existing profiles may be significant and hard to predict. Since the training periods are often long, e.g. to learn the seasonal variations, it is also not often desirable to completely re-learn the profiles after such a change. So, the features used for detection should be selected and designed such that they are robust against the re-configurations and other disruptive changes in the environment. This relates to the discussion in Section 9.2.2.

The number of features used for detection should often be minimized and they should be the ones that are strong indicators for the kind of events to be detected, which makes the method more robust against re-configurations in each context. In case of more dramatic reconfigurations, it may be unavoidable that the profiles are, at least, partly invalidated and need to be re-learned. In these cases, transfer learning methods may be useful in re-using the knowledge from the old profiles in the new context. The transfer learning methods and knowledge sharing are discussed in Section 9.5.

9.3.5 Anomaly-Level Calculation

The anomalousness of a given observation can be calculated against the profiles. This is done for each profiled feature to get the *anomaly level* of the feature in that observation. If the feature is a single KPI that is normally distributed and profiled with a diurnal pattern as shown in the example in Figure 9.7, the anomaly level can be simply the z-score of the KPI value in that hour of day, i.e. normalized by the profiled mean and standard deviation values for the hour of day. In this case, the detection threshold is often defined as $z = 2.5$ [14], the so-called *2.5 sigma certainty*, implying that 98.8% of the observations fall within the normal range. Any observations outside this range, or at least the feature is the observation, are considered as anomalies [10].

Many KPIs in RAN, however, have, more complicated distributions, e.g. with several peaks in the probability density function. Also, it is sometimes necessary to capture the multivariate distributions amongst a set of KPIs. Finding a suitable general multi-dimensional distribution is not straightforward, especially considering computational limitations. A possible approach to handle this complexity is to cluster the multi-dimensional space into smaller parts, within

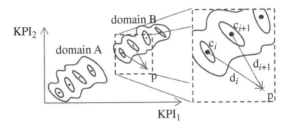

Figure 9.9 Compound profile.

which a multi-variate normal distribution can better approximate the data. A two-dimensional example is shown in Figure 9.9, where the cross-correlation of two KPIs is profiled. Firstly, the observations in the training data are clustered to find the higher-level structure in the data, i.e. to find the 'denser' areas. Then, each cluster is approximated by a multivariate normal distribution, which is represented by its vector-valued mean (which are also the cluster centroids) and its covariance matrix. In Figure 9.9, the mean values are depicted as the bullets and the covariance matrices as the ellipses having axis lengths equal to the standard deviations of the marginals, i.e. the diagonals of the covariance matrix [6].

A good candidate for a multi-dimensional anomaly level value of the feature (the probability density function), e.g. for a KPI pair as in Figure 9.9, is the weighted sum of the Mahalanobis distances [15] (the multi-dimensional generalization of the normal distribution) from the profile centroids, where the weights depend on the Euclidean distance of the observation from each cluster centroid. The value can be normalized and compared against a set threshold as in the one-dimensional time series case.

Other anomaly detection methods based on clustering such as topic modelling [16] or Growing Neural Gas (GNG) [17, 18], may also be used. In GNG, for example, the number of profiled states do not need to be defined upfront, but the algorithm keeps adding new states until a stopping criterion is reached. Ensemble methods combining several detection algorithms can also be used [13].

9.3.6 Anomaly Event Detection

Given the anomaly levels, the next question is, which observations are to be classified as anomalous and if certain anomalous observations belong together, i.e. may have the same root cause? This process is called anomaly classification or anomaly event detection. The simplest approach thereof is to set a fixed threshold for the classifying based on the feature's anomaly level, like the 2.5 sigma approach described in the previous section. But how about the anomalousness of the whole observation, considering all KPIs and features? Is it classified as anomalous if any

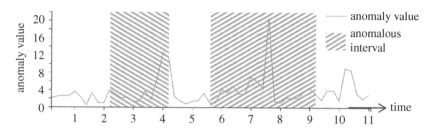

Figure 9.10 Anomaly value development in time.

of the features is above the threshold? How about if no threshold is exceeded, but the system remains close to an anomalous state for an extended period? How about unusual cyclical behaviour of anomaly levels, where no instantaneous threshold is crossed?

Without labelled training data available, from where the anomaly classification could be learned, models able to capture diverse anomaly classes need to be designed. One example is shown in Figure 9.10, where a detection method is based on a Density-Based Spatial Clustering of Applications with Noise (DBSCAN) algorithm. The y-axis charts the anomaly level of KPI in time. The anomaly level could be from either a diurnal or a multi-dimensional KPI correlation profile. DBSCAN does a clustering of the data points and calculates the anomaly value for a point in time t by, in effect, integrating the area created by the anomaly level curve within $t - \varepsilon$ and $t + \varepsilon$, where the parameter ε defines the window length. If this area is larger than a specified threshold *MinPts*, the observation at time t is labelled as anomalous. By adjusting the attributes ε and *MinPts*, the compromise between detection sensitivity for longer lasting but less severe anomalies can be adjusted. Additionally, a simple threshold can be used to detect instantaneous anomalous KPI anomaly levels [19].

9.4 Diagnosis and Remediation in Radio Access Networks

The term diagnosis is defined as 'the identification of the nature of an illness or other problem by examination of the symptoms' [20]. In the context of RAN fault management, diagnosis has typically been done by a human operator. The process is often triggered by an alarm event or exceeded by an PM KPI threshold. The target of the diagnosis process is to understand the severity and the root causes of the problem to then be connected to corrective actions.

The anomaly detection function can simply act as an additional trigger to the existing fault management processes, in which case, the detected anomalies are either raised as alarms or otherwise reported to the operator. The aim of cognitive self-healing, however, is to automate the complete self-healing process as much as possible. What is desired, is also to automate the diagnosis function, the selection of corrective actions to take and the deployment of those actions. This, however, is a much more complicated problem than the anomaly detection. This section discusses some of the key questions in closing the automation loop in RAN self-healing use cases.

9.4.1 Symptom Collection

As the definition for the word *diagnosis* already suggests, the process starts with the collection of observed symptoms. In case of RAN, this can mean the PM KPIs, FM alarms, system logs of the managed network elements or functions and so on. Additional context data, such as the configuration of the functions and any other data sources that might be relevant (weather, road traffic, etc.), can also be collected. As discussed in Section 9.2.3, the feature set is typically much wider in diagnosis than what is required for the detection.

The question of scope is also relevant in determining, where and from which period the symptom features should be collected. The detected anomaly event may already give an indication of the temporal, spatial (i.e. which RAN cells are impacted) and logical scope (i.e. which network functions), as well as which features to look for. For example, depending on the type of KPIs, it might be important to collect the average or minimum or maximum values, or even to detect unusual fluctuation in time (jitter) even if the extreme values are not anomalous.

As a result of the symptom collection, a so-called *anomaly pattern* is created. It contains the features describing and anomaly event. A relatively simple example is shown in Figure 9.11. In this case, the anomaly pattern consists of the values of 20 PM KPIs, averaged over the duration and the scope of the anomaly event. The darker grey circle depicts the profiled, expected value for the KPI. The blue and orange (superimposed) areas represent the description of two detected anomaly events. The approach is rather simple and averaging the KPI values over the whole scope loses a lot of information, but it offers a quick and intuitive way to visually compare the similarity of detected anomaly events.

9.4.2 Diagnosis

Perhaps the most typical example of an automated diagnosis function in a RAN self-healing context is alarm correlation, where from the stream of alarm events the correlation function tries to group the ones that share a common root cause. Instead of reporting all the alarm events, only the root cause may be reported. The diagnosis in self-healing functions the same way. The simplest diagnosis

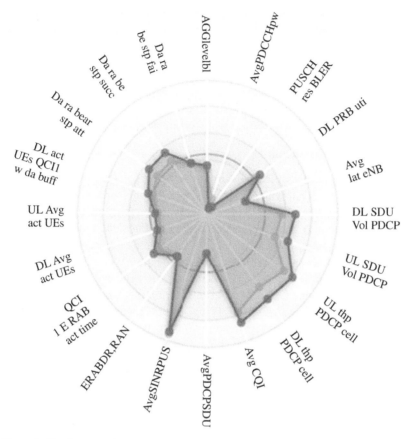

Figure 9.11 An anomaly pattern depicted as a radar chart [6].

automation is similarly based on rules, which connect the presence of certain symptoms and the absence of others to a given root cause diagnosis. Often, the rules are represented as Bayesian networks for a probabilistic representation of different states and their relations to the symptoms [21]. The collection of a rule base or creation of a Bayesian network are heavy and time-consuming tasks, however. They also need to be adapted, whenever there are significant changes in the environment.

Other approaches use Markov Logic Networks (MLNs) for the diagnosis [16]. MLNs are well-suited for diagnosis in self-healing, because they can be used to determine the most likely explanation for an event from noisy, incomplete, or even contradictory data. Probabilistic parameter values, also called weights, can be learned through experience and user feedback. However, MLNs also require upfront design of the rule set and labelled training data. To be able to adapt to any changes in the context and to new problems with agility, requires the process to be as dynamic as possible.

In Case Based Reasoning (CBR), the system reasons based on examples that have been stored into a knowledge base. The knowledge base can be extended at runtime improving the quality of CBR-based diagnosis. In self-healing, this means that the anomaly patterns of analysed anomalies are added to the diagnosis knowledge base with an attached label indicating the diagnosed root cause. Subsequently, detected anomalies can then be diagnosed by finding the closest in similarity amongst the already analysed examples in the knowledge base. If no reasonably similar example is found, the anomaly is raised for evaluation by a human operator whose findings are then added to the diagnosis knowledgebase. As such, the size of the knowledge base increases as more and more examples are analysed and added.

The performance of the diagnosis depends heavily on the features that are selected in the anomaly patterns, the distance metric used to find the most similar analysed example in the diagnosis knowledgebase and the quality of the knowledge base itself. One outcome of the diagnosis process can also be that more symptoms need to be collected, or from a longer and wider scope, i.e. it can be an iterative process.

9.4.3 Augmented Diagnosis

In self-healing, the ultimate target is a fully autonomous system. For the anomaly detection step this is usually possible, but as we've seen, the diagnosis is a more complicated problem and required more human involvement. Human knowledge and intuition are required to understand and give meaning to the collected symptom data. On the other hand, machines are much better at tirelessly processing large amounts of high-dimensional data. Thus, one of the key questions in designing a diagnosis function is, how to best combine the strengths of both people and the computing power of cognitive diagnosis functions.

The man-machine interface becomes an important part of the self-healing diagnosis function, as well as the question, how to best collect the operator knowledge into the diagnosis knowledge base. On their own, both have difficulty diagnosing large sets of anomalies, but by leveraging the knowledge and capabilities of the other agent, it is possible to greatly increase the accuracy and speed of the diagnosis process. The interworking between human and machine, illustrated in Figure 9.12, is called *augmented diagnosis* [6].

In augmented diagnosis, active learning methods are utilized to optimize the anomaly labelling effort [6]. Only those anomalies, which improve quality of the automated diagnosis process the most, are raised to the human operator for analysis. These could be either anomalies, for which the function has only

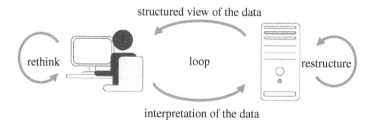

structured view of the data

Figure 9.12 Augmented diagnosis with active learning.

a low confidence level for an automated diagnosis, or anomalies which are on the borderline between two or more diagnoses. When new anomaly events are received, the diagnosis function matches them to available diagnoses in the knowledge base and identifies the ones, where the confidence is low.

Additionally, the diagnosis function can, in an unsupervised way, cluster the anomaly events and find structure in the data. Only an example of each anomaly event cluster or the cluster centroid may be raised to the operator, to reduce the workload of repetitively analysing and labelling similar anomalies. Any added labelled anomalies are then considered in the next iterations of diagnosis and can be used to improve the clustering in a semi-supervised way. Through this iterative process, depicted in Figure 9.13, both the insights from human analysis and the improved clustering of the augmented learning function incrementally improve each other.

A similar approach can be used for a simple transfer of diagnosis knowledge from one self-healing deployment to another, e.g. to bootstrap a diagnosis knowledge base in a new and fresh self-healing function instance [6]. This can be achieved by comparing the structure detected by clustering of the anomaly

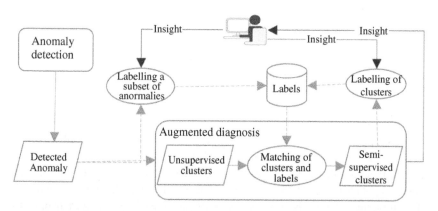

Figure 9.13 The augmented diagnosis process [6].

events in the new system to an existing diagnosis knowledge base of another, existing diagnosis function, and attempting to connect the anomaly clusters of the new system to those diagnoses. This, of course, requires that the feature sets in the anomaly patterns are similar. The use of transfer learning in self-healing is further discussed in Section 9.5.

9.4.4 Deploying Corrective Actions

The most common corrective action in RAN self-healing is a cell reset, since it resolves many of the software issues [21]. Typically, a restriction is applied that another reset is not allowed for a given period to prevent continuous resets in case the problem persists. Other automated corrective actions can include switching over to a standby unit in case a hardware failure is suspected by the diagnosis process and redundancy has been built in the system, e.g. a spare unit. The automated post-action configuration change verification method described in Section 11.3 presents an example, how misconfigurations may be resolved with a rollback to previous stable configuration and also discusses the complexities of the seemingly simple fix.

Typically, the correction that can be deployed automatically has to be relatively non-intrusive, like an automatic reset in case the cell is down. If these do not solve the issue and a new failure occurs during the blocked period, manual intervention is often required. In this case, the initial analysis done by the self-healing process can already help to point the diagnosis in the right direction and the collected data can be useful.

9.5 Knowledge Sharing in Cognitive Self-Healing

This section introduces the concept of information sharing amongst different domains, i.e. how the *knowledge* learned in one specific domain can be *transferred* to a different domain. This is particularly relevant for cases in which the data available in one domain is limited, which might compromise the effectiveness of training. Anomaly detection is a typical example in which few samples are collected, since anomalies do not often appear in a system. In this regard, transfer-learning models and techniques are introduced, and specific use cases presented. Finally, a specific implementation (i.e. knowledge cloud) is detailed, providing a numerical quantification of the *sharing gain*.

9.5.1 Information Sharing in Mobile Networks

Information sharing is not trivial since the domains (such as network management or orchestration) usually have their own data models (i.e. different KPIs collected with different levels of granularity), use cases and requirements tailored to their own needs. So, interpretation is typically required to transfer knowledge between domains. The main features of such sharing information principle should delineate:

- **What** information needs to be shared, e.g. which format, level of abstraction.
- **How** information should at best be shared (e.g. which learning algorithm should be developed).
- **When** information needs to be shared (e.g. periodically, on-demand, etc.).
- **Where** and **by whom** information should be stored (e.g. data collection vs data sharing).

The transfer-learning concepts have been developed to answer such problems [22]. Note that transfer learning typically refers to the transfer of *knowledge*, which differs from transfer of raw data. It implies processing of data, such as data mining, compression, and information extraction. Transfer learning, between different domains and tasks, can potentially greatly improve the performance of learning by avoiding or reducing expensive data-collection and data-labelling efforts.

Formally, a *domain* D consists of two components: a feature space $\mathcal{X}$ and a marginal probability distribution $P(X)$, with $X \in \mathcal{X}$. Given a specific domain D, a *task* $\mathcal{T}$ consists of two components: a label space $\mathcal{Y}$ and an objective predictive function $f(\cdot)$, which is not observed, but can be learned from the training data, i.e. pairs $\{x_i \in \mathcal{X}; y_i \in \mathcal{Y}\}$. Given a source domain D_S and learning task $\mathcal{T}_S$, a target domain D_T and learning task $\mathcal{T}_T$, transfer learning aims to help improve the learning of the target predictive function $f_T(\cdot)$ in D_T using the knowledge in D_S and $\mathcal{T}_S$, where $D_S \neq D_T$ or $\mathcal{T}_S \neq \mathcal{T}_T$.

The condition above, i.e. $D_S \neq D_T$, implies that either the two domains have different features space, $\mathcal{X}_S \neq \mathcal{X}_T$, or different marginal probability distributions, $P_S(X) \neq P_T(X)$. Similarly, $\mathcal{T}_S \neq \mathcal{T}_T$ implies that either $\mathcal{Y}_S \neq \mathcal{Y}_T$ or $P(Y_S | X_S) \neq P(Y_T | X_T)$.

In general, when the target and source domains are the same as well as the corresponding tasks, the learning problem becomes a traditional machine-learning problem. In contrast, when either the tasks or the domains are different, the supervised transfer learning can then be categorized as:

- **Inductive transfer learning:** When the tasks are different and labelled data are available at least for the target domain.
- **Transductive transfer learning:** When the task is the same, but domains are different. In this case, labelled data are assumed to be available only for the source domain.

Both classes are classified as *supervised learning* since it is assumed that some labelled data is available in at least one domain.

9.5.2 Transfer Learning and Self-Healing for Mobile Networks

Sharing knowledge allows the network to fully exploit the potential of learning mechanisms in finding intrinsic generalized patterns. However, knowledge sharing is not trivial, since it requires the capability to generalize knowledge from one domain and apply it to another, which is typically difficult in artificial intelligence (AI). This requires providing an intelligent knowledge-sharing framework that is able to identify what, when, where, and how to share. Methods for modelling and realizing such knowledge sharing usually go under the term of *transfer learning*, which is a very active area of research in the machine-learning community. Knowledge can be shared amongst different domains of the same network, such as network management, orchestration, and QoE management. The goal is, for instance, to provide prognostic cross-domain anomaly detection, where root cause analysis and proactive mitigation are jointly performed with an end-to-end perspective. Moreover, with the introduction of network slicing, knowledge can be shared amongst different slices to facilitate e2e cognitive slice life-cycle management which is able to learn from experience. Knowledge sharing is also envisioned to happen amongst different networks (e.g. technologies, operators), ultimately to be able to provide generalized models and provide a 'theory of networking' [23] which would solve the current problem that each network has to be 'learned' separately.

Transfer learning intrinsically helps to answer the four questions on information sharing [22].

What to transfer: Knowledge amongst domains can be transferred as follows:

- *Instance-transfer* assumes that parts of the labelled data in the source domain are reused for learning in the target domain by *reweighting*.
- *Feature-representation-transfer* finds a 'good' feature representation that reduces the divergence between source and target domains and improves performance of the target task.
- *Parameter-transfer* discovers shared parameters between the two domains.
- *Relational-knowledge-transfer* builds a mapping of relation knowledge between source and target domain.

While inductive transfer learning has been studied for each transfer setting, the transductive approach has only been investigated for the first two classes of transfer.

How to transfer: Specific learning algorithms need to be developed based on the considered use case, i.e. domains and tasks, as well as based on the available data. Furthermore, the problem of 'how' to transfer knowledge should also address the question 'by whom' data should be stored (i.e. **where**). Two approaches are possible:

- A *centralized* approach, which has the advantages of a global picture of source and target domains and tasks but requires a lot of effort and cost for transferring the data.
- A *distributed* approach, which is more complex to implement, but can be less expensive in terms of computational effort and data transfer.

When to transfer: Transfer learning should be done only when it will *actually* benefit the target domain and improve the target-task performance. In some situations, transfer learning may not be needed, nor indeed helpful, or in some cases even harmful. Those cases are referred to as *negative* transfers, and it is important to understand and avoid them. Moreover, knowledge transfer could happen both in a *periodic* manner or *on-demand,* i.e. when needed. The decision between 'push-based' or 'pull-based' data exchange depends on the specific task and considered use case.

9.5.3 Applying Transfer Learning to Self-Healing

Applied transfer learning in the context of network and service management aims at fully exploiting correlations and dependencies amongst domains and providing *cognitive* support to decision-making and automation. In this respect, data analysis and learning play a crucial role. The entities (or functions) designed to take the task of collecting data, implementing machine learning as well as transfer learning techniques are the so-called *cognitive agents.* The subsequent sections consider two use cases and related possible architectures, supported by inductive and transductive transfer learning, respectively.

9.5.4 Prognostic Cross-Domain Anomaly Detection and Diagnosis

Cross-domain anomaly detection and diagnosis refers to the possibility of exploiting cross-domain information sharing to ensure that upcoming faults are detected as early as possible, and even anticipated, so that countermeasures can be adopted in time. This scenario considers the case where two different domains have different tasks (to detect anomalies within its own domain), but they share knowledge to improve the performance of their own task. Formally, the domains are denoted as D_1 and $D_2, (D_1 \neq D_2)$, and their respective (domain-specific) tasks as T_1 and T_2, respectively. This case deals with inductive transfer learning, where:

- The **domains** D_1, D_2 are characterized by different features space, $\mathcal{X}_1 \neq \mathcal{X}_2$.

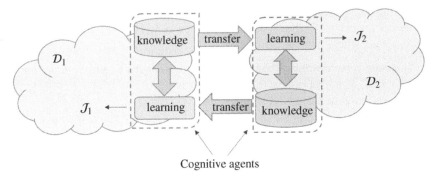

Cognitive agents

Figure 9.14 Inductive transfer learning architecture for e2e prognostic cross-domain anomaly detection and diagnosis.

- The **tasks** are different, i.e. $\mathcal{T}_1 \neq \mathcal{T}_2$.
- The **aim** is to help improve the learning of the target predictive function $f(\cdot)$ of one domain using the knowledge from the other domain (and task).

When assuming that labelled data is available in both domains, this type of learning is usually referred to as *multi-task learning*. Also, note that in this case both domains can either play the role of source domain as well as target domain, and improve the performance of their own task by exploiting the knowledge of the other domain. Figure 9.14 shows the two considered domains, assuming a cognitive agent in each domain (*distributed architecture*). Each cognitive agent has data from its own domain (knowledge database) and acquires knowledge from the other domain (via transfer learning). Both sets of data are used for learning and performing the specific task.

9.5.5 Cognitive Slice Lifecycle Management

End-to-end slice instantiation (and management) requires that resources in various domains are allocated, managed, and orchestrated in a coordinated manner. In this case, information sharing amongst different instances of the same slice type (in principle both within a single network or even amongst different networks) is necessary to enable coordination and to guarantee that the slice manager has a holistic view of the end-to-end network and can properly react to changes. When considering a single network, the coordination and managing of the different slice instances is done by a centralized entity, i.e. the slice orchestrator, which is also responsible for enabling knowledge sharing among slices.

In particular, consider the case where different slices within a single network exchange knowledge towards the same task. To represent this scenario, assume that slices S_1 and S_2 play the role of source domain, while a central cognitive agent

C (within the slice orchestrator) plays the role of target domain and has a specific (network-level) task $\mathcal{T}$. The problem can then be formulated as transductive transfer learning, where:

- There are three **domains**, namely two source domains S_1 and S_2 and a target domain C, which represents the cognitive agent.
- There is a unique **task** $\mathcal{T}$.
- The **aim** is to help improve the learning of the target predictive function $f(\cdot)$ of domain C using the knowledge from the both domains S_1 and S_2.

Furthermore, assume that the target domain has no available data, while labelled data is available at the source domains. In Figure 9.15, a *centralized architecture* is shown. S_1 and S_2 collects labelled data and transfer them to the central cognitive agent. The combination of this information is then used to perform a common task $\mathcal{T}$.

9.5.6 Diagnosis Knowledge Cloud

Experience from network operation is highly important for efficient network management. For any machine learning based management automation method it is especially so. After all, machine learning is learning from examples, from experience. Obviously, experience needs to be collected over time and this needs to be repeated for each deployment. In the case of self-healing, it means profiling the normal network behaviour and collecting a diagnosis knowledge base. Especially collecting a comprehensive diagnosis knowledge base can be a long process,

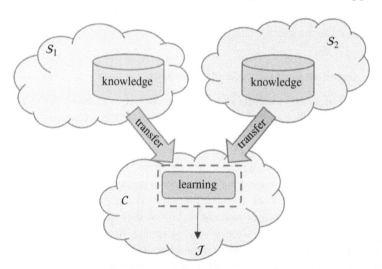

Figure 9.15 Transductive transfer learning architecture for slice life-cycle management.

because anomalies are, by definition, rare events which also holds true for network problems. Furthermore, it would be desirable to be able to detect and diagnose issues even when they are experienced for the first time in a given deployment. So, self-healing and especially the diagnosis step could benefit greatly from being able to use diagnosis information from other networks. Such an approach of diagnosis knowledge sharing across different networks has been pursued in a joint collaboration between Nokia and SRI International and has been disclosed in [24]. Hereby, only selected aspects of this joint work are highlighted, whereas further details are available in [24].

The diagnosis knowledge sharing framework described in detail in [24] and summarized hereby, called *diagnosis cloud*, collects the diagnosis knowledge from individual networks, based on which it derives the new knowledge to be applied to different networks. Such knowledge is used either for facilitating the management of newly deployed networks or new behaviour of already existing networks.

In order to execute the abovementioned steps, specific functions need to be introduced within the diagnosis cloud framework. Such functions are responsible for data anonymization, data collection, joint processing of collected data for knowledge extraction, etc. The next section highlights the foreseen procedures in the diagnosis cloud framework as well as the functional components needed for their execution.

9.5.7 Diagnosis Cloud Components

The diagnosis cloud infers the diagnosis knowledge based on the local diagnosis process executed at individual networks. Such knowledge is abstracted in a form that is network non-specific and further processed in order to derive new diagnosis knowledge. Finally, the diagnosis cloud uses such derived knowledge in different networks by using the parameters of that specific network as described in [24].

The main functional entities of the diagnosis cloud are Local Diagnosis Agent (LDA) which is responsible for local diagnosis process of individual network, Gateway Diagnostic Agent (GDA) which performs the data abstraction, and Central Diagnostic Agent (CDA) which includes the global knowledge about the network. The detailed descriptions of the three main functional entities are available in [24] and are summarized as following.

The **LDA** issues a diagnosis request to the cloud/central entity once it experiences a new or unknown problem. The LDA includes the local diagnosis models which identify an anomalous behaviour and are derived by individual local networks using, for example, machine learning approach and or rule-based methods. The LDA stores the models that were either obtained locally or retrieved from the cloud but are used in a local network in a local database. Furthermore, the LDA performs the model assessment estimating the quality of the model according to defined local criteria.

The **GDA** is responsible for translating data during the exchange between the local network/entity and the cloud/central entity, and in a way that no information is lost and that the privacy is preserved. The data translation and data generalization in the GDA also enables the data to be anonymized. The GDA also maps the related parameters between the cloud and the local network, e.g. mapping between network-specific and network-agnostic parameters.

The **CDA** maintains the global knowledge base which contains all local network models. The CDA collects the models from individual local networks, ranks them, and, when required, provides the relevant models to each local network, i.e. to the LDA, through the GDA. The CDA includes functionalities such as model ranking, model similarity check, model database, and its updates, global knowledge along with its analytics, e.g. extracting the statistics on the model utilization, estimating the accuracy of the model and combining different models.

9.5.8 Diagnosis Cloud Evaluation

The diagnosis cloud framework has been evaluated in [24], this section only highlights the major outcomes for the case where new cells are deployed in already existing networks. In such a case, the existing local models are not sufficient for identifying the network state and the anomaly cause for the new cells. An approach based on topic modelling [25] is used for capturing the network state. The most likely explanation for captured network state is then derived using the MLN [26] graphical models that enable probabilistic inference.

In order to evaluate the benefits of sharing knowledge between two networks using the diagnosis cloud, the evaluation of the abnormal state of newly added cells using the initial local model as well as baseline model for comparison has been done. Without using diagnosis cloud, the initial local model erroneously indicates a network state that is more anomalous than measured with the baseline model.

When diagnosis cloud framework is used, the LDA determines the quality of its local topic models. If the quality is not satisfactory, the LDA issues the model retrieval request from CDA. The CDA ranks all available models and sends back to LDA only the most relevant models. After receiving the models, the LDA estimates its detection capability and ranks the models accordingly. The model with the highest rank is finally selected to be used for detection. The model derived from the diagnosis cloud follows the baseline model far better, thus compared to the case where no diagnosis cloud is used, the considerable improvements in detection capability are achieved.

In further evaluation using the mean entropy measure of the diagnosis MLN quality (lower entropy means higher quality), the local MLN resulted in a mean entropy of 0.666 without knowledge sharing from the cloud. Applying the diagnosis cloud MLN to the same time series, results in a mean entropy of 0.5, thus the cloud MLN shows better performance. The detailed evaluation results and description is available in [24].

In general, all these results show that knowledge cloud framework can be leveraged to achieve better results in both the detection and diagnosis of faults by considering knowledge from self-healing deployments in other networks.

9.6 The Future of Self-Healing in Cognitive Mobile Networks

The previous sections have shown that developing a self-healing solution for a mobile network is a complicated process with lots of tweaking required to make it work. In many fields, deep learning methods have enabled more of the model development to be learned from less processed features. Can this apply to the self-healing use case as well? One requirement for deep learning methods is the availability of very large amounts of training data, and often labelled data for supervised learning. Typically, this has not been available for mobile network self-healing. This section will shortly discuss the future of self-healing, – the prospects of improving the performance of the current methods, for example, towards more predictive and preventive self-healing, and what special limitations might apply to it.

9.6.1 Predictive and Preventive Self-Healing

The ultimate goal of resilience is, of course, not only to be able to recover from a degradation, but to prevent it altogether by adapting to any changes before they cause a problem. In mobile network management this means that any faults would be detected or predicted before they lead to a service degradation. Here, the first question is, which problems are at all predictable. If a RAN backhaul cable is cut by construction work, this is probably something not realistically predictable by a self-healing function. There should be some early signs of the problem that the system can monitor and detect. In this sense, predictive self-healing is detection as well, but detection of the first symptoms of a cause, rather than detecting the symptoms of problem that it caused. A failing power amplifier could be an example of something predictable, because, for example, detected jitter or reduction in power output could be signs of an impending issue. If such early signs are known *a priori*, they may be configured in the detection and diagnosis system as rules. The early symptoms can, however, depend on the context, so it may be hard to provide general rules, and then there is of course the question of the faults, where the early signs are not known. The problem becomes, how to learn what are the symptoms of the causes that lead to given problems.

In an anomaly detection system that is based on monitoring time series data, such as the KPIs in a RAN, one approach is to apply time series prediction methods, such as ARIMA, to the KPI inputs, in context, and determine, if such predicted states are detected as an anomaly. However, many KPIs do not present an easily predictable behaviour on their own. Several KPIs may be considered together to make the prediction based on the wider context.

System behaviour leading to a fault can often be complex and not captured by simple time series prediction of the input features. In order to learn, without making too many assumptions, what the sequences of signs leading to a problem are, deep learning-oriented methods can be applied. For example, a Long-Short Term Memory (LSTM) Recurrent Neural Network (RNN) can be trained to predict from the input features, if a state labelled as degraded is probably in the future. The LSTM RNN can, in principle, learn the state transition paths that the system is taking and try to predict the future paths based on these models, including ones leading to degraded states. The difficulty, as is often in the case of self-healing, is that to learn the typical state transition paths that can lead to a degraded state, many examples of such events are required. Since faults are typically rare events, such training examples are not available, at least not in the limited context of a given network function, subnetwork, etc.

To mitigate the problem of insufficient examples to learn from, inherent to the diagnosis in self-healing, the knowledge sharing methods described in Section 9.5 can be important. Detecting and diagnosing a fault that has never occurred in a given network, based on knowledge transferred from elsewhere, where it has, is a kind of prediction. Another alternative is to use simulators and other such 'digital twins' to play out different scenarios to train the models, but the usability of the models depends heavily on the accuracy of the simulation. Often it is, however, the only alternative, since approaches like fault injection are not really an alternative in live deployments.

9.6.2 Predicting the Black Swan – Ludic Fallacy and Self-Healing

The United States Secretary of Defence Donald Rumsfeld famously coined the terms *known knowns, known unknowns,* and *unknown unknowns* in his statement [27]:

> -- there are known knowns; there are things we know we know. We also know there are known unknowns; that is to say we know there are some things we do not know. But there are also unknown unknowns—the ones we don't know we don't know. And if one looks throughout the history of

our country and other free countries, it is the latter category that tend to be the difficult ones.

And it is this latter difficult category that has been considered in this chapter. Starting with the definition of resiliency over robustness, the critical discussion for the network to be able to survive not only the known problems, but also the unknown ones. It is one thing to be able to detect the symptoms of these unknown problems and try to figure out some response to them, but as discussed in the previous section, the ultimate goal is to be able to predict and prevent them. But how realistic is that?

Nassim Nicholas Taleb developed what he calls the *black swan theory* to describe the often-inappropriate rationalization people have with hindsight on unexpected, unpredictable but catastrophic events, which gives rise to the demand to be able to predict them [28]. The name comes from common belief in Europe in the Middle Ages that there were no black swans, which was why the term 'black swan' was even used as an expression to refer to something impossible, until black swans were discovered by European explorers in 1697 in Western Australia. Taleb asserts three attributes on what he calls a proverbial *Black Swan* [28]:

> First, it is an outlier, as it lies outside the realm of regular expectations, because nothing in the past can convincingly point to its possibility. Second, it carries an extreme 'impact'. Third, in spite of its outlier status, human nature makes us concoct explanations for its occurrence after the fact, making it explainable and predictable.

From the black swan theory, Taleb concludes what he calls the ludic fallacy. It is an argument against applying naïve and simplified, game-like statistical models in complex domains. His argument centres on the idea that predictive models are based on 'platonified' ideal forms, which ignore the incredible complexity of reality:

- It is impossible to have the entirety of information available.
- Even the smallest variations can have significant effects (the chaos theory).
- Because of this, what is typically observed, post hoc, as a failure of the predictive capability of a model, may very well be a Black Swan, something that was real, but unpredictable (or *predictable only after the fact*).

It is exactly the completely unexpected and catastrophic failures, where it is anticipated that more cognition in the mobile network self-healing methods may be able to help predict and prevent them. But is the state and the ensuring of the QoS of an entire mobile network a complex domain? Absolutely. Is a universal predictive self-healing function, able to predict any unforeseen issues, then a Black

Swan? Probably. After a failure, and it is clear, which features to look at, it may very well have been predictable. However, if a similar failure has never occurred before (in a given context), the system has no way of knowing which features to look at for the prediction. But is there still something that can be done to improve the predictive capability of the self-healing functions? Very likely.

In recent years, deep learning methods have been able to combine knowledge in a way that can give new insights, but typically in less complex and chaotic contexts than mobile network self-healing. Still, if restricting the outcome from a truly universal predictive self-healing function to a somewhat more limited set of problems and managed domains, it may be possible to develop methods that are able to reason and learn based on huge amounts of data collected from the mobile communication that goes on around the world. This may require transfer learning methods and new ways of thinking about how to collectively share such knowledge. Deep learning methods can learn from less processed and feature-engineered data, and hopefully also learn the less idealistic factors of the model. How will the all-great predictive self-healing function of the future then look like? Well, that may be hard to say, because, to quote Niels Bohr: 'It is difficult to predict, especially the future'.

References

1 Sanneck, H., Hämäläinen, S., and Sartori, C. (2012). *LTE Self-Organizing Networks*. Wiley.

2 Zolli, A. and Healy, A.M. (2012). *Resilience – Why Things Bounce Back*. Headline Publishing Group.

3 Berns, A. and Ghosh, S. (2009).Dissecting Self-* Properties. IEEE International Conference on Self-Adaptive and Self-Organizing Systems, San Francisco.

4 Laster, S.S. and Olatunji, A.O. Automic Computing: Towards a Self-Healing System. Spring American Society for Engineering Education Illinois-Indiana Section Conference, Illinois.

5 Mwanje, S., Decarreau, G., Mannweiler, C. et al. (2016). Network management automation in 5G: Challenges and opportunities. PIMRC, Valencia.

6 Ali-Tolppa, J., Schultz, B., Kocsis, S. et al. (2018). Self-Healing and Resilience in Future 5G Cognitive Autonomous Networks. ITU Kaleidoscope, Santa Fe.

7 Michalopoulos, D., Gajic, B., Crespo, B. et al. (2017). Network Resilience in Virtualized Architectures. International Conference on Interactive Mobile Communication Technologies and Learning (IMCL), Thessaloniki.

8 Gajic, B., Mannweiler, C. and Michalopoulos, D. (2018). Cognitive Network Fault Management Approach for Improving Resilience in 5G Networks. EuCNC, Ljubljana.

9 Bajzik, L., Deak, C., Karasz, T. et al. (2016). QoE Driven SON for Mobile Backhaul Demo. International Workshop on Self-Organizing Networks (IWSON), Valencia.

10 Bodrog, L., Kajo, M. and Schultz, S.K.a.B. (2016). A robust algorithm for anomaly detection in mobile networks. IEEE Annual International Symposium on Personal, Indoor, and Mobile Radio Communications (PIMRC), Valencia.

11 Aytekin, C., Ni, X., Cricri, F. and Aksu, E. (2018). Clustering and Unsupervised Anomaly Detection with L2 Normalized Deep Auto-Encoder Representations. IJCNN, Rio de Janeiro.

12 Weldon, M. (2016). *The Future X Network – A Bell Labs Perspective*. CRC Press.

13 Ciocarlie, G., Novaczki, S. and Sanneck, H. (2013). Detecting Anomalies in Cellular Networks Using an Ensemble Method. CNSM, Zurich.

14 Brase, C.H. and Brase, C.P. (2008). *Understanding Basic Statistics*. Brooks/Cole.

15 Mahalanobis, P.C. (1936). *On the Generalized Distance in Statistics*. National Institute of Science of India.

16 Ciocarlie, G.F., Connolly, C., Cheng, C.-C. et al. (2014). Anomaly detection and diagnosis for automatic radio network verification. International Conference on Mobile Networks and Management, Würzburg.

17 Nováczki, S. and Gajic, B. (2015). Fixed-Resolution Growing Neural Gas for Clustering the Mobile Networks Data. International Conference on Engineering Applications of Neural Networks, Rhodes Island.

18 Nováczki, S., Tsvetkov, T., Sanneck, H. and Mwanje, S.S. (2015). A Scoring Method for the Verification of Configuration Changes in Self-Organizing Networks. International Conference on Mobile Networks and Management, Santander.

19 Ester, M., Kriegel, H.-P., Sander, J., and Xu, X. (1996). A density-based algorithm for discovering clusters in large spatial databases with noise. *Kdd* 96 (34): 226–231.

20 Lexico (2019). Oxford Dictionary. http://en.oxforddictionaries.com (accessed 16 January 2019).

21 Asghar, M.Z. (2016). *Design and Evaluation of Self-Healing Solutions for Future Wireless Networks*. Univeristy of Jyväskylä.

22 Pan, S.J. and Yang, Q. (2010). A survey on transfer learning. *IEEE Transactions on Knowledge and Data Engineering* 22 (10): 1345–1359.

23 Matsumoto, C. (2017). Why Machine Learning Is Hard to Apply to Networking.https://www.sdxcentral.com/articles/news/machine-learning-hard-apply-networking/2017/01 (accessed 12 February 2019).

24 Ciocarlie, G.F., Corbett, C., Yeh, E. et al. (2016). Diagnosis Cloud: Sharing Knowledge Across Cellular Networks. International Conference on Network and Service Management (CNSM), Montreal.

25 Blei, D., Carin, L., and Dunson, D. (2010). Probabilisitc topic models. *IEEE Signal Processing Magazine* 27 (6): 55–65.

26 Richardson, M. and Domingos, P. (2006). Markov logic networks. *Machine Learning* 63 (1–2): 107–136.

27 Rumsfeld, D. (2002). News Transcript: DoD News Briefing – Secretary Rumsfeld and Gen. Myers. United States Department of Defense. http://defense.gov (accessed 29 January 2020).

28 Taleb, N.N. (2007). Chapter 1: the impact of the highly improbable. In: *The Black Swan*. Random House.

10

Cognitive Autonomy in Cross-Domain Network Analytics

Szabolcs Nováczki, Péter Szilágyi and Csaba Vulkán

Nokia Bell Labs, Budapest, Hungary

Telecommunication networks exhibit complexity on many aspects: large topology, multiple interworking technologies, service variety, diverse, and mobile end-user demand. Given the scale and depth of complexity, to reach a high level of autonomy in network automation requires a paradigm shift in both system and interface design. Within the system, autonomous operation requires real-time awareness of end-to-end performance via correlated insight to user experience, application state, and network state across multiple network domains. Interfaces within the system should enable collection and sharing of insight to facilitate analytics and decisions. Interfaces should support the principles of intent, where the network receives high-level objectives and service definitions from the operator and returns system state in a transparent and humanly understandable way. This also redefines the role of the operator, as it means both delegation of responsibility and delegation of trust towards the system.

An enabler for automation is the large volume of real-time operational data generated in the network. Integrating machine learning (ML) methods into workflows enables leverage of this data for network automation and to unlock the potential of insight-driven decisions and action within the system. However, a core challenge for ML is the interpretation of the ML-generated output, and correlation with additional contextual data (including intents, business, and planning targets, cost analysis) to convert ML output to actionable insight. Moreover, to achieve closed-loop automation acting on the result of ML algorithms requires APIs exposed by the network management or user plane functions to deliver such actions. ML also transforms the operator interfaces, requiring novel ways to bring domain expertise into and insight out of the system. While this disrupts the traditional configuration, threshold driven PM/FM monitoring and reactive

Towards Cognitive Autonomous Networks: Network Management Automation for 5G and Beyond, First Edition. Edited by Stephen S. Mwanje and Christian Mannweiler.
© 2021 John Wiley & Sons Ltd. Published 2021 by John Wiley & Sons Ltd.

re-configuration and planning processes, it has the potential to create cognitive, self-managing systems, exposing lower complexity, and improving the operator's experience.

This chapter discusses these challenges, as well as the opportunities, from an end-to-end network operations perspective. The chapter starts with a general discussion of autonomous network operations and then dives into specific use cases, specifically the challenges on how to model network states, and how to leverage real-time information for analytics and customer experience management, ending with a discussion of mobile backhaul automation as a use case for such real-time analytics.

10.1 System State Modelling for Cognitive Automation

To improve the performance of a mobile network, network automation functions need to understand the state of the system not only to select the proper action to perform but, even more importantly, to notice that an is action needed in the first place. Unfortunately, as network entities are extremely complex, the possible states of aggregate parts of the networks are even more complex and cannot be determined by classical methods. There is no single measurement that could be used to determine the actual state of a network for any chosen aspect or dimension. On the contrary, these systems produce large amounts of data, each piece capturing some aspect of the system and in aggregate the pieces characterizing the actual system state. The goal of system state modelling is to analyse the available information received about a system and assemble the system's actual state. This section outlines the basic concepts of System State Modelling, gives some implementation alternatives and real-life examples of using these foundations.

10.1.1 Cognitive Context-Aware Assessment and Actioning

A key enabler of data-driven operation in Cognitive Autonomous Networks is insight collection. Without the system (in this case, the network) having detailed information about its own state, that of the users (humans or machines) and the facilitated communication or content transfer, automation becomes unfeasible. The critical learnings from Self-Organizing Networks (SONs) are that: (i) reducing complexity through automation requires a simplified operator interface – rather than exchanging one type of complexity for another, and (ii) obtaining a multi-domain end-to-end view for the autonomous functions is essential to achieve harmonized system operations. The multi-domain end-to-end view includes service/network measurements, network state, application level insight and others. To account for planning, cost, business strategy, revenue insight, and customer relations, actions should be aligned not only with the real-time end-to-end network/service/resource context but also with the high level, longer-term targets.

Collecting insight requires tapping into various data sources; including traffic measurements at various locations and abstraction/aggregation levels (application, radio, transport, etc.); direct information on resource availability and utilization from the network functions in various technology domains as well as explicit network-side context as insight from Operations Support Systems (OSSs), counters/alarms, event sequences, logs, etc. Combining these sources enables the estimation of the demand to be served by the network, to assess the state of the system detect or, in certain cases, even predict network or service degradations as well as to localize potential resource conflicts where that might require optimization actions. The different sources provide context for cognitive decision making. On the radio side, context includes location information, resource availability, load, channel quality, and various configuration parameters while on the mobile backhaul, it includes topology, transport services, and their parameters, routing and traffic management policies, etc. On the user and traffic level, context describes simultaneous users of shared resources, the type of applications they use, their communication requirements and, the traffic mix they generate, while on the application and service level, context includes details of specific application sessions (application state, progress of data transmission, responsiveness, etc.). For QoE and QoS, context includes measurement or estimation of end-user perceived quality.

In general, context may be seen as the set of measurable or observable parameters that define the unity of network configuration, the users of the network and the initiated communications. Such context includes both instantaneous information and collected or aggregated information that gives a historical view of the characteristic behaviour (both on user and system level) that can be used to automatically estimate the expected demand and the reaction of the system. As such, a context-aware system should collect up-to-date context information and have a model about its own operation, which when correlated with the context, can be used to evaluate the need for any actions and analyse the impact of potential actions. There is great potential in applying ML algorithms for such context data driven state modelling, as discussed later in detail.

10.1.2 State Modelling and Abstraction

The managed elements of a communication network are complex hardware/software systems with sensory units that provide performance measurements about their operation. Since the network (or elements thereof) consist of cooperating submodules, the captured measurements are correlated in some way. It follows then that the valid state of the network can only be captured by a system that aggregates these partial insights into a single system-wide observation.

This challenge is similar to capturing the state of a modern car, which also has multiple sensors that produce correlated measurements, that are then aggregated by some central unit. As illustrated in Figure 10.1a for example, the fuel consumption is related to the engine speed measured in revolution per minute (rpm) [1]. During a typical driving session, the car traverses several operation modes: acceleration, breaking, tuning, go reverse, or even some combination of these. In general, it is not possible to directly figure out the actual operation mode but looking at the car's sensor measurements, it is visible that each operation mode produces different measurement patterns, as illustrated by Figure 10.1b. Once each state is associated with its typical measurement pattern, it is possible to reason about the actual system state or to detect the operation mode by looking at the measurements. The importance of this concept is clearer by considering states for sub-optimal or degraded operations. Once the pattern of a faulty operation mode is known, the fault can be identified by the analytics module without need for a trained mechanic to check the car.

As shown in Figure 10.1a, traditional performance monitoring solutions handle the managed network elements as if they are a set of independent and uncorrelated data sources. Each data source is analysed independent of others (typically as a function of time), without considering that they are strongly bound together by the source system's implementation rules.

Instead, performance monitoring should handle the managed network elements as complex, partially observable states machines; and use the correlation structure of the data sources to reason about the actual state of the system (see Figure 10.2b). A partially observable state machine implies that: (i) the modelling abstraction is

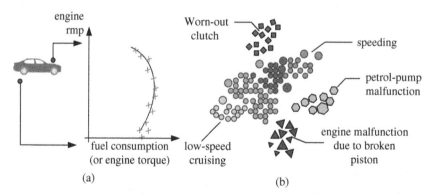

Figure 10.1 (a) Fuel consumption is proportional to the engine rpm; (b) Operation modes have unique data fingerprints.

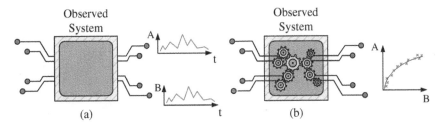

Figure 10.2 System State Modelling contrasted to the traditional approach that handles measurements independent of each other.

that, at any point in time, the observed system is in one of its possible states or operation modes; (ii) it is, in general, not possible to directly access the actual state as it is characterized by the combination of states of its components and; (iii) only indirect reasoning can be made by analysing information on the system. The complete set of all the possible states may not be known or there may be no formal mapping (e.g. defined by domain experts) between the observable information and the states. A common format of information is a set of numeric values where each element represents the actual value of a physical or logical/derived measurement, a specific property of the observed system. At each point in time, the received information defines a state within the state space of the system measurements, which when performed at regular time intervals provides a time series.

The main objective of System State Modelling is to discover all possible operation modes of the observed system by finding patterns in the measurement data and associating them with the operation modes. In other words, find the data fingerprints of the operation modes so that later, each operation mode can be identified by the associated fingerprint. Therein, the modes and states are differentiated as follows:

- The operation mode is the qualitative human-level description of the system state, e.g. the car is in acceleration, breaking, turning mode of operation.
- The system-state is the quantitative description of an operation mode. It defines an area in the multi-dimensional space spanned by the measurements when the system is in the associated operation mode.
- The micro-state is a single point in the multi-dimensional space spanned by the measurements. A system-state is reflected by a set of micro-states which are likely to be observed when the system is in the associated operation mode.

10.1.3 Deriving the System-State Model

To discover all the modes, the system needs to be observed for a long enough period but there is no limitation for such a period as there is no guarantee that all states

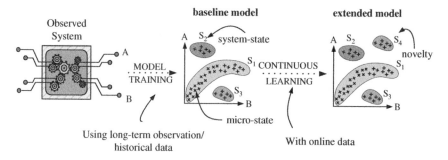

Figure 10.3 Continuous system state learning.

will have been manifested in any such period. Thus, it is necessary to develop a continuous learning approach that is capable of: (i) automatically detecting when a novel state is manifested and, (ii) learning its' pattern (Figure 10.3). Typically, a certain amount of historical data is needed to build a baseline model to apply in the continuous learning with new data. This history or observation window can be shortened if multiple entities of the same type of system can be concurrently observed, e.g. using data of several hundreds of Long-Term-Evolution (LTE) cells increases the chance of observing a significant fraction of the operation modes of any cell in a small time-interval.

10.1.3.1 The Static-State Model

The first step is to create a static- (i.e. time-invariant-) state model learned over a long-term observation of the system. In general, a network can have an extensive amount of possible states but, in practice, it is only useful to distinguish states that are significantly different from each other, i.e. to learn a model with only a finite number of states. This can be achieved by a vector quantization learning process.

During the training, a set of code vectors are created, each representing a discrete micro-state, i.e. a point in the multi-dimensional state spanned by the considered measurements. The training samples are submitted to the training process in a sequential manner. Each sample modifies the code vectors to represent the training data with a slightly lower quantization error. As each dataset is different, it is not possible to define (*a priori*) a fixed number of code-vectors that guarantees a specific quantization error for the dataset. Instead, an incremental model training procedure, similar to Figure 10.4, must be used. Therein, the number of code vectors is gradually increased if this improves the overall quantization error and the process stops when the error does not improve any further, indicating that the model has converged.

MODEL TRAINING

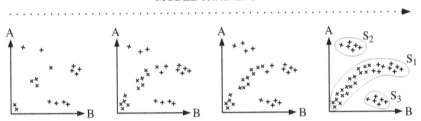

Figure 10.4 Incremental training of the static state model.

The static state model is essentially a quantization function, which associates any measurement vector (i.e. any point in the possibly continuous multi-dimensional measurement space) to one of the micro-states. A measurement vector is assigned to the closest micro-state according to some selected distance metric, e.g. Euclidian, supremum, cosine, etc. In other words, the micro states divide the space into regions, where the region associated to a micro state is the set of points for which that is the closest micro-state. The closest micro-state is often referred to as the best matching unit and the associated area as the coverage area of the micro-state.

10.1.3.2 State Trajectory Modelling and State Clustering

The static state model captures all the micro-states manifested in the training data, i.e. the static state model contains the data fingerprints of the different operation modes like accelerating, decelerating, etc. for the car analogy. The system may manifest multiple micro states during a certain operation mode, e.g. the operation mode acceleration produces different combinations of engine rpm and fuel consumption measurements. As such, micro states need to be grouped according to the operation modes, i.e. micro states need to be sorted in such a way that they get into the same group if they are manifested during the same operation mode.

Over a certain period of time, the manifested micro states can be drawn on a state trajectory similar to Figure 10.5. The state trajectory shows that the possible micro states do not follow each other in random order but, in most cases, the current state determines with a non-zero probability what the next state will be. Moreover, empirical experience suggests that micro states that appear close to each other in the micro state trajectory tend to belong to the same system state or operation mode. This suggests that micro states should rather be grouped by a metric that defines their temporal proximity than by their distance (e.g. Euclidian) in the measurement space.

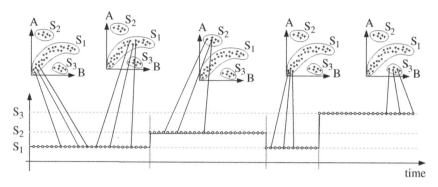

Figure 10.5   An example state trajectory.

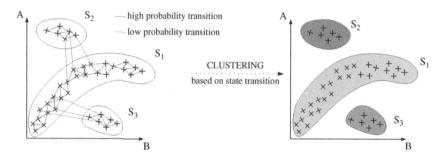

Figure 10.6   State clustering based on transition probabilities.

A possible solution is to create a state transition graph based on the individual micro-state trajectories of the observed entities. The nodes of the graph are the micro states while the edges represent transitions between the micro states. Each edge is assigned a weight that is proportional to the relative frequency of the transition between the connected micro states as showed in Figure 10.6.

10.1.4   Symptom Attribution and Interpretation

System State Modelling automatically finds patterns in the measurement data, but their association to actual, human interpretable operation-mode descriptions remains a manual task. However, the human effort can be lessened by attributing the patterns with symptoms: interpretable characteristics of the patterns that best-distinguish one pattern from the rest. In general, a pattern in the System State Model is a multi-dimensional object with dimensionality equal to the number of the considered measurements, which makes it practically impossible for humans to interpret the patterns. As such, the pattern information needs to be translated into a smaller dimension, i.e. to provide a one-or two-dimensional

(1D/2D) projection of the original multi-dimensional space, which captures the characteristic difference of a certain state from the other states.

The 1D projection may be obtained by selecting a single measurement that has extremely high or low values when the system is within the studied system-state. This delivers quick and obvious information about the system-state, e.g. in a certain, anomalous system-state, the Uplink interference in an LTE cell is significantly higher than in the rest of the cells, so uplink interference may be used as the single projection for identifying an anomalous cell.

For a 2Dprojection, the relation between two measurement values may be considered: none of the component measurements have extreme values that can singularly characterize the state, but the constellation of the measurement-value pair in the state is distinct from the constellations in the other states. A typical 2Dsymptom for LTE cells is the relation between the cell's Physical Resource Block (PRB) utilization and the cell level downlink (DL) throughput. The two values are proportional to each other under good radio conditions: the more PRBs used, the higher is the cell level DL throughput. The relation might, however, change in poor radio conditions, e.g. due to interference, the observed cell level throughput at a certain level of PRB utilization is smaller than in the rest of the cells.

In theory, a pattern can have multiple 1D or 2D symptoms and one could define three or higher dimensional symptoms. However, there is no guarantee that multiple lower dimensional symptoms must directly translate into a higher-dimensional symptom, yet any increase in the dimensionality sharply decreases the interpretability of symptoms.

Selecting 1D symptoms is relatively simple: for each cluster, the average values of each measurement are computed, and a measurement qualifies itself as a symptom if its average is at extreme quantiles of the global distribution of that measurement. However, more considerations must be made to identify appropriate 2D symptoms. In general, having n measurements, $0.5n * (n-1)$ pairs are possible. However, for scalability reasons, not all of them should be used. A second reason to reduce their numbers is redundancy due to strong correlation amongst measurements. Taking the correlation structure into account, a process called relation discovery finds a minimum set of measurement pairs that ensures the retention of the essential information while minimizing redundancy.

Relation discovery computes the pairwise correlations between the n measurements to produce a correlation graph (full graph with undirected edges), with $0.5n * (n-1)$ edges, which are weighted according to the absolute value of pairwise correlations. The graph is pruned to identify a subset, using the maximum weight spanning tree of the correlation graph. The required 2Dsymptoms for each state are those which show significant difference from every other state i.e. over some quantitative measure that captures the degree of deviation. A good measure is a clustering separation index for each state in each 2D projection. Thus, the challenge is to identify those 2D projections where this separation is (almost) equally good as the clusters' separation in n-dimensions. For a given state, a separation measure, called the symptom strength can be computed for each selected pair of measurements (defined by the edges of the relation graph). The pairs with the highest symptom strengths deliver the most important information about the state. This becomes especially important when the state in question belongs to a degradation or failure

10.1.5 Remediation and Self-Monitoring of Actions

Generally, the executed actions may be regarded as network configuration, policy, or packet forwarding changes that directly or indirectly impact the network behaviour. Changes may be delivered as traditional parameter configuration (such as tuning the radio parameters of various sites) or through reconfiguring transport services (creating, managing, re-routing, etc.) but, increasingly, changes are made through API calls in software or through the management (deployment, scaling, relocation) of virtualized functions or micro-services. Although actions may focus at a specific impact area (e.g. specific sites, transport devices, network functions) or technology domain (e.g. radio, transport, cloud), their impact may be across the wider network and may affect both the end-user perceived quality and the efficiency (in terms of resource utilization and cost) of the system. A considerable real-time impact is exercised by actions targeting flow management and traffic management (such as QoS/QoE enforcement), which require intervention in the handling of packets within a time scale of one end-to-end round-trip time. The scope of such actions is to maximize the user experience and provide the best service for most users within the constraints provided by the momentary network configuration. The configuration itself is also managed automatically through actions on a longer time scale to adapt the service to changes in demand or trends (Figure 10.7).

Note that every time a network fulfils a demand (i.e. traffic is generated), many low-level autonomous functions are involved. These include functions that execute at the lowest abstraction level to implement the high level (service level, network level, QoE level, etc.) targets. Cognitive actions may reprogram these underlying enforcement mechanisms so that their operation is aligned with that of the services and the network. This requires self-programming capabilities, realized as interfaces and APIs that enable one network function to call procedures in other network functions without having humans in the loop.

As networks require automation on multiple levels (from packet-level localized real-time actions to breaking down high-level context-free intents to context-based

Selection process (implicit harmonization):

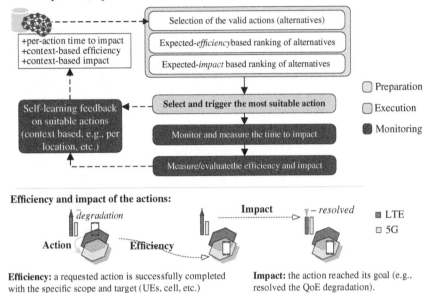

Figure 10.7 Self-monitoring of actions: efficiency and impact.

configuration changes), the network functions should also exercise their own insight-based automation. This can be implemented via embedded cognitive Functions within Radio Access Networks (RANs) and within the mobile backhaul or within the network-side functionalities for handling the specific requirements of verticals such as V2X, IoT, AR/VR, tactile, and their combinations.

To learn the best action for a specific network operation within a given context, context-based self-monitoring is required to build a knowledge base of the context and goal of each action as well as the consequence thereof characterized by the efficiency and impact of the action.

Efficiency describes the degree to which the action successfully achieved the intended changes. In general, some actions may fail, e.g. those that require the compliance or support of external devices (such as user equipment [UEs]) or those that are opportunity driven (e.g. whether coverage really exists to execute a radio traffic steering action). Such failures should be noted in the efficiency knowledge base, which thus grows as the network is operated for longer and as Cognitive Functions collect data over the scenarios. The actions are ranked for the different contexts so that non-efficient actions can be blacklisted, and the most efficient ones identified for use when needed. The filtration must also consider the latency of the actions – the time until the associated changes propagate into the metrics that characterize the goal of the action. The action may have been effective but too late for realizing the goal in time. The corresponding lead-time (which may also be context dependent) should thus also be learnt and utilized by the decisions.

Impact, the other metric for self-evaluating the action, evaluates whether the action reached its intended goal. Only efficient actions (those that successfully re-configure the system in time) can be impactful, but not all efficient actions move the network state or service quality to the intended goal(s).

Accumulated data on efficiency and impact enables the Cognitive Autonomous Network to build a model of the network in which context, action, and consequences are linked to each other. This is mandatory to make meaningful decisions that are most likely to converge the network in the right direction (closer to fulfilling the intents, farther from operational anomalies or failures and back to normal working conditions). The concepts on multi-dimensional end-to-end data analytics can be applied to many use cases, amongst them being the three use cases in the subsequent sections on real-time network analytics for user plane data, customer experience analytics or transport automation.

10.2   Real-Time User-Plane Analytics

The service provided by the network at a given time is a superposition of many factors, including the physical resource availability, topology, configuration, and parameters of network functions, radio propagation, etc. Having real-time up-to-date information about these aspects is critical for handling short-term user demand. As the demand is quite dynamic and the end-user perception or machine-to-machine vertical applications are impacted by short time scales, corresponding decisions/actions should also be dynamic and real-time, and so also should the collection of the enabling insight be. A major source of real-time information is available from the U-plane traffic itself since the traffic flows are modulated by the resources, the interaction with other flows and the services. This makes techniques for analysing U-plane data an indispensable component that real-time context-aware Cognitive Autonomous Networks should maintain.

10.2.1   Levels of User Behaviour and Traffic Patterns

The observable U-plane traffic patterns are highly related with the user. During an application session, and depending on the time scale of any online activity, micro-level and macro-level patterns are generated. Micro-level patterns are realized on a few seconds time scale and are defined by the momentary interaction between the user, the application, the initiated (e.g. requested or produced) content, the state of the network, the coupling of multiple traffic flows served by shared network resources, etc. These patterns may indicate how well a particular application session is served and they are important indicators of how much pressure the network resources have to handle (characterized by the demand-to-load comparison). Macro-level patterns are generated on much greater time scales (minutes, hours, days) and they are indicative of the user's overall approach in using digital services. Such patterns are defined by the overall user interest, lifestyle, social network, mobility/commute routes, home/work environment, whether the user is performing the usual daily routine or is on holiday, etc. The occurrence of application sessions within a longer time span depend on the user and the application (e.g. how frequently the user checks emails or news, what is the usual interaction with friends over social networks). The service received by such sessions depends on the context of other flows, network state, etc. and is observable via micro-level behaviour and patterns.

The micro- and macro-level behaviour are worth considering as they indicate two key measures: the demand that the network must serve; and whether the right level or service is provided. Understanding the difference between load and demand is key when looking for the right measures to quantify user plane performance and create actionable insights. Load is usually a metric between 0% and 100% describing how much of a given available resource is utilized. The resource itself may be dynamically changing (e.g. the capacity of a radio link) or fixed (e.g. the physical bandwidth of a wired link or number of vCPUs within a cloud computer node). High load means that the resources are utilized well, and 100% load means that all resources are utilized. However, even 100% load does not explain whether the resources are sufficient for good service. If the utilized or available resources are sufficient for good service, there is still no guarantee that good service is indeed achieved with the current service and thus no action is needed. And when the available resources are indeed insufficient to properly handle the traffic, it is still important to know what other actions (e.g. traffic steering/offload, bandwidth reconfiguration, etc.) could be used on which UEs/flows, to which destination cells, on which transport services, etc. in order to improve the service; and how many more resources are needed, if capacity reallocation is possible. These considerations do not imply that load is a useless metric to consider; however, it is inadequate.

Measuring the demand, and correlating it with the load, is a rather more efficient approach to generate quantitative and actionable insights to the U-plane network functions (Figure 10.8). Demand is a given measurable quantity (amount of physical resource, bandwidth, throughput, delay, etc.) that is needed for a flow or set of flows to maximize the perceived quality of the application they serve.

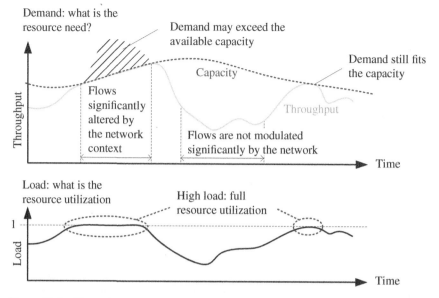

Figure 10.8 Illustration of the concept of demand vs load.

10.2.2 Monitoring and Insight Collection

There are multiple interactions points between the traffic flows and the network, such as RAN, mobile backhaul, link aggregations, etc. It is possible to increase the details of insight and to obtain localized measurements by combining the information obtained from multiple monitoring points (Figure 10.9). Multiple monitoring points enable the measurement of the same flow (even same packets) in greater detail, such as measuring the end-to-end delay and that between measurement points to infer which part of the system (including end devices and content servers) contribute the most to the end-to-end latency. Localizing where packets are discarded is also possible by tracking individual packets at multiple locations and observing the segment between monitoring points where they disappear. Such basic correlation between the measurements of multiple monitoring points already provides a rough insight into the U-plane operation and how the network modulates the traffic it serves. Correlating the information in real-time from multiple monitoring points requires that the monitoring points cooperate, i.e. that they exchange measurements regarding the same U-plane aggregations (packets, flows, etc.). U-plane insight derived from the observation of packets and flows at different locations provides different levels of insight depending on the protocol stack, traffic engineering, aggregation, and the applicable encryption at the monitoring point (Figure 10.9). For example, monitoring at RAN level provides insight at the granularity of each flow, application, UE, bearer,

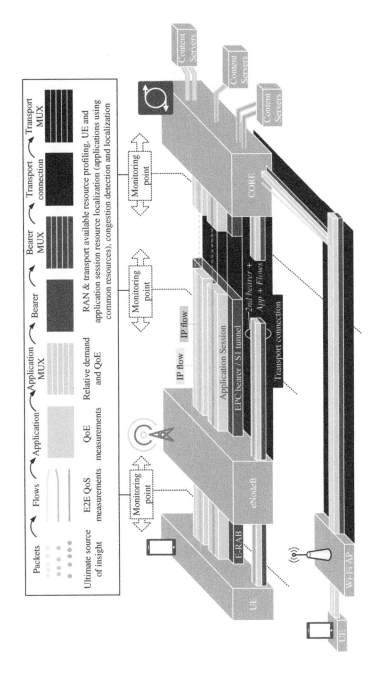

Figure 10.9 Insight aggregation from multiple monitoring points.

or QoS flow. At a transport link, traffic is multiplexed on various aggregates (e.g. observable per eNB, or per transport service, depending on the protocol level and encryption) that limits the granularity of the measurements.

The insight collected from packets can be used to detect the state of the flow at the protocol level (such as TCP), at the QoS level (delays, losses, etc.) and at the application or QoE level. Importantly, in many cases, the demand is also inferable from such measurements by profiling the amount of traffic usually transferred by the application and anticipating the usual context-based behaviour. For user and application state profiling, it is necessary to aggregate insight from packet level to higher abstraction levels, mainly flows and applications. Additionally, the U-plane measurements can describe part of the network state, including resource utilization at various locations, bottlenecks, individual (per UE) or more widespread anomalies, and even provide useful information for their localization and characterization [2].

For network state profiling via user plane insight, it is only necessary to aggregate insight from packet level to higher aggregations and abstraction levels along shared resources and service points; such as radio bearer, radio function, transport service, transport Per Hop Behaviour (PHB), virtual network function, etc. In addition to aggregation points, it is also beneficial to consider the model of resource schedulers to understand the reasons for the observed behaviour (e.g. to recognize particular types of congestion from the way the packet scheduler divides the resources amongst competing flows) and to calculate potential actions (e.g. how to re-configure the schedulers to perform a different kind of resource partitioning that better suits the high-level needs). These principles will be discussed further along use cases and examples in this chapter.

10.2.3 Sources of U-Plane Insight

Measurements derived from packet monitoring enable the generation of network-side and user or application-side insight on various aggregation levels. A non-comprehensive list of Key Performance Indicators (KPIs) that can be obtained on per flow, bearer, radio cell, radio site, transport service, or other aggregation levels is listed below:

- **Throughput and round-trip time (RTT)/delays:** these detect congestion build-up, localize bottlenecks and are also relevant to measure end-to-end performance visible to the users.
- **Buffer status (e.g. Packet Data Convergence Protocol (PDCP) buffer, transport buffer):** these inferred by increased delay and discard pattern.
- **Packet inter-arrival pattern:** this indicates how the network modulates the packet level flow behaviour (micro-behaviour) due to its real-time resource management and enables the inference of load on certain elements.

- **Flow internal status and events:** these (e.g. handshake, slow start, ongoing transmission, retransmission problems, etc.) enable understanding of whether the observed behaviour is in-line with that expected from the protocol model (e.g. observing TCP slow-start after connection establishment) and anticipation of the short term behaviour of the flow (e.g. exponential increase, or decrease).

The above measurements are most accurate if taken on the original traffic, although measurements may also be taken on injected traffic. Additional measurements are also required to better understand the model of the traffic, the application behind it, as well as the provided user experience. These, generated for each flow or group of flows serving the same application session, include amongst others:

- **Application type or category:** the basis of contextualizing and interpreting measurements is to understand the demand of the applications. The same QoS may mean significantly different end user experiences, depending on whether the traffic is generated by a browser, chat, streaming video, social media, or productivity tool.
- **Application state and session parameters:** this is specific to application type: e.g. for a web browser the start of page request and size of web page respectively; or for a video player in state ongoing video with session parameters progress of download, video momentary media rate, etc. They may both be inferred from the observed traffic pattern if it is characteristic to certain phases or use cases of the application.
- **QoE indicators:** these are also specific to the application type, corresponding to measurable quantities that are observable to the users. For example, the total web page download time during browsing and blog/news reading, video pre-buffering and stalling, etc.

10.2.4 Insight Analytics from Correlated Measurements

Some of the above most basic of measurements, including e.g. the throughput, application types, QoE, etc. are attributes of the end-to-end flows and can be obtained at any location. Other items (e.g. RTT, loss) provide different values depending on the location of measurement and thus enable correlation of the same KPI between the location-specific and the end-to-end measurements.

Correlation between different KPIs also enables the detection of a complex network/resource state that is defined by the superposition of multiple events/patterns (as illustrated by Figure 10.10). An example is to measure the momentary capacity on dynamic links (such as radio) using the correlation of throughput and delay. Such capacity is the momentary throughput at the times when the resources are fully driven, characterized by increased delay due to

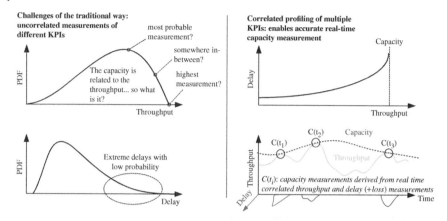

Figure 10.10 Challenges of correlating traditionally collected measurements; correlated profiling of multiple KPIs.

resource competition and slight queuing. With uncorrelated throughput and delay measurement, the relevant points in time are not available as illustrated in Figure 10.10. Correlating the measurements enables the grip of the relevant points and measurement of the capacity through observing the throughput at the right time.

The rich context of U-plane data not only consists of flow-based numerical measurements but also patterns, states, and dynamic transition profiles. Advanced profiling of the network state through U-plane measurements requires suitable methods and analytical apparatus, even up to the point of using apparatus trained via deep learning.

10.2.5 Insight Analytics from Packet Patterns

In the highly dynamic U-plane context where demand may shift rapidly, awareness of potential bottlenecks and their characteristics is important for a self-programming network to make informed decisions for real-time traffic and flow management. This motivates the next example shown in Figure 10.11 that goes beyond numerical measurements and infers the bottleneck type based on the discard pattern of the packets. The different observed patterns imply different urgency on resolving the bottleneck, if needed, as well as on the potential action (or its extent) that may be used successfully.

If all applications are properly served, there is no motivation for changing the network state and the insight is used by the network to keep track of the status. If, however, the QoE or related intents are not fulfilled, additional analytics may help to decide whether a network-side action may improve the situation and, if so,

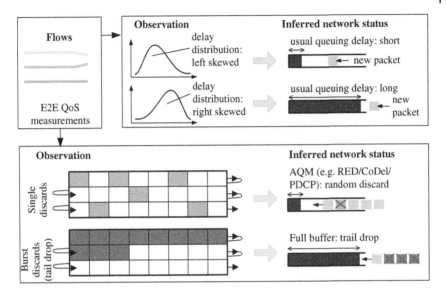

Figure 10.11 Inference of bottleneck characteristics via packet discard pattern classification.

how to parameterize the action. For such a decision, the network needs to infer two things:

1) the location and cause of the problem (if it is within the network, at the radio or Mobile Backhaul (MBH); whether it is due to congestion or link degradation, etc.), and
2) the gap between the current and the optimal state (i.e. how much resource would be needed for the task, how much is available, etc.).

The combination of these network-side information and the state of the applications and flows provides a context of the action. According to the self-learning and self-monitoring principles, the context is consulted with the existing database of actions that are efficient and have good impact within the given context to help pre-select an action. Properly parameterizing the action also requires numerical calculations based on the latest measurements.

The above concept is illustrated in more detail in Figure 10.12, where the QoE (or more generally, intent target) driven context is shown, with the additional consideration of existing resources vs demand. These together define the QoE driven congestion situation, which can be roughly classified into three stages: low load, resource conflict, and demand conflict.

Low load means that resources are not fully utilized; therefore, no network-side action is needed. If there was degradation from QoE or intent perspective, the source of degradation is not resource shortage or suboptimal resource allocation.

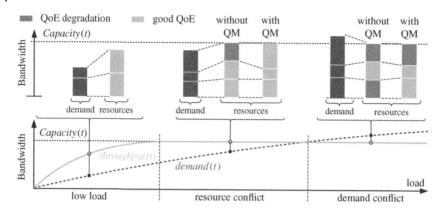

Figure 10.12 Congestion detection.

To create insight for the cause, the numerical and localized measurements on delay, loss, throughput may be used to locate the bottleneck segment (which in this case may be the external Internet part or the device itself). Observing simultaneous external network limitations at various locations may further strengthen such hypothesis.

Resource conflict means that resources are fully utilized but the cumulative demand is below the capacity. This is usually a normal working condition due to the greedy nature of transport protocols (e.g. TCP, Quick UDP Internet Connections [QUIC]) that take all available resources. Good QoE or meeting the intents is achievable given that the resource allocation is aligned with the intents. As the resource allocation needs to follow the changes in traffic mix and the individual applications or services, re-programming packet level resource allocation should be a routine part of cognitive U-plane management.

Demand conflict describes the state when the cumulative demand is higher than the available capacity, so it is not possible to simultaneously satisfy all demands. In that case, selective degradation (e.g. fall-back to best effort service, gradually compromising QoE targets) may be selectively used, to decrease the effective demand of the applications to a manageable level. The selection of sessions subject to such limitations should be performed according to the intents that may also describe elasticity in such matters. If available within the context-based action database, actions such as traffic steering or load balancing should be considered to reduce the load. Note that if demand conflict is a repeating network state, it should be a trigger for longer-term network optimization (potentially outside of the scope of automation) and it should be indicated to the operator via suitable interfaces.

The localization of congestion is performed by correlating the level of QoE (or level of intent fulfilment), radio RTT and radio packet loss at a radio cell as well as for an individual UE served by the radio cell. If the individual UE's metrics show degradation compared to the cell level aggregation, the QoE problem is localized to the UE. This is further reinforced if the radio KPI degradations are only present at the particular UE. Such analysis also illustrates the need for granular (user, bearer level) measurements. The location and scope of the degradation gives indication to the type and extent of action that has the potential to deliver improvement. For example, when a user's QoE is limited by the poor quality of its UE's own radio channel, redistributing resources from other UEs to the limited UE would not deliver the right impact (only marginally higher throughput is expected at the expense of degrading others). Therefore, for a Cognitive Function, this action would be present with a recorded low impact action, not to be recommended. The low impact may be self-learned by the system through applying it and monitoring the impact afterwards; or it could be a good application of expert knowledge infusion (since the reasons why this action is not impactful are well known and determined by the laws of the physical system). Other potentially more impactful actions could be traffic steering (e.g. inter-frequency handover or to offload different radio access technology) or fall-back to best effort service to avoid wasting physical resources without delivering acceptable user experience.

10.3 Real-Time Customer Experience Management

Customer experience management refers to a set of analytics and actions that are executed autonomously to improve the perceived experience of applications (hosted or Over the Top [OTT]) and services. The experience may be related to the human perception (e.g. for Internet applications, or AR/VR) or for use case dependent machine/application requirements (for machine to machine, IoT, or other machine type communication). In that sense, the metrics to be optimized are those that relate to the usage of the information (being rendered for immediate view, played back as video stream, controls a mechanical appliance, stored in background database, etc.). These metrics are defined with terms close to the applications logic and thus it may be distant in semantic from the terms used to define packet level transmission within the network. The former is referred to as quality of experience (QoE) named after the set of applications intended for human use (mostly vision), such as web browsing, video streaming, etc. but may also cover machine type communication requirements. The latter is referred to as quality of service (QoS), carrying a well-established set of metrics within the networking domain. As the QoE metrics depend on the application (which may be diverse in the use of the information transferred over the network), there is no universal set of QoE metrics; they vary depending on the application. The QoS metrics, on the other hand, are rather well established, at least at the packet level (i.e. delay, loss, throughput, etc.).

The scope of implementing customer experience in a Cognitive Autonomous Network is to autonomously adjust the network configuration to ensure good QoE while, at the same time, the system efficiency (e.g. amount of traffic sent, resource utilization, sessions served per resource amount) is also maximized. The intended QoE is to be communicated by the operator as intents so that the network has a freedom to exercise various optimizations to provide the expected observable quality at the lowest possible cost. As network traffic is highly dynamic, the adjustments made to the network configuration should also be dynamic and adaptive to the changes in context (such as the traffic or user mix to be served at a given area, or the radio conditions changing due to mobility). In fact, in order to enable the prevention of a visible degradation due to rapid changes, QoE management should be real-time (capable of analysing, deciding, and acting on the time scale of a round-trip time). The analytics, decision, and action in the context of the QoE management needs to bridge the gaps between intents, application level metrics and QoS/packet level actionable targets [3] – as described in this section.

10.3.1 Intent Contextualization and QoE Policy Automation

The contextualization of intents, i.e. the transition from context-free high-level targets to low-level context-based actions or micro-programs has been mentioned before as a general concept. Within the scope of QoE management, the concept is heavily applied to autonomously bridge the gap between the intents and application-specific QoE metrics and between the QoE metrics and packet/flow level QoS targets that are enforced. This process is policy automation, where policies are defined on three abstraction levels: intent level, customer experience level, and network level.

The highest level is the intent level, defined exclusively by the operator (i.e. not derived automatically). These policies define implicit targets, such as relative priorities between users/services/applications or specify intent-based rules such as a list of important applications for which QoE should be enforced (e.g. YouTube playbacks should not stall). The policies do not contain any numerical targets, QoS parameters, or details on how to achieve the policy target – these are left for the network to derive automatically through policy specialization.

The second level is the customer experience level, describing specific QoE targets for applications (e.g. web browsing) or their type (e.g. interactive download). The targets (e.g. download time, bandwidth), are either explicit ('low level intent' defined by the operator) or implicit and derived automatically from the intents by the system.

The third and lowest level, network level policies are QoS parameters (with numerical values) suitable to be passed directly to low level U-plane traffic or resource management entities such as packet schedulers. The parameters and their values are derived by the system based on the QoE targets or, in special cases, may be defined directly by the operator (such as default QoS parameters to be used for applications that are not recognized or not managed by the system, or non-dynamic service verticals such as native conversational voice with constant and well known QoS requirements, Service Level Agreements (SLAs), non-QoE specific user level limits such as aggregated maximum bitrate).

The process of policy automation for QoE management is illustrated in Figure 10.13. The example shows the possibility for the operator to define intents (highest level) or directly lower level policies, showing the flexibility of the concept. The goal for the system is to reach the network level policies starting from the level that was defined by the operator in each case.

In the leftmost example, the policy supplied by the operator comes with explicit QoS parameters already on the network level, thus there is no need for policy derivation in that case; the given QoS parameters can be enforced directly. The network still has to automatically detect when the policy should be activated, i.e. when the designated application usage is detected.

In the middle example, the policy from the operator comes from the customer experience level and defines a numerical target for the QoE parameter download time. When web traffic is detected, this policy needs to be compiled into a network level policy that can be executed on the packet level. The resulting policy is a bandwidth QoS parameter, based on the QoE requirement of downloading a specific amount of data (the size of the web page) within a given target download time (five seconds in the example). Since each web page has a different size and structure, the value of the bandwidth QoS parameter is defined on-the-fly by the network at the time the web page download actually starts and it detects attributes from the U-plane flows. Additionally, as the web page download progresses (i.e. the amount of downloaded/remaining data changes) and the time left from the five seconds budget decreases, the network level bandwidth target is updated so that the QoE target is met.

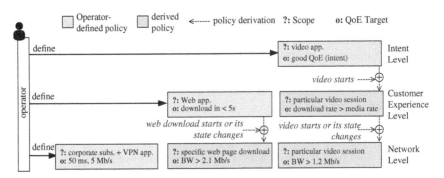

Figure 10.13 Interpretation of intent in the context of QoE management.

In the rightmost example, the policy is defined at the intent level without any context – only the high-level desire to appropriately serve a type of application. So, the system must first derive a QoE target (that describes the critical application level parameters for the application type) and then compile it further to numerical network level targets that are directly executable. The customer-experience-level policy defines that: to provide good QoE for the video download (i.e. implement the corresponding intent), the QoE target for the video is achieved via a bandwidth that is sufficient for the video to pre-buffer and start the playback quickly and higher than the media rate of the video so that it can be played back continuously without stalling (after the pre-buffering is finished). The policy's scope is narrowed compared to the general intent, here to apply only to the particular video because the amount of bandwidth and the momentary state of the video (pre-buffering, playing, etc.) is specific to each video. Then, the network derives a network-level policy that quantifies the specific bandwidth value using real-time application-level insights such as a QoS parameter (1.2 Mbps in the example) that is valid at a given time and can be directly enforced by the system (e.g. by programming a packet scheduler, as discussed later). And, as the video state changes, the network-level policy is updated with higher or lower bandwidth values.

10.3.2 QoE Descriptors and QoE Target Definition

The first level of intent contextualization the Cognitive Autonomous Network has to perform during QoE management is to derive QoE descriptors (which are the application level metrics or measurements) and targets (their numerical values) that apply to specific application sessions from general intents that apply to applications in general. In order to derive the QoE descriptors, the network has to know what the KPIs or measurements are that characterize the user perceived quality or state of various applications. This knowledge is a good example of expert or domain knowledge, so that it is often (at least for popular well-established applications) not something to be indirectly discovered but could be incorporated directly. Therefore, once the network has detected a certain type of application, the corresponding QoE descriptors may be retrieved from a knowledge database. The QoE targets are then dynamically calculated for each descriptor based on the numerical network measurements and state (e.g. available resources and congestion state), context (e.g. traffic mix) and the intent (e.g. what is the precedence of different applications or per user limits).

As the QoE descriptors are application specific, information about the type of the applications should be available for the function that performs the QoE management. This information may be provided by external sources and conveyed through interfaces (or packet marking) or it can be built-on to the QoE manager itself. The techniques of application detection are numerous and widespread thus they are outside of the scope of this chapter. Note, however, that for the detection of application type, a great challenge comes with the encryption of almost all traffic. Therefore, instead of deep-packet inspection and content-based application detection, there is an increasing need to apply pattern analytics on top of packet and flow level measurements (discussed earlier). This is the aspect where ML algorithms become important in this respect.

Defining QoE descriptors is possible by associating quantitatively measurable parameters to key requirements. As a prominent example, for stored video download and playback (constituting the majority of mobile data traffic) the key requirement is data availability at the client's playout buffer – if that is satisfied, the playback is guaranteed to be smooth and free of errors (as the underlying transport protocols deliver an error-free multimedia stream) [4]. Therefore, the QoE descriptor is the target throughput that has to be derived from the state of the application, so that the achieved throughput is not limiting the application's logic to deliver seamless experience to the user. The potential states of such multimedia applications basically consist of an initial pre-buffering state, where the application fetches enough data into its buffer to start the playback; and a playback state, when buffer works as a First-in-first-out (FIFO), depleting data through playback and adding new data via download. The playback and the download rates should be balanced so that the buffer is neither inflating (which would not be efficient) nor completely depleting (which would cause visible playback degradation). The state of the application is observable from the network side by detecting active and inactive download periods: during active periods, new multimedia chunks are fetched; during inactive periods, there is no data transfer as playback is done from the buffer. The balance between the playback rate and download rate is reflected in the pattern of active and inactive periods: as long as they alternate, the throughput achieved by the application is sufficient to keep the buffer loaded. Once the inactive periods start to disappear, it means that the momentary achieved throughput has become too low and that the target throughput needs to be increased. If the inactive periods are too long, the application performs well but it could present the same visible experience having less resources (shorter inactive periods). Therefore, there is headroom in the target throughput for optimization. The headroom may be claimed if other applications transit to a state that requires more resources for them (e.g. as in Figure 10.12 'resource conflict' state) or it may be left at the application if there is no need to alter the network's behaviour (e.g. as in Figure 10.12 'low load' state).

Application sessions of the same type expose similar behaviour as they are implemented along similar principles (or even run the same code). However, the sessions are unique due to the exact content they move around: watching two different videos with the same application can well be described with the same QoE descriptor but require different numerical targets.

For interactive applications (e.g. web page downloads), a good QoE descriptor is the amount of time needed to respond to a user input (e.g. click, scroll) by delivering the data needed to update the application's view. Such time may be quite independent of the network conditions thus qualifies as a good customer experience level policy or intent. How to achieve such time requires quantification of the throughput needed by the application which, in turn, requires constantly monitoring the progress of the download (amount of transmitted data) and the leftover data (still expected to be transmitted). The leftover data may be learned via ML profiling the usual session behaviour. As with the video downloads, increasing the target throughput alone stops improving the visible experience at some point (i.e. stops reducing the time to content) as there are other influencing factors to it (such as the intrinsic latency of the network, transport protocol behaviour, structure of the content, etc.). Unlike video sessions, however, decreasing the bandwidth does not cause sharp QoE degradation (like stalling); the experience is degrading more gracefully (longer experienced download times). Therefore, if the intent is to keep as many sessions with good experience as possible during a resource conflict network state, resources may be borrowed from such sessions to keep less gracefully degrading ones from exposing visible impairments to the end users.

10.3.3 QoE Enforcement

The enforcing of QoE targets on the very short term (in real-time, within one end-to-end RTT) is possible by self-programming packet or flow-level user plane functions and schedulers. Such mechanisms are able to change how packets are serviced immediately if required. Other actions such as traffic steering, load balancing and re-routing may also be triggered via QoE targets, however, their time to act is usually longer. Therefore, the most efficient QoE enforcement instrument lies on the packet scheduler level.

Enforcement means to schedule packets at the proper rate and order that implement the numerical targets derived from QoE descriptors and targets. These numerical targets are on the packet or flow level, such as target throughput, or latency, which are enforceable via a delay sensitive packet scheduler. Between packet level schedulers and QoE targets, however, there is a mismatch in aggregation: QoE targets are defined for applications which are potentially served by multiple flows, whereas schedulers usually operate on packet class level where a class is defined based on logical groups other than the application (such as radio data bearer on connectivity basis, or IP QoS class based on Differentiated Services Code Point (DSCP) aggregate). The gap may be bridged by an additional layer or scheduler, which implements application level traffic separation and resource allocation granularity. This is referred to as the application scheduler. The application scheduler dynamically maintains virtual buffers to separate packets of applications (not flows or classes) and maintains numerical service targets (such as bandwidth or delay urgency) per application buffer. The packets are transferred from the application buffers to lower layers. When the application scheduler is deployed at the RAN, it forwards packets to traditional radio schedulers so that they take care of maximizing radio efficiency on a traffic mix that has already been optimized for QoE. Therefore, the dual layer of schedulers (application and radio) jointly implement a resource allocation scheme that is optimal from both QoE and radio efficiency points of view [5].

The QoE management requires context and insight-based self-programming as changes in the traffic mix and other conditions require updates to lower level scheduling principles. Therefore, the application scheduler interfaces with measurement and analytical functions that supply the necessary context information. Important measurements include throughput, RTT, as well as congestion detection and capacity measurement per resource aggregate (e.g. cell, radio scheduler). Application detection and QoE target definition are also needed to provide the numerical targets for application aggregates within the scheduler. Using the combination of these insights, the service of each application buffer is programmed (continuously updating as the application mix or context changes), achieving standalone operation always providing optimal experience within the limit of the available resources and according to the high-level intents.

10.4 Mobile Backhaul Automation

MBH is the transport network that provides the connectivity between the Radio Access and core network elements for all traffic including user-, control and management-plane. Typically, MBHs are complex and heterogeneous systems (multiple technology layers from multiple vendors) spanning over large areas (whole countries), collecting and aggregating the traffic of a whole mobile network. Accordingly, their CAPEX and OPEX have significant share in the total cost of mobile network ownership. Keeping CAPEX low mandates careful planning and sequential commissioning, whereas OPEX can be reduced through automating the network management and operation. The current practice of manual planning is costly and inaccurate itself, whereas the sequential commissioning leads to suboptimal resource utilization.

The MBH automation concept attempts to introduce cognitive techniques in the planning and operation of MBH. This section describes a framework for MBH automation, that simplifies the planning and configuration process and maximizes the efficiency of the network through enhanced optimization and automation techniques [6, 7].

10.4.1 The Opportunities of MBH Automation

Network planning is typically a process with low accuracy as the resource requirement is calculated based on traffic forecasts (inaccurate by nature) that, at best, are based on historical measurements and simplified network models. Simplifications are needed to keep the numerical complexity and processing requirements at a reasonably low level. The parameter configurations of each network element calculated during the planning are eventually downloaded to the network elements. Due to the inaccurate inputs and simplified models these parameters will most likely not provide optimal system operation under the real traffic load.

The common characteristic of all the alternative/complementary transport solutions is that, in each case, long-lasting transport tunnels are configured between the RAN and core network nodes, whereas QoS schemes are applied at the packet level. These transport tunnels are configured either manually or via provisioning tools (that provide some level of automation) when the network is extended with new network elements or when new services are introduced. During the provisioning process, the transport tunnels are configured on top of the existing ones one by one (as the network is extended) having their path calculated in a locally optimal way. The result is suboptimal on the system level and inefficient compared to the case when all the existing and new transport tunnels over a system are known in advance and the configuration is calculated by considering the whole network, the total demand, the granularity of the allocations vs the granularity of the resources and by considering the multiple technology layers of the transport network (e.g. optical, Ethernet, IP/Multi-Protocol Label Switching [MPLS]). The efficiency gap increases with the size of the transport network and the amount of configured connections, making room for significant CAPEX savings.

Statically configured resource allocations and transport QoS parameters are not optimal as there is no globally suitable node configuration especially because in mobile networks the traffic demand changes dynamically. Due to the complexity of the system and the high number of parameters that must be configured in each case, reconfigurations and re-parameterizations with traditional processes are rarely executed. Efficient operation requires the self-adaptation of the transport parameters to the continuously changing traffic.

As the MBH has a significant role in end-to-end performance, its inefficient operation directly translates into high total cost of ownership (TCO), poor QoS/QoE, reduced system efficiency, waste of resources and energy, frequent degradations, difficult troubleshooting, and negative user experience all of which affects the revenues the operators can realize. Introducing MBH automation is important to reduce OPEX and to increase the resource usage efficiency that reduces the CAPEX. Specifically, MBH automation addresses the following challenges:

1. Increased system complexity: large amount of network elements, complex topologies, multi-technology environment in the MBH that means that the system is difficult to dimension, commission, and manage.
2. Complex and dynamic traffic mix that cannot be handled efficiently with static parameter sets, that is, there is no globally valid configuration that once downloaded to each network element would provide efficient end-to-end network operation. Adaptive and context specific parameterization is needed instead.
3. Inaccurate planning that is based on traffic forecast and simplified network models; even the most accurate traffic forecast becomes obsolete rather quickly. To prevent the planning (not only the configurations but the physical capacity and the resource allocation) becoming obsolete, over-dimensioning is needed.
4. Non-aligned transport and radio QoS. The former is aggregate centric whereas the latter radio service centric. This means that upon transport congestion, the end-to-end QoS targets are not met.
5. Time consuming, failure prone node and network commissioning, operation, and management.
6. The need for rapid delivery of mobile services (coverage, connectivity, capacity, access to the applications).

10.4.2 Architecture of the Automated MBH Management

MBH automation applies the concepts of plug-and-play, self-configuration and self-optimization mechanisms combined with advanced measurements (KPI collection), automated anomaly and degradation detection, failure localization, and cognitive decision making. In this context, Network Function Virtualization (NFV) and Software Defined Networking (SDN) are enabler technologies. The Cognitive Function is a new paradigm that offers the capability that the network or the network elements can profile and learn their own operation, to find their suboptimal states and define their own operating point by adapting to the actual network and traffic context. Each Cognitive Function monitors the impact and the efficiency of its self-optimizing/configuring actions (Figure 10.14); feeding the results of the monitoring back to its decision engine to learn the best action for a given context. The impact is considered high if the action reached its goal (e.g. resolved the degradation); while the efficiency is considered high if the originated action is successfully completed within the specific scope and target. Then, low impact and/or inefficient actions are filtered out by the Cognitive Function for the given context.

Despite targeting full MBH automation, it should be supervised by the network operator (Figure 10.13) by defining the intents, objectives, policies, and rules for

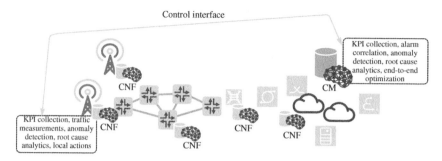

Figure 10.14 MBH automation architecture.

its operation; providing learning feedback by annotating the detected anomalies, providing the ground truth, block, or force actions, etc. as may be needed.

MBH automation applies a hybrid architecture with distributed and centralized elements as illustrated in Figure 10.14. The embedded cognitive network functions (cognitive network functions [CNF], embedded in the network elements/functions such radio nodes, core elements, clouds) enable closed-loop actions when the detected anomaly has only local relevance and the problem can be resolved by re-configuring or optimizing the local parameters. Then, for transport degradations that impact multiple network elements or end-to-end contexts, a centralized decision and action engine called the Cognitive Management Function (CM) is used.

The CNFs maintain end-to-end QoS and optimal transport configuration of the network element to which it is attached, and also completes the commissioning of the RAN nodes by creating and configuring their transport. The CNF executes continuous profiling of the network element to learn its capabilities, identify its status and quantify its performance. It utilized the learned profile to detect anomalies in the operation of the network element that are either due to wrong parameterization or suboptimal transport resource allocation. Since anomalies at a network element can also manifest themselves in the control or user traffic, the CNF monitors and profiles the whole traffic of the radio node. When an anomaly/degradation is detected, the CNF first identifies the corresponding reason, initially separating radio and side problems. It then narrows down the detected transport-related problems to a transport segment by involving the insight collected by other CNFs and the CM, exchanged via in-band or out-of-band (e.g. JSON/REST) interfaces. Finally, if needed (e.g. to resolve the degradation or to maintain the QoS), it initiates reconfigurations and optimizations, which can target either the transport interface, the site router of the node (if such device exists) or the end-to-end transport service. Through its operation, the CNF is a measurement point that extracts real time QoS and QoE KPIs and intercepts/detects relevant events from/within the user, control, and management plane traffic of the radio node. Moreover, it monitors the control plane traffic/messages of the transport network to detect any relevant change in the status of the network/link/path, etc.

The CM is an OSS/Central Management entity that acts as a system level optimization and configuration engine responsible for resolving the degradations that are related to end-to-end transport services or that affect multiple network elements, e.g. congestion on a transport link shared by the traffic of multiple radio nodes. Additionally, the CM analyses the relevant KPIs collected by the CNFs (and delivered via out of band interfaces such as JSON/REST) to predict the associated trends. This allows preventive operation, e.g. to trigger re-configuration before the negative tendencies would cause hard failures. The CNFs trigger the CM whenever a detected anomaly cannot be resolved through local configuration or if it is due to a problem with the transport service (e.g. improper resource allocation) or if it is clearly within the MBH, i.e. it might affect other radio nodes as well. The CM collects these triggers, consolidates them to filter out false positives and finally correlates them to identify incidents reported by distinct CNFs that are caused by the same failure/problem. The CM can operate in real time, implementing a closed control loop, e.g. as an entity that resolves single or transient incidents on the MBH and/or through a long control loop, e.g. as an entity that resolves persistent degradations or prevents failures caused by negative trends. Additionally, it can proactively configure the system to follow the typical traffic profile of the radio nodes and the shifting load between network domains (caused by the mobility routing of the users).

Besides resolving transient or pathological failures, the CM is responsible for maintaining the system efficiency and keeping the system operation and resource usage/allocations at an optimal working point. Accordingly, it can trigger actions by itself whenever it detects inefficient resource usage/allocation or system operation by analysing the collected KPIs. This mandates that the CM is integrated with the existing OSS and customer experience management (CEM) tools and has access to their KPI database. Depending on the environment (the existence of SDN controller, Path Computation Element (PCE), or other transport provisioning tools) the CM either connects directly to some or all of the transport nodes through the existing/standard management and configuration interfaces or acts as a user of the respective transport provisioning tool. If the reach of an SDN controller is limited to a certain domain, or there are several SDN controllers managing distinct network domains, the CM acts as integrator, that is, it triggers reconfiguration or optimization by separately addressing the individual SDN controllers or even directly configuring the network elements out of the reach of the SDN controller(s) to achieve a coherent end-to-end system configuration/status.

10.4.3 MBH Automation Use Cases

Plug and play commissioning: This use case simplifies the transport planning to a great extent and increases the level of automation of the radio node commissioning process. Traditionally, the network element commissioning requires that transport services are pre-planned and pre-configured before the provisioning process. In contrast, the MBH automation framework eliminates the need of pre-planning and pre-configuring the transport connectivity of each and every radio node. The radio node side CNF detects the start of the commissioning process and automatically triggers the transport configuration through the CM. By the time the first user plane connection is established, the transport connectivity is available. In continuation, the transport configuration and resources allocated to the network element are dynamically adapted/optimized to the traffic demand and networking context by the MBH automation during its whole lifetime.

RAN node transport self-configuration and parameter optimization: This is a use case that not only simplifies the planning and configuration process but also enables harmonized end-to-end QoS enforcement and guarantees coherent end-to-end transport configuration. Through the CNFs, MBH automation enables the dynamic and adaptive configuration of the transport parameters that best serve a given traffic mix. The CFs monitor the traffic at a radio node to detect if the target QoS/QoE is not met (e.g. due to the non-harmonized radio and transport configuration) or if the configured resources are insufficient for proper QoS/QoE. Upon such cases, the relevant transport parameters are redefined in a self-configuring process or the CM is triggered for action as previously described. This allows the deployment of network elements without detailed configuration as the solution itself takes care that the best parameters for a given radio node are found.

Transport service self-optimisation: This use case enables automated detection and reconfiguration of end-to-end transport services with suboptimal transport resource allocations. CNFs trigger the CM on detecting degradation due to poor transport service configuration. Subsequently, the CM selects one of the following actions: (i) increase of the bandwidth allocation in case of enough free resources on the existing path; (ii) in case of inadequate transport resources, identify under-utilized transport services that share resources with the service having inadequate bandwidth allocation, and downsize them to allow for more resources towards the problematic transport service; (iii) reroute transport services (while maintaining even load and optimal system resource utilization) in case of inadequate resources on the path of the congested transport service and no possibility to free up enough bandwidth by downsizing other services; (iv) trigger the operator with an alarm message coupled with a recommended configuration in case all the above actions are not possible.

Detecting transport improperly-configured elements: The CNF(s) continuously perform end-to-end measurements based on which transport elements with configurations that degrade the QoS or the system efficiency can be detected. Depending on the set-up, the CM can even identify these elements and correct their configuration.

Traffic trend analysis and preventive optimisation: The CM continuously monitors the traffic to detect trends and identify the potential future resource limitations/narrow points within the system. Then, it updates the system configuration to prevent these incidents. For example, to leverage the gain of the geographical diversity and the daily users commuting routines whenever the topology allows it, the CM can, during the business hours, free-up resources allocated to serve suburban cells and increase the resource allocation for downtown areas. It may then revert the configuration after the business hours are over. Moreover, the CM can delay the reconfiguration in order to not prevent planned network extensions or to operate based on allocation and retention priorities, cost functions, etc.

Maintaining optimal system level configuration and resource allocation: At network deployment, when the system is extended upon the commissioning of radio nodes, the system state can drift from the optimal status as transport resources are provisioned one-by-one to the new network elements. This is because transport services of a newly provisioned network elements are created by considering the actual status, i.e. the already configured services and the resource requirement of the new transport services being established. This approach provides a local optimum but as the number of newly configured radio nodes increases it leads to suboptimal configuration at system level. MBH automation CFs continuously monitor the existing status (resource allocations vs available capacity, the paths of the transport tunnels vs the topology, etc.) and where enough gain can be achieved by a system level optimisation, it calculates the optimal path for each service, creates a step-wise plan for reconfiguration and finally triggers the reconfiguration(s) according to this plan.

SLA monitoring: The MBH automation framework can monitor the SLA when the transport services are provided over leased lines; identify underutilized resources; predict resource limitations; and quantify the required resources.

Measurements, KPIs: The MBH automation framework can act as a measurement mechanism that provides detailed and accurate KPIs, identifying and localizing failures thus improving the network monitoring, management, and troubleshooting capabilities.

10.5 Summary

Network automation is an essential requirement for telecommunication networks to provide dynamic and versatile services over multi-layered complex technology domains, software architectures and large topologies. The automation has a transformational impact on service and network management, operator roles and interfaces, delegation and trust, the design and cooperation of individual network functions, and more. ML is a key technology enabler to introduce cognitive capabilities into the networks, such as self-learning, cognitive decision, reinforcement and verification of actions, enabling novel or better services with cost efficiency and reduced operational overhead. Embedding ML into the network, especially to drive closed-loop automation, has its specific requirement on data collection, context and insight generation, training, validation, transparency of automated decisions, augmentation with domain expert knowledge. Utilizing the capabilities of ML is a promising way to leverage the data available within networks, make data-driven decisions and deliver on the promises of the telecommunication industry.

References

1 Stark, A. (2017). Brake Specific Fuel Consumption (BSFC). https://x-engineer .org/automotive-engineering/internal-combustion-engines/performance/brake-specific-fuel-consumption-bsfc (accessed 28 January 2020).

2 Szilágyi, P. and Vulkán, C. (2015). LTE User Plane Congestion Detection and Analysis. IEEE 26th International Symposium on Personal, Indoor and Mobile Radio Communications (PIMRC), Hong Kong.

3 Radics, N., Szilágyi, P. and Vulkán, C. (2015). Insight Based Dynamic QoE Management in LTE. IEEE 26th International Symposium on Personal, Indoor and Mobile Radio Communications (PIMRC), Hong Kong.

4 Szilágyi, P. and Vulkán, C. (2015). Network side lightweight and scalable YouTube QoE estimation. In: *International Conference on Communication*, 3100–3106. London: ICC.

5 Héder, B., Szilágyi, P., and Vulkán, C. (2016). Dynamic and adaptive QoE management for OTT application sessions in LTE. In: *IEEE 27th Annual International Symposium on Personal, Indoor, and Mobile Radio Communications*. Valencia: PIMRC https://doi.org/10.1109/PIMRC.2016.7794857.

6 Bajzik, L., Deák, C., Kárász, T. et al. (2016). QoE driven SON for mobile backhaul demo. In: *6th International Workshop on Self-Organizing Networks*, 1–2. Valencia: IWSON https://doi.org/10.1109/PIMRC.2016.7794576.

7 Bajzik, L., Kárász, T., Vincze, Z. et al. (2017). SON for mobile backhaul. In: *7th International Workshop on Self-Organizing Networks*. Barcelona: IWSON.

11

System Aspects for Cognitive Autonomous Networks

Stephen S. Mwanje[1], Janne Ali-Tolppa[1] and Ilaria Malanchini[2]

[1] *Nokia Bell Labs, Munich, Germany*
[2] *Nokia Bell Labs, Stuttgart, Germany*

Network Management (NM) is a system of devices and software functions with the respective interfaces to allow for monitoring, control, and configuration of network devices. Network Management Automation (NMA) extends this system by adding automation software functions both within the network devices and in the network control devices. Beyond the individual functionalities for configuration, optimization, and healing (as discussed in Chapters 7–10), there are system-wide challenges that also need to be addressed. The most widely discussed challenge here is self-organizing networks (SON) coordination for which early solutions focused on prioritization of SON functions (SFs) at run-time. Cognitive Network Management (CNM) takes the perspective that such system-wide challenges can be addressed using cognition as the basis for decision making.

This chapter discusses these system-wide challenges in the CNM environment where cognition is the basis for functional development. It seeks to address two core questions: (i) what are the core ideas that can be added to the SON environment to advance it towards more cognition and autonomy? (ii) What are the system-wide challenges that need to be addressed when the functions are themselves cognitive and how should such a cognition system be managed or leveraged? The chapter starts with a baseline discussion that summarizes the SON framework highlighting the two SON layers – the functional and the SON management layers. Thereafter, the two core major parts of the chapter are presented: (i) the transient phase with the advancements to SON and the forward-looking Cognitive Autonomous Networks (CAN) system challenges and solution ideas.

The advancements to SON highlight the value of Augmented automation, where rule-based SON functions are complimented with management-layer

Towards Cognitive Autonomous Networks: Network Management Automation for 5G and Beyond,
First Edition. Edited by Stephen S. Mwanje and Christian Mannweiler.
© 2021 John Wiley & Sons Ltd. Published 2021 by John Wiley & Sons Ltd.

cognitive capability. The discussion focusses on the verification of actions and its interaction with SON coordination. Then, the forward-looking discussion presents a new framework for the CNM/CAN system with a focus on cognition as the baseline for decision making even in the individual functions – hereinafter called Cognitive Functions (CFs). Subsequent sections discuss solution ideas to the specific challenges in the CNM/CAN scenario – specifically, the abstraction and learning of network states, Multi-agent coordination of non-deterministic heterogeneous agents through synchronized cooperative learning (SCL) and the coordination of multi-vendor functions over thin standardization interfaces.

11.1 The SON Network Management Automation System

SON, as the first step towards automated mobile networks operations, administration, and maintenance (OAM), focused on addressing specific OAM use cases, e.g. Mobility Load Balancing (MLB), Mobility Robustness Optimization (MRO), or Automatic Neighbour Relations (ANR) [1]. SON seeks to improve overall network performance while minimizing operational expenditures by reducing human network operations [1]. This section characterizes the SON NMA paradigm by describing the framework within which SON functions were developed and mechanisms for decision making.

11.1.1 SON Framework for Network Management Automation

In the SON paradigm, each automation use case is accomplished by a specific SON Function (SF) which ensures scalability since the whole SON system does not need to be implemented at once. To achieve the automation objectives, the SON system, shown in Figure 11.1, was conceptualized to have two layers – the algorithms layer which implements the individual SON Functions and the SON operation layer which implements system-wide functions.

The SON framework, shown in Figure 11.1, operates on input characterized by a defined set of Key Performance Indicators (KPIs) which 'model' the operating point, i.e. the static and dynamic characteristics of the network environment. The 'ideal' operating point depends partly on the network characteristics (like architecture, network function properties and configuration, current load, etc.), and partly on the requirements and properties of the services and applications to be supported. This points to the first SON limitation: by always using the same KPI set, SFs have a static and restricted view of the environment. Consequently, their performance is limited by the degree to which KPIs accurately measure, represent, and abstract/model the environment. And although the SF may also

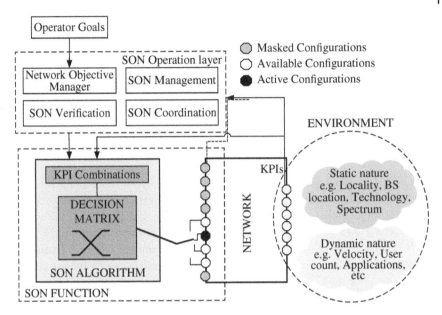

Figure 11.1 The SON framework.

include the active Network Configuration parameter (NCP) values as input to the decision-making process, it always has only one active NCP-value set (chosen from a limited set of possible NCP value sets). The ultimate solution requires that a much wider set of available information is used to evaluate the network state for each decision that is taken. This, however, complicates the design of both the SON Functions and their operations layer. New ideas on how multi-agent systems (MASs) approaches could be used for implementing network automation functions and systems need to be considered. This is the subject of the generic discussion in Section 11.2.

11.1.2 SON as Closed-Loop Control

Each SF, as a closed-loop control algorithm for the specific use case(s), acquires data from the network elements, typically KPI values, raw counters, timers, or alarms. With the data, it autonomically determines or computes new NCP values according to a set of algorithm-internal static rules or policies. The algorithm can be seen, thereby, as a decision matrix (see Figure 11.1) that matches inputs (combinations of input data) to outputs (NCP values), i.e. the SF is a state machine that derives its output (the NCP values) from a combination of inputs and function-internal states. This decision matrix enforces a fixed behaviour in that its input-output relationship, or the path thereto is predesigned into the solution through the rules (states and state transitions) of the algorithm.

As networks become even more complex, e.g. with the addition of 5G, the automation functions need to be more flexible which has led to the push for Cognitive Functions which apply the cognitive technologies presented in Chapter 6 to achieve better outcomes. Correspondingly the complete framework will need to be revised as is presented in Section 11.5.

11.1.3 SON Operation – The Rule-Based Multi-Agent Control

The SON operation layer provides extra functionality beyond that in the SON functions that ensures the entire SON system delivers the expected objectives.

Firstly, for a conflict-free joint operation of multiple, independent SF instances, concepts for SON coordination and SON management were introduced [27, 28]. SON coordination [2–4] is the run-time detection and resolution of conflicts between SF instances, e.g. if two instances simultaneously modify the same NCP, or one instance modifies an NCP that influences (thus corrupts) measurements used by another instance. SON management enables the operator to define performance objectives as target KPI values or ranges [3, 4], while verification may be added to ensure that set targets are always met for all SFs [5]. The objectives, combining the KPI's relative importance and the cell's context, such as location, time of day, cell type, etc. [3], enable SON management to influence the behaviour of a given SF by modifying its SF Configuration Parameters (SCPs). Accordingly, different sets of SCP values lead to a different KPI-to-NCP value mapping, i.e. a different decision matrix for each SCP value set. These matrices must, however, be pre-generated by the SF vendor prior to deployment in a production environment, e.g. through simulations.

Like SFs, both SON coordination and SON management rely on operator or SON-vendor defined fixed rules and policies leading to another SON limitation: while minor modifications to the network environment, context, or objectives can be autonomously handled, the underlying algorithms (i.e. state machines and transitions) remain unchanged. This hinders adaptation to major changes of cell density, network technology, architecture, context definitions, or to newly defined operator business and service models. As such, the revised approach to automation to leverage cognitive technologies requires a revision as well to the function coordination mechanism as is discussed in Section 11.6.

Beyond SON coordination and SON management, other studies have proposed the need for and the design of SON verification solutions whose focus is to ensure that conflicts amongst SON Functions that cannot be resolved by pre-action coordination can still be resolved post-action or that the effects thereof can at least be minimized. Although verification implementation can also be rule based, it has been shown that more complex solutions are possible especially in computing the extents of the different effects on the network. Details of these will be discussed in Section 11.3.

11.2 NMA Systems as Multi-Agent Systems

Extending the definition in [6], an MAS may be defined as a collection of agents in a common environment with the agents co-operating or competing to fulfil common or individual goals. It is evident then, that NMA systems are MASs in which the individual automation functions (the SON Function in SON or the Cognitive Functions in CANs) are the agents. However, the agents may also be the instances of the automation functions be it in the cells or the OAM domains.

There is a large body of knowledge on the development of MASs, e.g. agent models, coordination, data collection, interaction amongst agents and system architecture. The biggest challenge, however, always remains the coordination and control of these agents. The solution thereof are the four options presented [7] and illustrated by Figure 11.2: (i) Single-Agent System (SAS) decomposition or simply Separation, (ii) Single coordinator or Team learning, (iii) Team modelling and (iv) Concurrent games. Correspondingly for NMA, these coordination and control mechanisms for MASs provide the alternatives for architecting the system. Their relative merits and demerits for NMA are summarized here.

11.2.1 Single-Agent System (SAS) Decomposition

Where the interactions amongst agents are not strong, the MAS problem may be decomposed into separate SAS problems as shown in Figure 11.2a. This is done

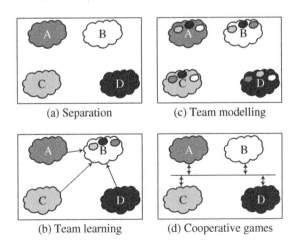

(a) Separation   (c) Team modelling

(b) Team learning   (d) Cooperative games

Figure 11.2 Multi-agent coordination.

with the assumption that the optimum solution can still be found despite the interactions or that the suboptimal solution is also appropriate for the application. Correspondingly, functions are scheduled to be independent and with no effects on each other's learning. Owing to the simplicity of dealing with separate SAS problems, many state-of-the-art SON coordination solutions have applied this kind of approach.

11.2.2 Single Coordinator or Multi-Agent Team Learning

Here, a single agent, called a coordinator or a team learner in learning problems, decides the behaviour for a team of agents including triggering activity of the individual functions and managing the effects amongst those functions. The coordinator agent (e.g. agent B in Figure 11.2b) decides when and which agents can take actions; the effects of these actions and the appropriate responses to such actions. Thereby, the coordinator requires the behavioural models of all team members (the dark bubbles in Figure 11.2b) which it uses for coordination.

The learner may be homogeneous, in that it learns a single agent's behaviour for all the agents in the team which can easily offer better performance with low complexity even if the agents may have different capabilities. It is, however, only applicable if the heterogeneous solution-space is not feasible to explore, so the search space is drastically reduced by homogeneity. On the other hand, a heterogeneous team learner allows for agent specialization by learning different behaviours for different members of the team. Examples of the two forms of learning are respectively a SON Function that learns a single behaviour for all cells in a network vs one that learns different behaviour for different cells. However, Hybrid Team Learning is also possible, in which case, the team is divided into multiple squads, with each agent belonging to only one squad and each squad taken as an agent within the team. Then, behaviour will be similar amongst agents in a squad and different amongst squads, maximizing the benefits of both homogeneous and heterogeneous team learning squads, i.e. simplicity that achieves specialized characteristics.

Team learning has the merits that the single coordinator can utilize the better understood SAS techniques with good convergence and stability characteristics and that it tries to improve the performance of the entire team and not only for a single agent. However, team learning suffers scalability challenges: (i) If the agents are not all implemented at once, as is the case with SON, the coordinator will have to be revised and/or reimplemented each time a new agent is added; (ii) it may not be feasible to maintain the state-value function as the number of agents increase, or at the least, the learning process is significantly dumped – its centralized nature implies collecting information from multiple sources which also increases the signalling rate.

11.2.3 Team Modelling

Here, each agent focuses on optimizing its objective but models the behaviour of its peers to account for their actions (Figure 11.2c). Using the models (the dark bubbles in Figure 11.2c), the agent evaluates its actions and determines the effects that such actions would have on the peers. It then predicts how peers are also likely to behave in response to its actions. The agents could be competitive or cooperative. The competitive agent focuses only on maximizing its objective with the expectation that the other agents are doing the same for their respective objectives. A cooperative agent, however, tries to select actions that concurrently maximize its benefits and, if possible, also maximizes the other agents' benefits.

Using peer modelling in network automation would also suffer scalability challenges since models in all SFs must be updated each time a new SF is added to the system. Besides, such models are very complex owing to the complexity of the individual SON functions. Consequently, the modelling processes would make each SF very complex, even as complex as the heterogeneous team learner.

11.2.4 Concurrent Games/Concurrent Learning

In concurrent games, multiple learners try to partly solve the MAS problem, especially where some decomposition is possible and where each sub-problem can, to some degree, be independently solved. Concurrent games project the large team-wide search space onto smaller separate search spaces thereby reducing computational complexity of the individual agents. However, learning is more difficult because concurrently interacting with the environment makes it non-stationary, i.e. each change by one agent can make the assumptions of other learning agents obsolete, ruining their learned behaviour.

Concurrent learning may be categorized as cooperative games, competitive games or a mixture of the two. Fully cooperative games utilize global reward to divide the reinforcement equally amongst all the learning agents with the same goal of maximizing the common utility. Where no single utility exists, either a further coordination structure is required to decompose observed rewards into the different utilities or cooperation must be enforced through the sharing of information during the optimization process as shown in Figure 11.2d. This exchange of information results in what are called Concurrent Cooperative Games, where the agents compete for the shared parameter or metric but are willing to cooperate on what the best compromise value should be.

Competitive games are winner-takes-all games where agents compete in a way that one agent's reward is a penalty for the other agents. Such games, which encourage agents to learn how to avoid losing strategies against other agents, are inapplicable in networks where all automation functions to must 'win'. Instead, mixed games, which are neither fully cooperative nor fully competitive, may be the applicable ones but accordingly, the applicable degree of competition remains an open challenge.

11.3 Post-Action Verification of Automation Functions Effects

The SON coordination challenge has been clearly justified and studied, i.e. to coordinate the actions of multiple network automation functions and/or instances and ensure system-level operational goals are achieved. The individual automation functions may, however, have undesired and unexpected side-effects that cannot be resolved by the pre-action coordination mechanisms which only resolve potential conflicts that are known *a priori*. To detect and rectify such issues, the concept of automated post-action SON verification has been developed.

The SON verification function [8–11] monitors the relevant KPIs after changes have been introduced in the network, runs anomaly detection algorithms on them to detect degradations and, based on the outcome, decides if corrective actions in the form of rollback of the changes need to be applied. The function can, in principle, be applied to different network functions, but for a demonstration of its usage, this section only uses verification in the Radio Access Network (RAN).

In general, the verification process is triggered by a Configuration Management (CM) change as proposed either by a SON function or a human operator. The subsequent process, as shown in Figure 11.3, includes five steps [8, 10]:

1) Scope generation, which determines, which network functions may be impacted by the change.
2) The assessment interval, the verification process monitors the network performance. The length of the assessment interval can depend on the type of change that triggered the process and other factors that influence how long it takes before the impact of the changes can be observed, and statistically significant data collected.

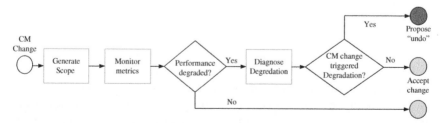

Figure 11.3 An overview of the SON verification process.

3) Based on the observations, a detection step determines if the changes can be accepted or if there is a degradation for which further action may be required.
4) In the case of a degradation, a diagnosis step is triggered to establish if the degradation is the result of the monitored CM changes,
5) Finally, if the diagnosis matches the degradation to the CM changes, corrective actions are undertaken, most commonly an undo of the configuration changes that led to the degradation.

The subsequent sub-sections will discuss the different steps of the process in detail. Note that besides verifying the actions of SON functions, verification functionality can also be applicable in other network use cases, amongst them in Network acceptance [2] and service level agreement (SLA) verification. In this case, network acceptance uses fixed performance thresholds, fixed scope (area around new network element) and simple action (either alarm the network operator or not) while the period of the verification is bound to the acceptance period and deadline. SLA verification can be applied in the same way as in network acceptance, but dependent on the SLA definition, some profiling may be required in the continuous verification process.

11.3.1 Scope Generation

In the verification scope generation, the verification process analyses which network functions or elements are affected by the CM change that triggered the verification process, i.e. the so-called *verification area*. For example, when antenna tilts are optimized by the Coverage and Capacity Optimization (CCO) function, the verification area could be the reconfigured cell, the so-called *target cell*, and all its geographical neighbours either to first or to second degree (to account for possible overshooting), called the *target extension set* [8, 12]. In RAN, it is therefore typical to include all the reconfigured network elements and the first-level neighbours (in terms of handovers) although, ideally, the area should be determined based on the particular CM change and should include other factors, like the network function containment hierarchy or the service-a function chain.

Another decision is the time required to monitor the network performance to collect a statistically relevant amount of data and reliably assess the impact of the performed change. This monitoring period is called the *observation window* and its length may also depend on the kind of change that triggered the verification process and, for example, when the change was made [12]. For certain changes, the impact can be observed rather quickly, but for others, several days of data are required to observe the impact of the change in different network traffic conditions. Correspondingly, as is discussed in later subsections, there can be overlapping verification operations with overlapping verification areas and observation windows that require coordination.

11.3.2 Performance Assessment

In the assessment interval, the verification function needs to monitor the performance of the verification scope and compare it to the performance before the change. Since the decision to either accept or reject the CM change will be based on the comparison, the feature set selected for the monitoring phase should be such that the decision can actually be made [8, 10, 12].

Typically, verification is done by monitoring the Performance Management (PM), Key Performance Indicators (KPIs) and the Fault Management (FM) alarms. For some KPIs, e.g. the failure KPIs like dropped call ratios, the operator's policies may already define fixed acceptable thresholds e.g. for the minimum or maximum KPI value. For such KPIs, it may be enough to verify that the KPI value is within this acceptable range as defined in the policies. In general, however, it is desirable to avoid such fixed, manually defined acceptance thresholds because for many KPIs, the acceptable values may depend on the verified network function instance and so a global threshold would not work well. Rather, it is best to learn how the network function typically performs and define acceptable changes in comparison to that typical performance. In general, this typical performance, called the KPI profile, defines a statistical description of the normal variation of the KPI.

The profiles, which may be of different types depending on KPI, form the basis against which the KPI is compared during performance monitoring. For the performance assessment, the KPI levels need to be normalized in a way that the quality or goodness of the change can be evaluated. Specifically, the normalization process must also take the KPI type into account, e.g. to capture the facts that success indicators are unacceptable if too low; failure indicators are unacceptable when too high and that neutral indicators have specific low and high threshold values [10].

As illustrated by Figure 11.4, the profiles need to be created against a specified *context*, like time and/or another KPI-like load. For dynamic KPIs that exhibit a seasonal variation, e.g. especially the traffic-dependent KPIs, the KPI values need to be normalized against the normal daily and seasonal patterns. Then, any observed changes will be more likely to be due to the configuration change and not part of the normal fluctuation of the KPI. Time dependence may also consider a different context for each period, e.g. for each hour of day and perhaps separately for weekdays, weekends, and public holidays [10].

Using other KPIs as context implies profiling and monitoring the correlation between the monitored KPI and the context KPI. The dynamic profile will then track the profile characteristics (e.g. the expected minimum, maximum, mean ...) against the different contexts as shown in the inset of Figure 11.4. For example, considering the Call Drop Rate (CDR) against load as context, the profile could state that: the target range of the CDR should be (1%, 3%) in low-load scenarios, but

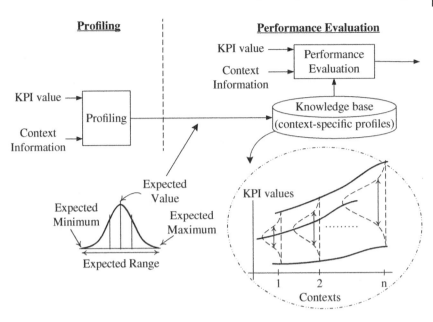

Figure 11.4 Generation and use of dynamic (context-dependent) KPI profiles.

that as the load increases, the two thresholds gradually increase to some maxima (e.g. [4%, 8%] respectively).

11.3.3 Degradation Detection, Scoring and Diagnosis

For the verification decision, the normalized KPI levels are typically aggregated to higher level performance indicators, to which the verification thresholds are applied. First the KPI-level anomaly profiles are calculated and then aggregated on cell-level, to give a cell-level verification performance indicator. These can be further aggregated to give a similar measure for the whole verification area [10].

Similarly, instead of simply defining one so-called 'super KPI' to represent the performance of the verification area, indicators may be aggregated for different performance measures, such as availability, accessibility, retainability, quality of service, mobility, etc. These may also be combined with a set of rules of what is acceptable performance level or performance change. The accuracy and reliability of the detection can also be improved by: (i) applying a hysteresis function instead of a single value and (ii) applying a Time-to-trigger, in which case, the detector raises/seizes an alarm only if the Super-KPI value is above/below the detection threshold for a certain time equivalent to the specified time-to-trigger value.

Together with the change in the performance, the verification decision also needs to consider the context, especially the performance of the cell before the

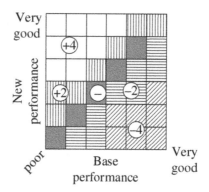

Figure 11.5 Verification assessment scoring function.

CM change. For example, different levels of freedom should be accorded to a CM change intended to optimize a stable, functioning network vs the CM change to repair an already degraded cell – more freedom should be accorded when the verification area is already in an unstable or a degraded state [10, 12].

Figure 11.5 shows an example SON verification scoring function, which accounts for both the assessment score from the verification degradation process (the absolute change in performance) as well as the change in the cell's performance compared to other similar cells. For well-performing cells, only changes that clearly improve the performance are accepted, whereas for worse performing cells more flexibility may be given in the grey and yellow zones [10].

Beyond simply detecting the degradation, it is necessary to diagnose that the degradation was really caused by the CM change. For example, it may be that the reconfiguration was done to prepare for some changes in the network function's environment and that, without the change, the degradation would have been even worse. So, undoing the changes would make the situation only worse.

As discussed in Chapter 9, diagnosis is a complicated problem and it is often not possible to have a reliable diagnosis. For this reason, the verification function tries to minimize the impact of external changes in the as-above-described high-level KPIs that are used in the verification decision. The simple assumption is often that after this process, all the observed changes in the performance (by comparing performance before and after the CM change) are caused by the re-configuration. This assumption may potentially lead to false positive rollback decisions, but the deployment of the undo operations should be done in as non-intrusive a way as possible. However, if one plans to learn from the verification decisions and to block rejected configuration changes, the impact of a false negative verification decision becomes more significant.

The diagnosis process can be improved by incorporating other known facts into the decision, e.g. the relevant (severe) alarms and the Cell status information (e.g. administrative lock). This allows the differentiation between causes and thus to take different decisions besides just 'undo' for each cause, e.g. to simply do nothing. However, further advanced diagnosis methods, such as those described in Chapters 9 and 10 may also be incorporated into the SON verification process.

11.3.4 Deploying Corrective Actions – The Deployment Plan

When the SON verification function has detected a degradation and has determined that it has been likely caused by the configuration change, the next step is to decide and create a plan of action regarding how the degradation could be corrected. In a simple scenario, the verification function can simply trigger a CM undo to roll the changes back and then, considering the rollback as another CM change, it re-executes the verification process for that change. The re-verification is necessary because the rollback may degrade the performance more than the initial CM change, for example, in case the environment of the verified network function has changed, in which case, it may be better to re-introduce the initial CM changes [5, 8, 9].

Automation functions may require multiple cycles to reach their optimization goals, each cycle requiring long observation windows to monitor the impacts of each step. So, several function instances may run in parallel, each optimizing the network according to its objectives. Correspondingly, several verification operations may run in parallel, possibly with overlapping *verification areas* and *observation windows* which results in *verification collisions* as depicted in Figure 11.6 [13, 14].

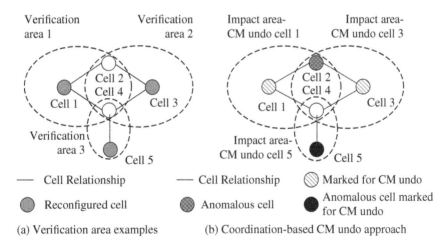

Figure 11.6 An example of a verification collision.

In a verification collision, if a degradation is detected in a cell that is included in more than one such verification areas, it is often not possible for the verification function to determine which CM change led to the degradation. For example, consider the five cells in Figure 11.6 where three cells (cells 1, 3, and 5) have been reconfigured and two (cells 2 and 5) have degraded after the CM changes. Configuring the verification areas to include the reconfigured target cell and its neighbour cells, leads to the verification areas labelled 1, 3, and 5 (respective to the target cells). However, this knowledge is inadequate to determine which change led to the degradation in cells 2 and 5.

An appropriate rollback mechanism is required for which two options are possible – an aggressive or a sequential mechanism, as described here.

1) **Aggressive rollback**: The aggressive rollback approach would be to undo the CM changes in all three re-configured cells 1, 3, and 5. Reasoning for such an approach could be that it would be the fastest and most certain way to return the network to a previously stable non-degraded state. However, it is often not that simple. A rollback is a further change in the network and not without a risk of its own. There is always the risk that the rollback may make things even worse. The more changes undertaken at once, in this case undo, the higher the chance. Furthermore, as with any changes, the more parallel changes there are, the harder it is to diagnose the one that caused the degradation if one occurs. Also, since there can be many overlapping verification areas, overlapping only partially in verification areas and observation windows (time), it may be that the combined undo scope becomes very wide. As such, it is critical to undo only the changes that are most likely to have caused the degradation.

2) **Sequential deployment of undo actions**: The other extreme would be to deploy all undo operations sequentially, i.e. one by one. However, this would also be very inefficient in a network, where there are frequent CM changes. The verification function may not be able to keep up and would become a bottleneck.

It is rarely the case that all verification areas are at verification collision with each other simultaneously. Therefore, it is possible to form an undo operation deployment plan, where only verification areas that are not in collision together are rolled back simultaneously. Figure 11.7 depicts the process of forming the undo operation deployment plan. The example shows a network of 12 cells, with their corresponding adjacency graph, in which cells 4, 7, and 10 are reconfigured. In each case, each of the verification areas v1, v2, and v3 consists of the reconfigured cell and its first-degree neighbours. A verification collision graph is constructed with the edges connecting the colliding verification areas, in this case only areas v1 and v2.

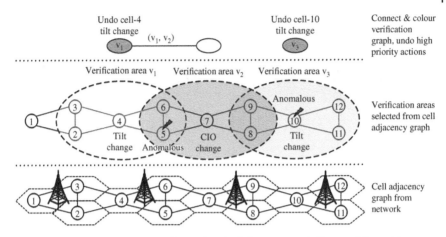

Figure 11.7 An example of the CM undo scheduling approach. Source: Adapted from [13].

A Graph-colouring algorithm is applied to the verification collision graph, where each node of the graph is assigned a colour in such a way that no connected nodes share the same colour. The undo deployment plan is then formed so that only verification areas of a certain colour are deployed simultaneously and thus avoiding simultaneous undo actions amongst colliding verification areas [13].

The undo operations can be prioritized, for example, so that the colour containing the verification areas with the highest total number of degraded cells are always rolled back first. On using this rule for the example in Figure 11.7, verification areas v1 and v3 would be undone first. Then, v2 is rolled back in the subsequent correction window, but only in case the undo of the change in cell 4 has not already corrected the degradation [13].

11.3.5 Resolving False Verification Collisions

Deploying an undo operation can be a time-consuming task since the impact of the undo also needs to be verified. In a network with lots of reconfigurations there might be insufficient correction timeslots available for deploying the corrective actions even utilizing the graph-colouring approach described in the previous subsection. Therefore, it is desirable to detect the verification collisions that could be resolved before deploying the corrective actions [14].

Figure 11.8 shows a *false verification collision* in a network of 10 cells. Cells 1, 2, and 3 are reconfigured leading to the depicted verification areas, as identified by the respective target cell. Now, consider that cells 5, 8, and 9 are degraded and that the verification mechanism is unaware that cell 2 is not responsible

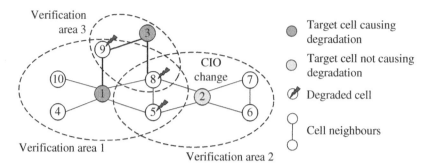

Figure 11.8 Example of a verification collision [14].

for any of the degradation. Since the degraded cell number 8 is included in all three verification areas, three correction deployment slots are required for the undo operation. However, since cell 2 isn't causing any degradation, the collision between verification area 2 and verification areas 1 and 3 are *false verification collisions*.

One approach for resolving such false verification collisions is by employing a behavioural graph, which indicates the degree of similarity amongst the performance in several cells. To demonstrate its usage, consider the example shown in Figure 11.9 with two KPIs a_1 and a_2 as the features to be used in the verification process. With the anomaly level of the features as the dimensions of the graph (see step 4), each cell is placed on the graph according to its anomaly levels for the respective features. For simplicity, some cells are completely overlapping in this example as indicated by the numbers in the graph vertices (e.g. cells 5, 9, and 12 at vertex $V_{5,9,12}$).

A fully connected graph is constructed amongst all the cells, with the weight of each edge as the Euclidian distance between the cells on the feature map. Then, removing a configured number of longest edges or edges longer than a set maximum edge-weight, the behaviour graph is transformed into an anomaly graph that clusters similarly performing cells together.

The result that cells belonging to a certain cluster exhibit similar anomalous behaviour in the verification process can be utilized to try to detect false verification collisions. Collisions between *weak verification areas* are eliminated by removing those extended-target-set cells from the verification area, which do not belong to the same cluster with the target cell of that verification area. For example, removing the edge ($V_{3,7} - V_{6,11}$) indicates that cells 6 and 7 do not need to be in the same verification area, so cell 7 is removed from verification area V_6 creating the smaller 'weak verification area 6' shown in step 5. This process eliminates the weak collisions between the verification areas allowing multiple simultaneous undo actions to be deployed.

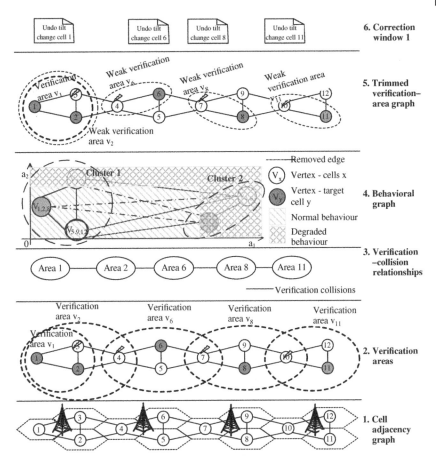

Figure 11.9 An approach for detecting false verification collisions [14].

In the example, the process results in a requirement for only two correction windows for the four verification collisions, i.e. only verification areas 1 and 2 in Figure 11.9 overlap necessitating separate two windows. The other areas can be deployed concurrently to verification area 1 in correction window 1.

The combination of the above processes ensures that network automation functions can be supervised to ensure that only their positive influences are accepted to the network and any negative influences addressed through these undo operations. The verification concepts, although developed for the SON framework, expect to be usable even when the functions are cognitive since they provide means of redress or at least a feedback mechanism to the cognitive function to re-evaluate their actions and minimize negative influences. The next section shows one such usage of the verification concept.

11.4   Optimistic Concurrency Control Using Verification

A major requirement for NMA systems is concurrency control, i.e. ensuring that network performance is not degraded because of multiple automation functions acting concurrently on the network. The simplest response in the SON paradigm has been to apply SON coordination mechanisms as the solution. Therein, the safest SON coordination scheme executes only one SON function instance at a time in the whole managed network, but this would be very inefficient. Another approach is to only allow SON functions with non-overlapping impact areas and times to run at the same time. Although more efficient than network-wide serialization, this approach also restricts the number of active SON function instances significantly, especially due to the long impact times of many SON functions. The execution of one function can last for several granularity periods (GPs) (the smallest periods of data collection) and the result can be the same function requesting to execute again. On the other hand, it is imperative that conflicts between SON function instances can be avoided. A combination of pre- and post-action coordination, namely combining SON coordination and SON verification, can be used to optimize the coordination performance and to implement an optimistic concurrency control (OCC) strategy [9].

11.4.1   Optimistic Concurrency Control in Distributed Systems

OCC assumes that multiple transactions can often complete without interfering with each other [15], so running transactions use data resources without locking the resources. Before committing, each transaction verifies that no other transaction has modified the data it has read. In case of conflicting modifications, the committing transaction rolls back and can be restarted. In systems, where the data contention is low, this offers better performance, since managing locking mechanisms is not needed and excessive serialization can be avoided [16].

In performance-critical distributed applications, data is additionally often processed in batches, to avoid the performance penalties of numerous remote procedure calls. When OCC is used, parallel batch operations are not synchronized, and this can lead to race conditions between some data elements in the batches. Database constraints, for example, can be used to check the consistency of the stored data and to catch such conflicts. In case of a constraint violation, the transaction for the whole batch is rolled back. The operation is then retried by the application with stricter concurrency control and possibly one by one for each of the batch elements. Using this method, performance remains good, when most of the batch write operations are successful and conflicts are rare but, at the same time, it ensures that one invalid element does not prevent processing the whole batch.

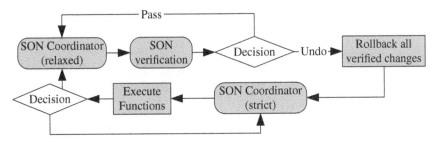

Figure 11.10 Optimistic concurrency control using SON coordinator and verification.

11.4.2 Optimistic Concurrency Control in SON Coordination

Figure 11.10 highlights how OCC can be implemented with the SON coordination and verification concepts. The basic idea is that the SON coordinator can allow for more parallel execution of SON function instances, if the result of the optimization actions taken by the functions is verified by the SON verification function. SON verification will ensure that any possible degradations that are a result of the conflicts are quickly resolved. This can be further optimized by changing the SON coordination policy based on the verification results. In case verification detects a degradation, the SON coordinator switches to a stricter coordination policy *in the specific verification area*. It can, for example, completely serialize the SON function instance execution until all functions have run at least once, after which the original more parallelized SON coordination scheme can be continued [16].

A more complicated organization of the Functions could also be considered when SON verification is combined to the coordination mechanism. For example, the conflicting SON function instances may from the SON-coordination perspective be placed into two categories: the hard conflicts, which must never be run in parallel and the soft conflicts, i.e. coupled function instances, which can only be run in parallel if the outcome is verified by SON verification. Correspondingly, the coupled instances can mostly be run simultaneously with the expectation of some infrequent race conditions.

11.4.3 Extending the Coordination Transaction with Verification

To enable this opportunistic concurrency control mechanism, the coordination transactions need to be extended with SON verification as depicted in Figure 11.11. When a SON function instance wants to optimize certain Network parameters, it sends an execution request to the SON coordinator which then initiates a new coordination transaction. Through this transaction, the coordinator coordinates the CM changes in the transaction area which includes all the Network Functions that are within the impact area of the SON function instance [16].

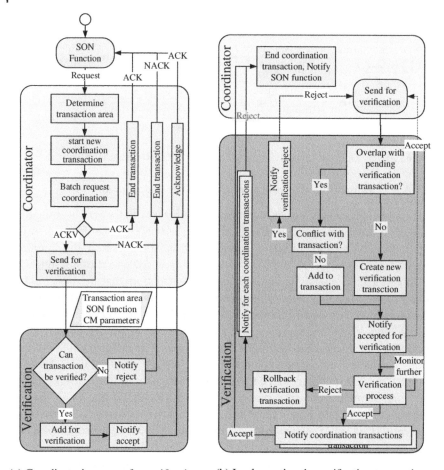

(a) Coordinator's request for verification   (b) Implementing the verification transaction

Figure 11.11 The extended SON coordination transaction implementing OCC.

In [1, 17, 18], the coordinator had only two decisions – either **acknowledge** (**ACK**) or **reject** (**NACK**). Extending these concepts, a third decision is now added. With an **ACK** decision, the network automation function instance is allowed to provision its CM changes in the network and the transaction ends (commit). Here, the coordinator does not require the changes to be verified, but they may be independently verified depending on the system configuration and the operator policies. For the **NACK** decision on the other hand, the SON function instance is not allowed to provision its CM changes in the network and the transaction ends. Opportunistic concurrency control introduces the **Acknowledge with Verification** (**ACKV**) decision, where the SON function instance is allowed to provision its CM changes in the network, but the coordinator keeps the transaction open and marks it for verification. The ACKV decision is signalled to the verification with at least three information elements, i.e.: (i) Transaction area, (ii) the originating SON function instance, and (iii) the updated CM parameters (see Figure 11.11a).

Verification wraps the coordination transaction in a higher-level verification transaction, which in the case of parallel verification requests with overlapping transaction areas may contain several coordination transactions. If the verification function is not able to verify the changes, for example, due to conflicting ongoing verification operation, it will reject the request for verification and the transaction must be rejected. Alternatively, the SON coordinator can execute batch coordination for the transactions, to check if some of the requests can still be acknowledged.

If the transaction area of the coordination transaction overlaps with another ongoing verification operation, the verification mechanism must decide, if the coordination transaction can be added to the existing verification transaction. Otherwise, it must be rejected, and the coordinator must be notified about the rejection.

As in [9], at each granularity period (GP), the verification monitors the KPIs and decides for each verification transaction either to 'pass', 'undo', or 'continue monitoring'. A **'Pass'** implies that all coordination transactions with their CM changes contained in the verification transaction are acknowledged and closed. An **'Undo'** implies that all the coordination transactions contained in the verification transaction are rejected, their CM changes are rolled back, and the transactions are closed. Finally, a **'Continue monitoring'** decision implies that the SON verification mechanism will monitor the verification area performance for at least one more GP and that the verification transaction and all contained coordination transactions remain open during this period.

The open challenge then is how to group requests into either ACK, NACK, or ACKV. This can be statically configured, i.e. according to the network automation function models provided by the function vendor. The coordinator may, for example, statically decide that conflicts between specified SON functions would not lead to acknowledging one and rejecting the other, but both would be acknowledged with verification. However, such a static approach can lead to situations where the function instances get caught in a conflict causing degradation, and a rollback by the SON verification mechanism, only to restart the same cycle from beginning.

The principles of OCC avoid rollback loops through an opportunistic coordinator. As before, the coordinator acknowledges requests with verification whenever possible instead of rejecting them. However, if verification rejects a coordination transaction, the coordinator switches into strict concurrency control in the specific transaction area, i.e. allowing only one function instance to run at a time or only function instances that are known not to conflict with each other, as in [18, 19]. Strict concurrency control continues until all automation function instances have run at least once or until a pre-configured time threshold is reached, after which, the coordinator reverts to relaxed concurrency control. This allows the function instances to reach their targets without conflicts from race conditions.

11.5 A Framework for Cognitive Automation in Networks

As has been discussed in Section 11.1, SON is unable to autonomously adapt to complex and volatile network environments, e.g. due to frequently changing operating points resulting from high cell-density as in UDNs, virtualization and network slicing in 5G RANs [20], or from frequently changing services/ applications requirements and characteristics. The solution to this challenge is the use of Cognition in Network Management, i.e. the use of Cognitive Functions as the intelligent OAM functions that can automatically modify their state machines through learning algorithms [21]. Correspondingly, in the CAN paradigm, these complex multi-RAT, multi-layer, multi-service networks shall remain operable with high (cost) efficiency of network management and with considerably lower necessity for manual OAM tasks.

11.5.1 Leveraging CFs in the Functional Decomposition of CAN Systems

The CAN paradigm advances the use of cognition in networks to: (i) infer environment states instead of just reading KPIs and (ii) allow for adaptive selection and changes of NCPs depending on previous actions and operator goals. The general idea, for example, in [21, 22], is that CFs will: (i) use data analytics and unsupervised learning techniques to abstract, contextualize, and learn their environment, and then, (ii) use Reinforcement Learning (RL) techniques like Q-learning to either learn the effects of their actions within the specific defined or learned contexts [21, 22] or simply to learn how to act in such environmental contexts. The design of the CAN system needs to take advantage of the CFs' capabilities not only to account for the behaviour of each CF both individually and collaboratively besides other CFs but also to allow for flexible deployments – be in centralized, distributed, or hybrid scenarios.

The proposed blueprint for the CAN system decomposes the system into smaller inter-related functions that leverage cognition at each step/function. The design leverages 'active testing' benefits that are inherent in machine learning (ML), i.e. knowledge build-up requires CFs to execute unknown configurations and evaluate

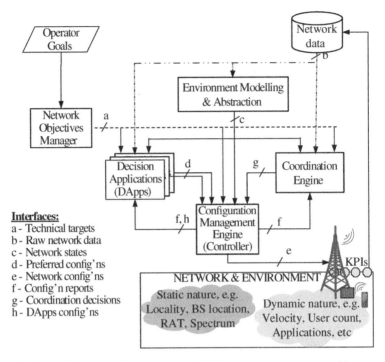

Figure 11.12 CAN framework – Functions of CAN system and related cognitive functions. [29]

how good or bad they perform in each context, which is the central feature of 'active testing'.

The proposed CF framework comprises five major components as shown in Figure 11.12 [29]: the Network Objectives Manager (NOM), the Environment Modelling and Abstraction (EMA), the Configuration Management Engine (CME), the Decision Applications (DApps), and the Coordination Engine (CE). These components include all the functionality required by a CF to learn and improve from previous actions, as well as to learn and interpret its environment and the operator's goals. While the CF operates in the same environment as SON functions, it deals differently with the KPIs' limited representation of the environment. Instead of simply matching network configurations to observed KPIs, the CF infers its context from the combination of KPIs and other information (like counters, timers, alarms, the prevailing network configuration, and the set of operator objectives) to adjust its behaviour in line with the inferences and goals. The subsequent sections describe the roles and interworking of the five CF components. Note that the discussion here is independent of the component implementation architecture. i.e. Figure 11.12 only depicts the required interfacing be it in a centralized, distributed, or hybrid implementation.

11.5.2 Network Objectives and Context

The generic input to the system is provided by the NOM and EMA, which respectively provide the operator or service context and the network environment and performance context.

The NOM interprets operator service and application goals for the CAN or the specific CF to ensure that the CF adjusts its behaviour in line with those goals. The other components take this interpretation as input and accordingly adjust their internal processes and subsequently their behaviour. In principle, each CF needs to be configured with the desired KPI targets and their relative importance, which the CF attempts to achieve by learning the effects of different NCP values. Without the NOM, such targets would be manually set by the operator who analyses the service and application goals (or KQIs) to derive the network KPI targets and their relative priorities. In this design, the NOM replaces this manual operation by using cognitive algorithms to break down the input KQIs (which are at a higher level of abstraction) into the output which are the prioritized KPI targets at a lower abstraction level.

The EMA on the other hand abstracts measurements into environment states which are used for subsequent decision making. Such environment abstractions (or 'external states') that represent different contexts in which the CF operates are built from different combinations of quantitative KPIs, abstract (semantic) state labels, and operational scenarios like the prevailing network or function configurations. Note that although SON also uses KPIs and current network configurations in the decision, it does not make further inference about the environment but instead responds directly and only to the observed KPIs. Even where contexts may be abstracted, the set of possible external states (usually, the considered KPIs) is fixed since it must be accounted for in the algorithm rules and the underlying decision matrix. The CF uses the EMA module to create new or change (modify, split, delete, etc.) existing quantitative or abstract external states as and when needed. These abstract states are then used by the further CF sub-functions – the DApp and CME, which may optionally also individually specify the KPIs, the level of abstraction, the frequency of updates, etc. that they require.

The simplest EMA engine is an ML classifier that clusters KPIs or combinations thereof into logically distinguishable sets. Such a classifier could apply a Neural Network, a Support Vector Machine (SVM), or similar learning algorithms to mine through KPI history and mark out logical groupings of the KPIs. Each group represents one environmental abstraction – requiring a specific configuration. Note, however, that an advanced version of the EMA may add an RL agent that selects the appropriate abstractions and (preferably) reclassifies them for the specific CF.

A centralized EMA provides the advantage of working with a wider dataset (network performance measurements, KPIs, context, etc.) across multiple cells or the entire network. While individual CFs can only have a limited view of the network context, a centralized EMA collects data across a defined network domain. This does not mean that it provides KPIs with the same level of abstraction to all CFs. Rather, depending on the CF and its feedback on the KPIs and context, the EMA dynamically adapts/changes its output. This will generally include multiple scales of measure, e.g. from ratio-scale KPIs to interval-scale metrics and to semantically enriched nominal-scale state descriptions. Further, the level of precision and accuracy can be modified dynamically.

11.5.3 Decision Applications (DApps)

The DApp matches the current abstract state (as derived by the EMA module) to the appropriate network configuration ('active configuration') selected from the set of legal/acceptable candidate network configurations. The DApp has the logic to search through or reason over the candidate network configurations, and to select the one that maximizes the probability of achieving the CF's set objectives for that context. In the SON paradigm, such an engine was the core of the SON function and the network configurations were selected based on a predefined set of static rules or policies (the decision matrix in Figure 11.1). In a CF, such an engine will learn (i) the quality of different network configurations, (ii) in different contexts (as defined by the EMA), (iii) from the application of the different legal network configurations, and (iv) towards different operator business and service models and associated KQIs. It will then select the best network configuration for the different network contexts, making the mapping (matching abstract state to network configuration) more dynamic and adaptive.

For this, the internal state space and state transition logic of the DApp (replacing the SON function's decision matrix) must also be flexible. Since there are no rules here (to be changed), changes in the DApp internal states (and transitions) are triggered through the learning. For example, using a neural network for selecting configurations, the DApp may be considered as a set of neurons with connections amongst them, in which neurons fire and activate connections depending on the context and objectives.

Besides the examples in Chapters 7–10, there are multiple ways in which the DApp may be implemented, typically as supervised learning or RL agents. The neural network example above could be an example instantiation of the supervised learning agent which is trained using historical network statistics data. It then employs the learned structure to decide how to behave in new scenarios. Using RL, the DApp could be a single-objective RL agent that learns the best network configurations for specific abstract states and CF requirements. The single objective hereby is optimizing the CF's requirements which may, in fact. consist of multiple targets for different technical objectives or KPIs as set by the NOM. As an example, for the MRO use case, the single objective of optimizing handover performance translates into the multiple technical objectives of minimizing radio link failures while simultaneously minimizing handover oscillations. Also, since there may not always be a specific network configuration that perfectly matches specific contexts, fuzzy logic (where truth values are not binary but continuous over the range [0,1]) may be added to RL to allow for further configuration flexibility network.

For most use cases, the DApp will be distributed (i.e. at the network function/element, like the base station) to allow for a more scalable solution even with many network elements. However, a centralized implementation is also possible either for a small number of network elements or for use cases with infrequent actions, e.g. network self-configuration scenarios like cell-identity management. Optionally, the CME may be integrated into the DApp, which is especially beneficial if both functions are co-located.

11.5.4 Coordination and Control

The Cognitive NMA system requires means to control the individual functions and to coordinate amongst their behaviour. This responsibility is undertaken by two units – the Configuration Management Engine (CME) and the Coordination Engine (CE) – which may, in some implementations, be combined into one.

11.5.4.1 Configuration Management Engine (CME)

Based on the abstract states as inferred by the EMA module, the CME defines and refines the legal candidate network configurations for the different contexts of the CF. In the simplest form, the CME masks a subset of the possible network configurations as being unnecessary (or unreachable) within a specific abstract state. In that case, the set of possible network configurations is fixed, and the CF only selects from within this fixed set when in the specific abstract state. However, a more cognitive CME could also add, remove or modify (e.g. split or combine) the network configurations based on the learning of how or if the network configurations are useful.

The CME is a multi-input multi-objective ML agent that gets input from the EMA and CE to determine the set of legal configuration candidates, i.e. the active configuration set. It learns the set of configurations that ensure accurate/effective fast-to-compute solutions for the CF's objective(s), the operator's objectives, and the CE requirements/commands. The simplest CME is a supervised ML agent (applying a Neural Network, an SVM or similar algorithm) that evaluates historical data about the quality of the configurations in different contexts (environmental states, peer functions, etc.) to select the legal configurations. However, an online ML CME could apply reinforcement learning to continuously re-adjust the legal set as different configurations are applied and their performance evaluated.

A centralized CME manages the internal state space for all CFs by constantly monitoring and updating (i.e. modifies, splits, deletes, etc.) the set of legal configurations available for each CF. Again, the advantage here consists of taking more informed decisions due to a broader dataset and sharing state-space modelling knowledge across multiple CFs. However, to manage scalability, e.g. for the case where multiple different CFs are implemented, a feasible centralized CME (and the most likely implementation) will only manage network configuration sets for CFs and not for CF instances. In that case, final NCP selection will be left to the DApp decisions/learning of each instance.

11.5.4.2 Coordination Engine (CE)

Like SON, the CNM paradigm also requires a coordination function, albeit of a different kind. Since the CFs will be learning, the CE needs to differently coordinate the CFs whose behaviour is non-deterministic owing to the learning. Specifically, the CE detects and resolves any possible conflicting network configurations as set by the different CFs. Additionally, (for selected cases) it defines the rules for 'fast track' peer-to-peer coordination amongst CFs, i.e. it allows some CFs to bypass its coordination but sets the rules for such by-pass actions. It also enables cross-domain knowledge and information sharing (i.e. across different vendor/network/operator domains). This may include environment and network models as well as the relevance and performance of KPIs and CF configurations in different contexts. Moreover, it supports the EMA and CME by identifying CFs with similar context in environment and legal configuration sets.

As stated earlier, the CE needs to (i) learn the effects of different CF decisions on other CFs; (ii) interpret the learned knowledge; and (iii) optionally suggest modifications to the CME and DApp on how to minimize these effects. Thereby, it undertakes the supervisory function over the CME and DApp, e.g. it may request the CME to re-optimize the legal configurations list but may also directly grade the DApp actions to enable the DApp to learn configurations that have minimal effects on other CFs.

If deployed in a distributed manner (i.e. at the network function), the CE becomes a member of a MAS of learning agents, where each agent learns if and how much the actions of its associated DApp affect other CFs or CF instances. It then appropriately instructs the DApp to act in a way that minimizes these negative effects. For this, the CE instances would have to communicate such effects with one another as suggested in [22, 23]. Otherwise, in a centralized CNM approach, the CE is also centralized to allow for a multi-CF view in aligning the behaviour of the CFs.

11.5.5 Interfacing Among Functions

The individual CF components described above interact via the interfaces depicted in Figure 11.12. Interface a from the NOM towards the CE, CME, and DApp is used to convey the KPI targets to each CF. The latter three components then read raw network state information like KPIs over interface b, and also receive individually customized representations of the current environment from the EMA component via interface c. A configuration change proposal computed by any of the DApps is transmitted to the CME for implementation via interface d while the CME activates via interface e. The activated network configurations are reported by the CME to the DApps and CE via interface f, which may also be extended to the EMA if the EMA's descriptions of the environment states also include the active network configuration values. The CE uses interface g to convey information on the impact of a CF's NCPs to the CME, e.g. by notifying which NCPs (or values thereof) have shown an adverse effect on the objectives of other CFs. Finally, interface h is used by the CME to update the set of legal configurations of the DApps. Further details on the information content of these interfaces are given in Table 11.1.

11.6 Synchronized Cooperative Learning in CANs

The CAN framework successfully justifies the use of ML for developing Cognitive Functions. More especially, using RL, each CF can learn the optimal to behave for all possible states in a particular environment. The assumption here is that the CF is afforded an environment to learn the independent effects of its actions and to determine the best actions to apply in a given state. However, even when acting alone, CFs can affect each other's metrics. For example, an MLB-triggered Cell Individual Offset (CIO) change in a cell can affect MRO metrics right after the change and at later points in time. Synchronous separation of the function execution (be it time or space) can thus not solve the challenge in this case, yet a coordinator would be too complex for learning-based CFs since it must account for the non-deterministic nature of the CFs. A good alternative in this case is SCL where the complex coordinator is replaced by an implicit mechanism that allows the CFs to communicate their effects to one another as described here.

Table 11.1 Descriptions of the required interfaces between CF component blocks.

Type	From	To	Information provided	Remarks
a	NOM	CE CME DApps	KPI target (values) to be achieved by the CFs *Optional:* target interpretations to distinguish their respective relevance; weights, priorities or utilities of the different KPIs	KPI targets may be universally sent to all so that each entity filters out its functionally relevant targets, or differentiated per recipient, e.g. the CME and DApp get only CF specific targets, CE gets the targets for all CFs.
b	Network, OAM, ...	EMA	Current network state parameter and KPI values	Allows DApps, CME and CE to evaluate how good certain actions are for the CF at hand, requiring
c	EMA	CE, CME DApps	Abstract environment states or contexts as created by the EMA.	States may be generic or specific to each CF, provided they have a common reference. But recipients may also specify the abstraction level required for their operation.
d	DApps	CME	Proposed network (re)configurations; Reports on the quality of the action(s) per context	May also be implemented directly between DApps to exchange reports on actions taken
e	CME	network	Activation of selected and approved network configuration values	
f	CME	CE, EMA, DApps	Reports on CME's network configurations and quality of the action(s) per context	*Optional for the EMA:* needed only if its state abstraction includes current configurations
g	CE	CME	CE configuration or report for each CF *Simple:* description of effects of CF action *Optional:* CE decisions/recommendations/	The CME uses the input to (re)configure the CF's control-parameter spaces. Recommendations/decisions may e.g. specify actions that should never be re-used
h	CME	DApps	KPI report on the DApp's action(s), CME configuration of the CF's action space (set of legal network configurations)	If the configurations database is part of the DApp, such configurations are sent to the DApps. Otherwise, the CME independently edits the standalone database.

11.6.1 The SCL Principle

Consider a network with C cells where each has F CF instances (also herein referred to as the learning agents) indexed as $CF_i \forall i \in [1, f]$. If, in a particular network state CF instance i in cell j (CF_{ji}) takes an action, that action will affect peers $CF_{jl} \forall l \in [1, f]$; $l \neq i$ which are the other CFs in cell j as well as the peers $CF_{ci} \forall c \in [1, k]$; $c \neq j$ & $\forall i \in [1, f]$, which are all the CFs in the other cells. For optimal network-wide performance in that state, CF_{ji} needs to act in a way that its action has: (i) the best performance considering CF_{ji}'s metrics and (ii) the least effects on the peers. Thereby, CF_{ji} needs to know the likely effects of its actions on the peers which requires that it must learn not only over its metrics but also over those peer effects.

The SCL concept enables this learning across multiple CF metrics through the three-step process illustrated in Figure 11.13, which involves:

1) After executing an action, CF_{ji} informs all its peers about that action triggering them to initiate measurements on their performance metrics.
2) At the end of a specified monitoring period which may either be preset and fixed or may be communicated as part of step 1, the peers report their observed effects to the initiating CF (here CF_{ji}).
3) The initiating CF (CF_{ji}) then aggregates the effects across all the reporting peers and uses that aggregate to evaluate the quality of its action and update its learned policy function.

In Figure 11.13a, for example, A informs B whenever it (A) takes an action, prompting B to monitor its metrics. At the end of the observation period, B informs A of the corresponding effects on B's metrics, with which A derives a penalty that qualifies the action. If the action was acceptable or good to B, A may not penalize

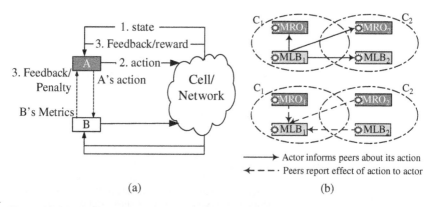

(a) (b)

Figure 11.13 Synchronized cooperative learning: (a) the SCL concept for two cognitive functions and (b) example message exchanges in two cells.

that action in a way encouraging the action to be reused in future. Otherwise A may heavily penalize the action to ensure it is blacklisted.

In a scenario with multiple CF instances, each CF instance that receives the active CF's message must report its observed effect, so that the reward/penalty is derived from an aggregation of all of the reports. For example, for the two cells in Figure 11.13b, both with instances of MLB and MRO, after taking an action in C_1, MLB_1 the MLB instance in cell C_1, receives responses from MRO instances in both cells C_1 and C_2 as well as from the MLB instance in cell C_2.

With the possibility of having multiple cells either as actors or as peers, two challenges must be addressed to guarantee optimal results:

- How to manage concurrency amongst CF instances within or across cell boundaries.
- How to aggregate the received information in a way that ensures effective learning.

These are described in the subsequent sections as are ideas on how to address them.

11.6.2 Managing Concurrency: Spatial-Temporal Scheduling (STS)

After CF_{ji} has taken an action, for it to receive an accurate report from a peer CF_{cl}, it is important that during the observation interval, CF_{cl}'s metrics are not affected by any other agent except CF_{ji}. Otherwise, since CF_{cl} is not able to differentiate actions from multiple CFs or instances thereof, its report will be misleading. It is appropriate to assume that CF_{ji} only affects CF instances in its cell or those in its first-tier neighbour cells and not in any other cells further out. This is a justified assumption except for a few radio-propagation-related automation functions, like interference management, which can easily be affected by the propagation of the radio signal beyond the first-tier neighbour cells. Even then, compared to effects in first-tier neighbour cells, effects in second and higher-tier neighbours are typically so small that they can be neglected.

Correspondingly, the assumption of having effects only in the cell and its first-tier neighbours and the requirement that CF_{cl} is only affected by CF_{ji} (at least during learning), imply that, CFs should only be scheduled so that no two CFs concurrently affect the same space-time coordinate, especially during the learning phase. This resulting mechanism, called Spatial-Temporal Scheduling ensures that each metric measurement scope (a space for a given time) is affected by only one CF, i.e. CFs are scheduled in a way that each CF acts alone in a chosen space-time coordinate.

For the learning-time spatial scheduling, consider the subnetwork of Figure 11.14 and a CF A with an instance executed in cell 14 (i.e. A_{14}). For A_{14} to

Figure 11.14 Spatial cell scheduling for concurrent actions in a hexagonal-grid.

learn the independent effects of its actions on the critical peers – CFs in 14 and its tier 1 neighbours (e.g. 12, 16), it is necessary that during A_{14}'s observation interval:

- only A_{14} is executed in cell A and no CF is executed in any of the neighbour cells to cell 14, otherwise, effects observed in cell 14 would not be unique to A_{14}.
- No CF is executed in any of cell 14's tier-2 neighbours (e.g. 23, 15), otherwise, the effects in cell 14's tier-1 neighbours (e.g. in 12, 16) would not be unique to A_{14}.

This implies that the nearest concurrent action to A_{14} should be in cell 14's tier-3 neighbours (e.g. cell 10), which are outside A_{14}'s reporting area. The result is the reuse-7 clustering profile of Figure 11.14 where concurrent activity is only allowed within a Cluster/set of cells with the same colour in the Figure 11.14. By doing this, each CF can measure its effects on the critical CFs without influence from any action in the potential conflict cells and CFs. To achieve this, SCL applies the time division multi-frame of Figure 11.15 with 7-frames per multi-frame which ensures that each of the seven clusters gets 1 frame of the 7-frame multi-frame.

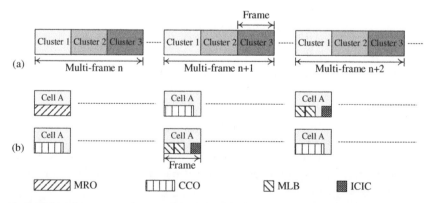

Figure 11.15 Space and Time separation of CF execution: (a) Cluster Frames in a Multi-frame and (b) STS scheduling in two cells during different multi-frames.

The clustering can be configured by the reuse-7 graph colouring scheme which allocates every one of any seven neighbour cells to a different Cluster creating the mapping illustrated by Figure 11.14. Given a seed cell, the graph-colouring algorithm first allocates to the seed's immediate neighbours ensuring there are two cells between any two identically-coloured cells, effectively allocating each of the seed's neighbours to a different cluster. Then, starting with any of the seed's second-tier neighbours, the algorithm again allocates cells to clusters following the same rule, i.e. ensuring there are two cells between any two identically-coloured cells.

11.6.3 Aggregating Peer Information

The space-time scheduling of CFs enables the CFs to make accurate observations of their environment and give accurate reports of their observation thereof. The simplest report of the observations is the hash of the important KPIs and their values for the respective peer CFs. This minimizes the need for coordinating the design of the CFs, i.e. no prior agreement on the structure and semantics of the report is necessary although it raises the concern of different computations (thus meanings) of the KPIs. However, to understand the concepts however, it is adequate to assume that this simple reporting mechanism is used.

Given the multiple metric reports of KPI-name to KPI-value hashes from the different peers, for the initiating CF (CF_{ji}) to learn actions with the least effects on those peers, CF_{ji} requires an appropriate objective function for aggregating those metrics. Thereby, for each peer that sends a report, CF_{ji} requires a peer-specific local 'goodness model' of the metrics i.e. a specific model that describes which metric values are good or otherwise. The model must be specific for each actor-peer CF pair since each CF affects each peer CF in a way that is specific to the acting CF and the peer. In Figure 11.13b for example, both MRO instances report the PP and RLFs rates while the MLB in cell B reports the dissatisfied user rate over the measurement interval, so the initiating CF (MLB_1) must account for the MRO effects different from the MLB effects when deriving the associated quality of the action taken.

Moreover, the initiating CF may also require an effect model for the different peers to describe the extent to which it should account for a given peer's observations. This model, which will typically be different for different CF instances would, for example, differentiate peers in the same cell as the initiator CF from those in neighbouring cells. For the MRO-MLB case, for example, the MRO effect model may consider neighbour cells' MLB effects as being insignificant, yet the MLB effect-model may consider neighbour cells' MRO effects as critical to the performance evaluation. CF_{ji} must then use the combination of goodness and effect models to evaluate the aggregate effect of its actions on the peers. Ideally, this will translate into a generic operator-policy aggregation function, typically as a weighted multi-objective optimization function, for which the operator sets the weights.

11.6.4 SCL for MRO-MLB Conflicts

The MRO-MLB conflicts are the most widely discussed conflict in SON, so it makes sense to use that to demonstrate the SCL ideas. This assumes that the two Cognitive Functions are implemented as Reinforcement learning (Q-learning) based agents as described in Chapter 9 (Sections 9.3 and 9.4). The characteristic behaviour of the two functions (hereafter respectively referred to as QMRO and QLB) are:

1) Each CF observes a state (mobility for QMRO and load distribution for QLB).
2) The CF selects and activates an action on the network (the Hys-TTT tuple for QMRO and the CIO for QLB).
3) The CF evaluates a rewards function that describes how good the action was for the network.

Considering a network of 21 cells with wraparound as in Figure 11.14, each cell is availed an instance each of QMRO and QLB. However, to improve the speed of convergence, the different instances of single CF (e.g. QMRO) learn a single-shared policy function. The cells and CF instances are clustered by a graph-colouring algorithm and configured with execution time slots are described in Section 11.6.2. To implement the SCL mechanism, steps 2 and 3 above are adjusted such that:

2) after activating the action, the CF informs the peers of the action and requests them to evaluate their metrics for an interval that is specific to the active CF.
3) The affected peer CFs report their metric values for the specified period and the initiating CF aggregates these in the reward used to evaluate the action.

The critical aspect then is how to design the rewards functions that enable each CF to learn based on the aggregate of the received information and its own measurement. As described above, the appropriate function needs to account for differences in CF types and instances. This, however, can be complex so to reduce complexity yet still evaluate SCL's benefits, the evaluations here only consider CF instances within the same cell and neglect intra-cell effects, i.e. a single-effect model is used with effect $= 1$ for intra-site peers and effect $= 0$ all other CFs. Considering Figure 11.13b, for example, this mechanism implies that after its action in cell C_1, MLB_1 only considers the feedback from MRO_1, and neglects the effects on MRO and MLB in cell C_2. This may not be enough to account for all effects, but the complete aggregation function requires a detailed study that quantifies the actual cross-effects among the CFs. The MLB and MRO reward functions are designed as described here:

11.6.4.1 QMRO Rewards

Alongside minimizing RLFs and PPs, QMRO needs to account for MRO effects on load, to minimize the N_{us} by, for example, reducing the load in an overloaded cell. To account for MRO effects on load, the reward derived from the Handover Aggregate Performance (HOAP) metric is scaled by a *Loadbonus* = 0.9 as given in Eq. (11.1), but only if the overload significantly reduces after MRO. Otherwise, the default *Loadbonus* = 1 is applicable.

$$r_{x,a} = -Loadbonus * (w_1 P + w_2 F_E + w_3 F_L) \tag{11.1}$$

'Significant load change' occurs if the cell was overloaded at the time of MRO action but the offered load subsequently reduced by more than 20%. The 20% reduction applied in all cells is heuristically obtained, but with two straightforward principles. Firstly, actions leading to load reduction will have better quality, if their respective *loadbonus* are less than the default. Secondly, the reward should not over emphasize load reduction to the extent that the initial MRO focus is lost, i.e. the MRO goal of minimizing RLFs and PPs should be paramount. These principles ensure that QMRO prioritizes MRO but with an eye to reducing any observed overloaded.

11.6.4.2 QLB Reward Function

To account for MRO effects while pursuing the QLB objective of instantaneously removing overload, QLB also now requires the penalization of actions that cause excessive degradation of HO performance. For each load scenario, Γ as defined in section QLB, the revised reward function adds an HO-related penalty (*HOpenalty*) as in Eq. (11.2).

$$r = \begin{cases} \Delta\rho_s + 1 - HOpenalty & ; & \Delta\rho_n = 2, \Gamma < 3 \\ \Delta\rho_s - HOpenalty & ; & \Delta\rho_n < 1 \\ \Delta\rho_s - 1 - HOpenalty & ; & otherwise \end{cases} \tag{11.2}$$

where $\Delta\rho_s$ is the achieved reduction in the serving cell offered load ρ_s and $\Delta\rho_n$ is the change in average target cells offered load. For the MLB decision, a *HOcost* can be estimated as in Eq. (11.3). Using this *HOcost*, the HO penalty can be derived from a penalty function like that in Figure 11.16.

$$HOcost = \frac{P + 2 * F_E}{H + F_E + F_L} \tag{11.3}$$

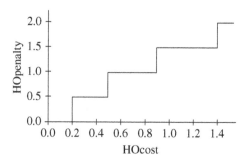

Figure 11.16 Example handover penalty function.

11.6.4.3 Performance Evaluation

Performance of SCL is evaluated by comparing: (i) to *Ref* the reference network without any CNM solutions but having manually optimized HO settings, (ii) to *QBOTH*, which is the operation of the two functions without any coordination solution (here denoted by), and (iii) *STS* which is the operation with only STS as an example implementation of Single Agent Systems decomposition solution. In all cases, two perspectives are considered:

1) **Transient performance**: With gains as the percent reductions in HOAP and *N*us for each batch relative to *Ref*, the transient results track the Simple Moving Average (SMA) of the gains in each metric using a window of 10 values (i.e. 10 batches). For example, if at t = τ the *Ref* and SCL *N*us are respectively $N^t_{us,Ref}$ and $N^t_{us,SCL}$, the transient *N*us is

$$
N_{us} = \mathop{E}_{t=\tau-9}^{\tau} \left\{ \frac{100(N^t_{us,Ref} - N^t_{us,SCL})}{N^t_{us,Ref}} \right\} \tag{11.4}
$$

2) **Steady state performance:** In the results, all solutions complete learning between 20 000 and 30 000 s. So, steady-state performance plots on a 2D grid the metric averages for the last 20 batches (which are all after 30 000 s). Note that *N*us here indicates the average number of dissatisfied events per second evaluated over the measurement period (e.g. a batch) in the network.

11.6.4.4 Observed Performance

The performance of SCL can be summarized by Figure 11.17 which evaluates the transient performance in terms of the HOAP metric [22] and the number of dissatisfied users (N_{us}). It is visible from the Figure 11.17 that compared to QBOTH, SCL achieves good compromise between the two CFs, although may not always be better than STS. During learning, SCL and STS performance is equivalent since the same CFs are active and exploring in the same way. After learning is completed, however, the performance differs because the learned

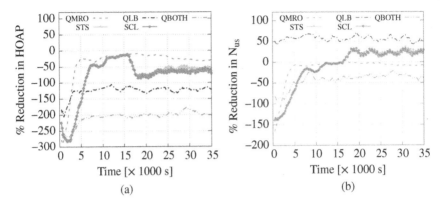

Figure 11.17 Transient performance of SCL-based CF coordination in a network with a 50 m wide hot-spot in a 60 km/h environment.

policy functions are different. For SCL, each CF is not only given a chance to learn its independent policy function but is also required to learn the solutions that have minimal effect on the peers. The cooperative competition minimizes trigger oscillations, i.e. CFs need not be triggered in response to a peer's action since the peer ensures that its actions have minimal negative effects on other CFs.

The relative performances of SCL and STS will differ depending on the scenario, mainly owing to the very dynamic nature of mobile communications. Major variations in network conditions are mainly expected in user numbers, velocity profiles and distribution; hotspot locations and sizes, as well as in the physical characteristics like Shadowing. User count may only have a small impact on HO performance, where it only changes the rate of HO of events and not their relative comparison. It may, however, have a more pronounced effect on MLB since it directly affects the offered load, just like hotspot size and velocity which are expected to have major effects on both load and HO metrics.

Regardless of scenario variations, coordination solutions should, at the very least, not degrade the performance. To evaluate this, Figure 11.18 shows the results of re-executing the STS and SCL studies in two variations of the initial scenario. The figure shows the respective performances of the initial scenario (i.e. 50 m hotspot and 60 kmph velocity) in Figure 11.18a; a lower velocity scenario (i.e. 50 m hotspot, 30 kmph velocity) in Figure 11.18b; and a smaller-hotspot scenario (i.e. 20 m hotspot, 60 kmph velocity) in Figure 11.18c. In each case, the sub-Figure 11.s compare the steady-state performance of the coordination solutions against the reference network, the independent CFs, and the combined but uncoordinated operation (QBOTH).

It is visible in both revised scenarios (Figure 11.18b,c), that owing to reduced mobility and load the QBOTH-induced degradation in N_{us} reduces. Thus, both

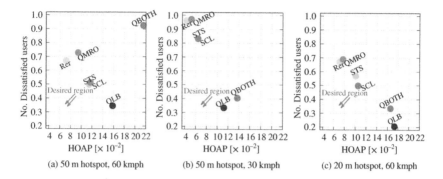

Figure 11.18 Performance of the CF coordination solutions in various network hotspot and velocity scenarios.

STS and SCL achieve good compromise since they significantly improve HO performance (reduce HOAP) although this is achieved at the cost of giving up some of the QLB gains. Meanwhile, with lower velocity, both solutions achieve the exact same performance, which may indicate that STS is already at the Pareto front of the performance compromise between the two CFs. However, in case of a small hotspot (Figure 11.18c), SCL achieves significantly better compromise compared to STS.

11.6.4.5 Challenges and Limitations
As presented here, SCL is quite limited although it demonstrates promise. Firstly, mixing metrics from different network perspectives implies comparing dissimilar items, e.g. comparing PPs to dissatisfied users. Without having a clear basis for such a comparison, the possible kinds of actions that can be taken by SCL get limited. An alternative, for example, using the KPI goodness scale is a necessary improvement.

In conclusion, STS achieves good performance since it allows CFs to independently act in their environments. It has been shown, however, that this could be improved using SCL by allowing the CFs to communicate their effects to one another as they learn their policy functions. Further improvement would, nevertheless, be expected if all effects can be accounted for as described in the discussion on limits and constraints in the next section.

11.7 Inter-Function Coopetition – A Game Theoretic Opportunity

Game theory can be defined as 'the study of rational decision-making in situations of conflict and/or cooperation'. The decision is a player's choice of what action to take, amongst a fixed set of alternatives, given some information about the state of the world. The consequences of a player's decision will be a function of her action, the actions of other players (if applicable) and the current state of the world. A rational player will choose the action which she expects to give the best outcome/consequences, where 'best' is according to her set of preferences. Each possible outcome is associated with a real number – its utility, which can be subjective (how much the outcome is desired) or objective (how good the outcome actually is for the player).

In a CNM system, several cognitive functions (CFs) need to simultaneously make decisions. However, as highlighted in the previous section, such CFs might affect one another, their control regions might overlap, and the respective decisions might collide leading to unexpected effects. Thus, game theory is a powerful tool that can be used in such situations where the decision of an entity influences the decisions of others as well as the achieved outcome (or utility). Namely, to improve (or maximize) the overall outcome, coordinated strategies might be adopted by the different agents (or functions) to handle the control requests and simultaneously converge to optimal equilibria. In particular, when non-cooperative game theory is used, the goal is to predict the individual strategies and utility achieved by each player (i.e. cognitive functions) and to find the set of stable solutions, i.e. equilibria, to which the system will converge to. In contrast, cooperative games investigate how different players can form coalitions in order to maximize their overall utility. Finally, 'coopetition' incorporates both concepts towards a 'cooperative competition'. In this case, players cooperate with each other to reach a higher utility value when compared to the case of no cooperation, but still compete to achieve advantage.

11.7.1 A Distributed Intelligence Challenge

Although deployment of Cognitive Functions in CAN promises to further minimize human effort in both the design and operation of the functions, in principle, it moves some of the complexity to another point, namely the function coordination layer. Each function being cognitive (an independent learning agent) and the CAN a set of concurrently and independently learning functions, raises two problems for function coordination. Firstly, the functions adjust the same (or at least related) parameters and use the same measurements, yet unlike in SON the response of the functions to any given state may not be consistent due to the learning. Most important however, is that functions observe and need to respond to dynamic, inconsistent network states, especially during the exploration of the candidate behaviour policies. Specifically, each function's view of the network states to which it is supposed to respond is only a partial description of the true network state. The effect of any action on the network is not fully bounded to a given space or time region, i.e. the effect can reach regions that may not have been anticipated by the function (due to the partial state view). Yet, the set of function cannot simply be replaced by a single function with a complete view of the network state and effect region since, in that case, the state space explodes.

This then requires a complex coordination function which must track the multiple dynamic agents each having non-deterministic behaviour owing to the learning. To simplify, the complexity at the coordination layer, the agents need to be 'smart enough' to not only learn optimal behaviour for their partial states, but also to learn to minimize the conflicts. In other words, the learning process should happen within a dynamic learning environment but with minimal or no explicit coordinators.

The SCL as described in [22, 23] and summarized in Section 11.6 is a first step to the solution but it still has limitations. The need for synchronization, in time and space makes the coordination very inefficient. The space-time separation solutions take a long time to converge since the Cognitive Functions are only allocated small portions of the space-time resources. The optimal solution, therefore, lies in a coordinated competitive-cooperation of the functions for which game-theoretic approaches are a good promise. The expectation in this case is that the functions learn concurrently within the environment but in a way that, as they update their policy functions for each action taken, they account for the existing of other learners who influence the same observation space. Ideally then, the functions would concurrently improve their performance over time (as shown in Figure 11.19) having learned to each behave in a way that minimizes the effects on the peer functions. The next sections explore the likely structure of such a game theoretic solution.

11.7.2 Game Theory and Bayesian Games

Game theory deals with situations in which multiple decision-makers interact, where each player's consequences are affected by her choice as well as the

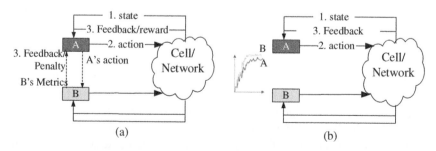

Figure 11.19 Synchronized learning in comparison to distributed learning.

choices of other players. The resulting game may be classified as a zero-sum or a non-zero-sum game.

In zero sum games (e.g. chess), one player's gain is the other's loss. The most important result here is the minimax theorem, which states that under common assumptions of rationality, each player will make the choice that maximizes her minimum expected utility. This choice may be a pure or a mixed strategy. A pure strategy is a fixed definition of the player's moves for all game states and other player's moves while a mixed strategy is a probability distribution over all possible pure strategies, i.e. it is a random choice between pure strategies.

In non-zero-sum games (e.g. business agreements), it is possible for both players to simultaneously gain or lose. Non-zero-sum games may be cooperative in that the players make enforceable agreements and each player holds true the understanding that 'I will cooperate if you do'). The games may, however, also be non-cooperative, in which no prior agreements can be enforced. Consequently, an agreement in non-cooperative games must be self-enforcing to ensure that players have no incentive to deviate from it. The most important concept here is the Nash Equilibrium (NE) which is the combination of strategy choices such that no player can increase her utility by changing strategies.

The basic assumption in game theory is that a rational player will make the decision that maximizes her expected utility. Three decision types can be expected:

1) **Decisions under certainty**: The consequences $C(a)$ of each action a are known. A rational agent chooses the action with the highest utility $u(C(a))$.
2) **Decisions under risk**: For each action, a probability distribution over possible consequences $P(C | a)$ is known. A rational agent chooses the action with highest expected utility, $\sum P(C | a)u(C) \forall C$.
3) **Decisions under uncertainty**: Agents are assumed to have a subjective probability distribution over possible states of nature $P(X)$. The consequence of an action is assumed to be a deterministic function $C(a, X)$ of the action a and the state S. A rational agent chooses the action with the highest subjective expected utility, $\sum P(X)u(C(a, X)) \forall X$. This is the capability that can be exploited for multi-agent learning by using Bayesian inference.

11.7.2.1 Formal Definitions

A *game in normal form* consists of:

1) A list of players $I = [1 \ldots n]$
2) A finite set of strategies S_i for each player $i \in I$
3) A utility function u_i for each player $i \in I$,

Where $u_i : S_1 \times S_2 \times \ldots \times S_n \rightarrow R_i$, i.e. u_i maps a combination of players' pure strategies to the payoff for player $i \in I$. The normal form gives no indication of the order of players' moves, so it is sensible to assume that all players choose strategies simultaneously.

A **Nash equilibrium** is a set of strategies $s_1 \in S_1, \ldots s_n \in S_n$, for players $i \in I$ such that for each player i:

$$s_i = \arg \max_s \{u_i(s_1, \ldots s_{i-1}, s_i, s_{i+1}, \ldots s_n)\} \tag{11.5}$$

For each player i, s_i is a strategy that maximizes her payoff, given the other players' strategies; no player can do better by switching strategies.

A **mixed strategy** σ_i is a probability distribution over i's pure strategies s_i; e.g. if A, B are the pure strategies for P1, $\sigma 1$ might be $\{p(A)\ A, p(B)\ B\}$. Then the utility for i given strategies $\sigma_1 \ldots \sigma_n$ of others is

$$u_i(\sigma_1 \ldots \sigma_n) = \sum_{s \in S} u_i\{s_1 \ldots s_n\} \prod_{j=1}^{n} \Pr_{\sigma_j}(s_j) \tag{11.6}$$

The Nash Equilibrium in mixed strategies is a set of mixed strategies $\sigma_1 \ldots \sigma_n$ such that for each player i:

$$\sigma_i = \arg \max_\sigma \{u_i(\sigma_1, \ldots \sigma_{i-1}, \sigma_i, \sigma_{i+1}, \ldots \sigma_n)\} \tag{11.7}$$

The Nash's Theorem [24] states that for every game, there always exists a mixed Nash equilibrium whereas a pure strategy equilibrium may or may not exist.

An **extensive form game** is a 'game tree': a rooted tree where each non-terminal node represents a choice that a player must make, and each terminal node gives payoffs for all players. In games with perfect information, at each node in the tree, the player knows exactly where in the tree she is. In games with imperfect information, this may not be true.

A pure strategy s_i for player i consists of a choice for each of player i's information sets. In a game with perfect information, each information set consists of a single decision node. For imperfect information, the information set is characterized by a *belief*, i.e. an assignment of probabilities to every node such that the sum of probabilities for any information set is 1. Games with imperfect information are also known as Bayesian games. Note that mixed strategies and the related mixed Nash equilibria as defined as above are also applicable to extensive form games.

11.7.2.2 Bayesian Games

Different from games with perfect information formalized above, i.e. where players have common knowledge of the game structure and payoffs, Bayesian games model the case in which some of the parameters are unknown. In other words, players have incomplete information on the available strategies and payoffs from the other players and, in contrast, base their decision on beliefs with given probability distributions.

An example of a Bayesian game could be the case in which the probability distribution over possible states *P(X) is not known*, but the consequence of an action is still a deterministic function *C(a, X)* of the action *a* and the state *X*. It is, however, also possible that both P(*X*) and C(a, *X*) are not known.

The fact that some of the information is unknown, it is modelled by defining *types* for the different players and by associating probability distributions over the type space. The type of each player determines the specific player's payoff function and/or associated strategies.

Formally, a Bayesian game consists of [25]:

1) A list of players i = 1 ... n.
2) A finite set of types Θ_i for each player i. where the type captures all unknown information about the state of nature in which the player finds herself, including all information about other players that she may not know.
3) A finite set of actions A_i for each player i.
4) A payoff function u_i for each player i, where u_i: $A \times \Theta \to R$. i.e. u_i maps a combination of a player's actions and types to the payoff for player i.
5) A prior distribution over types, P(θ) for θ in Θ.

The set of pure strategies S_i for each player i are defined as S_i: $\Theta_i \to A_i$ i.e. A strategy is a mapping from types to actions.

A Bayesian Nash equilibrium (BNE) is a combination of (pure) strategies $s_1 \in S_1$ ... $s_n \in S_n$, such that for each player i and each possible type θ_i in Θ_i:

$$s_i(\theta_i) = \arg \max_s \sum u_i(S_{NE}(\theta_i))P(\theta_{-i}|\theta_i) \tag{11.8}$$

where S_{NE} is the set of these pure strategies for all players

$$S_{NE}(\theta_i) = s_1(\theta_1), \ldots s_{i-1}(\theta_{i-1}), s_i(\theta_i), s_{i+1}(\theta_{i+1}), \ldots s_n(\theta_n) \tag{11.9}$$

At a BNE, no player type can increase her expected payoff (over the distribution on of possible opponents' types) by changing strategies.

Similarly, to what is above, mixed strategies can be defined as probability distributions over the available strategies, and the corresponding BNE in mixed strategies can be derived. A mixed strategies BNE always exists.

11.7.3 Learning in Bayesian Games

What happens when the probability distribution over the possible types for a player is not known? Or even worse, what happens when the player's utility for each state of nature is not known. The simplest answer is that the players should act in their world/nature and learn the distributions and Utility. It has been argued that 'Nash equilibrium might arise from learning and adaptation' [26]. Multiple models for learning in games have been proposed with varying levels of sophistication. In general, these are classified as either passive or active (see [26]) as summarized below.

In static simultaneous move games, the strategies are simply choices of actions and learning is 'passive', i.e. because what players do have no impact on what they see, players have no incentive to change their actions to gain additional information. There are two main models of passive learning – *fictitious play* and *reinforcement learning models.*

In *fictitious play,* the players begin by making arbitrary choices as they cannot learn without receiving any data. Subsequently, players keep track of the frequency with which their opponent has played different actions. However, fictitious play can open a player up to exploitation by a clever opponent because of the dependence on a deterministic best response based on the information collected. Besides, from a purely descriptive point of view, the exact best response implies that a small change in beliefs can lead to a discontinuous change in response probabilities, which seems implausible.

On the other hand, *reinforcement learning* models do not deal with beliefs but rather directly update a measure of the utility of each action – called a 'propensity' – and derive probabilities to reinforce actions with higher propensities and make them more likely to be played. In the earliest studies, only the selected action was updated according to how well it performed, i.e. utility weights are updated only for the action that was chosen. Recent studies have instead now proposed 'self-tuning experience-weighted attraction', where weights are updated for every action that would have done at least as well as the action that was chosen and the utility of actions that are not used is depreciated.

Both passive learning model types can be improved with the application of recency considerations, i.e. that recent observations might get more weight than older observations. The reasons and conditions thereof are that if the process that is generating observations undergoes unobserved changes, then older observations may indeed be less informative than recent ones. Most models of recency have focused on using simple rules of thumb, e.g. by specifying that older observations receive exponentially less weight. A more accurate, albeit more complex, method is to develop explicit Bayesian models of changing environment. This, however, may in some cases lead to a distribution of play that is very different than any Nash equilibrium [22].

Active learning is necessary mainly because a passive learning approach will not gather information about the outcomes of alternative strategies, as much of the game is 'off the equilibrium path'. Instead, players in a dynamic game may choose to experiment with actions that they think might be suboptimal, but the agents do so to learn more about consequences of those actions.

Active learning, however, needs to address four key issues. Firstly, the patience of the players in terms of time preference or discounting is crucial, since a patient player will be more willing to risk short-term mistakes in pursuit of better long-term performance. Secondly, random play may be incorporated as a mechanism for learning about off-path play. And the effects of the suboptimal actions are crucial in that if the potential risks from experimentation are large and negative, then less of it will occur. Finally, games that include many information sets may potentially require a lot of experimentation to learn the off-path effects. This is especially crucial case in the case of repeated games.

Games that are repeated over time raise the possibility of creating incentives through rewards and punishments as means of encouraging particular behaviour. However, learning in repeated games is complicated by the need to infer causality 'off the equilibrium path'. This is described as the 'Folk Theorem': any payoff vector that is individually rational for the players is a subgame perfect equilibrium provided that players are sufficiently patient. This implies that repeated games allow the possibility of cooperation using incentives that are established through future rewards.

It is evident from this summary that learning in games in a generalization of models of RL and adaptive control. In general, the simple RL models evaluated in optimal control (e.g. as summarized in Chapter 6) are a special form of learning in games, i.e. ones in which adaptive learning models of the agents do not incorporate beliefs about opponent's strategies or do not require players to have a 'model' of the game. So, the case of coordinating multiple network automation functions with each of them as a learning agent is equivalent to the challenge of a game of learning agents. The challenge then is on how to map the expected game to the appropriate learning model.

11.7.4 CF Coordination as Learning Over Bayesian Games

Bayesian games can then be used to model and optimize CF coordination when different CFs affect each other, e.g. in terms of the achieved reward or payoff. In this case, it is justified to assume that the different players are the different CFs, and the set of strategies and payoffs are modelled using the definition of 'types', as highlighted above. Formally, the games may be defined as:

- A list of players i = 1 ... n, which may for example be the Q-Learning based CFs.
- A finite set of types Θ_i for each player i, equivalent to the states as used in QL. These capture all information about the state of nature in which the player finds herself including all information about other players that she may not know.
- A finite set of actions A_i for each player i.
- A utility function u_i for each player i, where; $u_i : A \times \Theta \to R$. i.e. ui maps a combination of a player's actions and types to the payoff for player i.

In this case, the game under consideration is more complicated than traditional Bayesian game, since it assumes that the prior distribution on types, $P(\theta)$ for θ in Θ, is unknown, but also all possible types are not known at the beginning, and must be learned over time, to eventually improve the outcome of the game. Other elements of the formulation are, however, applicable to the CF coordination problem. The information sets or types capture the different views of the games for an agent. These are private to the agent and could be equivalent to the individual states as observed by the CF (as the agent). For example, consider the two global states {HO state = 2, LB state = 6} and {HO state = 2, LB state = 8}. For a mobility optimization agent, these are the same type/information set i.e. {HO state = 2, whatever all else there may be}.

This private view may be contrasted to the global view of the game. Therein, each game state or game equilibrium represents a specific combination of the players' possible views with a utility attached for each player to the actions taken when that particular combination of views is observed. This is equivalent to a complete state combination of multiple Cognitive Functions, e.g. one game state could be the tuple {mobility state = 30 kpmh, Load state = highload, Interreference state = low, …}.

Although this formulation is logical, it needs to be studied further with prototypes that demonstrate its application. A critical challenge at this early state is the design of the utility functions. Since more agents are expected to be added over time (i.e. more cognitive functions may be introduced on the network), the utility needs to be set up in a way that it can be extended to learn over new action spaces as the new agents are added. This is expected to be starting point of such a study.

11.8 Summary and Open Challenges

The SON control loop relied a lot on human design, but this becomes inadequate as the function becomes more cognitive. The critical challenge is the coordination amongst the functions. Although multiple approaches are possible if the system is viewed as a MAS, the initial approaches, as expected, have taken a simplistic command and control approach. In particular, the system management has focused on supervising the functions and adding mechanisms to verify that the actions taken are positive for the whole network. These are the roles shared between the coordinator and the verification function.

11.8.1 System Supervision

By enabling the network to react quickly to any perceived degradations after a configuration change, verification can make the system more robust and solve unforeseeable conflicts between independent automation function instances. The risk, however, is that at the same time it makes the system more resistant to change. It makes the network more robust, yet fragile against changes in the context or in the environment. In other words, it doesn't necessarily make the system more resilient.

The scoring method and the diagnosis process introduced in Section 11.3.3 can mitigate this problem and ensure that verification allows the system to adapt when changes in the environment so require. A further concept extending this principle introduces the concept of Network Element Virtual Temperature (NEVT), which indicates the state of stability of the element and its context and environment at a given moment. The idea behind the NEVT is that for network elements or functions with higher NEVT values, in other words functions with more 'unstable' context, SON verification is more likely to accept reconfigurations that do not improve or may even (slightly) degrade the performance to allow for better adaptation to the changes. The NEVT is increased by changes in the context or discontinuities like software upgrades, but over time it 'cools down', to ensure that the system converges to a stable, well-performing state. The NEVT can also be distributed to dependent network functions, for example, to the geographical neighbour cells in RAN, because the instability may also propagate in the network.

Besides supervising individual function, verification can be added to other functions to enhance the functionality. In particular, it was shown that verification can be combined with coordination to implement an opportunistic concurrent control mechanism. Accordingly, a coordinator takes a laissez fare attitude to mild conflicts by allowing verification to manage the degradations arising from such conflicts. The coordinator can then switch to a strict mode that rejects conflicting requests if such requests have been discarded by the verification process as unresolvable. The strictness is maintained until some terminal condition is fulfilled e.g. either until all automation function instances have run at least once or until a pre-configured time threshold is reached. It is only after this that the coordinator reverts to relaxed or laissez fare concurrency control mode.

11.8.2 The New Paradigm

Coordination and verification will remain critical even as more cognition is added to the functions albeit in a new framework. The new framework required new functionality e.g. to characterize the observed state of the network in a way that it consistently labelled for all of the automation function. However, old functionality will also need to be adapted. For example, the configuration and management of the network key performance targets will require more automation to keep the targets synchronized with the now non-deterministic functions. Similarly, the mechanisms for coordination will need to be changed since the basic serialization to avoid concurrency will no longer adequate.

One such candidate mechanism was presented, one through which the functions keep track of each other's actions and collaborate in minimizing their negative effects to one another. Although the presented solutions have only been narrowly tested and are not proved to work in all conditions, they provide a starting point on which an advanced version can be developed. For this, it has been argued that game theory provides a good theoretical framework in which such a solution may be realized. In particular, it was proposed to consider a game state as each network state in which multiple functions need to make individually and globally optimal decisions. The required equilibrium can then be developed through learning.

11.8.3 Old Problems with New Faces?

It is worth noting that besides solving the function coordination, other critical challenges remain outstanding. Firstly, how will the system keep track of what is or has happened given the indirection of control? For example, the network state is labelled at the point different from where the state is used for making decisions. And if the objectives are not static but are adjusted according to context, which unit would be held responsible any observed negative outcome. Yet even designing the entire system as a single monolithic unit is also not a candidate solution. How to troubleshoot the entire system in case of unexpected system-wide failure. And, if multiple actions are taken and a degradation observed, to which time point or actor should such a degradation be accorded? These are critical challenges that need to be addressed and it may be the case that they will not be solved technically but institutionally. For example, the DevOps framework (discussed in the next chapter), may enable the solutions to be developed and implemented incrementally even without fully developed solution concepts. As such, it is important to consider the institutional ideas alongside the technical and functional ideas that have so far been presented.

References

1 Hamalainen, S., Sanneck, H., and Sartori, C. (eds.) (2011). *LTE Self-Organising Networks (SON): Network Management Automation for Operational Efficiency.* Wiley.

2 Hahn, S. Gotz, D., Lohmüller, S. et al. (2015). Classification of Cells Based on Mobile Network Context. 81st Vehicular Technology Conference (IWSON Workshop), Glasgow.

3 Frenzel, C., Lohmüller, S., Schmelz, L.C. et al. (2014). Dynamic, Context-Specific SON Management Driven by Operator Objectives. Network Operations and Management Symposium (NOMS), Krakow.

4 Lohmüller, S., Schmelz, L.S., Hahn, S. et al. (2016). Adaptive SON Management Using KPI Measurements. Network Operations and Management Symposium (NOMS), Istanbul.

5 Tsvetko, T., Ali-Tolppa, J., Sanneck, H. and George, C. (2016). Verification of Configuration Management Changes in Self-Organizing Networks. IEEE Transactions on Network and Service Management (TNSM).

6 Weiss, G. (ed.) (1999). *Multiagent Systems: A Modern Approach to Distributed Artificial Intelligence.* MIT Press.

7 Panait, L. and Luke, S. (2005). Cooperative multi-agent learning: the state of the art. *Autonomous Agents and Multi-Agent Systems* 11 (3): 387–434.

8 Tsvetkov, T.I. (2017). Verification of Autonomic Actions in Mobile Communication Networks. Dissertation. Technical University of Munich.

9 Tsvetkov, T., Novaczki, S. and Sanneck, H.C.G. (2014). A Post-Action Verification Approach for Automatic Configuration Parameter Changes in Self-Organizing Networks. International Conference on Mobile Networks and Management, Würzburg.

10 Novaczki, S., Tsvetkov, T., Sanneck, H. and Mwanje, S. (2015). A Scoring Method for the Verification of Configuration Changes in Self-Organizing Networks. MONAMI, Santander.

11 Novaczki, S. (2013). An Improved Anomaly Detection and Diagnosis Framework for Mobile. International Conference on Design of Reliable, Budapest.

12 Ali-Tolppa, J. and Tsvetkov, T. (2016). Network Element Stability Aware Method for Verifying Configuration Changes in Mobile Communication Networks. International Conference on Autonomous Infrastructure, Management and Security (AIMS), Munich.

13 Tsvetkov, T., Sanneck, H. and Carle, G. (2015). A Graph Coloring Approach for Scheduling Undo Actions in Self-Organizing Networks. IFIP/IEEE International Symposium on Integrated Network Management (IM 2015), Ottawa.

14 Tsvetkov, T., Ali-Tolppa, J., Sanneck, H. and Carle, G. (2016). A Minimum Spanning Tree-Based Approach for Reducing Verification Collisions in Self-Organizing Networks. IEEE/IFIP Network Operations and Management Symposium NOMS, Istanbul.

15 Johnson, R. (2002). *Expert One-on-One J2EE Design and Development.* Wrox Press.

16 Ali-Tolppa, J. and Tsvetkov, T. (2016). Optimistic Concurrency Control in Self-Organizing Networks Using Automated Coordination and Verification. IEEE/IFIP Network Operations and Management Symposium, Istanbul.

17 Bandh, T. (2013). Coordination of autonomic function execution in Self-Organizing Networks. Ph.D. dissertation. Technical University of Munich.

18 Romeikat, R., Sanneck, H. and Thobias, B. (2013). Efficient, Dynamic Coordination of Request Batches in C-SON Systems. IEEE Vehicle Technology Conference, Dresden.

19 Räisänen, V. and Tang, H. (2011). Knowledge Modeling for Conflict Detection in Self-Organized Networks. International Conference on Mobile Networks and Management (MONAMI), Aveiro.

20 Mwanje, S.S., Mannweiler, C., Schmelz, L. and ul-Islam, M. (2016). Network Management Automation in 5G: Challenges and Opportunities. IEEE International Symposium on Personal, Indoor and Mobile Radio Communications (PIMRC), Valenica.

21 Mwanje, S.S., Schmelz, L.C., and Mitschele-Thiel, A. (2016). Cognitive cellular networks: a Q-learning framework for self-organizing networks. *IEEE Transactions on Network and Service Management* 13 (1): 85–98.

22 Mwanje, S.S. (2015). Coordinating Coupled Self-Organized Network Functions in Cellular Radio Networks. Dissertation. Technical University Ilmenau.

23 Mwanje, S.S., Sanneck, H., and Mitschele-Thiel, A. (2017). Synchronized cooperative learning for coordinating cognitive network management functions. *IEEE Transactions on Cognitive Communications and Networking* 4 (2): 244–256.

24 Nash, J.F. (1951). Non-cooperative games. *Annals of Mathematics* 54 (2): 286–295.

25 Levin, J. (2001). Dynamic games with incomplete information. *Nature* 2 (2): 2.

26 Fudenberg, D. and Levine, D.K. (2016). Whither game theory? Towards a theory of learning in games. *Journal of Economic Perspectives* 30 (4): 151–170.

27 3GPP (2011). *Evolved Universal Terrestrial Radio Access Network (EUTRAN); Self-configuring and Self-Optimizing Network (SON) use cases and solutions.* Sophia Antipolis: 3GPP.

28 SOCRATES (2010). Deliverable d5.9: Final report on self-organisation and its implications in wireless access networks. Eu Strep Socrates (INFSOICT-216284).

29 Mwanje, S.S., Mannweiler, C. and Schmelz, C. (2016). Method and Apparatus for Providing Cognitive Functions and Facilitating management in Cognitive Network Management Systems. Patent PCT/IB2016/055288, filed 02 September 2016 and issued 08 March 2018.

12

Towards Actualizing Network Autonomy

Stephen S. Mwanje[1], Jürgen Goerge[1], Janne Ali-Tolppa[1], Kimmo Hatonen[2], Harald Bender[1], Csaba Rotter[3], Ilaria Malanchini[4] and Henning Sanneck[1]

[1] *Nokia Bell Labs, Munich, Germany*
[2] *Nokia Bell Labs, Espoo, Finland*
[3] *Nokia Bell labs, Budapest, Hungary*
[4] *Nokia Bell Labs, Stuttgart, Germany*

The previous chapters of this book have presented the technical challenges in managing communication networks and solutions and solution concepts that can be used to achieve cognitive autonomy for networks. Effectively achieving that outcome, however, requires more than technical solutions. An appropriate environment or operations framework is needed. This chapter seeks to answer two questions:

1) What is the framework and non-technical concerns that need to be resolved to ensure that cognitive autonomy can be achieved?
2) How should these paradigms be leveraged towards achieving cognitive autonomy for networks?

In general, the book does not claim that the network automation challenge has been fully resolved. Rather, it has laid down solutions and ideas, which – when combined – can move network operations closer to cognitive autonomy. In the same respect, this chapter attempts to look towards that future with a focus on how the cognitive autonomous vision can be impacted by the operations framework/environment. The chapter lays out a vision on the expectations of cognitive autonomy, the challenges of network modelling and creating ever more abstract representations of the network, as well as the implications of cognition for the architectural and functional design of networks. It describes what would be expected from a network management perspective and relates it to other high-level goals including intent-based operations. Meanwhile, as an attempt to highlight a critical solution idea, the chapter discusses the concepts of DevOps and their inter-relations with cognitive management. Finally, given the central role of data to the entire cognitive management paradigm, the chapter closes with a discussion of the value and implication of data in (cognitive) network operations. This is crucial because it will determine the degree to which cognitive solutions can be appropriately trained and to what extent knowledge can be transferred from one operation to the other.

12.1 Cognitive Autonomous Networks – The Vision

Chapter 1 characterized the levels of network automation distinguishing between scripted, self-organizing, autonomous, and cognitive networks. Specifically, it defines a scripted network as the human-controlled network that allows scenario-specific execution of automation scripts and contrasts it with a Self-Organizing Network (SON), which automates the interpretation of events under different contexts as well as the selection and execution of actions. An autonomous network is able to act on its own even although it may not be able to appropriately reason when selecting actions, while a cognitive network is able to reason and formulate recommendations for behaviour, even when it may not be independent in execution. It is evident, as such, that the desired system is a cognitive autonomous network, i.e. one that uses reasoning to formulate recommendations for action (cognition) and, subsequently, independently executes the derived actions (autonomy).

Chapters 7–10 focused on solutions towards cognitive autonomy, but it is still necessary to answer questions on the underlying principles and concepts that will guarantee the success of Cognitive Autonomous Network (CAN). Amongst the questions to answer are following:

1) How do the different cognitive techniques (as presented in Chapters 5 and 6) compare to one another, in particular, for network management automation tasks? In general, the recent success of machine learning (ML) hints at its foremost position in the pack, but one still needs to contrast it against other solutions to ensure that the right solution will be picked for any given job.
2) What are the factors to guarantee the successful implementation and deployment of ML applications and algorithms? ML here is used as a test case with the understanding that principles picked for ML will also be of interest for the technologies.
3) What changes are needed on the entire Network Management (NM) architecture? It is expected that the new data intensive cognitive approaches will require new components in the NM system, and these must be identified to determine if CAN developments are justified amidst these new changes. Moreover, the question on how data can be enriched to be enable ML is also critical to answer.

4) What are the implications on Key Performance Indicator (KPI) design and event/data management? Essentially, given the capability to process large amounts of data at different points in the network, should the KPIs be redesigned or should the collection of events be revised both to ease and leverage these data processing capabilities?

Obviously, there can be many other questions, but the four questions above are amongst the most crucial to address. The rest of this section will attempt to discuss these and how they could be addressed.

12.1.1 Cognitive Techniques in Network Automation

As discussed in Chapters 5 and 6, it is clear there are many techniques that can be used for decision making with each providing different cognitive capabilities. On the one hand, the successful usage of cognitive techniques will highly depend on the operational conditions including data management capabilities and the culture mechanisms underpinning proper application of cognitive tools especially ML techniques. On the other hand, however, there remains a critical challenge on the choice of technique for a given network automation requirement.

12.1.1.1 Matching Problem Requirements

The most important requirement is that the capabilities of the selected technology achieve the requirements of the problem at hand. Figure 12.1 summarizes the comparative capabilities of the technologies, as presented in Chapters 5 and 6, when gauged on the data-processing model of cognition presented in Chapter 4. Many technologies offer inference capabilities and only a few offer perceptive capabilities, so it is important to identify the problem requirements in order to choose the right technology.

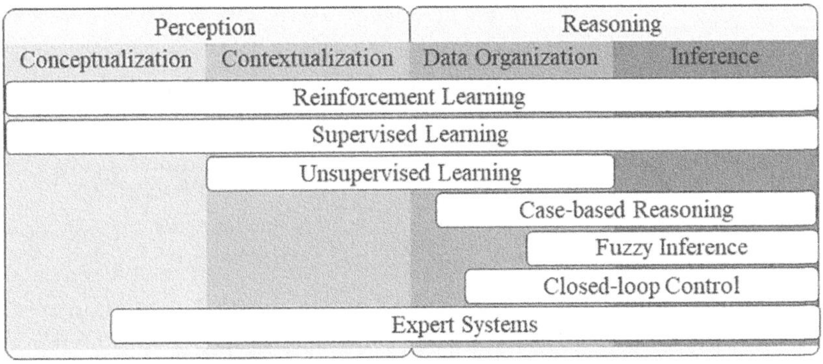

Figure 12.1 Capabilities of the different automation techniques.

However, besides matching the problem requirements, there is also a challenge in selecting the technology when there are multiple techniques that achieve the problem requirements. In principle, this breaks down to how these technologies map to the cognitive dimensions. The choice lies in two dimensions:

1) Weighing the need for human development and configuration effort against the data and training requirements for the selected solution.
2) Weighing the development effort against training time.

12.1.1.2 Development Effort vs Required Data

One observation is that the techniques tend to trade human effort for intensity of training (data) as illustrated by Figure 12.2. Expert systems have the most intensive requirement for human design especially in the knowledge needed to derive the rules. However, it has the least training needs; the developed solution is applied directly to solving the problem without need for training. Closed-loop control systems simplify the human design challenges given their ability to use feedback to improve the solutions. This however comes with a need for training; the system requires some time before reaching its steady state. Fuzzy inference systems also require intense human design needs owing to the rules that must be evaluated by the system. Otherwise, Case-Based Reasoning (CBR) has the least need for human design as the system independently captures the observed states. However, it does have a high need for training as it takes time to learn all the applicable possible

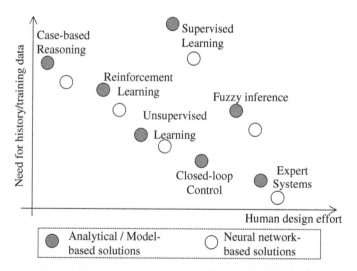

Figure 12.2 A comparative map of the design effort against training data requirements for the different automation techniques – in each case, neural network-based solutions require less data for the same level of accuracy.

states. Note that in the cases where one wishes to use CBR to identify *a priori* known cases from the input data stream and make the system to react to those cases, a lot of human effort will be required to identify, analyse, select, and annotate or attach actions to those cases needed to be found. This, however, is true for all cases where the human has special or specific requirements on the used data. The tasks of identifying, analysing, selecting, and annotating or attaching actions to specific data will be required in all such special tasks.

ML systems are moderate on human design effort – at least for the system operator. For example, supervised learning cost functions are easier to design compared, for example, to rules in expert systems and unsupervised learning only needs to design the measures of accuracy, e.g. the distance measure in clustering problems. However, the training requirements of ML solutions differ depending on the problem. Unsupervised learning generally requires less training data compared to supervised learning and Reinforcement learning (RL), which has the most confusing characteristics. Although the science behind the machine learning solutions is complex, especially for RL, it is easily transferred from one problem to another when clearly understood. Similarly, RL is also different in that no data is required to be accumulated offline but the solutions require a lot of time for the online learning solution to converge.

The individual technologies also differ in terms of the applied computing engine, either an analytical model or a neural network (NN) model. In general, neural networks are used for ML solutions, but it should also be possible to apply them for the classic artificial intelligence (AI) solutions, e.g. for fuzzy inference or CBR. In CBR, for example, the NN can be used to make hypotheses and to guide the CBR module in the search for a similar previous case that supports one of the hypotheses. If used for diagnosis, for instance, the knowledge acquired by the network is interpreted and mapped into symbolic diagnosis descriptors, whose structure is stored in the NN. This can then be used by the system to determine whether a given answer is credible for a given observation.

Comparing the analytical and NN solutions, although neural network-based solutions achieve better accuracy and can solve more complicated problems, they also come with more design challenges, mainly in selecting the NN's hyperparameters. For a specific problem, however, their ability to achieve better performance for the same amount of training data implies that they may require less data for the same level accuracy owing to their ability to easily learn more complex structures [1].

12.1.1.3 Training Time vs Development Effort
Another dimension is the contrast between development effort and training time. For most technologies, inference time is infinitesimally small, i.e. after the model is adequately trained, to compute a solution is quasi instantaneous for any

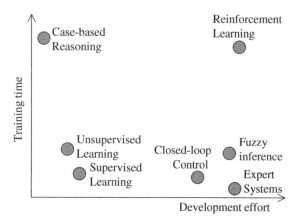

Figure 12.3 A comparative map of the design effort against the required training time for the different automation techniques.

observation or scenario. The training time, however, is significant and varies for the different technologies as illustrated by Figure 12.3.

Besides CBR which requires minimal rules to be applicable to a problem, all classic AI (expert systems, Fuzzy inference, closed-loop control) requires significant development effort owing to high reliance on rules that must be developed by the human experts. CBR, on the other hand, requires little development effort, only requires a data base for tracking past cases, but requires a significant amount of time to train because it must observe the events and compare them one by one with previous events.

Supervised and unsupervised learning require less development effort compared with classic AI systems since they only require a general understanding of the concepts and little solution-specific details. Some development effort is, however, still needed to ensure that the designed solution meets the requirements of the problem at hand. Moreover, there may be some meta design needed, e.g. to select the most appropriate hyper parameters for model.

Since the data for training is collected/availed beforehand, training takes a short time, at least in comparison to CBR and RL. Such a training time is only limited by the capabilities of the available computation capacity (for a detailed discussion see Chapter 6).

RL is the significant outlier. The development effort is significant owing to the complexity of the concepts and, mainly, to the need to develop appropriate reward mechanisms. The learning is highly dependent on the reward mechanism and rewards need to be designed to reinforce the desired outcomes, a process that is not obvious to all. Then, the need to learn in an online format implies a long training period that highly depends on how quickly the underlying system takes to effectuate suggested actions and on how fast it generates subsequent observations.

In general, these large differences in the development vs training characteristics imply that the technologies are not all equally appropriate for all problems. There should be some effort expected to match the problem to most appropriate technology.

12.1.2 Success Factors in Implementing CAN Projects

Besides developing systems for partial or full autonomy, AI/ML can be used to gain insights, information, and knowledge from large datasets or systems with complex dynamics and to augment the capabilities of human personnel by creating enhanced tools, models, processes, etc. This is the capability expected to be at the core of CANs and through it, CANs will deliver solutions that make it easier to configure and operate as well as improve the performance of systems through optimization and/or better algorithms. To achieve these outcomes, some specific means relating this to the successful implementation of ML projects must be afforded – specifically as regards to the objectives, the data and the resources.

First, a **mathematically expressible objective** is paramount. Although ML is able to 'choose' amongst many models, the objective function must be expressible mathematically and must appropriately represent the goal for ML to work its magic. As a conceptual descendant of curve-fitting, ML, selects the best model (the best fit for curve-fitting) because it minimizes the objective function, which measures the 'goodness-of-fit'. It follows then, that the appropriate objective function must be known otherwise an unwanted result may ensue.

Similarly, it is important to use the **appropriate class of models**, i.e. the models must fit the goal. Taking the curve-fitting example, if a class of linear models is considered, the curve-fitting solution will only find the best learn model, which may not actually be accurate as possible. Instead, by changing to another class of models – say polynomial models, a better result may be observed. Note that the reverse is also true, i.e. utilizing a polynomial model where a linear model may have been appropriate may lead to overfitting the training data resulting in poor performance on the unseen data.

The third requirement is that any ML project starts with a **large dataset** as the essential raw material that shapes the project goals. It follows then that any conversation on applying ML to a problem is followed by the reply 'what does your data look like?'. The amount and kind of data needed for success is often more expensive to acquire and maintain than is anticipated. The kind of data (e.g. labelled) determines what kind of ML methods can be applied. For some applications, data can be generated artificially by, for example, simulation or test environments. This kind of data has some limitations that must be correctly accounted.

Finally, any project requires the right resources, i.e. the implementation team should have the right resourcing level with right skill sets and in the right proportions. The major expertise needed are: a domain expert, a data architect, a data engineer, a software coding expert, a ML expert (which is essentially a mathematical skill) as well as a visualization expert. Each of these is an engineering discipline in its own right and cannot simply be replaced by tools. A ML expert cannot, for example, be replaced by a ML tool such as TensorFlow just as a visualization expert cannot be replaced by a 'visualization tool' such as Tableau or plotly. Also, the right amount of computing resources must be allocated. Often, this is greater than anticipated. In general, a project that is not properly resourced will fail with the consequence that the failure will be interpreted as a failure of ML.

12.1.3 Implications on KPI Design and Event Logging

The application of ML in network operations will have a significant effect on the way network performance is evaluated. Allowing for maximizing the capabilities, for which machines outperform humans (e.g. handling scale and diversity), may guarantee better outcomes. Traditionally, network KPIs were designed for interpretation by humans so the design of KPIs needs to change to maximize these machine capabilities.

Firstly, KPIs are made concise as humans have limits on the amount of data they can handle simultaneously. The KPI could not be designed with so many specifics or specialties, otherwise the human operator can no longer evaluate that many specialties. However, the advent of advanced data processing capabilities should allow for more fine-grained evaluation of KPIs. This directly translates into the capability to process more data even for the same decisions as were taken by operators. Many more KPIs than what could be simultaneously handled by the human operator can now be added to give even more insight into what is happening, i.e. multiple views may be taken of the same event, which would allow for even better decisions.

Secondly, the number of KPIs for each event or measure in traditional operations were made fewer since humans can easily contextualize them. Humans are good at identifying subtle differences with few examples, so even when the number of different measures for an event are few, the human can still identify the differences. Moreover, the human operator needs to make a quick decision and is unable to take an infinite time evaluating KPIs before coming to a decision. All this needs to change for machine-centred CAN. With lack of intuition in machine decision making, yet with the capability to handle large amounts of data, context could be added even for the existing events/KPIs to enable the machine to make rich decisions. For example, instead of simply noting that a Radio Link failure (RLF) has occurred, it may make sense to record the RLF with all the context as to when it happened, e.g. how fast the user was moving, what the exact radio conditions were, etc. the Machine would then learn better how to diagnose RLFs or any other such KPIs.

This added meta information may also be wider than the specific KPI of interest but may also include other meta information that closely relates to the event. Instead of simply stating that an event x happened, information is added to answer the question on 'what else happened when event x occurred?', e.g. 'what was the weather when the RLF occurred?'

Most important, however, is the capability to process large amounts of data which allows for finer granular data to be used for decision making, especially in cases where that data is processed locally. This allows the recoding of events to be contextualized to enrich the data. One example is to increase the frequency in some contexts and reduce it in others. For example, if a session is dropped, the gNB may increase the collection of drive-test data or channel-state information to diagnose the cause of the drop. Such data collection may then lessen when the drop is corrected. Similarly, better or expanded KPI information may be collected/computed if a certain event is predicted and relaxed when such a threat is no longer likely.

Also, the capability for big data implies that data from different perspectives can be mixed when deriving decisions. Data may be processed locally to capture local insight and then sent to a remote entity to aggregate the different local insights with a broader perspective. This allows multi-dimensional views on network data and multi-perspective decisions, which may result in better operational decisions.

12.1.4 Network Function Centralization and Federation

The broad CAN capabilities, i.e. cognition and autonomy, reflect a technological evolution of NM solutions, but they must also track network characteristics, i.e. CAN capabilities must address specific networks features and capabilities – including the degree of centralization.

Future networks will be highly distributed as they are driven by concepts on ultra-dense HetNets. It is desired, however, that this should not be a challenge for CAN or, at most, should be a manageable challenge. This is mainly because being autonomous translates into the ease to scale the solutions – the required human effort does not increase with the number of network elements to be simultaneously managed. There is the concern that low-delay services (and the related demands on data granularity, algorithm execution intervals and prediction needs) may lead to distribution of CAN solutions and thereby reduce the capability to scale. This is a justified concern, but the distributed solution in such a case only addresses local scope and does not need to share large amounts of data with centralized instances. So, the scaling may only be a challenge on set up but will be contained during the operations.

Concurrent to distribution, there is also a big push for centralization, e.g. to centralize the Radio Access Networks (RANs) as much as possible to the extent of even centralizing the near-real-time part of the radio protocol stack [2]. As such, CANs must concurrently support distribution and centralization – be it for the core, RAN or otherwise. Of great importance in this centralization is the concept of cloudification. More and more network functions will be hosted in cloud instances – be it central, edge or other cloud instances – leaving only the bare minimum to the access site. This cloudification, however, should not only be a challenge but also the opportunity for validating cognitive autonomy. Given that network data will also be increasingly available at cloud locations, cognitive algorithms can exploit a broader dataset to make decisions quicker and better.

As the number and combinations of use cases supported by networks increase, the nature of the networks will also change. More importantly, networks cease to be homogeneous, becoming different and distinct for each owner, operator, and market. For example, instances of non-public networks will proliferate with both licensed and unlicensed spectrum scenarios or combinations thereof. The size of these networks will also greatly vary from campus networks in a small rural factory with tens of access points, for example, to dense public land mobile networks of large network operators with thousands to millions of access points. So, NM systems must fit the network and its owner or operator conditions. For example, management of campus networks bears the challenge that the enterprise has to manage a network with similar network elements and, in many cases, with comparable complexity as a mobile network operator, yet the enterprise does not have the scale to invest in detailed know-how on network planning and operation as is done by mobile network operators. In general, only very big companies will undertake the management on their own, whilst all others will only carry out the basic local maintenance and outsource the rest to a remote service where the management is fairly similar to the Communication Services Provider (CSP's) or a CSP-managed services case. For the enterprises who run small private networks, the local maintenance needs to be as zero touch as possible and autonomous network operation will become of special importance. In this case, the (Big)data-native character of CANs will again become indispensable, the CAN becomes both the driver and the requirement.

12.1.5   CAN Outlook on Architecture and Technology Evolution

As network slicing becomes the principle way of supporting logical separation amongst services, CANs can be expected to support additional slicing management requirements. Slicing may become normal even in non-public networks, for example, to differentiate the communication services for different production lines in a factory. Similarly, CANs must support slicing used to separate non-public and public services over a common substrate network. Moreover, it will become increasingly necessary to support dynamic slicing, i.e. to allow for flexible deployment, reconfiguration, and decommissioning of network slices both for end-to-end slices and slice subnets in individual network domains.

The maturity of cloud services and infrastructure combined with the successes of open-source software in the IT industry have motivated many in the industry to push NM towards these two realities, i.e. to cloud-native, open-source implementations. The NM framework and the corresponding cognitive automation must, as such, also evolve to meet these demands. The expected outcome is that there will most likely be a separation into three distinct parts (see Figure 12.4):

1) vendor-specific management and instrumentation of the network functions;
2) a unified (and possibly open-source) management platform to which vendors may contribute capabilities, and
3) a combination of vendor-specific and vendor-neutral (common) management applications, where the vendor's agnostic applications undertake the simple, common, or generic use cases leaving the specialized tasks to the vendor-specific applications.

Here, the CAN framework should mainly support the described separation but does not drive or demand such a split. Instead, the split places new design requirements on the CAN systems – the CAN system has to be designed in a way to leverage all the data and capabilities from the three subparts (i.e. from the infrastructure, the platform, and the applications) and from multiple vendors who may be supported in each of the three parts.

The NM architecture needs to evolve to a system that is able to leverage multiple solution approaches and technologies. Specifically, the NM architecture, represented by Figure 12.5, must allow for policy- and/or intent-based management fulfilled by multiple automation functions (as opposed to rule-based functions) that are hosted over multiple single- and/or multi-domain management and

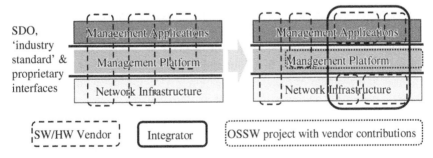

Figure 12.4 Cloudification as a drive to evolve the architecture of network management systems.

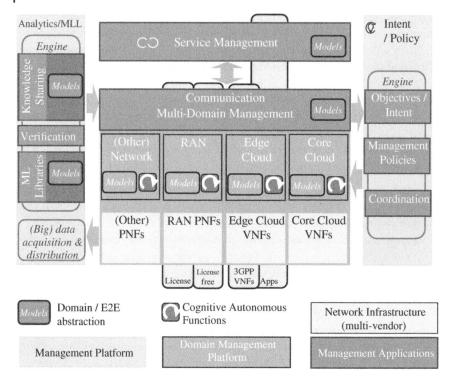

Figure 12.5 The components of the future CAN-ready network management system.

data/analytics platforms. Concurrently, the management functions must allow for better scalability and simpler standardization of interfaces. This requires the right level of abstraction, so the models assumed in NM must be harmonized to guarantee the right abstractions for automation, cf. Section 2.6.

As some networks will even be operated by non-experts, e.g. the non-public networks (NPNs) customized to a particular enterprise, the NM system needs to take more responsibility than it has taken to date. For example, leveraging the DevOps paradigm (described in Section 12.2), network diagnostics could go as far as editing the software for the network functions and automatically raising or reporting the changes to the vendor or the communication service provider without involving the NPN owner/operator who may not be in position to make sense of the changes or even the observed anomaly. Of course, this will not be desirable in all scenarios and will, in some cases, even be totally unacceptable to the NPN owner/operator, especially since it will not be easy to train the machine to achieve human-level expert skills in all NM aspects. However, it still demonstrate the need to expand the level of cognition exhibited by NM systems.

The key technologies for the basic building blocks of future NM systems are illustrated by Figure 12.5 and include pipelines of AI/ML Models, intent-based and multi-domain management. These have major impacts on the system design as well as on the resulting AI/ML architecture. The subsequent paragraphs and sections discuss the extent to which each of these will affect the evolved NM system architecture.

Intent-based management will allow human-level abstraction and reduce the need for a very high expertise level in operating networks. Multi-domain management for example, for slicing management will need to become a standard way of undertaking management actions. The management must support both software abstractions of logical network functions as well as the legacy physical network functions, both supported over vendor-specific and vendor neutral Domain management (DM) platforms. However, intent-based management still remains controversial as discussed in Section 12.2.

Pipelines of AI/ML models will allow different ML capabilities to be applied coherently to solve a certain use case. For example, in the development part of the process ML libraries and algorithms should be reused to create Certification Authority (CA) functions and in the operational part knowledge sharing amongst CA function instances will be possible. Similarly, trust in the actions proposed by ML algorithms can be developed when such algorithms are connected to standardized pipelines (e.g. for data pre-processing and sandboxing) that can verify a subset of the actions in predefined controlled scenarios.

ML is about utilizing the data that is available in the system and, therefore, data collection and processing is one of the key architectural questions in the development of CANs. This process can be modelled as a ML pipeline (ML-pipeline), which is in principle a service chain, that is collecting data from different sources, processing and, combining it and eventually producing higher level insights. Such ML-pipelines may span over several different domains (e.g. RAN, Core or outside 3rd Generation Partnership Project [3GPP] scope, etc.). For example, data collected in RAN may be processed by a service in the core network and the insights from this service used in the Operation, Administration, and Management systems.

To fully utilize the potential of the ML pipelines, they should be able to access different data sources in a flexible but consistent manner. This could include, for example, data sources with a very different granularity and timespan, for example, real-time reporting of user equipment (UE) measurements and NM data aggregated over several sites and long time periods, or different data types such as numerical timeseries KPIs or textual system logs. The existing hierarchical models in collecting counters and KPIs on different timeframes, domains, and abstraction levels can be too inflexible for setting up a variety of ML pipelines and, therefore, a Service-Based Architecture (SBA) is required. Also, due to the large amounts of data that can be utilized in ML and analytics, data sharing may work better than the more traditional data collection methods.

The common tools that are needed to store, share, and process data, form the ML platform. The use of ML may also set requirements on the hardware that the functions or services are running on. Graphics Processing Units (GPUs) or Tensor Processing Units (TPUs) may be required to accelerate both the training and the inference of the ML models. The requirements on the available hardware and also the available data, i.e. the volume of it or any labels for supervised learning, can be different for the training of the model to the inference. For example, it may be that the model is trained in a more centralized location, where more data and computational resources are available, and then deployed in a more distributed fashion to the edge for inference to meet latency requirements. Very low latency inference may also require specialized hardware to run on. To give a more concrete example, let us consider a scheduler that is optimized with ML to learn the optimal scheduling policies for the local context, i.e. the type of mobility pattern that is typical for a particular RAN cell. The model for the Radio Resource Management (RRM) optimization could be trained in a central cloud in the core network, where more data can be collected and stored for training and also data from many cells and sites can be combined for a more comprehensive training set. On the other hand, the optimization algorithm needs to be able to run in a very short and fast cycle, so for inference it needs to be deployed in the NodeB.

The initial development and training of a model for a certain use case should be distinguished from its training for a deployment to a new context, for example, in a new network. The initial development process would involve experts such as a domain expert, data scientist, and a ML algorithm specialist, who look at the requirements of a particular use case and, largely by iterative experimentation, find the best ML model to meet those requirements. This process requires training data to validate the model. But once the model design is finished, it may be deployed in several instances, where it needs to be trained to adapt to the local context. In our previous RRM optimization example, the same model could be deployed to several cells in several networks and learn the optimal model parametrization for each. It is also important to remember that after the initial training, the model may require re-training at a later point, to adapt to context changes, or, in some cases, e.g. with RL, the training process can be continuous during the inference. The different requirements on the network architecture for the training and the inference phase of the model are especially relevant for this operational phase training, re-training and inference. The operational phases of a ML-model or ML-pipeline could be called its *life-cycle*.

To manage the task of setting up a ML pipeline and to take care of the *ML-pipeline life-cycle management*, a component called Machine Learning

Orchestrator (MLO) can be introduced. It takes care of setting up the required service chains for the ML-pipeline, including placement decisions for each of the involved services to meet any hardware requirements. It is important to notice that these requirements and also the service chain required for different life-cycles of the ML pipeline can be different and so the MLO should be able to take care of all the required ML-pipeline life-cycle management operations as well, e.g. setting up the required service chains for each phase.

Having enough training data can often be an issue. Also, in case of ML-pipelines that include RL, it may not always be possible do the training phase in live network, since that can risk a service degradation. To address these issues, sandboxes can be used at least for the initial training of the models. A sandbox can be a limited part of the network, or a simulator or an emulator. The difficulty in this case is to make the simulations realistic enough. In case of campus networks, for example, in Industry 4.0 use cases, the simulations can be based on a more comprehensive digital model of the environment, where the network is deployed, the so-called *digital twin*. Such environments enable the ML methods to more freely experiment with different options, without endangering the service quality.

Since to fully utilize the potential of the ML-pipelines they should be able to access data on very different domains and timeframes, the pipelines may span over many domains (RAN, core, etc.) and planes (user, control, management). Also, the consumers for the ML-based insights might be in different domains or plains. Therefore, it could be argued that the ML-pipeline can be orthogonal to these traditional views on the network and even form a new *automation plane* of its own.

12.1.6 CAN Outlook on NM System Evolution

The network operations features and architecture in Sections 12.1.4 and 12.1.5 are part of the drivers, together with the complexity discussed in Chapter 1, for evolving NM, and it is evident that the CAN must be flexible to support all of these different features. In other words, even as it becomes more complex owing to increased cognitive capability and autonomy, the complexity of the CAN system should co-exist with and not erode flexibility.

To meet the future NM challenges, the NM system needs to evolve to the point where the operator ceases to be involved in the large majority of the operational activities and simply retain the supervisory and system planning roles. The large majority of operational activities will be undertaken by the automation system capable of responding to different network events with modules at different abstraction levels. In general, this evolution will follow the different levels of automation introduced in Section 1.3.1, i.e. from a system that can detect a single network event to one that correlates and diagnoses multiple events and eventually to one that contextualizes the events and anticipates both single- and multiple-correlated events.

The characterization of the different levels of automation hints at the stages, through which NM automation is expected to transit. This path, illustrated by Figure 12.6, starts with manual control transiting through automated and cognitive automated control, eventually to a cognitive autonomous network:

1) The manually controlled network may undertake some automation e.g. through scripts that automate specific routine actions to offer convenience to the operator.

2) The automated network, exemplified by SON, has human-designed control loops that undertake simple control through monitoring and (re)configuration of network parameters, but the functions are configured by the operator to optimize their performance. The majority of current network deployments are, at this stage, since they have deployed a number of automation functions in the networks, which are still largely being controlled by the operators.

3) The cognitive automated network adds cognitive capabilities over and above the automation functions. There is recent work that pushes the boundary towards stage three by proposing learning functions that take the responsibility of configuring the automation functions away from the operator. This has mainly been referred to as Cognitive Network Management (CNM). For a cognitive-automated network, the network automation system matches the prevailing contexts in the network environment to the manually defined context model, for which the cognitive functions then learn to identify the respective optimal or best configuration parameter values. This is, however, inadequate given that even the operator cannot enumerate all the possible context differences for any given context dimension. Specifically, it is impossible to detect and isolate the differentiating dimensions that would enable the operator to clearly differentiate between contexts.

4) Cognitive autonomous networks have full control and responsibility for the network. This final stage requires an end-to-end design of the CAN in a way that embeds the learning capabilities of AI/ML techniques into each part of the cognitive functions. The CAN will be characterized by functions that, besides being able to learn appropriate configuration parameter values for specific contexts in the network, are also able to learn the contexts and subsequently remap the learned structure to the context model. It is, therefore, necessary that the network automation system also automatically adjusts the context model during its operations. Consequently, the CAN system will have to learn the context dimensions and the granularities for each such dimension even as it learns the optimal network configurations for each such context.

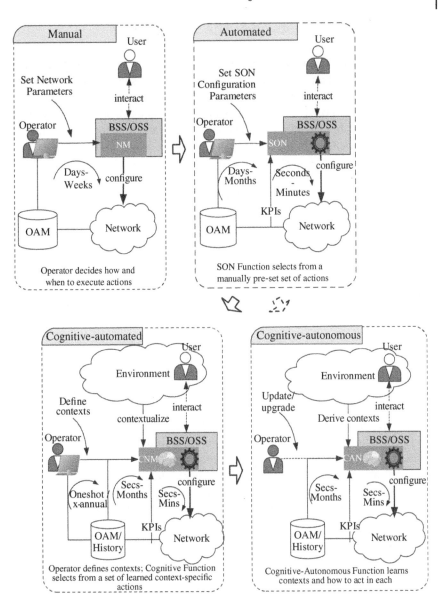

Figure 12.6 From a manually controlled to a cognitive autonomous network.

12.2 Modelling Networks: The System View

In traditional management, as described in Section 2.6, modelling the interfaces between management tools and the managed entities is critical since the interfaces support the human operator in transferring management data between the entities. NM, however, is more than pushing and pulling management data across some interfaces but includes the injection of the skilled human operators' knowledge and experience into the network processes. Modelling for network automation attempts to formalize the generation and transfer of this human knowledge to control the network.

Chapter 2 (specifically Section 2.6) has shown that based on current modelling approaches, it seems impossible to derive a *common* model that eases NM and automation. First, the telecommunication management network applications differ in their focus, scope of operations, timescale of operations and levels of abstraction. Similarly, managed entities have different management requirements, i.e. the requirements in the Radio, Core, Transport, and IT systems or 'Cloud' networks are different. And since these entities, managed and, managing, are provided by different vendors, they have too little in common that can be expressed in a common information model.

An alternative approach is to take a systems theory approach to the network's behaviour, to describe the relationship between the inputs and outputs of a network. This enables one to derive the relationship between the automation intents and the resulting policies and correspondingly allows more possibilities to further automate network control. Although the terms policy, intent, goals, and objectives have already been discussed in the literature e.g. by ETSI in [3] and IETF in [4], their understanding and usage nevertheless still causes a lot of misunderstandings. So, this section introduces a simple systems theory description of the network behaviour to clearly differentiate between the space of configurations and the space of observations. This separation enables the relationships between different strategies to control the network and is used to further clarify the terminology and motivate the study of intent-based approaches to network control. The section starts with general background of the system view before presenting the novel approach to system control starting with subsection 12.2.4.

12.2.1 System Description of a Mobile Network

A mobile network can be regarded a system that is described by a transfer function S, which translates the vector of controlling parameters $X = (x_1, x_2, \ldots x_n)$ in the configurations state-space into a resulting vector of observations $O = (o_1, o_2, \ldots o_m)$ in the observations -space:

$$S(X) = O \tag{12.1}$$

The parameters in the configurations state-space comprise all cells, their locations and directions, all parameters of the cells and their relations, as well as the configuration of core and transport network elements. In LTE this comprises thousands of parameters per cell and hundreds per neighbour relation. The configurations state-space also comprises the known and modelled part of the network's environment like buildings, mountains, etc. and the location of network elements from core and transport, which is relevant for the latency, for example.

The observations state-space represents the outcomes from the network that lead to how the user equipment (UE) experiences the mobile network performance. This space comprises, for example, air interface measurements (e.g. signal strength, interference, etc.), all events encountered at the many processing steps and subsystems (dropped calls, handover attempts, and failures), all alarms as well as the signalling traces between the network elements and the UE. The observations state-space also comprises end-to-end measurements of the user experience like bandwidth, latency, and jitter.

Unfortunately for mobile networks, the system transfer function S is not known in detail. Yet, as in any real-life system, a mobile radio network is disturbed by uncontrolled input D, which is, at least partly unknown as well:

$$S(X, D) = O \tag{12.2}$$

This disturbance is caused by, amongst others, the only approximately modelled environment of the network, the partly unknown traffic distribution, and the limited accuracy of simulating the propagation of electromagnetic waves in complex and partially unknown environments.

In general, the system's behaviour can be described either by the set of configurations state-space parameters X that control the behaviour or by the observations state-space output O that result from that behaviour. However, since neither the system transfer function S nor the disturbance D is known in detail, there is limited possibility to accurately predict how changes to the configuration will influence the observed output. Similarly, changes in the observed output may not always be traced back to specific changes of the configuration. It is, therefore, critical in the analysis of mobile radio networks to distinguish between the configuration and the observations state-spaces and their capability to describe the system's behaviour.

The following example might illustrate the problem: In the area between two (or more) cells, the UEs might experience a high number of dropped calls, which are typically assumed to be caused by a so-called coverage hole, i.e. an area between two (or more) cells that is not covered by any of the cells (Figure 12.7a). The typical solution here is to increase the transmission power of the cells to close the coverage hole. However, there is a chance that the dropped calls are caused by interference between the cells due to too much overlap of the two cells (Figure 12.7b). In this

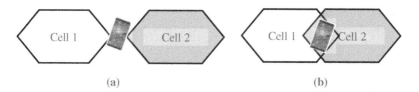

Figure 12.7 Coverage holes appear if the transmit power of the cells is to low (a), whereas interference of two cells happens if the transmit power of the cells is too high (b). In both case the probability of dropped calls is high.

case, increasing the transmission power of the cells would even worsen the problem as the required solution is to reduce the transmit power of the cells.

This example highlights the challenge with configurations-based control (i.e. deciding parameter changes) when the system transfer function S and the disturbance D are not well known. This then justifies the need for alternative control approaches that state the expected outcomes (e.g. to reduce the dropped calls) as opposed to stating changes in configuration (increasing the transmit power).

12.2.2 Describing Performance

By using a function F, the vector of raw observations O can be aggregated into one or several 'performance metrics' p_i.

$$p_f = F \cdot O \tag{12.3}$$

The overall performance p_f is as a combination (vector) of several components or metrics $p_f = (p_1, p_2, \ldots)$, like coverage, capacity, latency, rate of radio link failures, or rate of unsuccessful handovers. These metrics p_i are functions of subsets of the raw observables and are intended to represent an abstracted view to the observations. But selecting the subsets and functions is not trivial, especially since the observations state-space is not based on orthogonal, independent dimensions, and the observations are not independent from each other. Correspondingly, only a few simple functions to calculate performance metrics ('Key Performance Indicators') have been standardized, whereas the majority are defined individually by vendors and operators.

In many cases, the individual performance metrics p_i are used to express the quality of the network. However, there is a high probability that individual performance metrics as such cannot be used as optimization targets, because such targets would be mutually conflicting. For example, performance metrics for energy consumption and for network coverage can neither be interpreted in a mathematical sense as quality metrics nor used as optimization targets, since the absolute minimum of energy consumption is reached if all base stations are switched off, which means no coverage at all. The target to optimize the energy consumption is in conflict with the target to optimize the coverage of the network. Even the KPIs 'energy per transferred bit' or 'energy per covered area' are not sufficient either, because optimizing energy per bit would result in networks that do not cover rural areas, whereas optimizing energy per area may provide insufficient bandwidth to the cities. Similarly, the performance metric 'cost of the network' reaches its minimum if no network is installed at all while in that case the quality of the resulting network is zero.

Hence, to use the abstracted performance metrics p_i as an objective in network optimization, the operator mandatorily has to define implicitly or explicitly a weighting function Q to balance between the conflicting targets of the individual performance metrics by using weights w:

$$Q: = Q(w_i, p_i) \tag{12.4}$$

The weighting of the different components $\boldsymbol{p_i}$ to result in an overall target or 'super KPI' $Q: = Q(w_i, p_i)$ depends on the situation and the business strategy of the network operator. Usually, the business strategy translates into the classical marketing mix that consists of product, price, distribution, and promotion. One operator might decide to deploy a network with few base stations, invest little in marketing, and offer the services at low subscription rates via retail chains. In contrast, another operator might invest more into the network and its operations, positioning themselves as a premium operator and offer the services at higher subscription rates. These basic strategic considerations will influence the weighting function Q for each operator.

With all these transformations, it is evident that the right side of the system description equation offers several levels of abstraction to describe the system: i.e. the vector of raw observations $O = (o_1, o_2, \dots o_m)$ (Eq. 12.2), the calculated performance metrics p_i (Eq. 12.3), and the quality metrics Q, which contain operator-specific weighting functions (Eq. 12.4).

12.2.3 Implications on Automation

Many currently known solutions to automate the management of networks are based on the notion of resources that are selected, configured, and composed ('orchestrated') by some management systems to provide the required end-to-end service. For transport and core networks and for the virtualization environment such automation is well known [5].

From a system description point of view, this is based on the fact that the system transfer function of these networks is reasonably well understood. For example, two routers connected by a cable or fibre provide a link with very well-defined behaviour that can be modelled as a well-described 'resource' independent from its environment. Due to the known system transfer function, any change in the configurations state-pace can be translated into the observations state-space and vice versa. At least in theory, an automation algorithm is able to derive the necessary configuration changes which result from the observations state-space changes that the service requires. In many cases, even 'if – then – else' rules might be able to derive the necessary configuration changes from the required observations state-space changes. This enables automation algorithms to compose a required end-to-end service by combining network functions ('resources') and to select configurations that will result in observations expected by the services and derived by the system transfer function.

As shown above for mobile RANs, neither the system transfer function nor the disturbance is known in detail. Correspondingly, an automation algorithm cannot guarantee that a change in the configurations state-space will result in the desired observation. Even in theory, an automation algorithm might not be able to derive by simple rules the necessary configuration changes to implement a new end-to-end service; In many cases, it requires multidimensional, non-linear optimization of several cost functions with inherently conflicting targets to derive the relationship between configurations and corresponding observations. Since the running network cannot be used for trial-and-error based optimizations, such optimizations can only be performed by simulating the network, which requires the corresponding computing resources and introduces further sources of errors.

This difference in observability and controllability is the reason why automation in the domain of RANs is far more complicated than in case of transport networks, core networks, or the virtualization infrastructure.

12.2.4 Control Strategies

In classical NM, the management functions are setting parameters from the configuration space to control the network. As discussed in Section 12.2.2, it is a precondition that the management functions and managed entities are using the same model for the dynamic behaviour, the format, and the semantics of the management data. Since the internal implementation of network functions depends on the vendor and the release, it is evident that the objects and parameters to configure the network functions will differ. As a consequence, common management functions will not be possible as long as they depend on a common model of the network functions in the configuration space. The following paragraphs explore how far the space of observations might be used to control the network.

12.2.4.1 Configuration vs Goal-Based Control

Using the system model, a controlling entity external to the system has two options to express its desired change to the system as illustrated by Figure 12.8.

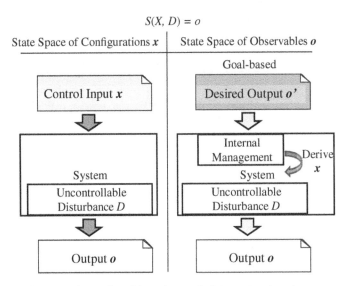

Figure 12.8 Configuration and goal-based control of the network system.

The controller can modify control parameters $X = (x_1, x_2, \ldots x_n)$ from the configurations state-space, expecting the system to result in proper behaviour and observations state-space outcome $O = (o_1, o_2, \ldots o_m)$. Alternatively, the controller sends a set of desired outcomes O, P or Q and expects some management entity within the system to find proper configuration X to reach the desired outcome. However, even in this case where the external controller articulates the control input as states from the observations state-space (right-side of Figure 12.8), the system still requires a concrete internal configuration, i.e. some vendor-/technology-specific adaptation needs to properly translate this observations state-space input into concrete parameter configurations.

In principle, the two approaches mirror the management principles of 'command and control' vs 'goal-based' management. In the first principle, the manager directs the employee to do specific tasks with the hope, that this will lead to the desired target, while in the second, the manager expresses the goal and leaves it to the employee to decide how to reach the target. Both approaches might be valid for any layer of the management hierarchy, with some examples of the options available at different layers listed in Table 12.1.

12.2.4.2 Command-Based vs State-Based Configuration Management

Many complex configuration changes of a managed entity require a sequence of actions to transition the managed entity from the source configuration state to the target configuration state. For example, from the 3GPP northbound interfaces as

Table 12.1 Examples of configuration and goal-based policies on different layers of management functions.

	Mode of management	
Layer of function	Configurations state-space	Goal-based/observations state-space
Business	Option 1: Roll out a network Option 2: Rent bandwidth from an existing network	Guarantee connectivity for UE in a large area
Service	Option 1: Install a new BTS Option 2: Optimize for coverage and capacity	Offer connectivity of 30 Mbps at a hotspot for 100 UE
Network	Option 1: Activate carrier aggregation Option 2: Optimize for coverage and capacity	Increase throughput at the cell border between two cells
Element management	Option 1: Set 'length of sliding window for load-based HO' to 10s Option 2: Change the size of the cells, change HO thresholds, etc.	The load between cell 1 and cell 2 shall be equally distributed.

well as from network elements the basic concepts of 'command-based' [6, 7] and 'state-based' [8, 9] Configuration Management (CM) are known:

In command-based (or statement-based) CM, the management system analyses how to transition the system from configurational source state to target state and derives an element-specific sequence of commands (e.g. in 'command line scripts'). This sequence of commands depends on the internal dynamic behaviour of the managed entity and, for example, the state of the ongoing call-processing. For the interface between network management and element manager, 3GPP's Basic CM [7] follows this approach. Further, this concept is the basis for the command line interfaces ('CLIs') of network elements.

In state-based CM, the management system provides the target state in configurations state-space to the managed entity and relies on the managed entity to autonomously perform all necessary steps to reach the desired target state. This concept is the basis for 3GPP's Bulk CM northbound interface between NM and element management [9] and is state-of-the art for the interface between element/domain managers and the network elements of the RAN. Many network elements from core and transport are adopting this concept, e.g. by the introduction of netconf [10, 11].

Both command- and state-based CM require expert knowledge about the allowed configurational states of a managed entity and the allowed state transitions. This implies that considerable effort must be expended to analyse, design, maintain, and test the corresponding generators of such error-free, optimal command sequences. In the command-based approach, this effort has to be invested by the management systems, whereas in the state-based CM, the corresponding functions are part of the managed entity. This has two major advantages: (i) the managed entity's R&D has more knowledge about the entity's internal behaviour, and (ii) the managed entity has better knowledge during run time about its current configuration and the ongoing call processing than does any management system.

From the management system perspective, the state-based approach represents an abstracted view that hides the details of the paths through the configuration space. This hides the internal and external dynamic behaviour of the managed entity from the management system. However, the management system needs to understand the detailed semantics of the standardized and implementation-specific parameters of the managed entities to select/derive the final state. From this point of view, state-based CM does not provide any abstraction.

Relatedly, one may be tempted to consider state-based CM as an example of goal-based control, i.e. because of the ability to hide the details of the configuration paths. However, even when that ability is considered abstraction, state-based CM cannot be regarded as a goal-based management strategy, because the final state is still within the configurations state-space, i.e. the management system does not state the goals for which the target configuration has been selected.

12.2.4.3 Benefits and Limitations of Configuration- and Goal-Based Control

In Table 12.1, there are many cases where multiple configurations need to be implemented to achieve a given goal. In the same way, there might be multiple reasons for a given configuration, i.e. a given configuration could be mapped to multiple goals as stated in Section 12.3.1. Specifically, because both the system transfer function S and the uncontrollable disturbance D are not known in detail, there is no one-to-one correspondence between configuration and the desired outcome, i.e. there is no guarantee that a particular configuration will fully achieve a particular stated goal for any of the functional layers. For example, the goal to 'guarantee connectivity for UE in a large area' might be implemented by rolling out a network, or by renting bandwidth from an existing network. Therefore, none of the two strategies is perfect, both having merits and limitations.

Configurations-based control imposes enormous maintenance work and cost since it must keep up with changes in the configuration parameters, a very tedious exercise especially at the network element level. The need to define values for all network element parameters requires the managing entity to understand the semantics of those configuration parameters, an error-prone and time-consuming task for the radio network elements, which have thousands of proprietary parameters. Moreover, because the parameters change with each new feature of the base station and have nothing in common with other vendors' implementation-specific parameters, it is very tedious to keep the external managing entity synchronized with the NEs. Specifically, the external controller need not only have release-specific, vendor-specific business logic, but must also have frequent business logic updates to cope with new the features of new releases.

Purely goal-based control avoids this problem since the goals are inherently independent of specific implementations and can be adequately expressed in terms of the observations state-space. The challenge, however, is that it still remains hard to develop mechanisms that are able to automatically derive the necessary configurations-state inputs from the desired outcomes.

Configurations-based management as well as goal-based management depend on both, the availability of proper information from all relevant entities and on the possibilities to control all affected entities. In the example in Figure 12.7 the system that derives the control input must be aware of the fact that the cells either leave a coverage hole or have too much overlap. In the given example, Cell-1 might even be able to infer from measurements that the interference is too high and that Cell-2 has to reduce its transmit power. But Cell-1 is not able to control the transmit power of Cell-2 if the cells are hosted by different base stations.

12.2.4.4 Implicit Mix of Strategies

Although there is a tendency to imply that management systems are moving from configuration-based management and towards goal-based management, it is worth noting that purely configuration-based management is not possible, i.e. there has not been any case of purely configurations-based control. Instead management systems have infused the two approaches even although with a more explicit focus on the configurations-based model.

Management systems positioned at higher layers of the management hierarchy typically have a more abstract view of the network elements: the higher the level in the management hierarchy, the less awareness of the details. At the business management layer, the operator defines a new service using a very abstract view to the network and its resources with the expectation that finally all the details for a valid, correct, and concrete configuration are available at network element layer. However, without further input, none of the management functions are by their own able to 'invent' sensible, detailed configurations. The management system requires additional input in the form of o, p, and Q (or combinations thereof), which is typically implicitly provided through the experience and background knowledge of the network engineer working with the management tools.

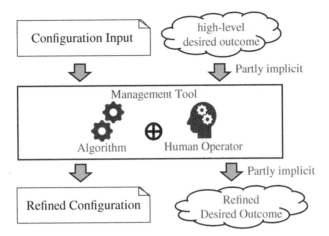

Figure 12.9 Each management layer receives configuration input, which is combined with the implicit operator input on desired outcome to generate a refined configuration for the next layer.

As illustrated by Figure 12.9, a typical layer of the management hierarchy receives abstract requests from a higher layer and outputs a refined, more specific (less abstract) configuration to management functions of the next lower level. Within the management layer, algorithms and the human operators evaluate the abstract control input and combine it with the (mostly) implicitly stated input on the desired outcomes, which is derived from the human operator's experience, background, and contextual knowledge.

For each domain, several machine-to-machine interfaces exist to transfer configuration data between the different management functions whilst highly-skilled network engineers control and implicitly provide goal-related input to the algorithms. For example, the operators provide thresholds, algorithm triggers, and the applicable parameter ranges for different contexts. Further automation of NM requires to add machine-to-machine interfaces to formalize the transfer of goal-based inputs. Additionally, it is necessary to find cognitive algorithms, which are able to combine the specified goals and outcomes with the necessary configuration changes and the current configurations to finally generate the optimized, specific configurations for the network as a whole and individual network functions.

12.2.5 Two-Dimensional Continuum of Control

The different abstraction of describing the network's behaviour, i.e. in terms of configuration or the desired outcomes, directly translate into different abstraction of control. This section derives these abstractions leading to clear definitions of the related terminology.

Table 12.2 Example network management control input differentiated along the dimensions of management functions layering and the abstract outcomes.

	Controlling input			
NM layer	Configuration X	Raw observation O	KPI P	Weighted quality metric Q
Business	`Not applicable in this example`	Support premium video service	Not applicable in this example	SLA for users in the large area shall be met.
Service	`Not applicable in this example`	Reduce video stalling, page download time	Average Bitrate per eNB	Throughput in given area shall be 10/40 Mbit/UE in rural/city.
Network	`Activate Carrier Aggregation`	Increase UE average throughput	Bitrate per UE per available resources	cell-edge Throughput shall be 3 Mbit/UE
Element management	`Set: activateCA=1; CellPairs: 1&4, 4&1, 2&5, 5&2, 3&6, 6&3; A3=offset; A5=thr1, thr2; A6=offset; ...`	Average bitrate cell = 1 per UE per 15 min	Bitrate per UE per available resources	If less than three UEs request CA, then switch off second carrier to save energy

Since the network's output can be described by raw measurements o, performance metrics p, and quality metrics Q, the desired outcome can also be expressed at any of these levels of abstractions. This applies to any of the management layers, which leads to a two-dimensional map of possibilities for controlling the system as illustrated by Table 12.2.

The two-dimensional continuum of control can be regarded as a generalization of the 'Policy Continuum' [3], which represents a 1D diagonal-section through Table 12.2, from upper-right (Weighted Quality Metrics/Business Management) to lower-left (Configurations-based/Element Management). However, although for many use cases the upper-left corner of the table is not applicable, some business management layer use cases do depend on raw metrics like latency. Similarly, although rare, the lower right corner of Table 12.2 might be applicable to express weighted quality metrics for individual network elements, e.g. to request an individual network element to balance resources between two user groups of different importance according to some given service policy or to control within one element the balance between energy consumption and number of available carriers.

Not only from theoretical point of view, the 2D continuum of control is a more generalized description. From a practical point of view, it does not restrict the overall management system to refine the control input in a linear (diagonal) way from top-right to bottom-left. It reflects the fact that depending on the use case the overall management system might derive the final, concrete configuration X by using different paths through the table.

The 2D continuum of control allows to describe management functions, which work in only one of the dimensions. A NM layer function might receive weighted quality metrics and as output produce policies expressed as expected raw observations, i.e. the function provides a more concrete control input within the same management layer of Table 12.2. Similarly, for purely vertical functions, a NM function may receive from service management layer a weighted quality metric that involves multiple network elements. The function at the NM layer then transforms the scope to individual managed entities by breaking the metric down into multiple weighted quality metrics that only depend on data from individual network elements, which the functions then pass on to the individual elements – i.e. to fill the lower right corner. An example here might be the request to a network element to balance throughput against energy consumption by activating or deactivating carrier aggregation depending on the number of carrier aggregation requests.

12.2.6 Levels of Policy Abstraction

Given the multiple levels at which system behaviour may be described, it follows that there are multiple levels at which polices for controlling network behaviour may be described. Network automation, which implies automating the process of generating the control input, can also possibly be implemented at multiple levels. Taking only the two extreme cases of configurations and goals is inadequate, in most cases 'if – then – else' rules will not be able to derive the necessary configuration changes from the stated observations state-space changes. Thus, besides the extreme means of control, the intermediate means also imply means of defining control policies. Considering the Handover optimization use case as the example in Table 12.3, these different means may be differentiated according the following definitions:

Raw observations: These are the lowest network outcomes and include network events, counters, and gauges. An event in this case is any network incident that

Table 12.3 Example abstractions of a mobility optimization policy.

			Controlling input	
Control abstraction	Configuration *X*	Raw observation *O*	KPI *P*	Weighted quality metric *Q*
Policy	Configuration policy		Target	Objective
Business goals				Maintain balance between QoE of mobile users and admission of new users
Service goals			• ensure zero session interruptions • ensure low cell congestion	Ensure seamless mobility across cells while balancing load distribution among the neighbour cells
Network management goals	`Reduce delay before triggering HO`		RLF rate <2%, Ping pong rate <5%	With high priority, RLF rate <2%, medium priority, RLF rate <2%, With low priority, RLF rate <2% while with high priority cell load <80%
Element Management	`Set: A3 offset = 2dB;`			

may be of interest to the network operator such as an RLF or a handover failure, while a counter is the number of any such events within a specified time. A gauge is the instantaneous measurement of a network variable, an example of which includes the instantaneous power consumption, CPU utilization, or latency.

Key performance Indicator (KPI): A KPI is a network performance statistic computed from counters of network events, i.e. a KPI is the function over one or more counters such as the ratio of handover (HO) failures per minute to the number of HO attempts per minute. Examples of KPIs may include the rates of RLF, Random Access Channel (RACH) failures or handover Ping pongs.

Metric: A metric is any abstract network quality measure, a weighted combination of two or more KPIs. This, as introduced in the previous sections, is intended to find a balance amongst conflicting KPIs. Example metrics are handover performance, energy consumption, or network coverage.

Policy: A policy is a statement of the course of action to be applied. Just as the system can be described on different levels of abstraction, policies can be expressed at multiple levels of abstraction. As such, a policy may be defined as any kind of control to the system in either the layering or abstraction dimensions. A policy may not be strictly configurations-based or goal-based but may be: (i) abstract with the different levels of abstraction between configurations-based and goal-based; and at any of the network control layers. For example, 'Guarantee coverage in area A' indicates that coverage has been prioritized against other metrics like energy savings. This, therefore, is no longer an abstract goal but an intermediate step between the goal and the configuration actions like 'deploying a new cell' or 'changing a parameter value like transmit power'.

The different levels of granularity represent multiple quasi-orthogonal levels of policy-abstraction between the configurations- and observations state-spaces. Correspondingly, given multiple parameters that affect a set of measurable metrics, policies can be expressed at each of four levels of abstraction between the configurations- and observations state-spaces. The four levels, as described below, can be distinguished in the case of a mobility-optimization policy using the examples in Table 12.3.

Objectives: At the highest level of abstraction are the objectives, which express the compromises that need to be made over multiple, possibly conflicting KPIs that are weighted to represent their relative importance. In other words, the objectives express the desired outcomes for a set of KPIs including the compromises to make amongst those KPIs. An example Objective would state that 'without affecting KPIs y1, y2, z1, z2 ..., attempt to ensure that KPI a1 achieves values a11, a12, ... with priority pa11, pa12, ...'. Here a1 describes a particular KPI for metric A such as RLF for the case of the HO performance metric. Thus, the objective is the policy that expresses expected metric outcomes.

Targets: The target specifies the desired outcomes on a specific KPI. An outcome, for which, the compromises have already been resolved. After the objectives are stated with their relative prioritizations, the compromises are resolved to generate the specific targets for each KPI. An example KPI target will be the statement that 'ensure that KPI a1 is less than /greater than/exactly av1 or lies between the values av1 and av2'. Thus, the target is the policy that expresses expected KPI outcomes.

Parameter configuration: The Parameter configuration is the lowest level of control, which is the command to set/change a specific parameter to some value. This is what is typically undertaken by the Automation function with the intention to achieve the stated targets. Thus, a configurations-based policy is the policy that expresses expected parameter configurations. Note that there is no policy defined in terms of raw observations simply because the raw observations are rarely used as a means of control.

Goals: The goal is any statement of expectation at a particular layer. KPI targets and objectives can be stated at any of the layers and so they need to be differentiated amongst the different layers. Goals state the policy or expected outcomes in each layer of control without clarifying what kind of expectation there may be. In other words, we can talk of business goals, service goals, or network management goals which, in each case, represent the outcomes at that layer. The goals do not explicitly state which kind of expectations are stated and so have to be refined into observations, KPI targets, or objectives. An example goal is the statement that 'Ensure seamless mobility while balancing load', which, in practice, is an objective to balance the mobility robustness and load-distribution metrics. The same goal can be stated differently, e.g. that: 'ensure zero session interruptions' and 'ensure low cell congestion' which may be interpreted as specific service-level KPI targets.

Note that there is a temptation to classify parameter configuration-based and goal-based control as imperative and declarative control respectively. This, however, confuses the terms since a goal may also be imperative. As an example, the command to 'rehome a cell' is declarative since there are a series of specific imperative commands that need to be executed to achieve that outcome. It is however, within the network configurations state-space and not within any of the network outcomes state-spaces (observations, KPIs, or weighted metrics), and so it cannot be considered a goal.

12.2.7 Implications on Optimization

Optimization means a variation of configuration parameters $X = (x_1, x_2, \ldots x_n)$ in order to optimize an overall cost function. As described in Section 12.2.4, neither raw observations nor calculated performance metrics are usable alone, as optimization cost function, often because neither of them is able to balance between conflicting targets. This balancing is mandatory input for any useful optimization and must be derived from the business strategy of the operator and the specific network environment and context.

12.2.7.1 Modelling Optimization

Since the optimization must include a function to balance between the potentially conflicting targets p_i, the optimization algorithm might use either as a cost function, the combined weighted Quality Metric $Q = Q(w_i, p_i)$ and optimize this combined function to result in configuration X, or the algorithm might split the overall cost function Q into components p_i, in order to optimize them in an iterative way individually until the weighted result Q reaches an optimum. The first approach follows a diagonal path through the map, while the split of the overall cost function into its components involves a dedicated step from right to left within the same layer.

In general, optimization is the mapping $F := (c, X_{SC}, P_{Sh}, O_{ex}) \rightarrow X_1$, i.e. a mapping from a vector $(c, X_{SC}, P_{Sh}, O_{ex})$ to a subset of the configurations state-space X_1, where c *is a set of* some internal parameters that control the optimization, P_{Sh} is the historical performance or quality metrics, X_{SC} is the current configuration, and O_{ex} is the expected outcome of the system. The optimization function f provides as output new values for a subset of the parameters X_1, without changing the rest of the parameters X_2:

$$X = X_1 + X_2$$

$$X_1 = f(c, X_{SC}, P_{Sh}, O_{ex})$$

$$S(f(c, X_{SC}, P_{Sh}, O_{ex}), X_2, D) = O_{real}$$

The system S combines contributions, that depend on the vector X_1 of parameters controlled by an optimization function and all other parameters X_2, that are not modified by the optimization function, and the uncontrolled disturbance D to generate the real outcome O_{real}. The aim of designing the optimization functions is to ensure that $O_{real} = O_{ex}$. This is not simple since the disturbance D may change with time owing, for example, to environmental influences like buildings, trees, weather, and especially the traffic, which all vary with time. Similarly, the configuration X_2 may also change with time, e.g. if modified by the operator to optimize other network aspects.

The optimization needs to separate the influence of X_1 from time-dependent contributions of $X_2(t)$ and $D(t)$, otherwise, useful optimization is not possible. This is, in fact, the main problem in real networks. In simulations, $X_2(t)$ and $D(t)$ are controlled by setting them to be constant during the simulations but yet the other modelled parts do not mirror all details of the real environment. The realistic solution must thus rely on real-network measurements that have filtered the uncertainties.

12.2.7.2 Dealing with Uncertainty

An acceptable solution for dealing with uncertainty in real networks is to collect data for a period that is 'long enough' for statistically significant data that is not skewed by temporal variations of the environment, the $X_2(t)$ and $D(t)$ components. One strategy to cope with the unknown temporal behaviour of $D(t)$ is to assume periodicity, e.g. of a day, week, or month. Then, averaging measurements for a whole period eliminates the time dependency and separates the influences of the optimization and the environment.

Naturally, this simple model has several drawbacks. Firstly, to ensure that data can be compared, samples need to be taken across the whole period, which means that data collection based on such a model cannot be faster than the period – no matter how fast measurements are provided. For example, 'real-time monitoring' does not improve latency of such a detector. Moreover, it does not eliminate effects like long-term trends or to isolate and optimize for parts of a period, e.g. the throughput during the peak or the energy consumption during a low-traffic period.

For faster reacting optimization functions, it is required to use better models for the environment. For example, to use daily profiles, which can e.g. be learned during some learning phase: Then, the function compares current against expected observations, and the detector raises a detection as soon as current observations deviate from the model. Also, data mining would be an approach, which compares actual data against certain 'patterns' to detect certain conditions in the network.

An alternative strategy of dealing with uncertainty is to assume that $D(t)$ is constant for a certain period of time. This has the challenge of defining the proper time window, which on the one hand must be long enough to collect sufficient data for the optimization target and on the other hand, must be short enough to ensure that the assumption '$D(t)$ is constant' remains valid.

Another strategy might even be to combine the above strategies, i.e. each observed period is partitioned into intervals, within which $D(t)$ is assumed constant and each interval is optimized by using data from multiple cycles. For example, assuming a periodicity of day and hourly intervals, the optimization function could optimize the heavy working hour (HWH) by evaluating the data of the HWH collected from multiple days. This, of course, does not fully address the challenges since: (i) cycles are not perfect as time is only one aspect of the changing environment and (ii) it assumes that network elements have data and can be optimized for each interval of the cycle.

12.2.8 The Promise of Intent-Based Network Control

There are many vendors with different networking equipment and different mechanisms for managing the equipment. Although standardization is intent to harmonize these it is not possible to harmonize each aspect of these systems, which also explains why there are many standards bodies, each promoting a different aspect of the management challenge. For the operators, this variability implies an ever-increasing complexity of managing their networks. A solution that is highly promoted is to develop mechanisms for 'intent-based management', i.e. solutions that derive the values of the control parameters based on the operator's intents, and thereby releasing the operator from having to deal with the details of individual parameters. The term 'intent' is, however, highly loaded and at least very confusing.

In general, intents are expected to be higher level outcomes expected by the operator. Example requests from the operator that may be considered as intents may include:

1) Set up a path for client CAN bank from their New York office to their London office capable of offering at least 20 Mbps and ensuring the minimization of the latency to no more than 100 ms and maximizing the availability to at least 99.92%.
2) Instantiate coverage at the Munich Oktoberfest capable of at least 40 000 simultaneous connections with on average 2 Mbps per connection.
3) Instantiate a network slice for an IoT customer capable of at least 2Kbps per second per connection for a total of 10 000 static devices that are evenly distributed throughout the city of Munich.
4) Re-home a cell from a source RAN Node to a destination RAN Node. As proposed in a 3GPP study [12].

As evident from these examples, there are so many possible statements that could be considered intents and so the discussion on intents always ends up going around in circles. There have been multiple definitions of intents as regards networking, but most are quite fuzzy and confusing. For example, the Android reference documentation defines an intent as an abstract description of an operation to be performed. This, however, assumes that there is only one abstraction of an operation, yet as shown in the previous sections, there can be many abstractions. Other definitions e.g. from CISCO [13], consider the intent to be the business goal that is intended to be achieved by the configuration action. Again, this assumes a simple translation from business goal/logic to configuration, which in case of the RAN is not given because of the inadequate knowledge on S and D. There has also been a tendency, in the case of sequential configurations (Section 12.4), for some authors to state that the final configuration is an intent and that the sequence of actions leading to that final configuration is the 'how' implementation of that intent. The challenge with this perspective, however, is how that definition should be mapped to intents if the required implementation is a single configuration step, i.e. if the implementation is simply a single configuration or multiple configurations in one bulk configuration action. Such an intent definition is thus faulty at the least or otherwise completely wrong. It is evident that all these definitions are applicable in domains of well-defined context only but cannot be generalized due to the heterogeneity and complexity of mobile networks. From that respect, it is better to derive the definition from basic linguistic usage of the term.

12.2.8.1 Definition

Linguistically, an intent is the aim of or the desired outcomes for doing something. Correspondingly, given the multiple levels of abstraction of network configuration policies, it follows that there should be an intent for each policy abstraction, which leads to the definition that:

An intent is the abstract description of a policy or action.

Such intents can be stated at each of the management layers as well as for each abstract policy level. Considering Table 12.4, for example, the business level goal to 'meet the Service Level Agreements (SLAs)' is the intent for the service-level KPI-objectives policy (say for ultra-reliable, low-latency communications [URLLCs] services) that 'Low latency has the highest priority of all metrics'. Then the service-level intent of 'Low latency has highest priority of all metrics'

Table 12.4 Intents are possible at multiple policy abstraction levels and intent translation occurs at each interface between layers and abstraction levels (the short arrows across the edges).

NM Layer	Controlling Input			
	Configuration, X	Raw Observation, O	KPI, P	Weighted Quality Metric, Q
Business				SLA for users in the large area shall be met
Service			Latency < 10ms	Low latency is highest priority compared to ES, etc
Network	Activate Carrier Aggregation	1. Reduce RLF 2. Increase throughput	RLF < 2% UE Bitrate > 2Mps	
Element Management	Set: activateCA=1; CellPairs: 1&4, 4&1, 2&5, 5&2, 3&6, 6&3; A3=offset; A5=thr1, thr2; A6=offset; ...			

may be interpreted as the setting a service-level KPI-target policy of 'Latency must be less than 10 ms'.

The intent may also be simultaneously translated along both dimensions, e.g. the service-level KPI-targets intent of 'latency <10 ms' may imply a NM-level outcomes policy to either 'reduce RLF' or 'increase user throughput'.

The NM level desired output to 'increase user throughput' then becomes the intent that leads to a NM level policy to 'activate carrier aggregation on a NE', which also becomes an intent for the low-level configuration of cell-level Carrier aggregation parameters.

12.2.8.2 Intent-Based Cognitive Autonomy

It would be wishful thinking to assume that the network will function completely independently even with the use of cognitive autonomous functions. In reality, the operators' actions will only move to a higher abstraction level, i.e. NM becomes cognitive and autonomous but based on the operator's intentions or guidance. This implies that concepts on intent-based Network operations are critical to the realization of CANs. Realizing intent-based network control, however, remains an open challenge.

Using Intent-based control, the operator informs the network about what he wishes to see happen either in terms of a configuration change or a desired outcome. The automation functions then derive the actions that must be implemented to realize the operator's intents. It follows then that the network automation system must provide an appropriate 'intent' interface through which the operator expresses the intents and the network responds with information about how effectively the respective intent has been achieved. It is through such an interface that the operator indicates the need to either upgrade a specific function or the complete CAN system.

With such an indication, e.g. to upgrade, the CAN system must then check for the appropriate pre-requisites (like licenses and dependencies) and make the necessary operational controls like saving function configurations or learned models.

Moreover, underpinning the Intent-based control challenge is the ability to translate the intents into lower level actions and mapping very many kinds of actions to only a few intents. However, with clear understanding of CANs, i.e. how they are able to learn both context and the context-configuration mapping, the challenge gets simplified.

Another challenge is the design of an appropriate intent expression language that is capable of capturing the right context dimension expressed within the intent, which are then matched by the cognitive functions to the learned context-configuration mapping. The successful implementation of CANs therefore promises a chance to transition to intent-based network control.

Another aspect of the 'intent-based management' challenge is describing the network in way that is dynamic. This may imply the inclusion of the semantics of the parameters into the network models or defining new abstract mechanisms that capture the semantics separate from the parameters. For example, the network models could be extended by providing a feature model of a network device like a base transceiver station (BTS), for which a Feature model may summarize the features in the BTS stating which ones are available, active or inactive. Using such semantic models, the optimization intents get to be described in terms of the semantics as opposed to doing it in terms of the parameters.

The above questions show that intent-based networking is both critical and yet not simple. It is desirable that CANs will be able to leverage their cognition to appropriately apply the intent framework and successfully doing so will inevitably increase the systems' operations and management autonomy. However, the implementation thereof can benefit from concepts on the new framework on combined development and operations as described in the next section.

12.3   The Development – Operations Interface in CANs

In a traditional network environment, there is a distinct separation between software development and the target operational environment (also typically called the production system). The software releases are typically of larger volume and deployed on a less frequent basis after heavy testing procedures and stages. One of the major shortcomings of this process model is the so-called 'wall of confusion' between the development stage and the production or operation stage. This is caused by a lack of communication between both sides and a mismatch of the targets: whereas the development is tuned towards the number of new features in the releases, the operation is more interested in stability and tries to avoid risky changes in the running system.

This system is not applicable for CAN, especially because of the likely increased risk. Instead, a DevOps paradigm would be much more suited owing to its emphasis on collaboration and the use of automation. With its intrinsic agility, DevOps provides the baseline to successfully implement ML and AI capabilities and put it in production for Cognitive Autonomous Network operations.

12.3.1   The DevOps Paradigm

The challenge of mismatched interests between development and production teams is being addressed by the DevOps paradigm, which emerged in the

2005–2010 time period. DevOps requires a very close cooperation, or even integration, of the development and operation teams. Unlike the large multi-month step-function changes in traditional delivery and deployment environments, DevOps is characterized by continuous delivery of small incremental changes to software applications via automation. Thus, DevOps delivers innovations to users rapidly and in a less risky manner.

DevOps is not just a methodology, it touches all parts of an organization over the whole lifecycle. The transformation towards DevOps comes along with both a technology and a culture change. On the technology side, illustrated by Figure 12.10, it requires continuous integration (CI), continuous testing, continuous delivery (CD), and continuous monitoring and, unlike traditional software development, DevOps relies on constant and immediate feedback at any phase of the lifecycle. On the culture side, DevOps represents the idea of tearing down organizational silos and building joint teams with shared responsibility over the complete lifecycle.

DevOps induces agility into the whole application lifecycle, where agile development focusses on the creation of software, the DevOps paradigm expands agility

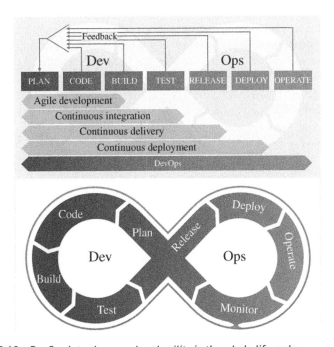

Figure 12.10 DevOps introduces end-end agility in the whole life cycle.

over all phases. A prerequisite to successfully applying the DevOps approach is to automate the processes and workflows to create the complete and closed-loop workflow (incl. feedback path) also known as the DevOps pipeline. The DevOps environment, based on agility and automation, provides an ideal framework to continuously deploy and update cognitive capabilities with integrated AI and ML techniques in the network, allowing for incremental evolution towards a cognitive autonomous network.

12.3.2 Requirements for Successful Adoption of DevOps

Leveraging the DevOps practices in the Telco ecosystem requires adaptation to the specifics of the supply and value chain in this industry. Unlike large webscale companies, where development and operation is under control of one party, the 'Dev' and 'Ops' roles in the Telco world are separated: the 'Dev' covered by the supplier/vendor and the 'Ops' role taken by an operator. This, as illustrated by Figure 12.11, not only requires DevOps readiness of both sides, but also a strict coordination between the two. Also, the introduced DevOps pipelines need to be compatible and interlinked to enable the automation of the closed-loop pipeline.

One option to successfully implementing DevOps in the Telco environment is to setup joint teams staffed by vendor and operator to manage and monitor paths of the interlinked DevOps pipeline between the vendor and operator.

From a technology standpoint, the DevOps paradigm requires a cloud native architecture, e.g. built on an abstraction and virtualization layer which might be implemented by hypervisor technology and Virtual Machines (VMs) that enable the execution of software in shared hardware environments. Hypervisor technology is based on emulating the underneath hardware's resources, typically memory and compute resources, and allowing each VM to run on top of the hardware like a

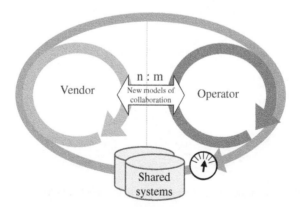

Figure 12.11 The target DevOps model for communication services providers.

separate computer. A major disadvantage of VMs is that because the VM acts as a separate computer, as such, it doubles the time needed for a full service to be operational. Containers minimize this by offering a way to share the host's resources instead of emulating them. This enables very fast start-up times, although usage is still not simple.

A number of approaches are developed, mainly within the open source community [14], to simplify the operational aspects. Docker [15] has made containers very popular by offering an easy way of running them that is similar to running VMs. Meanwhile, VM management has also matured with OpenStack as the de-facto standard software to manage VM-based services across multiple physical nodes. Kubernetes has also gained a lot of momentum, but it still has a lot of missing features that must be addressed to be able to manage large-scale multitenant applications in a distributed operation environment.

Adopting the DevOps paradigm into software (SW) design has an impact on the actual SW architecture as well. Microservice architectures offer convenience in software design by trading off easier development with ease of scaling to the cost and complexity in operation. However, the update of individual SW components become easier. Moreover, organization structure also needs to reflect the agility required to design and deliver a SW to the operation.

12.3.3 Benefits of DevOps for CAN

In general, DevOps is a powerful way for operators to accelerate innovation and to deliver new services to win business in the rapidly changing telco market. First of all, DevOps enables fast delivery and deployment of new features and provides early feedback from the market, which either results in a competitive edge in the market or in the potential to quickly react to new competitive challenges in the market. It also enables vendors and operators to quickly drop non-successful offerings before major costs are sunk therein.

Secondly, DevOps improves the quality of the software significantly as bug localization and bug fixing in small software increments will be much easier and faster than in big and complex releases. Moreover, DevOps improves the security of the software: security vulnerabilities can be detected and mitigated much faster due to the continuous monitoring of the software and the ability to quickly correct and update the affected parts of the software. Like a moving target, the continuous change of the deployed software provides less exposure to potential attackers.

In the case of NM, the continuous monitoring capability of a DevOps ready environment fits very well with the monitoring needs of learning functions that are inherent in cognitive autonomous networks. Moreover, the software specific benefits above are also critical in CAN. Specifically, since CAN Functions are new and not evolutions of old software, they may be prone to bugs, in which case, the quick bug fix promised by DevOps becomes important. Similarly, through the DevOps framework, vendors, and operators will be able to quickly and less expensively introduce new CAN functions into their operations and can smoothly upgrade and improve those functions with very little disruption to their operations.

12.4 CAN as Data Intensive Network Operations

The introduction of cognitive decision making for network operations implies a heavy reliance on network data. Consequently, this changes the operational environment – at the least into one where network data takes centre stage compared to previous network management paradigms. This then requires the value of data to be appropriately appreciated by the different entities. This appreciation must cover implications to the commercial interest of all the parties, its implications on the nature and structuring of operations, and the risks posed by this over reliance on data and automated data-based decisions. This section briefly discusses these challenges as open questions that need to be addressed before CANs can be realized.

12.4.1 Network Data: A New Network Asset

Modern society is surrounded and controlled by data, every day humans generate an incredible amount of data, more than ever before. The sixth edition of DOMO's report [16] states that 'over 2.5 quintillion bytes of data are created every single day' and that by 2020, 'it's estimated that 1.7 MB of data will be created every second for every person on earth'. Correspondingly, the data analytics market is incredibly valuable, with assessments ranging from 80 to 100 billion of Euros by 2025 [17].

Mobile network data are passively generated, from the end user and network devices, and easily collected by network operators. The value brought by this data has been increasingly recognized, as reported by Ovum's recent report [18], which found that '67% of brands already view mobile operators as good original sources for customer data insight, ahead of digital media companies, chat app platforms, and device manufacturers'. In this context, the challenge for network companies is not about creating data sources or collecting raw data – which is already done – rather being able to extract *information* and *create value* from the data, which is not straightforward. 'Vendors that are able to translate data into value will survive and thrive. Those that do not will be left behind' states a recent report by 451 Research [19]. And that is why data is the *ultimate asset* [19].

Data will be monetized and sold, but in the long-term owning data will not be enough. The sustainable value of the market will be given by the ability to transform data into profitable business actions. Investments need to be made to develop the right platforms and algorithms. And numbers show that this is where companies are going. In 2018, the AI funding worldwide reached about $20 billion, which is also one of the highest levels of investor financing amongst emerging technologies [20]. Several companies in the mobile ecosystem are increasingly deploying AI solutions in their products, services and, customer management with network operations as a major intended area of applying the AI algorithms mainly to cope with the ever-growing network complexity [20].

12.4.2 From Network Management to Data Management

As explained earlier, the cognitive automated network adds cognitive capabilities over and above the automation functions throughout the heavily interlinked NM and operation systems. This will provide a new set of requirements for data management within and between the systems.

CAN modules that control network functions (which together provide the end-to-end service to subscribers) need to share the same data and models of the service performance. Otherwise, they end up optimizing their limited territory, which can cause a very sub-optimal chain of functions, each reacting out for their local optimums. In practice, there are two ways of countering this challenge.

Firstly, each CAN module will be a data consumer with a continuous need for up-to-date data representing the current, yet continuously evolving network state. As such, there is need to share and route management plane data not only upwards towards Operations Support Systems (OSSs) and Business Support Systems (BSSs) systems but also downwards towards distributed CAN modules. This introduces a new type of data sharing and transfer tools based on streaming data and publish-subscribe paradigms [21].

Secondly, compared to classical NM systems, in addition to the standardized signalling between network functions, there will be a need for statistical summarisation and event generation layers. As new services appear, and old ones evolve, these layers must adapt to the changing information requirements of CAN modules and the new parameters carrying information about the performance of those new services. This then requires inbuilt mechanisms to modify data models, extend interfaces and, create new interface instances in such away, that the models, interfaces, and algorithms are able to support the evolving needs.

Most of the multi-dimensional ML methods require some form of normalization or scaling of parameters in order to make parameters and their value ranges comparable with each other. However, scaling strongly affects the results of the analysis, yet there are several heuristics that may be applied for scaling [22]. Therefore, the minimum requirement for a data management system serving CAN modules is that the same scaling method and corresponding parameters is used for model training and model execution in analysis.

Another problem, that is also often neglected in academic literature, is the varying quality of streaming data. As communication networks are very complex systems, it sometimes happens that due to different reasons, like local congestions or other random failures, some of the data points will either get lost or corrupt. This needs to be considered both in data-sharing systems and also in CAN modules to make them able to operate on partially missing, incomplete, or corrupted data as well.

In data-sharing systems, the problem can be mitigated by adding redundancy and assurance mechanisms to data flows. Unfortunately, they often slow down the data transfer and, therefore, their usage is always a trade-off between robustness and speed.

In CAN modules, multiple options may be used to qualify incoming data. These include, e.g. traditional methods, like definitions of possible value ranges, but also ML-based anomaly detection (AD) methods can be used to recognize data points that have already been seen. However, if AD methods are used, there must also be mechanisms to identify and adapt CAN modules to newly appearing network states that produce previously unseen value combinations to data points.

Incomplete data for a CAN module, is yet another type of challenge. What should a module decide, if suddenly half of parameter values in data points are missing? Should it continue normally using existing values and ignoring or replacing missing ones with something? And if replacing, then what is that something? Or should it stop working totally? The answer depends on the purpose, use, and place of each CAN module and must be tailored specifically thereto. However, that answer may need to be defined and already given when usage and role of a CAN module is defined.

12.4.3 Managing Failure in CANs

A general trend in applying analytic technologies described in this book are that they can succeed in tasks and contexts, where an incorrect decision does not cost too much while the correct action can pay back many-fold. For example, in tailored advertising, if you see an advertisement that does not interest you, you skip it without thinking twice, yet if you see a good bargain of a product you have wanted for ages, you will happily click the advert and buy the product. This will also, most probably, be the case with usage of CAN modules. They will be beneficially implemented to perform such tasks, where there are a high number of simple decisions to make and where the penalty of a wrong decision is low. In all other scenarios, extra measures are required, i.e. new capabilities need to be developed for CANs to be deployed in roles, where the likelihood increases that an incorrect action can shut down large portion of a network.

In a traditional network operation set-up, all decisions that change the state of the network, are done by a human expert. In his training and agreed processes guiding his actions, there are several practices, whose purpose is either to guarantee the correctness and quality of the applied measures and changes or to at least mitigate and limit effects of these possible errors. Such practices should also be defined for and included in systems operated by autonomous AI-based entities like CAN functions. This is quite a tedious task to do as it needs to cover all the points and tasks where such functions may run. The challenge is even greater in systems-like communications networks, where: (i) phenomena and performance on one side of the system can propagate through the network as a service is established; and (ii) traffic is transferred end-to-end from one side of the network to another in a very short time.

A good starting point for consideration and implementation of such practices is to provide a monitoring system that is able to provide information about – not only services – but also about performance and actions taken by CAN modules. By monitoring and learning to understand what, why and in what context the CAN function do certain actions, a system owner can gradually develop a covering monitoring system and train meta CAN modules to monitor the other modules. This may lead to a layered architecture, where upper layers are controlling and guiding lower ones.

In building such a monitoring system, one should not limit the data collection to internal reporting of CAN modules about what a module was doing and why. Internal data does not provide any information about, e.g. programming errors like memory leaks or system errors that make CAN modules to stall or miss some needed resources or data. This kind of information can be only collected from lower implementation layers like the operating system, hypervisor, or communication gateway, to name a few.

The first step in developing such a mitigation mechanism is the use of risk assessment procedures [23]. In risk assessment, possible failures are identified, their probabilities of occurrence estimated, and the effects and value of their likely damage quantified. The process allows the risks to be prioritized according to their severity and for mitigation efforts to focus on the most severe risks. This same procedure could be applied to communications networks. However, there still remains the problem that because communications networks are highly redundant and robust systems, the effects and value of a malfunction can be hard to know before the malfunction is encountered.

References

1 Armenteros, J.J.A., Tsirigos, K.D., Sonderby, C.K. et al. (2019). SignalP 5.0 improves signal peptide predictions using deep neural networks. *Nature Biotechnology* 37 (4): 420.

2 O-RAN ALLIANCE (2018). *O-RAN: Towards an Open and Smart RAN*. O-RAN ALLIANCE.

3 ETSI GR ENI 003 (2018). *Experiential Networked Intelligence (ENI); Context-Aware Policy Management Gap Analysis*. Sophia Antipolis: ETSI.

4 RFC 3198 (2001). *Terminology for policy-based management*. IETF.

5 ETSI GR NGP 011 (2018). *Next Generation Protocols (NGP) E2E Network Slicing Reference Framework and information Model*. Sophia Antipolis: ETSI.

6 3GPP TS32.601-2 (2000). *Basic CM Integration Reference Point*. Sophia Antipolis: 3GPP.

7 3GPP T.32.602 (2018). *Basic CM Integration Reference Point: Information Service*. Sophia Antipolis: 3GPP.

8 3GPP TS32.602-2 (2000). *Bulk CM Integration Reference Point*. Sophia Antipolis: 3GPP.

9 3GPP TS32.612 (2004). *Bulk CM Integration Reference Point: Information Service*. Sophia antipolis: 3GPP.

10 RFC 4741 (2006). *NETCONF Configuration Protocol*. IETF.

11 RFC 6241 (2011). *Network Configuration Protocol (NETCONF)*. IETF.

12 3GPP TR 28.812 (2018). Intent driven management service for mobile networks; Study on scenarios for Intent driven management services for mobile networks. Sophia antipolis: 3GPP.

13 Cisco (2019). Cisco Intent-Based Networking At-a-Glance. https://www.cisco.com/c/en/us/solutions/collateral/enterprise-networks/digital-network-architecture/nb-06-intent-based-networking-aag-cte-en.html?oid=aagen016865 (accessed 17 June 2019).

14 Cloud Native (2019). Building Sustainable Ecosystems for Cloud Native Software. https://www.cncf.io (accessed 31 January 2020).

15 Docker (2019). Docker: Get started. https://www.docker.com (accessed 31 January 2020).

16 DOMO (2018). Data Never Sleeps 6.0. https://www.domo.com/learn/data-never-sleeps-6 (accessed 03 March 2019).

17 Accuray Research LLP (2017). Global Big Data Market Analysis & Trends - Industry Forecast to 2025.

18 OVUM (2017). Data Insights in Digital Advertising and the Role of Operators. https://ovum.informa.com/resources/product-content/data-insights-in-digital-advertising-and-the-role-of-operators (accessed 31 January 2020).

19 451 Research (2018). Empowering the digital revolution.

20 GSMA (2019). Embracing AI in the 5G era: How momentum is building across the mobile industry.

21 Kojola, V., Kapoor, S. and Hätönen, K. (2016). Distributed computing of management data in a telecommunications network. International Conference on Mobile Networks and Management, Abu Dhabi.

22 Scikit learn (2019). SCiKit Learn User Guide: 5.3. Preprocessing data. https://scikit-learn.org/stable/modules/preprocessing.html (accessed 17 June 2019).

23 Hopkin, P. (2018). *Fundamentals of Risk Management: Understanding Evaluating and Implementing Effective Risk Management*. Kogan Page Publishers.

Index

a

access and mobility management function   48, 53, 55, 57
accounting management   73
Activation Function   220
active learning   354, 369, 370
Actor-Critic   251
Adam   209
alarm correlation   358, 367
allocation and retention priority   38
almost blank subframe   329
anomaly event aggregation   352, 353, 365
anomaly level   352, 360, 364
anomaly pattern   353, 367
Ant Foraging   145, 152–153
ARIMA   363, 380
Attention   155–164
augmented diagnosis   369
Auto Regressive and Moving Average Models   198
Auto-Commissioning   98, 99
Auto-configuration   253
Auto-connectivity   98, 99

Autoencoder   238
Automatic Neighbour Relation   101, 129, 137
Automation   1, 5, 9, 15, 19, 20, 23, 93, 94, 95, 97, 98, 100, 101, 133, 136, 138, 140, 141
automation plane   483
autonomic   20
autonomous   1, 20, 22, 24
autonomous capability   22
Autonomous Reasoning   173
autonomy   18, 20, 21, 24

b

backhaul   64, 65, 66, 68–72
Backpropagation   222
Basic Cognitive Processes   155
Bayesian games   458, 460
Bayesian Network   192
beam management   62
beamforming   60, 61, 87
bearer   38, 63
behavioural graph   434
Belousov-Zhabotinsky reaction   150

Towards Cognitive Autonomous Networks: Network Management Automation for 5G and Beyond, First Edition. Edited by Stephen S. Mwanje and Christian Mannweiler.
© 2021 John Wiley & Sons Ltd. Published 2021 by John Wiley & Sons Ltd.

Bio-inspired Autonomy   146
Blind PCI auto configuration
    267
Bloom's Taxonomy   158–159
Boolean Logic   186, 187
Business Support System   12, 13

c
carrier Ethernet   69, 71
Case Based Reasoning   182, 369,
    472
Centralized SON   95, 96, 119
circuit-switched   31, 33, 34, 36
Classic AI   174
Classification   212
Closed Loop   95
Closed-loop Control   180, 421
Closed-loop Self-Optimization
    304
cloud-native networks   42, 45,
    46, 47
Clustering   215
CNM environment   419
cognition   18, 19, 20, 21, 24
cognitive   1, 20, 21
Cognitive autonomy   146, 154
Cognitive Function   18, 25, 405,
    413
cognitive network management
    18,
    25
Cognitive Processes in Learning
    158
cognitive self-healing   349
Cognitive Self-optimization   302
Common-Sense Knowledge
    Problem   174

communication service   29, 50
complexity   2, 3, 8
Conceptualization   159–166
Concurrent games   423, 425
concurrent learning   425
Concurrent Processing and
    Actioning   162
conditional eigenvalue problem
    332
configuration   13, 16, 25
Configuration Management   73,
    76–78, 426,441
context classes   263
Context Model   260
context properties   260
context region   263
context-aware configuration
    256
Contextualization   159–166
continuous delivery   507
continuous integration   507
control and user plane functions
    separation   52
control plane   34, 40, 49, 51, 53,
    58, 65, 68, 78
Convolutional Networks   231
Cooperative learning   308, 313
Coordination and Control   444
Coordination Transaction   437
core network   29, 33, 38, 39, 51
Coverage and Capacity
    Optimization   117, 129,
    327
customer experience level   406,
    407, 408, 410
customer experience
    management   13

d

DARN   284
data analytics   21
data management   471, 511
DBSCAN   366
Decision Applications   441, 443
decomposition   64, 67, 70, 87
Deep Learning   227
Deep Neural Network   227
Degradation detection   413, 429
degree of automation   145,156
demand   385, 387, 394, 396, 397,
          398, 399, 400, 401, 402,
          403, 404, 412, 416
deployment options   51, 63, 65,
          68
deployment plan   431
DevOps   469, 506
diagnosis   353, 366, 427,429
Dimensionality Reduction   213
Directed Acyclic Graph   193
Discrete Logic   186
Distributed Cell Activation   281
distributed intelligence   457
Distributed SON   96, 97, 121
Domain Manager   95
Dropout   225

e

ECGI   101, 102, 103
EDGE   32, 33
eNB deployment   327
Energy Saving   109, 115, 116,
          117, 139
eNode B   40, 41
Environment Modelling &
          Abstraction   441

Environment Modelling   441
E-UTRAN   39, 40, 41
evolved packet core   38, 39
evolved packet system   40
Expectation Maximization   207
Expert Knowledge Base   177
Expert systems   177, 472
extensive form game   459

f

fault management   73, 78
FCAPS   94
FDD   37
feature selection   351, 356, 357
Features of Self-Organization
          152
5G   93, 108, 136, 137, 138, 139,
          140
flexible frame structure   60, 61
Flexible Resource Assignment
          329
Foraging of ants   152–153
fronthaul 58   65, 67
Function Configuration
          Parameters Values   253
Functional Decomposition of
          CAN   440
Fuzzy Inference   186
Fuzzy Q-Learning Controller
          322
Fuzzy Q-Learning   310–313

g

generation of mobile network   6
goal-based control   490
GPRS 30   33
Gradient Descent   208

Graphics Processing Unit 482
Grid Search 207
GSM 30, 31, 33, 35
guaranteed bit rate 38

h
heterogeneous network 10
Het-Net 106, 107, 108, 109, 115, 118
Higher Processes 156–166
HSDPA 30, 33, 37
HSPA 33, 37
HSPA+ 33
Human Cognition 154–159
Hybrid SON 97
Hyper-Parameters 207

i
ICIC 102, 122, 123
Implicit Knowledge 174
Inductive transfer learning 372–375
Inference Engine 177
Inference 159–168
information model 75, 78–82, 85–87
infrastructure as a service 43
Input World and Action Model 176
Intelligence 156–157
intent level 406, 407, 408
intent-based management 481, 502
inter-cell interference 322
Inter-function coopetition 456

inter-mode interference 329, 337
Intra-frequency 102, 109, 113

k
Key Performance Indicators 302
K-means 215
KPI database 415
KPI profiles 429
KPI target 259

l
Language 157
latency 395, 398, 410
layer 4, 5
Learning Augmentation 289
Learning in Bayesian Games 461
levels of autonomy 24
licensed assisted access 10
Load Balancing 101, 108, 109, 110, 111, 139
localization 399, 400, 404, 413
localized measurements 398, 404
Long Short-Term Memory 236
LTE 97, 101, 102, 103, 108, 113, 114, 116, 118, 119, 122, 137, 138, 139

m
Machine Learning 9, 18, 21, 25, 203

Machine Learning Orchestrator
482
management interface 76–79,
82
management layer 31
management layer 74, 78
Markov Decision Process 243
Markov Logic Networks 368,
378, 379
Markovian models 306
Max Pooling 234
MBH automation 411, 412, 413,
414, 416, 417
Membership Functions 186
Memory 156, 162
mental processes 154, 164
Metric Clustering 291
midhaul 67
Minimization of Drive Test 133,
139
Mitola's cognitive cycle 162
mobility management entity 39,
40, 41
Mobility Robustness
Optimization 93, 111,
309,314
Model-based methods 306
Modelling Cognition 159
multi agent Q-Learning 308
Multi-agent coordination 420
Multi-agent systems 423
multi-task learning 375
multi-tenancy 44, 50
multi-vendor functions 420

n

Natural Stochasticity 146
Nesterov Momentum 209
Network Automation Function
253
network automation 385, 386,
417
network configuration parameter
17
network contexts 254
Network Element 93
network engineering 12, 13
network function virtualization
7
network level policies 406, 407
network management 11, 13,
15, 25
Network Management 93, 94,
95, 96, 115, 125, 133, 136,
140
Network Manager 95
Network Objectives Manager
441
Network Resource Model 101
network slice 53, 58, 60, 79, 88
network slicing 7
Neural Network 219
Neuron 220
NGNM 97, 98
node B 36
non-public networks 478, 480
non-standalone architecture 29,
51, 52, 62
Numerical Optimization 207

o

OAM   94, 96, 98, 99, 102, 103,
     119, 120, 123, 125, 135, 138
Objective Model   257
Observe-decide-act   164
observe-orient-plan-decide-act
     cycle   164
OFDM   39, 59, 60
OFDMA   41
OMC   94
Operations Support System   12,
     13
Operations, Administration, and
     Maintenance   43, 47, 72
Optimal Resource Allocation
     334
Optimistic Concurrency Control
     436
optimization   16, 25
Overfitting   209

p

packet scheduler   408, 410
packet-switched   33, 34, 39
Partial Resource Muting
     329–335
PCI Assignment Objectives   266
PCI auto-configuration   265
PCI update   270
PCI   101, 102, 103, 104, 105, 106,
     107, 108, 114, 117, 139
Perception-Reasoning pipeline
     159,
     164
performance management   73,
     78, 83
performance monitoring   388,
     400

Physical Cell Identifier   101
Ping-Pong   314
plane   4, 5
platform as a service   43
plug-and-play   413
Plug-and-Play   98
policy abstraction   497, 499, 504
policy continuum   496
Policy gradient Algorithm   251
Post-action Verification   426
Power Saving Groups   274
PRB   109, 110, 122, 123, 124
predictive self-healing   379
Principal Components Analysis
     214
priority measure   260
proactive Self-Optimization   305
Problem-solving   156–166
profiling   351, 361
programmability   46

q

Q-Learning based MRO   315
Q-Learning   245
Q-Learning   307
QoE descriptor   408, 409, 410
QoE indicators   401
QoE management   406, 407,
     408, 411
QoE target   404, 406–411
QoE   387, 394, 395, 399–408,
     411, 416
QoS class   32, 38
QoS   387, 394, 399–417

r

RACH   120, 121, 122, 138,
     139

radio access network   29, 35, 36, 40, 56

Radio Link Failure   314

radio network controller   36, 37

RAT   98, 102, 103, 108, 109, 111, 113, 116, 118, 139

Real-time Network Control   284

Recurrent neural networks   235

Regression   211

Regularization   211, 224

Reinforce Algorithm   249

Reinforcement Learning   241

ReLu   220

Resource Muting   329–340

RLF   111, 112, 114, 115, 121, 134, 138

RMSprop   209

RTT   400, 401, 404, 410, 411

Rule Based Fuzzy Inference   323

Rule-Based System   177

s

sandboxing   483

SC-FDMA   41

Scope Generation   427

Search and Planning   175

security management   73

self-adaptation   412

Self-Configuration   97, 98, 100

self-evaluating   396

Self-healing   327

Self-Healing   97, 124, 125, 126, 128

Self-optimization   16, 97, 108, 124, 301–307

Self-organization   21, 146–154

Self-Organizing human systems   151

Self-Organizing Networks   17, 93, 140

self-organizing systems   148–153

self-programming   394, 402, 410, 411

Service-Based Architecture   53, 54, 78, 481

service-based interface   42, 55

service-level agreement   50, 89

session management function   53, 56

shared data layer   48

Sigmoid   220

Simulated- Annealing   207

Single coordinator   424

Single-Agent System decomposition   423

Single-Agent System   423

SLA monitoring   417

smart network fabric   48, 49, 65

software as a service   43

SON Coordination   17, 94, 97, 129–132, 436,437

SON Framework   420

SON Function   95–97, 119, 121, 129–132

SON management   17

SON Operation   422

Spatial-Temporal Scheduling   449

spatio–temporal patterns   150

split RAN   58, 67

Standard interference functions   330

state model   386, 387, 389, 390, 391, 392

State-Based Configuration
Management   491
statistical learning   306
Stochastic Gradient Descent
209
Successive Approximation of
Fixed Point   335
Supervised Learning   204
Symbolic AI   174, 176
symptoms   392, 393
Synchronized cooperative
learning   420, 447
system architecture   33, 39, 51,
52
system state   386, 388, 389, 390,
391, 392, 393, 417

t
TDD   37
TDMA   31
Team learning   424
Team modelling   425
telco cloud   43, 48
telecommunication management
network   73, 74, 80
Tensor Processing Unit   482
Term
Thought   154–157
throughput   393, 397, 398, 400,
401, 402, 404, 405, 409,
410, 411
Tilt Optimization   322
Time Series Forecasting   196
time series   389
Traffic Steering   108, 109

Transductive transfer learning
372, 374, 376

u
ultra-dense network   10
Underfitting   210
Unsupervised Learning   204
user plane   34, 40, 51, 52, 53, 57,
58, 62, 65, 68
user plane function   52, 53
UTRAN   33, 36

v
Value and Policy Iteration   244
Vanishing Gradients   228
verification areas   427, 431–435
Verification assessment Scoring
function   430
Verification assessment   430
Verification Collisions   431–435
vertical industry   8
virtualization   44, 46, 58
voice over LTE   39

w
WCDMA   32, 37, 41
Weight-Decay   225

x
x-haul   68

z
z-score   364

Printed and bound by CPI Group (UK) Ltd, Croydon, CR0 4YY
27/09/2021

03084441-0001